Cancer Drug Discovery and Development

Series Editor
Beverly A. Teicher

For further volumes:
http://www.springer.com/series/7625

Giovanni Melillo
Editor

Hypoxia and Cancer

Biological Implications and Therapeutic Opportunities

Humana Press

Editor
Giovanni Melillo
Bristol-Myers Squibb
Princeton
New Jersey
USA

ISBN 978-1-4614-9166-8 ISBN 978-1-4614-9167-5 (eBook)
DOI 10.1007/978-1-4614-9167-5
Springer New York Heidelberg Dordrecht London

Library of Congress Control Number: 2013952769

Printed on acid-free paper

Springer is part of Springer Science+Business Media (www.springer.com)

To my wife Maria Giovanna and to my sons
Simone and Mattia

Preface

This is an exciting time in Oncology! The tremendous progress in our understanding of genetic and epigenetic alterations that drive the oncogenic process has provided a number of new opportunities for the development of more effective and less toxic anticancer therapies. More importantly, the emphasis that researchers have placed on the cross-talk between cancer cells and the host is finally paying off, with novel therapeutic strategies boosting the patient's immune system generating transformational results in diseases that were difficult to treat with available therapies. The interaction between cancer cells and immune cells takes place in the context of a tumor microenvironment that is profoundly impacted by many factors. This book emphasizes the role of a universal adaptive mechanism, i.e., adaptation to changes in oxygen levels, as a fundamental biological variable of the tumor microenvironment. The inevitable imbalance between oxygen demand and supply caused by a dysfunctional and inefficient tumor vasculature is a hallmark of human cancers. The unprecedented progress in elucidating the molecular mechanisms underlying tumor hypoxia has fueled enthusiasm and efforts in better understanding the biological implications and exploiting potential therapeutic opportunities. This book attempts to capture this wealth of information in a coherent and comprehensive collection of contributions by world renowned scientists and investigators, involved in basic, translational and clinical research. The goal of this book is to provide a framework to generate novel hypotheses and opportunities that can ultimately be translated to the clinic to benefit cancer patients. Despite the unquestionable progress in the development of novel cancer therapies, achieving cures in metastatic disease remains an elusive and sought-after goal. Only a comprehensive and systematic approach that takes into account all of the components that ultimately embody the "oncogenic paradigm" will allow to overcome this challenging disease.

It has been a privilege for me to have an opportunity to work with such a talented group of scientists and investigators and I would like to thank all of them for their outstanding contributions to this book. I also would like to thank all of the patients who participate in clinical trials and contribute to the progress in the development of novel and more effective cancer therapies. Lastly, I would like to thank my family; without them I would have never been able to do what I love.

Giovanni Melillo MD

Contents

Contributors

Jan Alsner Department of Experimental Clinical Oncology, Aarhus University Hospital, Aarhus C, Denmark

Maria Carla Bosco Laboratory of Molecular Biology, Gaslini Institute, Genoa, Italy

Mary-Keara Boss Department of Molecular Biomedical Sciences, North Carolina State University, Raleigh, NC, USA

Morten Busk Department of Experimental Clinical Oncology, Aarhus University Hospital, Aarhus C, Denmark

Denise A. Chan Department of Radiation Oncology, University of California, San Francisco, CA, USA

Yao Dai Department of Radiation Oncology, University of Florida Shands Cancer Center, Gainesville, FL, USA

Mark W. Dewhirst Department of Radiation Oncology, Duke University Medical Center, Durham, NC, USA

Souvik Dey Department of Radiation Oncology, Smilow Center for Translational Research, Perelman School of Medicine, University of Pennsylvania, Philadelphia, USA

Nicole D. Fer Tumor Microenvironment Laboratory, SAIC-Frederick Inc., Frederick National Laboratory for Cancer Research, Frederick, MD, USA

Agnes Görlach Experimental and Molecular Pediatric Cardiology, German Heart Center Munich at the TU Munich, Munich, Germany

Amato J. Giaccia Department of Radiation Oncology, Stanford University School of Medicine, Stanford, CA, USA

Simona Greco Molecular Cardiology Laboratory, IRCCS-Policlinico San Donato, Milan Italy

Ester M. Hammond The Department of Oncology, The Cancer Research UK/MRC Gray Institute for Radiation Oncology and Biology, The University of Oxford, Oxford, UK

Adrian L Harris Molecular Oncology Laboratories, Department of Oncology, Oxford University, Oxford, UK

Kevin O. Hicks Auckland Cancer Society Research Centre, The University of Auckland, Auckland, New Zealand

Geoff S Higgins Gray Institute for Radiation Oncology and Biology, Department of Oncology, Oxford University, Oxford, UK

Michael R. Horsman Department of Experimental Clinical Oncology, Aarhus University Hospital, Aarhus C, Denmark

Mircea Ivan Department of Microbiology and Immunology, Indiana University, Indianapolis, IN, USA

Department of Medicine, Indiana University, Indianapolis, IN, USA

Constantinos Koumenis Department of Radiation Oncology, Smilow Center for Translational Research, Perelman School of Medicine, University of Pennsylvania, Philadelphia, USA

Stacey L. Lehman Department of Radiation Oncology, Smilow Center for Translational Research, Perelman School of Medicine, University of Pennsylvania, Philadelphia, USA

Cell and Molecular Biology Graduate Group, Perelman School of Medicine, University of Pennsylvania, Philadelphia, USA

Raymond J. Louie Department of Radiation Oncology, University of California, San Francisco, CA, USA

Fabio Martelli Molecular Cardiology Laboratory, IRCCS-Policlinico San Donato, Milan, Italy

Thomas Nielsen Department of Experimental Clinical Oncology, Aarhus University Hospital, Aarhus C, Denmark

Marianne Nordsmark Department of Oncology, Aarhus University Hospital, Aarhus C, Denmark

Monica M. Olcina The Department of Oncology, The Cancer Research UK/MRC Gray Institute for Radiation Oncology and Biology, The University of Oxford, Oxford, UK

Jens Overgaard Department of Experimental Clinical Oncology, Aarhus University Hospital, Aarhus C, Denmark

Mercè Padró Department of Radiation Oncology, University of California, San Francisco, CA, USA

Gregory M. Palmer Department of Radiation Oncology, Duke University Medical Center, Durham, NC, USA

Silvia Pastorekova Department of Molecular Medicine, Institute of Virology, Slovak Academy of Sciences, Bratislava, Slovak Republic

Frederik B. Pruijn Auckland Cancer Society Research Centre, The University of Auckland, Auckland, New Zealand

Annamaria Rapisarda Tumor Microenvironment Laboratory, SAIC-Frederick Inc., Frederick National Laboratory for Cancer Research, Frederick, MD, USA

Carly M. Sayers Department of Radiation Oncology, Smilow Center for Translational Research, Perelman School of Medicine, University of Pennsylvania, Philadelphia, USA

Pharmacology Graduate Group, Perelman School of Medicine, University of Pennsylvania, Philadelphia, USA

Luana Schito Vascular Program, Institute for Cell Engineering, Johns Hopkins University School of Medicine, Baltimore, MD, USA

McKusick-Nathans Institute of Genetic Medicine, Johns Hopkins University School of Medicine, Baltimore, MD, USA

Gregg L. Semenza Vascular Program, Institute for Cell Engineering, Johns Hopkins University School of Medicine, Baltimore, MD, USA

Departments of Pediatrics, Medicine, Oncology, Radiation Oncology, and Biological Chemistry, Johns Hopkins University School of Medicine, Baltimore, MD, USA

McKusick-Nathans Institute of Genetic Medicine, Johns Hopkins University School of Medicine, Baltimore, MD, USA

Dietmar W. Siemann Department of Radiation Oncology, University of Florida Shands Cancer Center, Gainesville, FL, USA

Claudiu T. Supuran Polo Scientifico, NEUROFARBA Department, Section of Pharmaceutical and Nutriceutical Sciences, Università degli Studi di Firenze, Florence, Italy

Luigi Varesio Laboratory of Molecular Biology, Gaslini Institute, Genoa, Italy

Jingli Wang Auckland Cancer Society Research Centre, The University of Auckland, Auckland, New Zealand

William R. Wilson Auckland Cancer Society Research Centre, The University of Auckland, Auckland, New Zealand

Xiaoling Zhong Department of Medicine, Indiana University, Indianapolis, IN, USA

Part I
Biology

Chapter 1
Hypoxia and Breast Cancer Metastasis

Luana Schito and Gregg L. Semenza

Abstract Mortality in breast cancer patients ultimately results from the dissemination of breast cancer cells (BCCs) from the primary tumor to distant organs (metastasis). The proliferation of BCCs leads to areas of decreased oxygen availability (intratumoral hypoxia), which drives breast cancer pathogenesis through the activation of hypoxia-inducible factors (HIFs). In this review, we outline emerging HIF-dependent molecular mechanisms that promote the metastatic dissemination of hypoxic BCCs through the lymphatic and circulatory systems. We show that HIFs contribute to key aspects of metastatic progression through the transcriptional activation of target genes in BCCs, lymphatic vessels, blood vessels, and supporting stromal cells within the primary tumor. Finally, we discuss the pharmacological inhibition of HIFs as a novel therapeutic approach to block breast cancer metastasis and improve patient survival.

Keywords Bone marrow derived cells · EMT · Extracellular matrix · Extravasation · HIF-1 · Intravasation · Lymphangiogenesis · Lymphatic metastasis · Margination · Tumor stroma

1.1 Introduction

Breast cancer comprises 18 % of all female malignancies, with about one million new cases reported worldwide annually (McPherson et al. 2000). Metastasis, which is the dissemination of breast cancer cells (BCCs) from the primary tumor to distant organs, is the major cause of mortality in women with breast cancer (Eccles and Welch 2007). BCCs characteristically metastasize to regional lymph nodes, bone marrow,

---

G. L. Semenza (✉)
Departments of Pediatrics, Medicine, Oncology, Radiation Oncology, and Biological Chemistry, Johns Hopkins University School of Medicine, Baltimore, MD 21205, USA
e-mail: gsemenza@jhmi.edu

L. Schito · G. L. Semenza
Vascular Program, Institute for Cell Engineering, Johns Hopkins University School of Medicine, Baltimore, MD 21205, USA

McKusick-Nathans Institute of Genetic Medicine, Johns Hopkins University School of Medicine, Baltimore, MD 21205, USA

G. Melillo (ed.), *Hypoxia and Cancer,* Cancer Drug Discovery and Development, DOI 10.1007/978-1-4614-9167-5_1,

brain, liver, and lungs through lymphatic and hematogenous routes. In this review, we discuss the molecular mechanisms by which hypoxia-inducible factors (HIFs) promote metastasis of BCCs and emerging evidence that HIFs represent important therapeutic targets in breast cancer.

1.2 Intratumoral Hypoxia in Human Cancers

BCCs are characterized by increased cell proliferation and decreased cell death that result from multiple somatic mutations and epigenetic changes (Veronese and Gambacorta 1991). The increase in cell number results in increased O_2 consumption, which leads to hypoxia (decreased O_2 availability). Intratumoral hypoxia is observed in approximately half of all locally advanced cancers and is an adverse prognostic factor in multiple types of human cancer that is independent of established prognostic parameters such as tumor stage, histological grade, and lymph node status (Vaupel et al. 2004; Vaupel 2009). Mathematical modeling suggests that intratumoral hypoxia occurs early in breast cancer progression with severe hypoxia and acidosis developing in intraductal tumors at distances greater than 100 μm from the ductal basement membrane (Gatenby et al. 2007). Measurement of the partial pressure of O_2 (PO_2) in patients with breast cancer revealed that the median PO_2 was 10 mmHg (1.5 % [v/v] O2) and that in > 50 % of breast tumors the measured PO_2 was less than 2.5 mmHg; in contrast, normal breast tissue presented a median PO_2 of 65 mmHg (Vaupel et al. 2007). Most metastatic lesions were also hypoxic, with PO_2 values that were comparable to the primary tumor (Vaupel et al. 2007). Low PO_2 stimulates blood vessel formation (angiogenesis), but the new blood vessels that form within cancers are morphologically abnormal, which results in heterogeneous perfusion of the tumor with persistent intratumoral hypoxia (Vaupel et al. 2004).

1.3 Adaptive Responses to Hypoxia are Regulated by HIFs

In mammalian cells, the response to hypoxia depends in large part on the activation of HIF-1, a heterodimeric transcription factor consisting of a hypoxia-inducible HIF-1α subunit and a constitutively expressed HIF-1β subunit (Wang and Semenza 1995; Wang et al. 1995). HIF-1 transactivates target genes containing *cis*-acting hypoxia response elements that contain the HIF-1-binding site sequence 5′-[A/G]CGTG-3′, which was first identified in the human genes encoding erythropoietin (EPO) (Semenza and Wang 1992), the glycolytic enzymes aldolase A, enolase 1, phosphofructokinase L, phosphoglycerate kinase 1, and lactate dehydrogenase A (Semenza et al. 1994; Semenza et al. 1996), and vascular endothelial growth factor (VEGF)-A (Forsythe et al. 1996).

Database searches identified HIF-2α as a protein that has high sequence identity to HIF-1α, and subsequent studies demonstrated that HIF-2α is also hypoxia inducible, heterodimerizes with HIF-1α, and transactivates reporter genes containing hypoxia response elements (Ema et al. 1997; Flamme et al. 1997; Hogenesch et al. 1997; Tian et al. 1997; Wiesener et al. 1998). The set of target genes transactivated by HIF-2 includes *EPO* and *VEGF* but HIF-2 does not activate transcription of the genes encoding glycolytic enzymes (Rankin et al. 2007; Loboda et al. 2010). Altogether, >1,000 genes are regulated by binding of either HIF-1, HIF-2, or both.

HIF-1α protein levels are tightly regulated based on the cellular PO_2. When cultured cells are exposed to 1 % O_2 for 4 h to induce HIF-1α protein expression and then returned to the standard 20 % O_2 tissue culture environment, HIF-1α protein levels decay with a half-life of <5 min (Wang et al. 1995), which reflects the time required for O_2 equilibration in the media (Jewell et al. 2001), i.e., HIF-1α protein degradation is remarkably rapid upon reoxygenation. O_2 directly regulates the half-life of HIF-1α through the hydroxylation of two proline residues (Pro^{402} and Pro^{564} in human HIF-1α) by the prolyl hydroxylase domain proteins PHD1, PHD2, and PHD3 (Kaelin and Ratcliffe 2008; Chowdhury et al. 2009). Both proline residues are embedded within the conserved motif LXXLAP that is also present in HIF-2α. This posttranslational modification of HIF-1α creates a recognition motif for the von Hippel–Lindau tumor suppressor protein (VHL; Kaelin and Ratcliffe 2008).

PHDs require O_2 and α-ketoglutarate as substrates for their enzymatic activity, catalyzing a reaction in which one oxygen atom is inserted into HIF-1α and the other is inserted into α-ketoglutarate to generate succinate and CO_2 (Kaelin and Ratcliffe 2008; Chowdhury et al. 2009). Among the PHDs, PHD2 plays a critical role in degrading HIF-1α in many tissues under normoxic conditions. PHDs contain Fe^{2+} in their catalytic center, which is inhibited when cells are treated with iron chelators or with Co^{2+} compounds, thereby explaining how these agents induce HIF-1 activity (Wang and Semenza 1993). Competitive antagonists of α-ketoglutarate such as dimethyloxalylglycine (DMOG) also inhibit PHDs and induce HIF-1α and HIF-2α expression under normoxic conditions (Elvidge et al. 2006).

1.4 Multistep HIF-dependent Mechanisms of Breast Cancer Metastatic Progression

The initiation of the metastatic process involves the sequential acquisition of phenotypes by BCCs that include: (1) increased motility through transformation of the epithelial-derived cancer cells into a mesenchymal phenotype (epithelial–mesenchymal transition, EMT); (2) enhanced cellular ability to disrupt the integrity of the basement membrane and invade through the extracellular matrix (ECM) and into the lumen of lymphatic or blood vessels (intravasation); (3) increased ability to survive in suspension in the lymph or blood (i.e., inhibition of anoikis) and to exit from the vessel at the metastatic site (extravasation); and (4) increased ability to revert to a proliferative epithelial phenotype (i.e., mesenchymal–epithelial transition) for clonal expansion and colonization of regional lymph nodes or distant organs.

In addition, BCCs require support from other cell types in order to successfully metastasize. BCCs stimulate the mobilization of bone marrow-derived cells, which home to future sites of hematogenous metastasis prior to the arrival of cancer cells in a process called pre-metastatic niche formation (Kaplan et al. 2005). Furthermore, the growth of both primary tumor and metastases is dependent on tumor angiogenesis, whereas BCC dissemination through lymphatic vessels relies on lymphangiogenesis. Fibroblasts and inflammatory cells such as tumor-infiltrating macrophages also support tumor growth (Lin and Pollard 2004) and lymphangiogenesis (Schoppmann et al. 2002a). Pericyte depletion within the primary tumor results in hypoxia-induced EMT and enhanced metastasis by activation of the C-MET receptor tyrosine kinase signaling pathway (Cooke et al. 2012).

1.4.1 EMT, Increased Cell Motility, and Invasiveness

The epithelial phenotype of early-stage tumors progresses towards a mesenchymal phenotype in a process known as EMT. The mesenchymal phenotype is characterized by loss of cell polarity and loosening of intercellular adhesion junctions, facilitating the migration of BCCs (Yang et al. 2008). One of the key proteins involved in EMT is E-cadherin, a major component of the adherens junction complex. HIF-1 indirectly represses the expression of E-cadherin through the transactivation of SNAIL (SNAI1), TCF3 (E47), ZEB1, and ZEB2 (Imai et al. 2003; Krishnamachary et al. 2006) with concomitant gain of N-cadherin expression, which allows cells to escape anoikis (Kumar et al. 2011). Through upregulation of urokinase-type plasminogen activator receptor (PLAUR) expression, hypoxia enhances proteolytic activity at the invasive front and alters the interactions between integrins and components of the ECM, thereby enabling cellular invasion into the basement membrane and the underlying stroma (Yoon et al. 2006; Lester et al. 2007; Buchler et al. 2009). Cell motility is increased through hypoxia-induced hepatocyte growth factor (HGF)–MET receptor signaling, resulting in cell migration towards the blood or lymphatic microcirculation (Pennacchietti et al. 2003; Parr et al. 2004; Eckerich et al. 2007).

EMT is also modulated by the establishment of a positive feedback between HIF-1α and Notch which results in increased SNAIL expression (Peinado et al. 2005; Higgins et al. 2007; Sahlgren et al. 2008; Chen et al. 2010). HIF-1 also directly transactivates the gene encoding TWIST, which is a central regulator of gastrulation, mesoderm specification, and EMT (Yang et al. 2008). BCCs constitutively expressing HIF-1α showed increased TWIST messenger RNA (mRNA) and protein levels, whereas reduction of endogenous HIF-1α levels blocked hypoxia-induced expression of TWIST (Yang and Wu 2008; Yang et al. 2008). Inflammatory cytokines produced by BCCs, such as interleukin-6 (IL-6), also induce EMT by suppressing E-cadherin expression and inducing vimentin, N-cadherin, SNAIL, and TWIST (Sullivan et al. 2009), but the role of HIF-1 in the production of, or response to, IL-6 has not been determined.

1.4.2 HIF-dependent Mechanisms of BCC Intravasation and Extravasation

BCCs disrupt the integrity of the basement membrane and interact directly with the underlying ECM, eventually penetrating the wall of a blood or lymphatic vessel (intravasation) to become circulating or intralymphatic tumor cells, respectively. It has been estimated that 80 % of breast cancer metastases disseminate through the lymphatic system, while 20 % of metastases may occur through the blood vasculature or by direct seeding (Leong et al. 2011), but this is an area of considerable controversy (Ran et al. 2010).

Increased expression of several genes implicated in breast cancer metastasis to the lungs has been reported in primary tumors of women with lung metastases and in breast cancer subclones selected for lung metastasis in mice (Minn et al. 2005; Gupta et al. 2007). Breast cancers spontaneously arising in conditional knockout mice expressing polyoma middle T antigen and lacking HIF-1α expression in mammary epithelial cells showed significantly reduced lung metastasis compared to breast cancers arising in wild-type mice, demonstrating that HIF-1 promotes breast cancer metastasis (Liao et al. 2007). Loss of HIF activity as a result of HIF-1α and HIF-2α knockdown in human BCCs led to a dramatic reduction in spontaneous metastasis of BCCs to lungs following orthotopic transplantation into the mammary fat pad of immunodeficient mice (Zhang et al. 2012). In addition, exposure of BCCs to 1 % O_2 in tissue culture (to simulate hypoxia within the primary tumor) prior to tail vein injection increased the extravasation of control BCCs, whereas extravasation of HIF-knockdown BCCs was significantly impaired (Zhang et al. 2012). These data indicate an important contribution of HIFs to metastasis via blood vessels from the breast to the lungs.

1.4.2.1 Angiogenesis and Blood Vessel Intravasation

Metastatic BCCs arising from the primary tumor access distant sites through either the systemic circulation or the lymphatic system. HIF activation in hypoxic BCCs stimulates tumor vascularization by inducing transcription of the genes encoding VEGF-A, stromal cell-derived factor 1 (SDF-1), placental growth factor, platelet-derived growth factor (PDGF-B), and angiopoietin 1 and 2, which act upon vascular cells through autocrine or paracrine signaling by activating their cognate receptors (Semenza 2012a). Intravasation of BCCs is facilitated by the greatly increased permeability of tumor blood vessels. A pivotal HIF target gene product in this process is VEGF-A (Forsythe et al. 1996), which is a secreted factor that induces a potent increase in vascular permeability that is 50,000-fold more potent (on a molar basis) than histamine (Yeo et al. 1993). VEGF-A expression mediated by HIF-1α in BCCs results in tumoral vessels with higher permeability to cells and macromolecules, thereby facilitating intravasation (Lee et al. 2003; Eum et al. 2004).

1.4.2.2 Blood Vessel Margination

Interactions between BCCs and endothelial cells (ECs) are regulated by HIF-1 through BCC expression of the transmembrane L1 cell adhesion molecule (L1CAM), which mediates cell adhesion by homophilic interactions and by heterophilic binding to integrins (Semenza 2012b). Genetic or pharmacologic inhibition of HIF activity impaired hypoxia-induced adherence between BCCs and ECs (Zhang et al. 2012). Further analysis revealed that HIF-1α, but not HIF-2α, was required for hypoxia-induced L1CAM expression and that short hairpin RNA (shRNA) interference targeting L1CAM (i.e., L1CAM loss of function) impaired the adherence of both hypoxic and nonhypoxic BCCs to ECs (Zhang et al. 2012). Thus, L1CAM is a key HIF-responsive mediator of the margination of BCCs onto ECs of blood vessels in the lungs and possibly other metastatic sites.

1.4.2.3 Blood Vessel Extravasation

It has been estimated that less than 0.1 % of cancer cells that enter the vascular system establish a metastatic lesion (Fidler 1970), which suggests that the extravasation of cancer cells from the circulation into a tissue microenvironment that is favorable for survival and proliferation may be a limiting step in the metastatic process. Angiopoietin-like 4 (ANGPTL4) is a secreted protein that is expressed at increased levels in the primary breast cancers of women with lung metastases (Padua et al. 2008) and in BCCs exposed to hypoxia (Zhang et al. 2012). By disrupting EC–EC interactions, ANGPTL4 increases the permeability of lung capillaries and facilitates the transendothelial passage of cancer cells (Padua et al. 2008). ANGPTL4 mRNA and protein expression were shown to be hypoxia inducible in BCCs with intact HIF-1 and HIF-2 activity but not in BCCs lacking HIF-1 and HIF-2 activity (Zhang et al. 2012). Screening of human genome sequences revealed a hypoxia response element in the 5′-flanking region approximately 1.5 kb from the *ANGPTL4* gene translation start site. Following mammary fat pad implantation of ANGPTL4-knockdown cells, primary tumor growth was unaffected relative to control BCCs but metastasis to the lungs was almost completely eliminated (Zhang et al. 2012). Despite the fact that both intravasation and extravasation require loosening of EC–EC interactions, ANGPTL4 only promotes extravasation of BCCs (Padua et al. 2008), suggesting that the ANGPTL4-secreting cell needs to be located on the luminal side of the vascular endothelium in order for ANGPTL4 to access its putative receptor(s) on ECs, the identity of which has not been definitively established but may include integrins β_1 and β_5 (Tan et al. 2012).

1.4.2.4 Lymphangiogenesis and Lymphatic Vessel Intravasation

Dissemination through the lymphatic system begins at the regional lymph nodes adjacent to the primary lesion (Chua et al. 2001) and accordingly the extent of lymph node involvement is one of the strongest prognostic indicators for breast cancer (Schoppmann et al. 2002b). Strikingly, the abundant literature highlighting the role of hypoxia and HIF-1α overexpression with increased angiogenesis and hematogenous dissemination (Erler et al. 2006a; Chan and Giaccia 2007; Liao et al. 2007; Rundqvist and Johnson 2010; Cooke et al. 2012) contrasts with the lack of studies addressing the role of HIF-1α in lymphangiogenesis or lymphatic dissemination of cancer.

Lymphatic vessels are blind-ended with rudimentary or absent pericytes, a fenestrated or absent basement membrane, and a lack of specialized junctions between lymphatic endothelial cells (LECs), resulting in high permeability (Swartz and Skobe 2001; Ji 2006). Lymphangiogenic vessels (i.e., newly formed lymphatics) in tumors present wider lumina and increased numbers of intercellular spaces compared to their established normal counterparts (Alitalo 2011). It is postulated that lymphatic intravasation is facilitated by the absence of a basement membrane and widened LEC–LEC junctions. BCCs can be transported through lymphatic vessels to local lymph nodes. Chemokine receptors such as CXCR4 and CCR7 are expressed by BCCs and melanoma cells, whereas their ligands, SDF-1 and CCL21, are expressed by LECs (Muller et al. 2001). Numerous reports indicate that hypoxia induces CXCR4 (Staller et al. 2003) and SDF-1 (Ceradini et al. 2004) through HIF-dependent mechanisms. Moreover, CCR7 expression is increased as a result of endothelin-dependent HIF-1α stabilization (Wilson et al. 2006). Endothelin-1 is encoded by a direct HIF-1 target gene (Camenisch et al. 2001), which suggests the existence of a feed-forward mechanism to increase HIF activity. These reports support the hypothesis that LECs have a chemoattractant effect upon BCCs that may facilitate intravasation and migration through the lymphatics.

A significant overlap was shown between the spectrum of cytokines with an angiogenic function that also presented lymphangiogenic properties. VEGF-A induces proliferation of VEGF receptor-2-expressing tumor-associated lymphatic vessels and occasionally intratumoral lymphangiogenesis (Nagy et al. 2002; Bjorndahl et al. 2005b; Hirakawa et al. 2005). Also, IGF-1 and IGF-2 have been shown to stimulate proliferation and migration of primary LECs (Bjorndahl et al. 2005a). A predominantly lymphangiogenic role has been attributed to two members of the VEGF family, VEGF-C and VEGF-D, which signal through the VEGF receptor-3 (VEGFR3), increasing lymphangiogenesis, lymph node invasion, and distant metastases (Joukov et al. 1996; Kukk et al. 1996; Makinen et al. 2001; Skobe et al. 2001; Stacker et al. 2001). Indeed, treatment with an inactive soluble form of VEGFR-3 or a kinase inhibitor blocked these processes (Mattila et al. 2002). A clinical study demonstrated a correlation between HIF-1α and VEGF-C expression in 119 cases of lymph node-positive breast cancer biopsies (Schoppmann et al. 2006).

Various clinical studies identified an association between HIF-1α, lymphatic metastasis, and lymphatic vessel density in breast cancer (Bos et al. 2003; Bos et al. 2005; Schoppmann et al. 2006). However, the molecular mechanism by which hypoxia contributes to lymphangiogenesis and lymphatic metastasis in breast cancer was not known. Previous reports showed that VEGF-A, in addition to angiogenesis, induces lymphangiogenesis and lymphatic metastasis in animal models of breast cancer and inflammation, providing a potential mechanism for HIF-dependent induction of lymphatic metastasis (Cursiefen et al. 2004; Hirakawa et al. 2005; Halin et al. 2007; Whitehurst et al. 2007). In contrast, the prototypical lymphangiogenic factors VEGF-C and VEGF-D are not induced by hypoxia (Karpanen et al. 2001; Currie et al. 2004; He et al. 2004; Okada et al. 2005; Hirakawa et al. 2007; Kopfstein et al. 2007; Gu et al. 2008).

Previous work showed that PDGF-B overexpression was associated with a high proliferation index in tumors from breast cancer patients (Coltrera et al. 1995). PDGF-B overexpression induced intratumoral lymphangiogenesis and lymph node metastasis in a mouse fibrosarcoma model (Cao et al. 2004). Moreover, increased expression of PDGF-B and PDGF receptor β was associated with lymphatic metastasis in patients with gastric cancer (Kodama et al. 2010). Intriguingly, PDGF-B mRNA expression was induced by hypoxia in several cell types (Ulleras et al. 2001; Zhang et al. 2003) in a HIF-1-dependent manner (Kelly et al. 2003), but the precise mechanism of HIF-1-dependent induction was not known.

To address the role of HIF-1 and PDGF-B in lymphatic dissemination of hypoxic BCCs, BCCs bearing a genetic knockdown of HIF-1α, HIF-2α, or both were implanted into the mammary fat pad of immunodeficient mice. Compared to control tumors, lymphatic vessel density was decreased by 45 %, 24 %, and 27 %, respectively, in tumors deficient for HIF-1α, HIF-2α, or both (Schito et al. 2012). Histopathological and immunohistochemical analysis showed that knockdown of HIF-1α, HIF-2α, or both impaired metastasis of BCCs to axillary lymph nodes (Schito et al. 2012). The luminal space of lymphatic vessels from control tumor-bearing mice was occupied by BCCs, whereas lymphatic vessels from HIF-knockdown tumor-bearing mice appeared collapsed and rarely contained intravasated cells. In addition, lymphatic vessels were observed in the subcapsular space of metastasized lymph nodes (Schito et al. 2012). These findings showed for the first time that HIF activity in BCCs contributes to lymphovascular and nodal invasion *in vivo*. HIF-1 binding to an intronic hypoxia response element was shown to transactivate the *PDGFB* gene in hypoxic cancer cells. Moreover, PDGF-B secreted by hypoxic BCCs bound PDGF receptor β (PDGFRβ) on LECs and increased LEC proliferation and chemotaxis, thereby providing a molecular mechanism for LEC migration into hypoxic cancers and lymphangiogenesis, facilitating the subsequent intravasation of BCCs into newly formed lymphatic vessels (Schito et al. 2012). Neutralization of PDGF-B in conditioned media from hypoxic BCCs was sufficient to abolish HIF-1-dependent proliferative and chemotactic effects on LECs, indicating a pivotal role of HIF-1 → PDGF-B → PDGFRβ signaling upon BCC-induced lymphangiogenesis *in vitro* (Schito et al. 2012).

Screening of 3,120 Food and Drug Administration (FDA)-approved drugs contained in the Hopkins Drug Library (Chong et al. 2007) revealed that digoxin and other cardiac glycosides inhibited HIF-1α (and HIF-2α) protein expression and growth of prostate cancer cell xenografts (Zhang et al. 2008). Digoxin was subsequently shown to inhibit BCC orthograft growth and lung metastasis (Zhang et al. 2012). Systemic treatment of tumor-bearing mice with digoxin blocked lymphatic metastasis to axillary lymph nodes (Schito et al. 2012). Treatment with the tyrosine kinase inhibitor imatinib also inhibited primary tumor growth and lymphatic metastasis. Thus, HIF-1 inhibition or PDGFRβ blockade effectively decreases lymphatic dissemination of BCCs.

1.4.3 HIF-dependent Effects upon the Host Tissue Facilitating Metastasis

The local microenvironment, or niche, of a cancer cell plays important roles in cancer progression. A major component of the niche is the ECM, a complex network of glycoproteins, proteoglycans, and polysaccharides with different physical and biochemical properties (Ozbek et al. 2010). ECM remodeling is tightly regulated during development through the expression and activities of ECM enzymes at multiple levels (Page-McCaw et al. 2007) and ECM remodeling also occurs in cancer. Tumor stroma is stiffer than normal stroma; in the case of breast cancer, diseased tissue can be ten times stiffer than the normal breast (Levental et al. 2009; Lopez et al. 2011). HIFs are known to activate expression of genes encoding the secreted matrix metalloproteinases (MMPs) 2 and 9 (Krishnamachary et al. 2003; Manalo et al. 2005; Koh et al. 2011). The *MMP14* gene encoding membrane type 1 MMP is a direct HIF-2 target in renal cell carcinoma (Petrella et al. 2005). MMP14 catalyzes the proteolytic activation of MMPs 2 and 9 from their inactive zymogens.

A seminal study demonstrated that VEGF receptor 1-positive bone marrow-derived cells (BMDCs) home to metastatic sites and form cellular clusters before the arrival of cancer cells (Kaplan et al. 2005). One of the critical mediators of BMDC recruitment in the pre-metastatic niche was lysyl oxidase (LOX) (Erler et al. 2009). Increased LOX expression is associated with metastasis and poor survival in patients with breast or head and neck cancer (Erler et al. 2006). LOX is secreted by hypoxic breast cancer cells into the circulation and accumulates at pre-metastatic sites, where it crosslinks type IV collagen in the basement membrane, which is essential for $CD11b^+$ myeloid cell recruitment to the lungs. Once $CD11b^+$ cells adhere to crosslinked collagen, they produce MMP-2, which in turn cleaves collagen, enhancing invasion and recruitment of BMDCs and metastasizing cancer cells (Erler et al. 2009).

In addition to LOX, two LOX-like (LOXL) proteins, LOXL2 and LOXL4, were expressed in a HIF-dependent manner in human BCC lines subjected to hypoxia (Wong et al. 2011, 2012). Knockdown of HIF-1α, HIF-2α, or both blocked the hypoxia-induced expression of LOX, LOXL2, and LOXL4. Genetic ablation of LOX, LOXL2, or LOXL4 expression blocked metastatic niche formation, which also occurred with knockdown of HIF-1α, HIF-2α, or both. In line with these findings, human breast cancers overexpressed LOX, LOXL2, and LOXL4 relative to surrounding normal tissue (Wong et al. 2011).

Cancer progression is influenced by tumor stroma reactivity. In breast adenocarcinoma, cancer-associated fibroblasts are affected by oxidative stress-mediated activation of HIF-1α, which in turn activates the secretion of SDF-1 (Toullec et al.

2010). A functional contribution of stromal cells to cancer progression arises from the pericytes, which in normal tissues provide structural support to blood vessels and regulate tissue physiology by influencing vascular stability (Kim et al. 2006; Dore-Duffy and Cleary 2011). Low pericyte coverage of intratumoral blood vessels is associated with invasive breast cancer and correlated with decreased patient survival (Cooke et al. 2012). In a 4T1 mouse breast cancer model, low pericyte coverage in tumoral vessels resulted in intratumoral hypoxia (Cooke et al. 2012). In this scenario, increased expression of HIF-1α promoted EMT and tumor progression. Loss of intratumoral pericytes was coupled with high MET expression, which in turn was associated with decreased survival of breast cancer patients (Cooke et al. 2012).

1.5 Therapeutic Implications and Concluding Remarks

The pleiotropic effect of hypoxia and HIF activation promotes breast cancer progression at multiple steps leading to acquisition of the metastatic phenotype. HIF inhibition offers the therapeutic potential of blocking angiogenesis, BCC invasiveness, EMT, and metastatic niche formation, while also decreasing lymphangiogenic and chemotactic cytokines that facilitate lymphatic intravasation of BCCs. Most of the studies described above utilized BCCs derived from triple-negative breast cancers that lack expression of the estrogen, progesterone, and HER2 receptors, and respond poorly to currently available therapies (Pal et al. 2011). In this challenging context, inhibition of HIF activity may provide a means of inhibiting primary tumor growth and blocking hematogenous and lymphogenous metastasis in patients with breast cancer. Analysis of biopsies from patients with invasive breast cancer showed that expression of HIF-1α and PDGF-B is linearly correlated with lymphatic vessel area quantified by immunohistochemistry and digital morphometry (Schito et al. 2012). In turn, these parameters correlated with the Scarf–Bloom–Richardson grade of the tumors (Schito et al. 2012), which is a strong predictor of patient mortality (Le Doussal et al. 1989; Amat et al. 2002). These clinical data provide support for clinical trials in which HIF or PDGF inhibition is evaluated. Preclinical data suggest that HIF inhibitors, when added to current therapeutic regimens, may increase the survival of breast cancer patients, particularly those with triple negative disease and high HIF-1α levels in their tumor biopsies, by impairing metastasis as well as other critical steps in cancer progression.

References

Alitalo K (2011) The lymphatic vasculature in disease. Nat Med 17:1371–1380

Amat S, Penault-Llorca F, Cure H, Le Bouedec G, Achard JL, Van Praagh I, Feillel V, Mouret-Reynier MA, Dauplat J, Chollet P (2002) Scarff-Bloom-Richardson (SBR) grading: a pleiotropic marker of chemosensitivity in invasive ductal breast carcinomas treated by neoadjuvant chemotherapy. Int J Oncol 20:791–796

Bjorndahl M, Cao R, Nissen LJ, Clasper S, Johnson LA, Xue Y, Zhou Z, Jackson D, Hansen AJ, Cao Y (2005a) Insulin-like growth factors 1 and 2 induce lymphangiogenesis in vivo. Proc Natl Acad Sci U S A 102:15593–15598

Bjorndahl MA, Cao R, Burton JB, Brakenhielm E, Religa P, Galter D, Wu L, Cao Y (2005b) Vascular endothelial growth factor-A promotes peritumoral lymphangiogenesis and lymphatic metastasis. Cancer Res 65:9261–9268

Bos R, Groep P van der, Greijer AE, Shvarts A, Meijer S, Pinedo HM, Semenza GL, Diest PJ van, Wall E van der (2003) Levels of hypoxia-inducible factor-1α independently predict prognosis in patients with lymph node negative breast carcinoma. Cancer 97:1573–1581.

Bos R, Diest PJ van, Jong JS de, Groep P van der, Valk P van der, Wall E van der (2005) Hypoxia-inducible factor-1α is associated with angiogenesis, and expression of bFGF, PDGF-BB, and EGFR in invasive breast cancer. Histopathology 46:31–36.

Buchler P, Reber HA, Tomlinson JS, Hankinson O, Kallifatidis G, Friess H, Herr I, Hines OJ (2009) Transcriptional regulation of urokinase-type plasminogen activator receptor by hypoxia-inducible factor 1 is crucial for invasion of pancreatic and liver cancer. Neoplasia 11:196–206

Camenisch G, Stroka DM, Gassmann M, Wenger RH (2001) Attenuation of HIF-1 DNA-binding activity limits hypoxia-inducible endothelin-1 expression. Pflugers Arch 443:240–249

Cao R, Bjorndahl MA, Religa P, Clasper S, Garvin S, Galter D, Meister B, Ikomi F, Tritsaris K, Dissing S, Ohhashi T, Jackson DG, Cao Y (2004) PDGF-BB induces intratumoral lymphangiogenesis and promotes lymphatic metastasis. Cancer Cell 6:333–345

Ceradini DJ, Kulkarni AR, Callaghan MJ, Tepper OM, Bastidas N, Kleinman ME, Capla JM, Galiano RD, Levine JP, Gurtner GC (2004) Progenitor cell trafficking is regulated by hypoxic gradients through HIF-1 induction of SDF-1. Nat Med 10:858–864

Chan DA, Giaccia AJ (2007) Hypoxia, gene expression, and metastasis. Cancer Metastasis Rev 26:333–339

Chen J, Imanaka N, Chen J, Griffin JD (2010) Hypoxia potentiates Notch signaling in breast cancer leading to decreased E-cadherin expression and increased cell migration and invasion. Br J Cancer 102:351–360

Chong CR, Xu J, Lu J, Bhat S, Sullivan DJ Jr, Liu JO (2007) Inhibition of angiogenesis by the antifungal drug itraconazole. ACS Chem Biol 2:263–270

Chowdhury R, McDonough MA, Mecinovic J, Loenarz C, Flashman E, Hewitson KS, Domene C, Schofield CJ (2009) Structural basis for binding of hypoxia-inducible factor to the oxygen-sensing prolyl hydroxylases. Structure 17:981–989

Chua B, Ung O, Taylor R, Boyages J (2001) Frequency and predictors of axillary lymph node metastases in invasive breast cancer. ANZ J Surg 71:723–728

Coltrera MD, Wang J, Porter PL, Gown AM (1995) Expression of platelet-derived growth factor B-chain and the platelet-derived growth factor receptor beta subunit in human breast tissue and breast carcinoma. Cancer Res 55:2703–2708

Cooke VG, LeBleu VS, Keskin D, Khan Z, O'Connell JT, Teng Y, Duncan MB, Xie L, Maeda G, Vong S, Sugimoto H, Rocha RM, Damascena A, Brentani RR, Kalluri R (2012) Pericyte depletion results in hypoxia-associated epithelial-to-mesenchymal transition and metastasis mediated by met signaling pathway. Cancer Cell 21:66–81

Currie MJ, Hanrahan V, Gunningham SP, Morrin HR, Frampton C, Han C, Robinson BA, Fox SB (2004) Expression of vascular endothelial growth factor D is associated with hypoxia inducible factor (HIF-1α) and the HIF-1α target gene DEC1, but not lymph node metastasis in primary human breast carcinomas. J Clin Pathol 57:829–834

Cursiefen C, Chen L, Borges LP, Jackson D, Cao J, Radziejewski C, D'Amore PA, Dana MR, Wiegand SJ, Streilein JW (2004) VEGF-A stimulates lymphangiogenesis and hemangiogenesis in inflammatory neovascularization via macrophage recruitment. J Clin Invest 113:1040–1050

Dore-Duffy P, Cleary K (2011) Morphology and properties of pericytes. Methods Mol Biol 686:49–68

Eccles SA, Welch DR (2007) Metastasis: recent discoveries and novel treatment strategies. Lancet 369:1742–1757

Eckerich C, Zapf S, Fillbrandt R, Loges S, Westphal M, Lamszus K (2007) Hypoxia can induce c-Met expression in glioma cells and enhance SF/HGF-induced cell migration. Int J Cancer 121:276–283

Elvidge GP, Glenny L, Appelhoff RJ, Ratcliffe PJ, Ragoussis J, Gleadle JM (2006) Concordant regulation of gene expression by hypoxia and 2-oxoglutarate-dependent dioxygenase inhibition: the role of HIF-1α, HIF-2α, and other pathways. J Biol Chem 281:15215–15226

Ema M, Taya S, Yokotani N, Sogawa K, Matsuda Y, Fujii-Kuriyama Y (1997) A novel bHLH-PAS factor with close sequence similarity to hypoxia-inducible factor 1α regulates the VEGF expression and is potentially involved in lung and vascular development. Proc Natl Acad Sci U S A 94:4273–4278.

Erler JT, Bennewith KL, Nicolau M, Dornhofer N, Kong C, Le QT, Chi JT, Jeffrey SS, Giaccia AJ (2006) Lysyl oxidase is essential for hypoxia-induced metastasis. Nature 440:1222–1226

Erler JT, Bennewith KL, Cox TR, Lang G, Bird D, Koong A, Le QT, Giaccia AJ (2009) Hypoxia-induced lysyl oxidase is a critical mediator of bone marrow cell recruitment to form the premetastatic niche. Cancer Cell 15:35–44

Eum SY, Lee YW, Hennig B, Toborek M (2004) VEGF regulates PCB 104-mediated stimulation of permeability and transmigration of breast cancer cells in human microvascular endothelial cells. Exp Cell Res 296:231–244

Fidler IJ (1970) Metastasis: quantitative analysis of distribution and fate of tumor emboli labeled with 125 I-5-iodo-2′-deoxyuridine. J Natl Cancer Inst 45:773–782.

Flamme I, Frohlich T, Reutern M von, Kappel A, Damert A, Risau W (1997) HRF, a putative basic helix-loop-helix-PAS-domain transcription factor is closely related to hypoxia-inducible factor-1α and developmentally expressed in blood vessels. Mech Dev 63:51–60.

Forsythe JA, Jiang BH, Iyer NV, Agani F, Leung SW, Koos RD, Semenza GL (1996) Activation of vascular endothelial growth factor gene transcription by hypoxia-inducible factor 1. Mol Cell Biol 16:4604–4613

Gatenby RA, Smallbone K, Maini PK, Rose F, Averill J, Nagle RB, Worrall L, Gillies RJ (2007) Cellular adaptations to hypoxia and acidosis during somatic evolution of breast cancer. Br J Cancer 97:646–653

Gu Y, Qi X, Guo S (2008) Lymphangiogenesis induced by VEGF-C and VEGF-D promotes metastasis and a poor outcome in breast carcinoma: a retrospective study of 61 cases. Clin Exp Metastasis 25:717–725

Gupta GP, Nguyen DX, Chiang AC, Bos PD, Kim JY, Nadal C (2007) Mediators of vascular remodelling co-opted for sequential steps in lung metastasis. Nature 446:765–770

Halin C, Tobler NE, Vigl B, Brown LF, Detmar M (2007) VEGF-A produced by chronically inflamed tissue induces lymphangiogenesis in draining lymph nodes. Blood 110:3158–3167

He Y, Karpanen T, Alitalo K (2004) Role of lymphangiogenic factors in tumor metastasis. Biochim Biophys Acta 1654:3–12

Higgins DF, Kimura K, Bernhardt WM, Shrimanker N, Akai Y, Hohenstein B, Saito Y, Johnson RS, Kretzler M, Cohen CD, Eckardt KU, Iwano M, Haase VH (2007) Hypoxia promotes fibrogenesis in vivo via HIF-1 stimulation of epithelial-to-mesenchymal transition. J Clin Invest 117:3810–3820

Hirakawa S, Kodama S, Kunstfeld R, Kajiya K, Brown LF, Detmar M (2005) VEGF-A induces tumor and sentinel lymph node lymphangiogenesis and promotes lymphatic metastasis. J Exp Med 201:1089–1099

Hirakawa S, Brown LF, Kodama S, Paavonen K, Alitalo K, Detmar M (2007) VEGF-C-induced lymphangiogenesis in sentinel lymph nodes promotes tumor metastasis to distant sites. Blood 109:1010–1017

Hogenesch JB, Chan WK, Jackiw VH, Brown RC, Gu YZ, Pray-Grant M, Perdew GH, Bradfield CA (1997) Characterization of a subset of the basic-helix-loop-helix-PAS superfamily that interacts with components of the dioxin signaling pathway. J Biol Chem 272:8581–8593

Imai T, Horiuchi A, Wang C, Oka K, Ohira S, Nikaido T, Konishi I (2003) Hypoxia attenuates the expression of E-cadherin via up-regulation of SNAIL in ovarian carcinoma cells. Am J Pathol 163:1437–1447

Jewell UR, Kvietikova I, Scheid A, Bauer C, Wenger RH, Gassmann M (2001) Induction of HIF-1α in response to hypoxia is instantaneous. FASEB J 15:1312–1314

Ji R-C (2006) Lymphatic endothelial cells, lymphangiogenesis, and extracellular matrix. Lymphat Res Biol 4:83–8100

Joukov V, Pajusola K, Kaipainen A, Chilov D, Lahtinen I, Kukk E, Saksela O, Kalkkinen N, Alitalo K (1996) A novel vascular endothelial growth factor, VEGF-C, is a ligand for the Flt4 (VEGFR-3) and KDR (VEGFR-2) receptor tyrosine kinases. EMBO J 15:1751

Kaelin WG Jr, Ratcliffe PJ (2008) Oxygen sensing by metazoans: the central role of the HIF hydroxylase pathway. Mol Cell 30:393–402

Kaplan RN, Riba RD, Zacharoulis S, Bramley AH, Vincent L, Costa C, MacDonald DD, Jin DK, Shido K, Kerns SA, Zhu Z, Hicklin D, Wu Y, Port JL, Altorki N, Port ER, Ruggero D, Shmelkov SV, Jensen KK, Rafii S, Lyden D (2005) VEGFR1-positive haematopoietic bone marrow progenitors initiate the pre-metastatic niche. Nature 438:820–827

Karpanen T, Egeblad M, Karkkainen MJ, Kubo H, Yla-Herttuala S, Jaattela M, Alitalo K (2001) Vascular endothelial growth factor C promotes tumor lymphangiogenesis and intralymphatic tumor growth. Cancer Res 61:1786–1790

Kelly BD, Hackett SF, Hirota K, Oshima Y, Cai Z, Berg-Dixon S, Rowan A, Yan Z, Campochiaro PA, Semenza GL (2003) Cell type-specific regulation of angiogenic growth factor gene expression and induction of angiogenesis in nonischemic tissue by a constitutively active form of hypoxia-inducible factor 1. Circ Res 93:1074–1081

Kim JA, Tran ND, Li Z, Yang F, Zhou W, Fisher MJ (2006) Brain endothelial hemostasis regulation by pericytes. J Cereb Blood Flow Metab 26:209–217

Kodama M, Kitadai Y, Sumida T, Ohnishi M, Ohara E, Tanaka M, Shinagawa K, Tanaka S, Yasui W, Chayama K (2010) Expression of platelet-derived growth factor (PDGF)-B and PDGF-receptor beta is associated with lymphatic metastasis in human gastric carcinoma. Cancer Sci 101:1984–1989

Koh MY, Lemos R Jr, Liu X, Powis G (2011) The hypoxia-associated factor switches cells from HIF-1α- to HIF-2α-dependent signaling promoting stem cell characteristics, aggressive tumor growth and invasion. Cancer Res 71:4015–4027.

Kopfstein L, Veikkola T, Djonov VG, Baeriswyl V, Schomber T, Strittmatter K, Stacker SA, Achen MG, Alitalo K, Christofori G (2007) Distinct roles of vascular endothelial growth factor-D in lymphangiogenesis and metastasis. Am J Pathol 170:1348–1361

Krishnamachary B, Berg-Dixon S, Kelly B, Agani F, Feldser D, Ferreira G, Iyer N, LaRusch J, Pak B, Taghavi P, Semenza GL (2003) Regulation of colon carcinoma invasion by hypoxia-inducible factor 1. Cancer Res 63:1138–1143

Krishnamachary B, Zagzag D, Nagasawa H, Rainey K, Okuyama H, Baek JH, Semenza GL (2006) Hypoxia-inducible factor-1-dependent repression of E-cadherin in von Hippel-Lindau tumor suppressor-null renal cell carcinoma mediated by TCF3, ZFHX1A, and ZFHX1B. Cancer Res 66:2725–2731

Kukk E, Lymboussaki A, Taira S, Kaipainen A, Jeltsch M, Joukov V, Alitalo K (1996) VEGF-C receptor binding and pattern of expression with VEGFR-3 suggests a role in lymphatic vascular development. Development 122:3829–3837

Kumar S, Park SH, Cieply B, Schupp J, Killiam E, Zhang F, Rimm DL, Frisch SM (2011) A pathway for the control of anoikis sensitivity by E-cadherin and epithelial-to-mesenchymal transition. Mol Cell Biol 31:4036–4051

Le Doussal V, Tubiana-Hulin M, Friedman S, Hacene K, Spyratos F, Brunet M (1989) Prognostic value of histologic grade nuclear components of Scarff-Bloom-Richardson (SBR). An improved score modification based on a multivariate analysis of 1262 invasive ductal breast carcinomas. Cancer 64:1914–1921

Lee TH, Avraham HK, Jiang S, Avraham S (2003) Vascular endothelial growth factor modulates the transendothelial migration of MDA-MB-231 breast cancer cells through regulation of brain microvascular endothelial cell permeability. J Biol Chem 278:5277–5284
Leong SP, Nakakura EK, Pollock R, Choti MA, Morton DL, Henner WD, Lal A, Pillai R, Clark OH, Cady B (2011) Unique patterns of metastases in common and rare types of malignancy. J Surg Oncol 103:607–614
Lester RD, Jo M, Montel V, Takimoto S, Gonias SL (2007) uPAR induces epithelial-mesenchymal transition in hypoxic breast cancer cells. J Cell Biol 178:425–436
Levental KR, Yu H, Kass L, Lakins JN, Egeblad M, Erler JT, Fong SF, Csiszar K, Giaccia A, Weninger W, Yamauchi M, Gasser DL, Weaver VM (2009) Matrix crosslinking forces tumor progression by enhancing integrin signaling. Cell 139:891–906
Liao D, Corle C, Seagroves TN, Johnson RS (2007) Hypoxia-inducible factor-1α is a key regulator of metastasis in a transgenic model of cancer initiation and progression. Cancer Res 67:563–572.
Lin EY, Pollard JW (2004) Macrophages: modulators of breast cancer progression. Novartis Found Symp 256:158–168
Loboda A, Jozkowicz A, Dulak J (2010) HIF-1 and HIF-2 transcription factors–similar but not identical. Mol Cells 29:435–442
Lopez JI, Kang I, You WK, McDonald DM, Weaver VM (2011) In situ force mapping of mammary gland transformation. Integr Biol (Camb) 3:910–921
Makinen T, Veikkola T, Mustjoki S, Karpanen T, Catimel B, Nice EC, Wise L, Mercer A, Kowalski H, Kerjaschki D, Stacker SA, Achen MG, Alitalo K (2001) Isolated lymphatic endothelial cells transduce growth, survival and migratory signals via the VEGF-C/D receptor VEGFR-3. EMBO J 20:4762–4773
Manalo DJ, Rowan A, Lavoie T, Natarajan L, Kelly BD, Ye SQ, Garcia JG, Semenza GL (2005) Transcriptional regulation of vascular endothelial cell responses to hypoxia by HIF-1. Blood 105:659–669
Mattila MM, Ruohola JK, Karpanen T, Jackson DG, Alitalo K, Harkonen PL (2002) VEGF-C induced lymphangiogenesis is associated with lymph node metastasis in orthotopic MCF-7 tumors. Int J Cancer 98:946–951
McPherson K, Steel CM, Dixon JM (2000) ABC of breast diseases. Breast cancer-epidemiology, risk factors, and genetics. BMJ 321:624–628
Minn AJ, Gupta GP, Siegel PM, Bos PD, Shu W, Giri DD, Viale A, Olshen AB, Gerald WL, Massagué J (2005) Genes that mediate breast cancer metastasis to lung. Nature 436:518–524
Muller A, Homey B, Soto H, Ge N, Catron D, Buchanan ME, McClanahan T, Murphy E, Yuan W, Wagner SN, Barrera JL, Mohar A, Verastegui E, Zlotnik A (2001) Involvement of chemokine receptors in breast cancer metastasis. Nature 410:50–56
Nagy JA, Vasile E, Feng D, Sundberg C, Brown LF, Detmar MJ, Lawitts JA, Benjamin L, Tan X, Manseau EJ, Dvorak AM, Dvorak HF (2002) Vascular permeability factor/vascular endothelial growth factor induces lymphangiogenesis as well as angiogenesis. J Exp Med 196:1497–1506
Okada K, Osaki M, Araki K, Ishiguro K, Ito H, Ohgi S (2005) Expression of hypoxia-inducible factor (HIF-1α), VEGF-C and VEGF-D in non-invasive and invasive breast ductal carcinomas. Anticancer Res 25:3003–3009.
Ozbek S, Balasubramanian PG, Chiquet-Ehrismann R, Tucker RP, Adams JC (2010) The evolution of extracellular matrix. Mol Biol Cell 21:4300–4305
Padua D, Zhang XH, Wang Q, Nadal C, Gerald WL, Gomis RR, Massagué J (2008) TGFī′ primes breast tumors for lung metastasis seeding through angiopoietin-like 4. Cell 133:66–77
Page-McCaw A, Ewald AJ, Werb Z (2007) Matrix metalloproteinases and the regulation of tissue remodelling. Nat Rev Mol Cell Biol 8:221–233
Pal SK, Childs BH, Pegram M (2011) Triple negative breast cancer: unmet medical needs. Breast Cancer Res Treat 125:627–636
Parr C, Watkins G, Mansel RE, Jiang WG (2004) The hepatocyte growth factor regulatory factors in human breast cancer. Clin Cancer Res 10:202–211

Peinado H, Del CIglesias-delaCM, Olmeda D, Csiszar K, Fong KS, Vega S, Nieto MA, Cano A, Portillo F (2005) A molecular role for lysyl oxidase-like 2 enzyme in snail regulation and tumor progression. EMBO J 24:3446–3458
Pennacchietti S, Michieli P, Galluzzo M, Mazzone M, Giordano S, Comoglio PM (2003) Hypoxia promotes invasive growth by transcriptional activation of the met protooncogene. Cancer Cell 3:347–361
Petrella BL, Lohi J, Brinckerhoff CE (2005) Identification of membrane type-1 matrix metalloproteinase as a target of hypoxia-inducible factor 2α in von Hippel-Lindau renal cell carcinoma. Oncogene 24:1043–1052.
Ran S, Volk L, Hall K, Flister MJ (2010) Lymphangiogenesis and lymphatic metastasis in breast cancer. Pathophysiology 17:229–251
Rankin EB, Biju MP, Liu Q, Unger TL, Rha J, Johnson RS, Simon MC, Keith B, Haase VH (2007) Hypoxia-inducible factor-2 (HIF-2) regulates hepatic erythropoietin in vivo. J Clin Invest 117:1068–1077
Rundqvist H, Johnson RS (2010) Hypoxia and metastasis in breast cancer. Curr Top Microbiol Immunol 345:121–139
Sahlgren C, Gustafsson MV, Jin S, Poellinger L, Lendahl U (2008) Notch signaling mediates hypoxia-induced tumor cell migration and invasion. Proc Natl Acad Sci U S A 105:6392–6397
Schito L, Rey S, Tafani M, Zhang H, Wong CC, Russo A, Russo MA, Semenza GL (2012) Hypoxia-inducible factor 1-dependent expression of platelet-derived growth factor B promotes lymphatic metastasis of hypoxic breast cancer cells. Proc Natl Acad Sci U S A 109:E2707–E2716.
Schoppmann SF, Birner P, Stockl J, Kalt R, Ullrich R, Caucig C, Kriehuber E, Nagy K, Alitalo K, Kerjaschki D (2002a) Tumor-associated macrophages express lymphatic endothelial growth factors and are related to peritumoral lymphangiogenesis. Am J Pathol 161:947–956
Schoppmann SF, Horvat R, Birner P (2002b) Lymphatic vessels and lymphangiogenesis in female cancer: mechanisms, clinical impact and possible implications for anti-lymphangiogenic therapies. Oncol Rep 9:455–460
Schoppmann SF, Fenzl A, Schindl M, Bachleitner-Hofmann T, Nagy K, Gnant M, Horvat R, Jakesz R, Birner P (2006) Hypoxia inducible factor-1α correlates with VEGF-C expression and lymphangiogenesis in breast cancer. Breast Cancer Res Treat 99:135–141.
Semenza GL (2012a) Hypoxia-inducible factors: mediators of cancer progression and targets for cancer therapy. Trends Pharmacol Sci 33:207–214
Semenza GL (2012b) Molecular mechanisms mediating metastasis of hypoxic breast cancer cells. Trends Mol Med 18:534–543
Semenza GL, Wang GL (1992) A nuclear factor induced by hypoxia via de novo protein synthesis binds to the human erythropoietin gene enhancer at a site required for transcriptional activation. Mol Cell Biol 12:5447–5454
Semenza GL, Roth PH, Fang HM, Wang GL (1994) Transcriptional regulation of genes encoding glycolytic enzymes by hypoxia-inducible factor 1. J Biol Chem 269:23757–23763
Semenza GL, Jiang BH, Leung SW, Passantino R, Concordet JP, Maire P, Giallongo A (1996) Hypoxia response elements in the aldolase A, enolase 1, and lactate dehydrogenase A gene promoters contain essential binding sites for hypoxia-inducible factor 1. J Biol Chem 271:32529–32537
Skobe M, Hawighorst T, Jackson DG, Prevo R, Janes L, Velasco P, Riccardi L, Alitalo K, Claffey K, Detmar M (2001) Induction of tumor lymphangiogenesis by VEGF-C promotes breast cancer metastasis. Nat Med 7:192–198
Stacker SA, Caesar C, Baldwin ME, Thornton GE, Williams RA, Prevo R, Jackson DG, Nishikawa S, Kubo H, Achen MG (2001) VEGF-D promotes the metastatic spread of tumor cells via the lymphatics. Nat Med 7:186–191
Staller P, Sulitkova J, Lisztwan J, Moch H, Oakeley EJ, Krek W (2003) Chemokine receptor CXCR4 downregulated by von Hippel-Lindau tumour suppressor pVHL. Nature 425:307–311

Sullivan NJ, Sasser AK, Axel AE, Vesuna F, Raman V, Ramirez N, Oberyszyn TM, Hall BM (2009) Interleukin-6 induces an epithelial-mesenchymal transition phenotype in human breast cancer cells. Oncogene 28:2940–2947

Swartz MA, Skobe M (2001) Lymphatic function, lymphangiogenesis, and cancer metastasis. Microsc Res Tech 55:92–99

Tan MJ, Teo Z, Sng MK, Zhu P, Tan NS (2012) Emerging roles of angiopoietin-like 4 in human cancer. Mol Cancer Res 10:677–688

Tian H, McKnight SL, Russell DW (1997) Endothelial PAS domain protein 1 (EPAS1), a transcription factor selectively expressed in endothelial cells. Genes Dev 11:72–82

Toullec A, Gerald D, Despouy G, Bourachot B, Cardon M, Lefort S, Richardson M, Rigaill G, Parrini MC, Lucchesi C, Bellanger D, Stern MH, Dubois T, Sastre-Garau X, Delattre O, Vincent-Salomon A, Mechta-Grigoriou F (2010) Oxidative stress promotes myofibroblast differentiation and tumour spreading. EMBO Mol Med 2:211–230

Ulleras E, Wilcock A, Miller SJ, Franklin GC (2001) The sequential activation and repression of the human PDGF-B gene during chronic hypoxia reveals antagonistic roles for the depletion of oxygen and glucose. Growth Factors 19:233–245

Vaupel P (2009) Prognostic potential of the pre-therapeutic tumor oxygenation status. Adv Exp Med Biol 645:241–246

Vaupel P, Mayer A, Hockel M (2004) Tumor hypoxia and malignant progression. Methods Enzymol 381:335–354

Vaupel P, Hockel M, Mayer A (2007) Detection and characterization of tumor hypoxia using pO2 histography. Antioxid Redox Signal 9:1221–1235

Veronese SM, Gambacorta M (1991) Detection of Ki-67 proliferation rate in breast cancer. Correlation with clinical and pathologic features. Am J Clin Pathol 95:30–34

Wang GL, Semenza GL (1993) Desferrioxamine induces erythropoietin gene expression and hypoxia-inducible factor 1 DNA-binding activity: implications for models of hypoxia signal transduction. Blood 82:3610–3615

Wang GL, Semenza GL (1995) Purification and characterization of hypoxia-inducible factor 1. J Biol Chem 270:1230–1237

Wang GL, Jiang BH, Rue EA, Semenza GL (1995) Hypoxia-inducible factor 1 is a basic-helix-loop-helix-PAS heterodimer regulated by cellular O2 tension. Proc Natl Acad Sci U S A 92:5510–5514

Whitehurst B, Flister MJ, Bagaitkar J, Volk L, Bivens CM, Pickett B, Castro-Rivera E, Brekken RA, Gerard RD, Ran S (2007) Anti-VEGF-A therapy reduces lymphatic vessel density and expression of VEGFR-3 in an orthotopic breast tumor model. Int J Cancer 121:2181–2191

Wiesener MS, Turley H, Allen WE, Willam C, Eckardt KU, Talks KL, Wood SM, Gatter KC, Harris AL, Pugh CW, Ratcliffe PJ, Maxwell PH (1998) Induction of endothelial PAS domain protein-1 by hypoxia: characterization and comparison with hypoxia-inducible factor-1alpha. Blood 92:2260–2268

Wilson JL, Burchell J, Grimshaw MJ (2006) Endothelins induce CCR7 expression by breast tumor cells via endothelin receptor A and hypoxia-inducible factor-1. Cancer Res 66:11802–11807

Wong CC, Gilkes DM, Zhang H, Chen J, Wei H, Chaturvedi P, Fraley SI, Wong CM, Khoo US, Ng IO, Wirtz D, Semenza GL (2011) Hypoxia-inducible factor 1 is a master regulator of breast cancer metastatic niche formation. Proc Natl Acad Sci U S A 108:16369–16374

Wong CC, Zhang H, Gilkes DM, Chen J, Wei H, Chaturvedi P, Hubbi ME, Semenza GL (2012) Inhibitors of hypoxia-inducible factor 1 block breast cancer metastatic niche formation and lung metastasis. J Mol Med (Berl) 90:803–815

Yang MH, Wu KJ (2008) TWIST activation by hypoxia inducible factor-1 (HIF-1): implications in metastasis and development. Cell Cycle 7:2090–2096

Yang MH, Wu MZ, Chiou SH, Chen PM, Chang SY, Liu CJ, Teng SC, Wu KJ (2008) Direct regulation of TWIST by HIF-1α promotes metastasis. Nat Cell Biol 10:295–305

Yeo KT, Wang HH, Nagy JA, Sioussat TM, Ledbetter SR, Hoogewerf AJ, Zhou Y, Masse EM, Senger DR, Dvorak HF, et al (1993) Vascular permeability factor (vascular endothelial growth factor) in guinea pig and human tumor and inflammatory effusions. Cancer Res 53:2912–2918

Yoon SY, Lee YJ, Seo JH, Sung HJ, Park KH, Choi IK, Kim SJ, SC Oh, Choi CW, Kim BS, Shin SW, Kim YH, Kim JS (2006) uPAR expression under hypoxic conditions depends on iNOS modulated ERK phosphorylation in the MDA-MB-231 breast carcinoma cell line. Cell Res 16:75–81

Zhang H, Qian DZ, Tan YS, Lee K, Gao P, Ren YR, Rey S, Hammers H, Chang D, Pili R, Dang CV, Liu JO, Semenza GL (2008) Digoxin and other cardiac glycosides inhibit HIF-1α synthesis and block tumor growth. Proc Natl Acad Sci USA 105:19579–19586

Zhang H, Wong CC, Wei H, Gilkes DM, Korangath P, Chaturvedi P, Schito L, Chen J, Krishnamachary B, Winnard PT Jr, Raman V, Zhen L, Mitzner WA, Sukumar S, Semenza GL (2012) HIF-1-dependent expression of angiopoietin-like 4 and L1CAM mediates vascular metastasis of hypoxic breast cancer cells to the lungs. Oncogene 31:1757–1770

Zhang SX, Gozal D, Sachleben LR Jr, Rane M, Klein JB, Gozal E (2003) Hypoxia induces an autocrine-paracrine survival pathway via platelet-derived growth factor (PDGF)-B/PDGF-β receptor/phosphatidylinositol 3-kinase/Akt signaling in RN46A neuronal cells. FASEB J 17:1709–1711

Chapter 2
Hypoxia and the DNA Damage Response

Monica M. Olcina and Ester M. Hammond

Abstract Gradients of hypoxia occur in most solid tumors. Cells found in these regions are associated with the most aggressive and therapy-resistant fractions of the tumor. Severe levels of hypoxia ($< 0.1\ \%\ O_2$) have been found to induce a unique DNA damage response (DDR) that includes both ATR- and ATM-mediated signaling. The consequences of the hypoxia-induced DDR include p53-dependent apoptosis and maintenance of replication fork integrity. Interestingly, the hypoxia-induced DDR occurs in the absence of detectable single- or double-strand breaks and in a background of repressed DNA repair. Inhibition of DNA repair in hypoxic cells has been proposed as a mechanism contributing to the increased genomic instability found in hypoxia. Furthermore, an increasing number of novel agents that target the DDR have been described and some are already undergoing clinical testing. Evidence from preclinical studies suggests that the use of some of these agents would be effective at targeting tumor cells in hypoxic regions.

Keywords DNA damage response · Replication stress · Synthetic lethality · PARP · Chk1 · ATM · ATR · p53

2.1 The DNA Damage Response (DDR)

The integrity of the genome is constantly being threatened by a number of endogenous as well as exogenous insults that can lead to the generation of DNA damage (Lindadhl and Barnes 2000). This damage will generally be in the form of single- or double-strand breaks (DSBs), which can occur following replication stresses (RSs) such as ultraviolet (UV) radiation or oncogene activation, as well as in response to other genotoxic stresses including ionizing radiation (IR), or chemotherapy (Jackson and Bartek 2009; Ciccia and Elledge 2010). The classical response to DNA damage leads to the activation of a complicated kinase signaling cascade that includes sensing and

---

E. M. Hammond (✉) · M. M. Olcina
The Department of Oncology, The Cancer Research UK/MRC Gray Institute for Radiation Oncology and Biology, The University of Oxford, Oxford OX3 7DQ, UK
e-mail: ester.hammond@oncology.ox.ac.uk

G. Melillo (ed.), *Hypoxia and Cancer*, Cancer Drug Discovery and Development, DOI 10.1007/978-1-4614-9167-5_2,

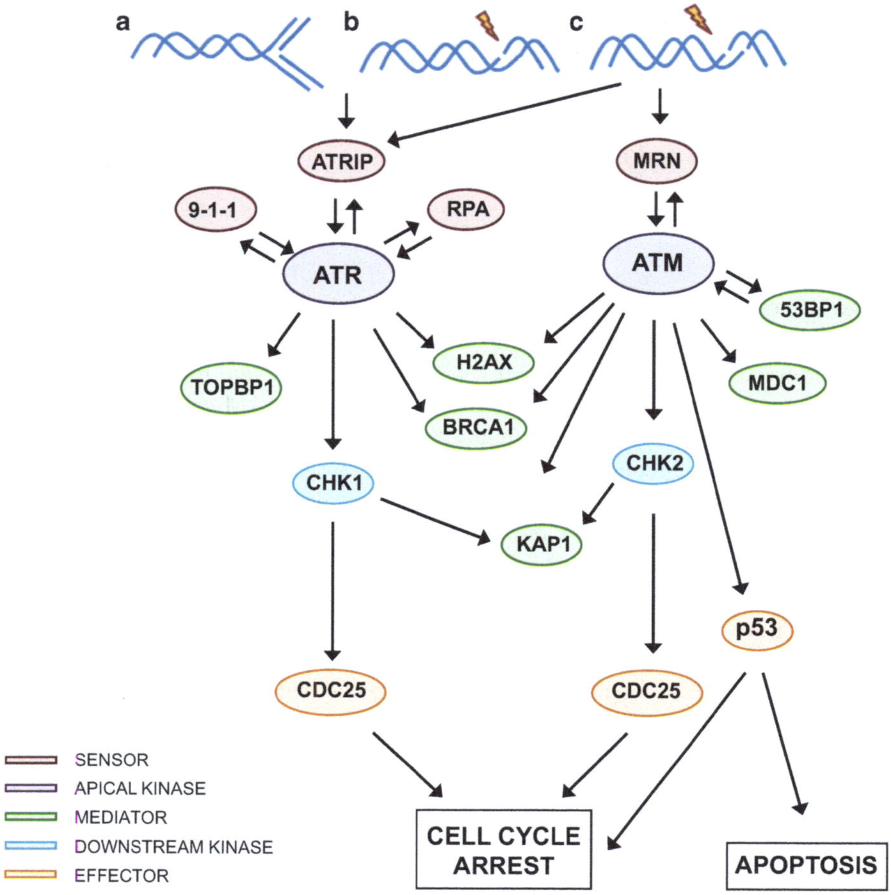

Fig. 2.1 Schematic representation of the DNA damage response. DNA lesions including stalled replication forks (**a**), single strand breaks (**b**), and double strand breaks (**c**) are detected by specific sensors. The apical kinases, *ATR* and *ATM*, are then activated and transduce the signal through mediators to downstream kinases and effectors

responding to the insult, as well as coordinating a number of downstream events generally resulting in cell-cycle arrest, DNA repair, senescence, or apoptosis (shown in Fig. 2.1 and recently reviewed in (Ciccia and Elledge 2010)). The cascade can be envisioned as having three principal kinases, although there is considerable cross-talk between them. If the stress activating the DNA damage response (DDR) leads to the generation of single-stranded DNA (ssDNA), the ATR PI-3K (phosphatidyl inositol-3-kinase) like kinase (PIKK) (Ataxia-telangiectasia and Rad3-related) will be activated (Cimprich and Cortez 2008). Alternatively, if the insult leads to the generation of a double-strand break (DSB), two other members of the PIKK family—Ataxia-telangiectasia mutated (ATM) and DNA-dependent protein kinase (DNA-PK)—become activated (Jackson and Bartek 2009). ATM is generally regarded as the main upstream kinase of the signaling pathway activated in response to DSB, while DNA-PK is mainly involved with the repair of DSB in the nonhomologous end-joining (NHEJ) pathway (Collis et al. 2005).

2.2 The DDR and its Relevance to Tumorigenesis

The action of the many players involved in the DDR will result in a coordinated defense mechanism against the original cellular insult. The mounting of an appropriate DDR is essential for the maintenance of genome integrity. Importantly, many studies now point to the importance of the DDR in cancer development, with the DDR acting as a barrier to tumorigenesis (Bartkova et al. 2005, 2006; Halazonetis et al. 2008). Factors involved in the DDR such as γH2AX are known to be expressed in early-stage tumors (Bartkova et al. 2005). Other studies have demonstrated that loss of DDR activation leads to loss of oncogene-induced senescence and the increase in cellular transformation (Di Micco et al. 2006). Furthermore, ATR haploinsufficiency together with endogenous levels of K-ras in p53 heterozygous mice result in enhanced incidence of lung adenocarcinoma, spindle cell sarcoma, and thymic lymphoma (Gilad et al. 2010). The protective function that might be exerted by the DDR in tumorigenesis may explain some of the common mutations in DDR genes found in cancers (Halazonetis et al. 2008; Gorgoulis et al. 2005; Kastan and Bartek 2004).

2.3 ATR-Mediated DDR

Following RS, areas of ssDNA become coated with replication protein A (RPA) (Cortez et al. 2001; Zou et al. 2003; Fanning et al. 2006). ATR-interacting protein (ATRIP) then associates with the RPA and ultimately recruits ATR to the sites of ssDNA. RAD17 will be directed to these sites where it will load the 9-1-1 complex (composed of RAD9-RAD1-HUS1). The activity of ATR is then enhanced by the phosphorylation of RAD17, 9-1-1, and a specific activator of ATR, topoisomerase II binding protein I (TOPBP1). ATR plays a critical role in the maintenance of replication fork integrity and checkpoint response. Accordingly, many studies have shown that if ATR function is lost, DNA repair and viability are severely compromised (Cliby et al. 1998; Hurley et al. 2007). One of the most characterized ATR substrates is checkpoint kinase 1 (CHK1). Phosphorylated and active CHK1 can direct cell cycle regulation by mediating a checkpoint response (Liu et al. 2000; Dai and Grant 2010; Sanchez et al. 1997). In response to a DNA damage signal (including aberrant replication structures), checkpoints act to halt the cell cycle to allow enough time for any damage incurred by the cell to be repaired. For example, the G_1/S checkpoint is maintained by p53, which is stabilized by the DDR. Phosphorylated p53 (at serine 15) activates cyclin-dependent kinase inhibitor 1A (p21). p21 interacts with CDK2 and CDK4 hindering the interaction between CDK2/cyclin B and CDK4/cyclin E (Sancar et al. 2004). This prevents key cellular events, such as chromosome segregation or DNA replication, being carried out in the presence of DNA damage or during repair (Cortez 2001).

2.4 ATM-Mediated DDR

A single DSB can be lethal to the cell so it is important that these lesions are detected and dealt with appropriately (Jeggo and Lobrich 2007). DSBs are generally sensed by the Mre11-RAD50-NBS1 (MRN) complex (Moreno-Herrero et al. 2005). Following the initial detection of the break, a number of signaling events will be coordinated. For instance, the ATM kinase undergoes autophosphorylation and becomes active (Bakkenist and Kastan 2003). Activation and retention of ATM at the sites of damage then leads to the phosphorylation and recruitment of other factors including mediator of DNA damage checkpoint 1 (MDC1) and 53BP1, which will further enhance ATM retention and signaling (Stewart et al. 2003; Goldberg et al. 2003). Specific chromatin changes also occur during the initiation of the DDR and serve to amplify the signaling and facilitate repair. Phosphorylation of the histone variant H2AX (γH2AX) occurs in an ATM-dependent manner following DSB generation, and as a result is often used as a marker of DNA damage (Rogakou et al. 1998). This modification is important for the recruitment of MDC1, which further enhances accumulation of γH2AX (Stucki et al. 2005). MDC1 has been shown to bind to γH2AX and initiates a feedback loop that encourages further MRN recruitment and sustained ATM activation (Stucki et al. 2005). Once active, ATM initiates the DDR signaling cascade leading to checkpoint activation, repair, or apoptosis (Lavi and Kozlov 2007). One of the key downstream targets of ATM is checkpoint kinase 2 (CHK2), which mediates cell-cycle arrest (Matsuoka et al. 2000). ATM signaling can also facilitate DNA repair by allowing the recruitment of repair factors. ATM-dependent phosphorylation of DNA repair protein BRCA1, for instance, is critical for the response to IR-induced DSB (Cortez et al. 1999). BRCA1-deficient cells have defective homologous recombination (HR)-mediated repair and display hypersensitivity to IR pointing to the importance of BRCA1 in directing repair following DSB formation (Chen et al. 1998).

2.5 Hypoxia-Induced p53-Dependent Apoptosis

One of the key effectors of the DDR is the tumor suppressor protein, p53, which can induce both cell-cycle arrest and apoptosis (Vousden and Lane 2007). p53 is, therefore, regarded as a critical tumor suppressor with significant roles to play in preventing cancer (Lane 1992). Hypoxia can also lead to the induction of the DDR, and p53 activation is indeed a principal consequence of the hypoxia-induced DDR (Riley et al. 2008; Green and Kroemer 2009; Yee and Vousden 2005; Harper and Elledge 2007). In response to hypoxia, p53 is phosphorylated at a number of residues including serine 15, which has been shown to be important in mediating nuclear accumulation of p53 (Hammond et al. 2002). The initial stabilization of p53 in hypoxia is also facilitated by phosphorylation on residues 6, 9, 20, 37, and 46 (Hammond and Giaccia et al. 2005). Interestingly, hypoxic conditions, which lead to the activation of p53, tend to induce apoptosis as opposed to a cell-cycle arrest in G_1. p53 activation is, in fact, critical for hypoxia-induced apoptosis, and loss of

a

	Hypoxia	Reox
COMET	No damage	damage
PFGE	No damage	damage
γH2AX	Rapid and robust but pan-nuclear	Unclear due to hypoxia signal
53BP1	No foci	foci

b

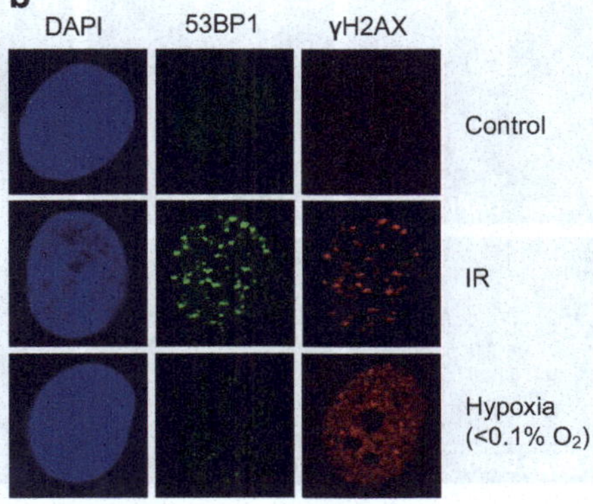

Fig. 2.2 Hypoxia does not induce detectable DNA damage. Standard assays for DNA damage detection demonstrate a lack of damage in hypoxic conditions. In contrast, reoxygenation after hypoxic exposure leads to abundant DNA damage (**a**). An example of cells exposed to either hypoxia or irradiation (*IR*) and stained for γH2AX or 53BP1 is shown (**b**)

p53 decreases hypoxia-induced apoptosis (Graeber et al. 1996). Regions of hypoxia have been elegantly shown to correlate with high apoptosis in tumors with wild-type p53, whereas low apoptosis levels are observed in hypoxic tumor regions that have lost p53 (Graeber et al. 1996). A strong selection pressure for p53 mutations and subsequent apoptosis has been described in hypoxic cells (1996; Levine 1997). The preferential selection of cells with apoptosis defects in hypoxia may provide an explanation for the increased resistance of many solid tumors to certain forms of cancer therapy, particularly those forms of chemotherapy that rely on p53-dependent apoptosis for their efficacy (Graeber et al. 1994).

2.6 Initiation of DDR Signaling in Hypoxia

As described, hypoxia triggers a DDR, characterized by p53 accumulation and apoptosis (Graeber et al. 1996). Initially, it was unclear exactly what the hypoxia-induced signal was that initiated the DDR. This was primarily because standard assays for DNA damage demonstrated that these hypoxic conditions did not induce DSBs (Hammond et al. 2002). For example, comet assays or staining for the presence of 53BP1 foci demonstrated that hypoxia, in the absence of reoxygenation, does not induce DNA damage or at least damage that is detectable using the assays indicated (Fig. 2.2). One complexity to this conclusion was provided by the observation that H2AX is robustly phosphorylated in response to hypoxia as this histone modification has been correlated with the presence of DSBs (Bencokova et al. 2009). However, as shown in Fig. 2.2, the γH2AX signal observed in hypoxic conditions is diffuse and pan-nuclear as opposed to forming discreet nuclear foci as seen in response to irradiation. This pattern of γH2AX staining is reminiscent of the response to RS

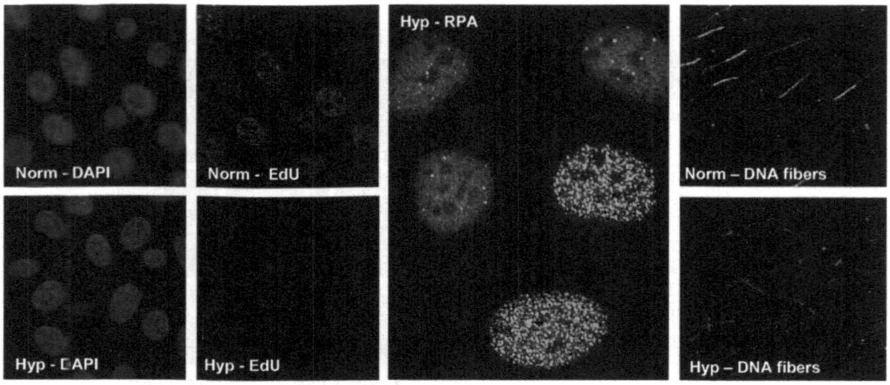

Fig. 2.3 Severe hypoxia induces an S-phase arrest, stalled replication, and a DDR. Representative images are shown for normoxic (*Norm*) and hypoxic (*Hyp*) cells. Hypoxic cells fail to accumulate EdU due to a lack of ongoing S-phase. RPA foci accumulate in the nuclei of hypoxia-treated S-phase cells. Replication stalls and stalled replication forks accumulate in hypoxia as shown by the accumulation of DNA fibers, which have not incorporated the second thymidine analog

for example as initiated by treatment with hydroxyurea or aphidicolin (Toledo et al. 2011) These data suggest that hypoxia induces a DDR, characterized by the robust induction of γH2AX but that this is more likely to occur as a result of RS rather than DSBs. In support of this a significant hypoxia-induced effect has been described on the S-phase population. In response to levels of hypoxia, which also induce p53 stabilization and γH2AX, cells rapidly undergo an S-phase arrest, i.e., they fail to incorporate thymidine analogs such as BrdU and EdU (Hammond et al. 2002) (as illustrated in Fig. 2.3).

More recently, hypoxia-induced replication arrest has been shown to correlate with a rapid decrease in the levels of available nucleotides (Pires et al. 2010a). This is predicted to be the result of decreased ribonucleotide reductase activity in these conditions. Ribonucleotide reductase catalyzes the conversion of ribonucleotides to deoxyribonucleotides in an oxygen-dependent manner (since molecular oxygen is required to regenerate a free radical at its active site) (Probst et al. 1989; Reichard and Ehrenberg 1983). Nucleotide levels have been measured *in vitro* and found to fall in response to severe hypoxia in parallel with the S-phase arrest. Importantly, nucleotide levels remain stable in mild hypoxia (for example, 2 % O_2) where no replication arrest is observed (Pires et al. 2010a,b). DNA fiber technology has been used to image individual ongoing or stalled forks as well as to measure the rate of ongoing replication fork speeds. This demonstrated a decrease in origin firing and fork speeds as well as stalling of replication forks in response to severe hypoxia. The occurrence of these aberrant replication structures correlates with the formation of RPA foci, which have been shown to form as a result of the coating of regions of single-stranded DNA (Pires et al. 2010a). Together, these data strongly support the hypothesis that, in response to severe hypoxia, nucleotide availability is limited, leading to stalled replication, the accumulation of regions of ssDNA and the DDR (Fig. 2.3).

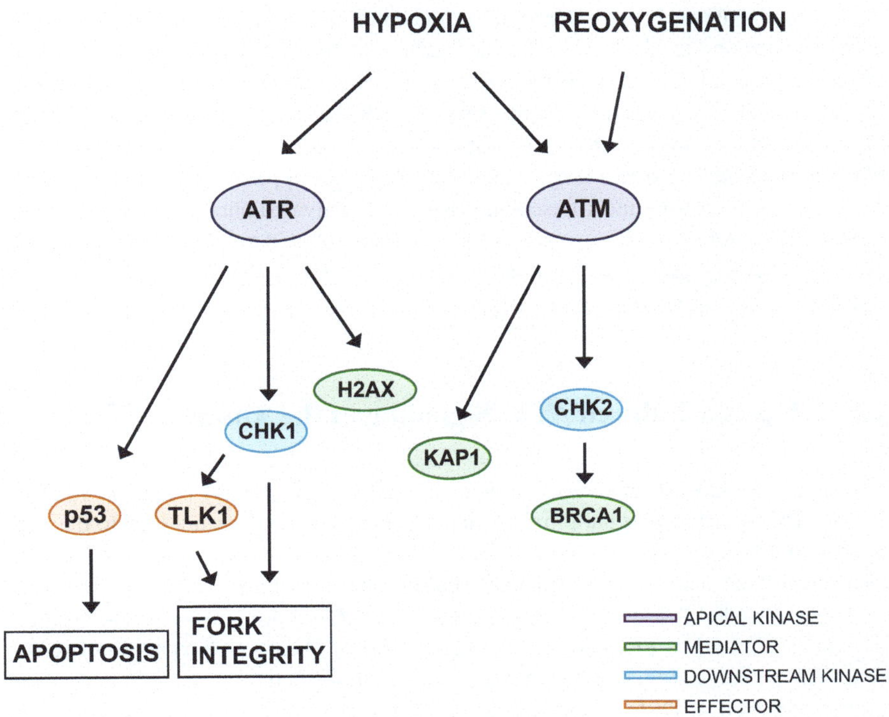

Fig. 2.4 Schematic representation of the hypoxia-induced DDR as described to date. In response to hypoxia, stalled replication initiates a DDR (both ATR- and ATM-dependent), while after reoxygenation DSBs are detected and lead to ATM signaling

2.7 Hypoxia-Induced CHK1 Signaling

ATR has been found to phosphorylate a number of downstream targets in hypoxia, including CHK1 (Hammond et al. 2002, 2003a, 2004) (Fig. 2.4). CHK1 is the key downstream kinase in hypoxia-induced ATR-mediated signaling. ATR and CHK1 have established roles in regulating normal replication and play critical roles in stabilizing stalled replication forks under severe hypoxia. This was demonstrated by showing that while hypoxia did not induce DNA damage, damage did accumulate if ATR or CHK1 were depleted/inhibited (Hammond et al. 2004). In support of the hypothesis that loss of nucleotides leads to replication arrest in hypoxia, loss or inhibition of CHK1 does not play a role in the initial hypoxia-induced replication arrest. If hypoxic cells are allowed to undergo reoxygenation and nucleotide pools are restored following acute hypoxia, replication forks will restart. In this context, an unscheduled increase in the number of new origins will occur if CHK1 is inhibited. CHK1, therefore, maintains genomic integrity following replication restart by delaying origin firing following reoxygenation-induced DNA damage to allow repair before replication can be resumed (Pires et al. 2010a,b). Furthermore, ATR/CHK1 inhibition or knockdown sensitizes cells to hypoxia/reoxygenation as assayed by colony formation assay (Hammond et al. 2004). The increased sensitivity to hypoxia/reoxygenation observed upon CHK1 loss is thought to be predominantly dependent upon inhibition of CHK1 during reoxygenation since loss of CHK1 activity during both reoxygenation and hypoxia does not lead to additional sensitization. Once active, CHK1 will phosphorylate downstream targets including TLK1 in hypoxia (Pires et al. 2010b). TLK1 is a serine/threonine kinase that is usually inactivated once phosphorylated by CHK1 following DNA damage and is thought to facilitate a number of processes, including DNA replication and chromatin remodeling (Groth 2003). The role of TLK1 in hypoxia has not been fully investigated.

2.8 Hypoxia-Induced ATM Signaling in the Absence of DSBs

The hypoxia-induced DDR also leads to the activation of ATM (Bencokova et al. 2009). This is somewhat paradoxical since severe hypoxia does not lead to the generation of detectable DNA damage (Fig. 2.2). Following hypoxia-induced RS ATM is phosphorylated at serine 1981 and has been shown to phosphorylate downstream targets including KAP-1 (Krüppel-associated box (KRAB) domain-associated protein 1), CHK2, and DNA-PKcs (Bencokova et al. 2009). The phosphorylation of KAP1 is somewhat surprising as it has previously been described to be entirely dependent on the presence of DSBs (Goodarzi et al. 2009). Furthermore, phosphorylation of p53 at serine 20 and BRCA1 at serine 988 occurs following CHK2 phosphorylation by ATM (Gibson et al. 2006). Hypoxia-induced ATM activity is independent of the MRN complex. As the principal role of the MRN complex is in the initial detection of DSBs, this fact is supportive of DSBs not being the signal which induces the DDR in hypoxia. Similarly to the classical DDR, the mediator protein MDC1 does amplify hypoxia-induced ATM activity and is required for maximal phosphorylation of ATM targets, for example, KAP-1. However, in contrast to the response to *bona fide* DNA damage, BRCA1, 53BP1, and RNF8 are not subsequently recruited to form nuclear foci by MDC1 in hypoxic conditions. ATM is a predominantly nuclear protein, although some reports indicate that it is also active in the cytoplasm (Watters et al. 1997; Ambrose et al. 2000). During hypoxia phosphorylated ATM is found in the nucleus but does not appear to be tightly associated with the chromatin. This is consistent with the fact that it is not activated in response to or recruited to sites of DNA breaks (Toledo et al. 2011). The role of ATM in severe hypoxia is unclear; however, ATM knockdown has been shown to cause sensitization to hypoxia/reoxygenation (Freiberg et al. 2006b).

The mechanism of ATM activation in response to severe hypoxia is still not well understood. It is possible that hypoxia-induced ATR may contribute to ATM phosphorylation and activation. This possibility is supported by the finding that in response to UV, ATR phosphorylates ATM (Stiff et al. 2006). Further investigation of this hypothesis in hypoxia is technically challenging since ATM is activated in response to pharmacological inhibition or genetic knockdown of ATR. As previously mentioned, this is believed to be due to the collapse of hypoxia-induced stalled replication forks (Toledo et al. 2011). Recently, we demonstrated that a potent ATR inhibitor (VE-821) induced DNA damage, detected by the presence of 53BP1 foci, in hypoxic cells (Pires et al. 2012). Hypoxia is not the only non-DNA damaging stress, which has been shown to induce ATM activity. For example, heat, high salt, and agents which modify chromatin have all been associated with an increase in ATM activity in the absence of DNA damage (Bakkenist and Kastan 2003). The induction of ATM by chromatin-modifying agents suggests the possibility that ATM responds directly to the stalled replication forks in hypoxia, as they would presumably present altered chromatin modifications. In addition, hypoxia is a strong modulator of the chromatin context, with many different chromatin modifications being induced in response to hypoxia (Johnson et al. 2008). Interestingly, the autophosphorylation (on serine 2056) and recruitment of DNA-PK in response to hypoxia (0.1–1 % O_2) have been associated with hypoxia-mediated changes in histone 3 acetylation. In these studies, the authors proposed that DSBs were not the signal for DNA-PK activation since recruitment of the XRCC4-DNA-ligase IV complex did not accompany such activation. DNA-PK was shown to modulate HIF-1-mediated signaling in this study (Bouquet et al. 2011) .

2.9 Two Pathways Collide—HIF-1 and the DDR

Recently, it has become clear that ATM is active (phosphorylated at 1981) in milder hypoxic conditions (0.2–1 % O_2) (Cam et al. 2010). In these conditions, ATM has been shown to phosphorylate and stabilize HIF-1α. ATM-dependent phosphorylation of HIF-1α leads to the stimulation of a negative mTORC1 regulator, REDD1, and a subsequent reduction in mTORC1 signaling in hypoxia. The activation of ATM in this context, however, was shown to be independent of DDR signaling. This is perhaps not particularly surprising since the milder hypoxic conditions used in this study are not associated with replication arrest, which is thought to be the activating signal for the hypoxia-induced DDR (Hammond, et al. 2002; Cam et al. 2010; Hammond et al. 2003a).

In addition, ATR has recently been shown to affect HIF-1α, pointing to a further link between these two signaling pathways (Pires et al. 2012). Cells grown as spheroids treated with the ATR inhibitor VE-821 were found to have lower levels of the HIF-1α target GLUT1 compared to untreated spheroids. This observation was expanded to show that in the presence of VE-821 HIF-1α stabilization was delayed in hypoxic conditions and that this manifested in delayed induction of HIF-1 target genes. This effect of ATR inhibition on HIF-1 stability and activity appeared transient and it is likely that this is due to the concomitant induction of ATM, which can also stabilize HIF-1α as described previously. Despite this, the role of ATR in stabilizing HIF-1α is of significant interest. An increasing number of human tumors have been shown to harbor ATM mutations, suggesting that ATR inhibition may have prolonged effects on HIF-1 activity in certain tumors. Interestingly, ATR inhibition was shown to slow down hypoxia-induced cell motility in a HIF-1α-dependent manner, suggesting that agents of this type might have unexpected impact on tumor spread. A similar role for ATR in regulating HIF-1 signaling has been demonstrated in a recent study where HIF-1α translation was shown to be regulated by ATR kinase activity at 0.1 % O_2 (Fallone 2012). In addition, Economopoulou et al. have shown that in response to mild hypoxia (1 % O_2), where no replication arrest is observed, H2AX is phosphorylated in an ATR-dependent manner in proliferating endothelial cells. The low levels of DNA damage arising during normal replication were thought to serve as the activating signal for DDR activation and H2AX phosphorylation in this case. These studies showed that γH2AX was required for pathological neovascularization in hypoxia. Notably, loss of γH2AX had no effect on developmental angiogenesis (Rankin et al. 2009; Economopoulou et al. 2009).

2.10 Reoxygenation and the DDR

As previously mentioned, reoxygenation following periods of hypoxia leads to the induction of DNA damage. This reoxygenation-induced damage signaling can be inhibited with the use of reactive oxygen species (ROS) scavengers, suggesting that ROS maintains the DDR after reoxygenation. In response to reoxygenation, ATM has been shown to phosphorylate downstream targets, including p53 and CHK2. ATM-dependent CHK2 phosphorylation leads to G_2 arrest (Gibson et al. 2005). In the absence of CHK2, cells undergo elevated levels of apoptosis and the G_2 arrest is abrogated (Freiberg et al. 2006a, 2006b; Hammond et al. 2004). If cells are exposed to acute periods of severe hypoxia (under 12 h) and then are allowed to undergo reoxygenation, they will undergo replication restart (Pires et al. 2010a,b). It is important to note that these cells will restart following the induction of ROS-induced DNA damage and in a context of reduced DNA repair (described below), potentially leading to increased genomic instability. Human tumor cells, with intact p53 function, will undergo p53-dependent apoptosis if exposed to periods of acute hypoxia followed by reoxygenation (Rzymski et al. 2010). When cells are exposed to longer periods of hypoxia (for example, over 12 h), replication does not restart in response to reoxygenation. Expression analysis determined that numerous key replication factors are repressed in hypoxic conditions. These include members of the MCM family, which act as a complex and are loaded onto chromatin for DNA replication. In addition, several members of this complex such as MCM3, MCM4, MCM5, and MCM6 have been shown to become unbound from chromatin in cells exposed to over 12 h of severe hypoxia. Replication and DNA synthesis restart cannot occur in these cases, even if reoxygenation occurs and nucleotide levels are restored (Pires et al. 2010a).

In addition after reoxygenation, ATM has been shown to be important in regulating pathological angiogenesis by limiting excessive ROS. Loss of ATM results in increased ROS and activation of the mitogen-activated kinase p38α and subsequent apoptosis in endothelial cells. This supports a distinct role of ATM in hypoxia/reoxygenation independent of its role in the DDR (Okuno et al. 2012).

2.11 Hypoxia Repression of DNA Repair

One prominent feature of tumor cells is genomic instability, now widely accepted as a hallmark of cancer (Negrini et al. 2010; Hanahan and Weinberg 2011). As mentioned above, the hypoxia-induced DDR can act as a barrier to tumorigenesis and could potentially act as protection mechanism against such genomic instability. Hypoxia, however, can also contribute to this hallmark in a number of ways. For instance, aberrant DNA replication (Young et al. 1988), gene amplification (Rice et al. 1986), as well as base substitutions and DNA deletions have all been linked to the hypoxic microenvironment (Reynolds 1996). Defects in sensing of DSBs have also been proposed to contribute to the hypoxia-mediated genomic instability. Irradiated cells in G_0-G_1 phase under chronic hypoxia or anoxia, for instance, have increased residual DSBs and chromosomal aberrations, pointing to a defect in DNA damage sensing in hypoxia (Kumareswaran et al. 2012). Finally, hypoxia can act as a driver of genomic instability since it can lead to the repression of a number of important DNA repair pathways (Klein and Glazer 2010). This final mechanism of driving genomic instability has been extensively studied in recent years and will be reviewed in detail below.

HR, a critical pathway involved in repair of breaks in the S and G_2 phases of the cell cycle, is functionally inhibited in hypoxia. Components of this pathway including RAD51, RAD52, BRCA1, and BRCA2 have been shown to be repressed in hypoxia by a number of mechanisms. These include binding of E2F4/p130 at the promoters of *RAD51* and *BRCA1* (Bindra et al. 2005, 2007; Halazonetis et al. 2008). Decreased RAD51 protein levels in moderate hypoxia have also been proposed to arise due to impairment of mRNA translation (Chan et al. 2008). Repression of *RAD51* and *BRCA1* also occurs by enrichment of repressive histone modifications such as H3K9me3 at specific loci on these genes (Lu et al. 2011).

The microRNA, miR-210, is a hypoxia-induced miR that also contributes to the repression of *RAD52*. This miR has been shown to bind the 3′ untranslated region (UTR) of RAD52. Use of an anti-miR-210 molecule prevented repression of RAD52 in hypoxia. Furthermore, *RAD23B*, a member of the nucleotide excision repair pathway, was also shown to be downregulated in hypoxia in this study. Overexpression of miR-373 resulted in repression of *RAD23B* as well as *RAD52*, whereas the use of anti-miR-373 in hypoxia partially prevented such repression. A binding site for miR-373 in the 3′ UTR of *RAD23B* was also described (Crosby 2009).

Functional assays using reporter plasmids capable of identifying simple repeated sequence instability, characteristic of mismatch repair (MMR) defects, have indicated an increase in genetic instability in hypoxia. Such instability was thought to occur due to the repression of *MLH1* in a HIF-1α-independent manner. Hypoxia-induced histone deacetylases (HDACs) were found to be important for this repression since the use of Trichostatin A (TSA), an HDAC inhibitor, prevented *MLH1* repression (Mihaylova et al. 2003). Epigenetic mechanisms of transcriptional repression of *MLH1* have also been proposed by Chen and colleagues (Chen et al. 2006). In this study, *MLH1* repression correlated with an increase in the activity of histone methyltransferase G9a and the subsequent enrichment of H3K9me2 at its promoter region, also in a HIF-1α-independent manner (Chen et al. 2006). Hypoxia-induced transcriptional repression of Myc has also been associated with *MLH1* and *MLH2* repression. In fact, a switch between activating c-Myc and repressive Mad/Mnt binding was proposed as a molecular mechanism for such repression and again occurred in a HIF-1α-independent manner (Bindra and Glazer 2007). In addition, HIF-1α-dependent mechanisms of MMR repression have also been reported. The transcription factors DEC1 and DEC2 are HIF-1α targets which are known to transcriptionally repress *MLH1* in response to hypoxia. Members of the MutSα, complex *MSH2* and *MSH6*, have been shown to be repressed in a HIF-1α-dependent manner in hypoxia. Again, a decrease in Myc binding at the promoters of these MMR genes was proposed to occur in response to hypoxia. In this case, Sp1, an interactor of both Myc and HIF-1α, is thought to preferentially recruit HIF-1α in response to hypoxia, displacing Myc binding from these sites. In these studies HIF-1α was, therefore, found to be a repressor of *MSH2* and *MSH6*, and interestingly such repression was also dependent on p53. The authors of this study also provided immunohistochemical evidence to show that low levels of MSH2 expression in tumors correlated with elevated HIF-1α expression, especially in those tumors where p53 could not be detected (Koshiji et al. 2005).

UBE2T is an E2 conjugating enzyme required for the activation of the Fanconi anemia pathway via FANCD2 ubiquitylation. The messenger RNA (mRNA) levels of UBE2T are rapidly decreased under hypoxic conditions in a HIF-1α-independent manner. *UBE2T* promoter activity in hypoxia was assessed in these studies by the use of a luciferase reporter construct composed of a 2 kb fragment upstream of *UBE2T*'s transcriptional start site. Promoter activity was found to be reduced in hypoxia and proposed as a possible mechanism for the reduction of *UBE2T* expression (Ramaekers et al. 2011).

Interestingly, nonhomologous end joining (NHEJ)-mediated repair is not compromised even though some members of the pathway are also repressed under hypoxic conditions (Bindra and Glazer 2005). It has been proposed that cells under hypoxic conditions may be forced to use this error-prone repair pathway preferentially given that most of the more accurate repair pathways are compromised. This may also lead to further genomic instability and potentially the aggressive phenotype of hypoxic tumor cells (Bristow and Hill 2008).

It is worth noting that repression of DNA repair pathways occurs in both mild as well as severe hypoxia suggesting that a large proportion of the hypoxic tumor may experience compromised repair. Moreover, the hypoxia-induced DDR is also repressed following prolonged exposures to severe hypoxia. The rapid hypoxia-induced phosphorylation of many of the members of the DDR that occurs following replication arrest has been shown to diminish following exposure to chronic hypoxia. This is most notable in the case of CHK1 phosphorylation, where the rapid and robust induction declines after prolonged hypoxia (Pires et al. 2010a,b).

It is also important to note that many of the DNA repair factors described above are also important facilitators of DNA damage sensing and signaling. For instance, BRCA1 forms a complex with MSH2 and MSH6, which is important for appropriate genomic stability maintenance following DNA damage (Wang et al. 2000). The appropriate activation of the S-phase checkpoint following IR also relies on the interaction between ATM and MLH1 and CHK2 and MSH2 (Brown et al. 2003). The involvement of many of these proteins in DNA damage sensing, signaling, and repair further highlights the impact of their repression in contributing to genomic instability.

All these factors taken in concert provide significant evidence for the role of hypoxia in contributing to genetic instability in tumors (Yuan et al. 2000). Of note, tumor cells often undergo cycles of hypoxia followed by reoxygenation. Reoxygenation can, for instance, occur following spontaneous vessel reopening or as a result of radiotherapy. Elegant studies have demonstrated that the hypoxic microenvironment is very dynamic with frequent changes in oxygen levels occurring within tumors (Dewhirst et al. 2008). As previously mentioned, even though hypoxia alone does not lead to the accumulation of DNA damage, reoxygenation leads to substantial damage. In the absence of fully functional repair pathways, the generation of DNA damage following cycles of hypoxia/reoxygenation can further compromise genetic integrity (Olcina et al. 2010).

2.12 Targeting the Hypoxia-Induced DDR

The hypoxia-induced DDR can be targeted in a number of ways leading to a potential increase in therapeutic window (Olcina et al. 2010). Reactivation of p53 has been attempted as a therapeutic strategy through the use of a variety of pharmacological agents. RITA (reactivation of p53 and induction of tumor cell apoptosis) activates p53, induces p53-dependent apoptosis, and can inhibit growth (Yang et al. 2009). This small molecule can also activate the DDR (leading to accumulation of p53 and γH2AX) and inhibit HIF-1α and its target, vascular endothelial growth factor (VEGF). Targeting the DDR in hypoxia by reactivating p53 may, therefore, increase cell kill as well as potentially reduce angiogenesis (Yang et al. 2009). Other compounds that can mediate p53 reactivation include Nutlin (an antagonist of p53's negative regulator, MDM2) and PRIMA (p53 reactivation and induction of massive apoptosis) (Di Cintio et al. 2010). The ability to accumulate functional p53 by any of these compounds may also provide a therapeutic advantage if these are combined with chemotherapy drugs that rely on p53-dependent apoptosis to exert their toxicity. This might provide a particularly good strategy to target hypoxic tumor cells under the strong selection pressure for loss apoptosis in the absence of p53 (Graeber et al. 1996; Kim et al. 1997; El-Deiry 2003).

The loss of components of the DDR such as ATR, ATM, CHK1, or CHK2 has been shown to sensitize cells to hypoxia/reoxygenation. This has been shown using several methodologies including gene silencing or the use of knockout or kinase dead cell lines (Hammond et al. 2004; Hammond and Giaccia 2004). More recently, the use of specific inhibitors of some of these kinases has also been shown to have promising sensitizing effects (Pires et al. 2012). Two specific ATR inhibitors, VE-821 and VE-822, were shown recently to reduce viability of a variety of tumor cell lines to a range of oxygen concentrations (Pires et al. 2012; Prevo et al. 2012; Fokas et al. 2012). Moreover, VE-821 was shown to enhance radiation-induced loss of viability of cells exposed to both normoxic as well as hypoxic conditions, showing the potential that these types of compounds could have in combination therapies (Pires et al. 2012). Importantly, the use of VE-821 either as a single agent or in combination with genotoxic agents only led to a reversible growth arrest in normal cells (Reaper et al. 2011; Charrier et al. 2011). Other ATR inhibitors, such as NU6027, have also been tested in combination with IR and temozolomide. This inhibitor was shown to sensitize breast and ovarian cell lines to both of these treatments (Peasland et al. 2011).

CHK1 inhibitors, such as compound AZD7762, are available and have been tested in combination with other forms of therapy such as nucleoside analog gemcitabine or topoisomerase inhibitors. Increased cell kill was observed when AZD7762 was used in combination with gemcitabine. The failure to stabilize replication forks together with the unscheduled activation of origin firing and entry into mitosis when cells are treated with gemcitabine under CHK1 inhibition is thought to be the cause of these effects. Furthermore, ATR/CHK1 inhibitors have been shown to be particularly effective in the treatment of tumors with high levels of oncogene-mediated RS. The promising effects seen with CHK1 inhibitors in tumors with high levels of endogenous or exogenous RS suggests they should also be effective for the targeting of severely hypoxic tumor cells (Gilad et al. 2010; Schoppy 2012). CHK1 inhibitors have also been used in combination with IR. The inability to maintain an appropriate checkpoint response and the repression of repair pathways such as HR are thought to play an important role in the increased radiosensitization observed when AZD7762 is used in combination with radiation (Morgan et al. 2010). Such radiosensitizaton has been observed both *in vitro* as well as in xenograft models *in vivo* (Mitchell et al. 2010; McNeely et al. 2010). Given that hypoxia leads to a repression of HR, it is possible that CHK1 inhibition in hypoxia may also lead to radiosensitization of hypoxic tumor cells through this mechanism.

Synthetic lethality describes a situation where a combination of mutations or inhibition of two genes or proteins leads to lethality whereas the loss of either one alone results in viability. This concept is becoming increasingly exploited as a potential means of therapeutically targeting the DDR. The most common example of synthetic lethality, aside from in yeast, is the use of inhibitors of poly (ADP-ribose) polymerase (PARP) in tumors with mutations in BRCA1/2 (McCabe et al. 2006). In these situations the loss of BRCA1 and the pharmacological inhibition of PARP lead to a reduced repair capability by the tumor and an increased cell kill (Morgan et al. 2010). PARP inhibitors have been in phase I, II, and III trials in the last

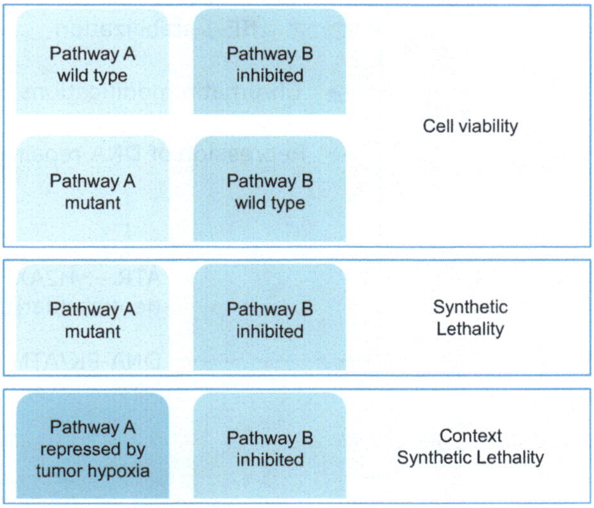

Fig. 2.5 Context synthetic lethality. When either pathway A or pathway B is functionally inhibited through mutation or inhibition cell, viability is unaffected. However, when this occurs in the same cell, synthetic lethality occurs. Context synthetic lethality occurs when one of the interacting pathways is repressed by the cellular context, in this case hypoxia

couple of years for a number of different indications (www.clinical trials.gov). Use of these inhibitors has been particularly encouraging as monotherapy in breast and ovarian cancers with BRCA mutations, although the emerging resistance to PARP inhibition has caused some concerns. The use of PARP inhibitors in combination therapy is also being investigated. Trials in non-small cell lung cancer with PARP inhibitors in combination with chemotherapy agents have been initiated. Synthetic lethality between ATR and PARP inhibition has also been demonstrated (Peasland et al. 2011). Furthermore, a growing number of tumor types have now been shown to harbor mutations or loss of functional ATM (Cam et al. 2010; Yu et al. 2004; Ye et al. 2007; Ding et al. 2008; Jiang et al. 2009). The use of ATR inhibitors in these tumors may provide a therapeutic advantage, particularly if these are used in combination with other treatments able to induce DNA damage. This has been proposed as yet another synthetic lethal opportunity (Reaper et al. 2011).

As hypoxia leads to the repression of a number of DNA repair pathways, the concept of synthetic lethality can be applied to any tumor where repair pathways are compromised. This is known as "context synthetic lethality" and has again been exemplified with the use of PARP inhibitors (Fig. 2.5). In hypoxic cells where members of the HR pathway, such as RAD51 or BRCA1, are repressed, the use of a PARP inhibitor, ABT-888, sensitized cells to hypoxia/reoxygenation (Chan et al. 2010). PARP inhibitors have also been shown to radiosensitize hypoxic tumor cells (Liu et al. 2008). Context synthetic lethality may, therefore, open up the possibility of using synthetically lethal interactions in a range of solid tumors where hypoxia can provide the context deregulation of particular pathways. It could be envisioned that the use of inhibitors to other parts of the DDR that work in synergy with DNA repair pathways would also render hypoxic tumor cells more sensitive to these kinds of agents while causing minimal effects on normal tissues (not present in hypoxic regions). These inhibitors could be used in combination with other agents causing

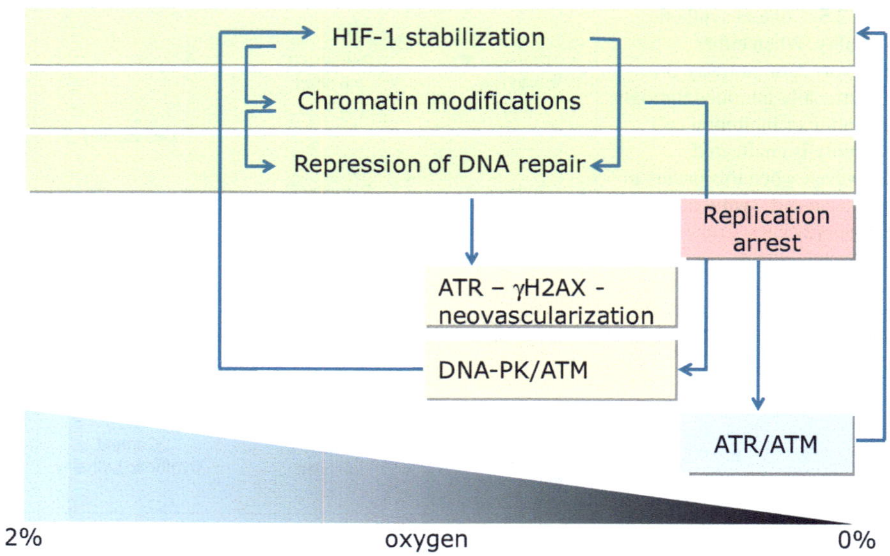

Fig. 2.6 Hypoxia-induced DDR signaling. The responses to a range of hypoxic conditions (0–2 % O_2) are illustrated. In response to these conditions, distinct chromatin marks are present and DNA repair is repressed (Johnson et al. 2008; Bindra et al. 2007; Chen et al. 2006; Bristow and Hill 2008). Replication arrest is induced in severe hypoxia and in turn initiates ATR/ATM signaling (Olcina et al. 2010). ATR/ATM and DNA PK have also been shown to be active in milder hypoxic conditions and have been shown to regulate HIF-1 stability under these conditions (Bouquet et al. 2011; Cam et al. 2010; Fallone et al. 2012)

DNA damage such as chemotherapy or radiotherapy, leading to an increase in the selectivity and efficacy of these agents. These combination therapies could form the basis of interesting therapies specifically targeting the tumor cells and, therefore, increasing the therapeutic window.

2.13 Conclusion

Regions of varying oxygen concentrations are commonly found within tumors (Vaupel et al. 1989). The activation and the consequence of the DDR are different at different oxygen tensions (as summarized in Fig. 2.6). One of the most constant features of the DDR is the repression of DNA repair, which is prevalent across oxygen tensions. The reasons for this phenomenon are unclear. Repression has been proposed to occur as an energy-saving measure in conditions of stress such as hypoxia (Klein and Glazer 2010). However, the apparent repression of repair pathways may simply be a reflection of the fact that elevated repair levels might result from cell lines adapting to tissue culture growth conditions, whereas hypoxic conditions may simply lead to a return to baseline physiological levels of repair. Severe hypoxia is an example of physiologically relevant RS that induces both

ATR- as well as ATM-mediated signaling. The apical PIKKs have all been shown to influence HIF-1 signaling also demonstrating a role for them in mild hypoxia. Importantly, inhibition of the hypoxia-induced DDR offers new and encouraging approaches to targeting some of the most aggressive fraction of solid tumors.

References

Ambrose M, Goldstine JV, Gatti RA (2000) Intrinsic mitochondrial dysfunction in ATM-deficient lymphoblastoid cells. Hum Mol Genet 16:2154–2164. doi:ddm166 [pii]10.1093/hmg/ddm166

Bakkenist CJ, Kastan MB (2003) DNA damage activates ATM through intermolecular autophosphorylation and dimer dissociation. Nature 421:499–506

Bartkova J et al (2005) DNA damage response as a candidate anti-cancer barrier in early human tumorigenesis. Nature 434:864–870

Bartkova J et al (2006) Oncogene-induced senescence is part of the tumorigenesis barrier imposed by DNA damage checkpoints. Nature 444:633–637. doi:nature05268 [pii]10.1038/nature05268

Bencokova Z et al (2009) ATM activation and signaling under hypoxic conditions. Mol Cell Biol 29:526–537. doi:10.1128/MCB.01301-08

Bindra RS, Glazer PM (2005) Genetic instability and the tumor microenvironment: towards the concept of microenvironment-induced mutagenesis. Mutat Res 569:75–85

Bindra RS, Glazer PM (2007) Co-repression of mismatch repair gene expression by hypoxia in cancer cells: role of the Myc/Max network. Cancer Lett 252(1):93–103

Bindra RS et al (2005) Hypoxia-induced down-regulation of BRCA1 expression by E2Fs. Cancer Res 65:11597–11604

Bindra RS, Crosby ME, Glazer PM (2007) Regulation of DNA repair in hypoxic cancer cells. Cancer Metastasis Rev 26:249–260. doi:10.1007/s10555-007-9061-3

Bouquet F et al (2011) A DNA-dependent stress response involving DNA-PK occurs in hypoxic cells and contributes to cellular adaptation to hypoxia. J Cell Sci 124:1943–1951. doi:10.1242/jcs.078030

Bristow RG, Hill RP (2008) Hypoxia and metabolism. Hypoxia, DNA repair and genetic instability. Nat Rev Cancer 8:180–192. doi:nrc2344 [pii]10.1038/nrc2344

Brown KD et al (2003) The mismatch repair system is required for S-phase checkpoint activation. Nat Genet 33:80–84. doi:10.1038/ng1052ng1052 [pii]

Cam H, Easton JB, High A, Houghton PJ (2010) mTORC1 signaling under hypoxic conditions is controlled by ATM-dependent phosphorylation of HIF-1alpha. Mol Cell 40:509–520. doi:10.1016/j.molcel.2010.10.030

Chan N et al (2008) Chronic hypoxia decreases synthesis of homologous recombination proteins to offset chemoresistance and radioresistance. Cancer Res 68:605–614. doi:68/2/605 [pii]10.1158/0008-5472.CAN-07-5472

Chan N et al (2010) Contextual synthetic lethality of cancer cell kill based on the tumor microenvironment. Cancer Res 70:8045–8054. doi:10.1158/0008-5472.CAN-10-2352

Charrier JD et al (2011) Discovery of potent and selective inhibitors of ataxia telangiectasia mutated and Rad3 related (ATR) protein kinase as potential anticancer agents. J Med Chem 54:2320–2330. doi:10.1021/jm101488z

Chen H, Yan Y, Davidson TL, Shinkai Y,Costa M (2006) Hypoxic stress induces dimethylated histone H3 lysine 9 through histone methyltransferase G9a in mammalian cells. Cancer Res 66:9009–9016. doi:66/18/9009[pii]10.1158/0008-5472.CAN-06-0101

Chen J et al (1998) Stable interaction between the products of the BRCA1 and BRCA2 tumor suppressor genes in mitotic and meiotic cells. Mol Cell 2:317–328

Ciccia A, Elledge SJ (2010) The DNA damage response: making it safe to play with knives. Mol Cell 40:179–204. doi:10.1016/j.molcel.2010.09.019 S1097–2765(10)00747-1 [pii]

Cimprich KA, Cortez D (2008) ATR: an essential regulator of genome integrity. Nat Rev Mol Cell Biol 9:616–627. doi:nrm2450 [pii]10.1038/nrm2450
Cliby WA et al (1998) Overexpression of a kinase-inactive ATR protein causes sensitivity to DNA-damaging agents and defects in cell cycle checkpoints. Embo J 17:159–169
Collis SJ, DeWeese TL, Jeggo PA, Parker AR (2005) The life and death of DNA-PK. Oncogene 24:949–961. doi:1208332 [pii]10.1038/sj.onc.1208332
Cortez D, Wang Y, Qin J, Elledge SJ (1999) Requirement of ATM-dependent phosphorylation of brca1 in the DNA damage response to double-strand breaks. Science 286:1162–1166
Cortez D, Guntuku S, Qin J, Elledge SJ (2001) ATR and ATRIP: partners in checkpoint signaling. Science 294:1713–1716
Crosby ME, Kulshreshtha R, Ivan M, Glazer PM (2009) MicroRNA regulation of DNA repair gene expression in hypoxic stress. Cancer Res 69:1221–1229. doi:0008-5472.CAN-08-2516 [pii]10.1158/0008-5472.CAN-08-2516
Dai Y, Grant S (2010) New insights into checkpoint kinase 1 in the DNA damage response signaling network. Clin Cancer Res 16:376–383. doi:1078-0432.CCR-09-1029 [pii]10.1158/1078-0432.CCR-09-1029
Dewhirst MW, Cao Y, Moeller B (2008) Cycling hypoxia and free radicals regulate angiogenesis and radiotherapy response. Nat Rev Cancer 8:425–437. doi:nrc2397 [pii]10.1038/nrc2397
Di Cintio A, Di Gennaro E, Budillon A (2010) Restoring p53 function in cancer: novel therapeutic approaches for applying the brakes to tumorigenesis. Recent Pat Anticancer Drug Discov 5:1–13. doi:PRA-ABS-Budillon-13 [pii]
Di Micco R et al (2006) Oncogene-induced senescence is a DNA damage response triggered by DNA hyper-replication. Nature 444:638–642. doi:10.1038/nature05327
Ding L et al (2008) Somatic mutations affect key pathways in lung adenocarcinoma. Nature 455:1069–1075. doi:10.1038/nature07423
Economopoulou M et al (2009) Histone H2AX is integral to hypoxia-driven neovascularization. Nat Med 15:553–558. doi:nm.1947 [pii]10.1038/nm.1947
El-Deiry WS (2003) The role of p53 in chemosensitivity and radiosensitivity. Oncogene 22:7486–7495. doi:10.1038/sj.onc.12069491206949 [pii]
Fallone F, Britton S, Nieto L, Salles B, Muller C (2012) ATR controls cellular adaptation to hypoxia through positive regulation of hypoxia-inducible factor 1 (HIF-1) expression. Oncogene. doi:10.1038/onc.2012.462onc2012462 [pii]
Fanning E, Klimovich V, Nager AR (2006) A dynamic model for replication protein A (RPA) function in DNA processing pathways. Nucleic Acids Res 34:4126–4137. doi:gkl550 [pii]10.1093/nar/gkl550
Fokas E et al (2012) Targeting ATR in vivo using the novel inhibitor VE-822 results in selective sensitization of pancreatic tumors to radiation. Cell Death Dis 3:e441. doi:10.1038/cddis.2012.181cddis2012181 [pii]
Freiberg RA, Hammond EM, Dorie MJ, Welford SM, Giaccia AJ (2006a) DNA damage during reoxygenation elicits a Chk2-dependent checkpoint response. Mol Cell Biol 26:1598–1609. doi:10.1128/MCB.26.5.1598-1609
Freiberg RA, Krieg AJ, Giaccia AJ, Hammond EM (2006b) Checking in on hypoxia/reoxygenation. Cell Cycle 5:1304–1307
Gibson SL, Bindra RS, Glazer PM (2005) Hypoxia-induced phosphorylation of Chk2 in an ataxia telangiectasia mutated-dependent manner. Cancer Res 65:10734–10741
Gibson SL, Bindra RS, Glazer PM (2006) CHK2-dependent phosphorylation of BRCA1 in hypoxia. Radiat Res 166:646–651. doi:RR0660 [pii]10.1667/RR0660.1
Gilad O et al (2010) Combining ATR suppression with oncogenic Ras synergistically increases genomic instability, causing synthetic lethality or tumorigenesis in a dosage-dependent manner. Cancer Res 70:9693–9702. doi:0008-5472.CAN-10-2286 [pii]10.1158/0008-5472.CAN-10-2286
Goldberg M et al (2003) MDC1 is required for the intra-S-phase DNA damage checkpoint. Nature 421:952–956, doi:10.1038/nature01445nature01445 [pii]

Goodarzi AA, Noon AT, Jeggo PA (2009) The impact of heterochromatin on DSB repair. Biochem Soc Trans 37:569–576. doi:BST0370569[pii]10.1042/BST0370569

Gorgoulis VG et al (2005) Activation of the DNA damage checkpoint and genomic instability in human precancerous lesions. Nature 434:907–913

Graeber TG et al (1994) Hypoxia induces accumulation of p53 protein, but activation of a G1- phase checkpoint by low-oxygen conditions is independent of p53 status. Mol Cell Biol 14:6264–6277

Graeber TG et al (1996) Hypoxia-mediated selection of cells with diminished apoptotic potential in solid tumours. Nature 379:88–91

Green DR, Kroemer G (2009) Cytoplasmic functions of the tumour suppressor p53. Nature 458:1127–1130. doi:nature07986 [pii]10.1038/nature07986

Groth A et al (2003) Human Tousled like kinases are targeted by an ATM- and Chk1-dependent DNA damage checkpoint. Embo J 22:1676–1687

Halazonetis TD, Gorgoulis VG, Bartek J (2008) An oncogene-induced DNA damage model for cancer development. Science 319:1352–1355. doi:319/5868/1352 [pii]10.1126/science.1140735

Hammond EM, Giaccia AJ (2004) The role of ATM and ATR in the cellular response to hypoxia and re-oxygenation. DNA Repair (Amst) 3:1117–1122. doi:10.1016/j.dnarep.2004.03.035

Hammond EM, Giaccia AJ (2005) The role of p53 in hypoxia-induced apoptosis. Biochem Biophys Res Commun 331:718–725. doi:10.1016/j.bbrc.2005.03.154

Hammond EM, Denko NC, Dorie MJ, Abraham RT, Giaccia AJ (2002) Hypoxia links ATR and p53 through replication arrest. Mol Cell Biol 22:1834–1843

Hammond EM, Dorie MJ, Giaccia AJ (2003a) ATR/ATM targets are phosphorylated by ATR in response to hypoxia and ATM in response to reoxygenation. J Biol Chem 278:12207–12213. doi:10.1074/jbc.M212360200

Hammond EM, Green SL, Giaccia AJ (2003b) Comparison of hypoxia-induced replication arrest with hydroxyurea and aphidicolin-induced arrest. Mutat Res 532:205–213

Hammond EM, Dorie MJ, Giaccia AJ (2004) Inhibition of ATR leads to increased sensitivity to hypoxia/reoxygenation. Cancer Res 64:6556–6562. doi:10.1158/0008-5472.CAN-04-1520

Hanahan D, Weinberg RA (2011) Hallmarks of cancer: the next generation. Cell 144:646–674, doi:10.1016/j.cell.2011.02.013S0092–8674(11)00127-9 [pii]

Harper JW, Elledge SJ (2007) The DNA damage response: ten years after. Mol Cell 28:739–745. doi:10.1016/j.molcel.2007.11.015

Hurley PJ, Wilsker D, Bunz F (2007) Human cancer cells require ATR for cell cycle progression following exposure to ionizing radiation. Oncogene 26:2535–2542. doi:10.1038/sj.onc.1210049

Jackson SP, Bartek J (2009) The DNA-damage response in human biology and disease. Nature 461:1071–1078. doi:nature08467 [pii] 10.1038/nature08467

Jeggo PA, Lobrich M (2007) DNA double-strand breaks: their cellular and clinical impact? Oncogene 26:7717–7719. doi:1210868 [pii]10.1038/sj.onc.1210868

Jiang H et al (2009) The combined status of ATM and p53 link tumor development with therapeutic response. Genes Dev 23:1895–1909. doi:gad.1815309 [pii]10.1101/gad.1815309

Johnson AB, Denko N, Barton MC (2008) Hypoxia induces a novel signature of chromatin modifications and global repression of transcription. Mutat Res 640:174–179. doi:S0027-5107(08)00017-1 [pii]10.1016/j.mrfmmm.2008.01.001

Kastan MB, Bartek J (2004) Cell-cycle checkpoints and cancer. Nature 432:316–323

Kim CY et al (1997) Selection of human cervical epithelial cells that possess reduced apoptotic potential to low-oxygen conditions. Cancer Res 57:4200–4204

Klein TJ, Glazer PM (2010) The tumor microenvironment and DNA repair. Semin Radiat Oncol 20:282–287. doi:S1053-4296(10)00040-8 [pii]10.1016/j.semradonc.2010.05.006

Koshiji M et al (2005) HIF-1alpha induces genetic instability by transcriptionally down-regulating MutSalpha expression. Mol Cell 17:793–803. doi:S1097-2765(05)01114-7 [pii]10.1016/j.molcel.2005.02.015

Kumareswaran R et al (2012) Chronic hypoxia compromises repair of DNA double-strand breaks to drive genetic instability. J Cell Sci 125:189–199. doi:10.1242/jcs.092262

Lane DP (1992) Cancer. p53, guardian of the genome. Nature 358:15–16. doi:10.1038/358015a0

Lavin MF, Kozlov S (2007) ATM activation and DNA damage response. Cell Cycle 6:931–942. doi:4180 [pii]

Levine AJ (1997) p53, the cellular gatekeeper for growth and division. Cell 88:323–331

Lindahl T, Barnes DE (2000) Repair of endogenous DNA damage. Cold Spring Harb Symp Quant Biol 65:127–133

Liu Q et al (2000) Chk1 is an essential kinase that is regulated by Atr and required for the G(2)/M DNA damage checkpoint. Genes Dev 14:1448–1459

Liu SK et al (2008) A novel poly(ADP-ribose) polymerase inhibitor, ABT-888, radiosensitizes malignant human cell lines under hypoxia. Radiother Oncol 88:258–268. doi:S0167-8140(08)00210-7 [pii]10.1016/j.radonc.2008.04.005

Lu Y, Chu A, Turker MS, Glazer PM (2011) Hypoxia-induced epigenetic regulation and silencing of the BRCA1 promoter. Mol Cell Biol 31:3339–3350. doi:MCB.01121-10 [pii]10.1128/MCB.01121-10

Matsuoka S et al (2000) Ataxia telangiectasia-mutated phosphorylates Chk2 in vivo and in vitro. Proc Natl Acad Sci U S A 97:10389–10394

McCabe N et al (2006) Deficiency in the repair of DNA damage by homologous recombination and sensitivity to poly(ADP-ribose) polymerase inhibition. Cancer Res 66:8109–8115. doi:66/16/8109 [pii]10.1158/0008-5472.CAN-06-0140

McNeely S et al (2010) Chk1 inhibition after replicative stress activates a double strand break response mediated by ATM and DNA-dependent protein kinase. Cell Cycle 9:995–1004. doi:10935 [pii]

Mihaylova VT et al (2003) Decreased expression of the DNA mismatch repair gene Mlh1 under hypoxic stress in mammalian cells. Mol Cell Biol 23:3265–3273

Mitchell JB et al (2010) In vitro and in vivo radiation sensitization of human tumor cells by a novel checkpoint kinase inhibitor, AZD7762. Clin Cancer Res 16:2076–2084. doi:1078-0432.CCR-09-3277 [pii]10.1158/1078-0432.CCR-09-3277

Moreno-Herrero F et al (2005) Mesoscale conformational changes in the DNA-repair complex Rad50/Mre11/Nbs1 upon binding DNA. Nature 437:440–443. doi:nature03927 [pii]10.1038/nature03927

Morgan MA et al (2010) Mechanism of Radiosensitization by the Chk1/2 Inhibitor AZD7762 Involves Abrogation of the G2 Checkpoint and Inhibition of Homologous Recombinational DNA Repair. Cancer Res doi:0008-5472.CAN-09-3573 [pii]10.1158/0008-5472.CAN-09-3573

Negrini S, Gorgoulis VG, Halazonetis TD (2010) Genomic instability—an evolving hallmark of cancer. Nat Rev Mol Cell Biol 11:220–228. doi:10.1038/nrm2858nrm2858 [pii]

Okuno Y, Nakamura-Ishizu A, Otsu K, Suda T, Kubota Y (2012) Pathological neoangiogenesis depends on oxidative stress regulation by ATM. Nat Med doi:10.1038/nm.2846

Olcina M, Lecane PS, Hammond EM (2010) Targeting hypoxic cells through the DNA damage response. Clin Cancer Res 16:5624–5629. doi:10.1158/1078-0432.CCR-10-0286

Peasland A et al (2011) Identification and evaluation of a potent novel ATR inhibitor, NU6027, in breast and ovarian cancer cell lines. Br J Cancer 105:372–381. doi:10.1038/bjc.2011.243

Pires IM et al (2010a) Effects of acute versus chronic hypoxia on DNA damage responses and genomic instability. Cancer Res 70:925–935. doi:10.1158/0008-5472.CAN-09-2715

Pires IM, Bencokova Z, McGurk C, Hammond EM (2010b) Exposure to acute hypoxia induces a transient DNA damage response which includes Chk1 and TLK1. Cell Cycle 9:2502–2507. doi:10.4161/cc.9.13.12059

Pires IM et al (2012) Targeting radiation-resistant hypoxic tumour cells through ATR inhibition. Br J Cancer 107:291–299. doi:10.1038/bjc.2012.265

Prevo R et al (2012) The novel ATR inhibitor VE-821 increases sensitivity of pancreatic cancer cells to radiation and chemotherapy. Cancer Biol Ther 13(11):1072–1081

Probst H, Schiffer H, Gekeler V, Scheffler K (1989) Oxygen dependent regulation of mammalian ribonucleotide reductase in vivo and possible significance for replicon initiation. Biochem Biophys Res Commun 163:334–340

Ramaekers CHMA et al (2011) Hypoxia disrupts the Fanconi anemia pathway and sensitizes cells to chemotherapy through regulation of UBE2T. Radiother Oncol 101:190–197. doi:DOI 10.1016/j.radonc.2011.05.059

Rankin EB, Giaccia AJ, Hammond EM (2009) Bringing H2AX into the angiogenesis family. Cancer Cell 15:459–461. doi:10.1016/j.ccr.2009.05.004

Reaper PM et al (2011) Selective killing of ATM- or p53-deficient cancer cells through inhibition of ATR. Nat Chem Biol 7:428–430. doi:10.1038/nchembio.573

Reichard P, Ehrenberg A (1983) Ribonucleotide reductase–a radical enzyme. Science 221:514–519

Reynolds TY, Rockwell S, Glazer PM (1996) Genetic instability induced by the tumor microenvironment. Cancer Res 56:5754–5757

Rice GC, Hoy C, Schimke RT (1986) Transient hypoxia enhances the frequency of dihydrofolate reductase gene amplification in Chinese hamster ovary cells. Proc Natl Acad Sci U S A 83: 5978–5982

Riley T, Sontag E, Chen P, Levine A (2008) Transcriptional control of human p53-regulated genes. Nat Rev Mol Cell Biol 9:402–412. doi:nrm2395 [pii]10.1038/nrm2395

Rogakou EP, Pilch DR, Orr AH, Ivanova VS, Bonner WM (1998) DNA double-stranded breaks induce histone H2AX phosphorylation on serine 139. J Biol Chem 273:5858–5868

Rzymski T et al (2010) Regulation of autophagy by ATF4 in response to severe hypoxia. Oncogene 29:4424–4435. doi:10.1038/onc.2010.191

Sancar A, Lindsey-Boltz LA, Unsal-Kacmaz K, Linn S (2004) Molecular mechanisms of mammalian DNA repair and the DNA damage checkpoints. Annu Rev Biochem 73:39–85

Sanchez Y et al (1997) Conservation of the Chk1 checkpoint pathway in mammals: linkage of DNA damage to Cdk regulation through Cdc25. Science 277:1497–1501

Schoppy DW et al (2012) Oncogenic stress sensitizes murine cancers to hypomorphic suppression of ATR. J Clin Invest 122:241–252. doi:58928 [pii]10.1172/JCI58928

Stewart GS, Wang B, Bignell CR, Taylor AM, Elledge SJ (2003) MDC1 is a mediator of the mammalian DNA damage checkpoint. Nature 421:961–966

Stiff T et al (2006) ATR-dependent phosphorylation and activation of ATM in response to UV treatment or replication fork stalling. Embo J 25:5775–5782

Stucki M et al (2005) MDC1 directly binds phosphorylated histone H2AX to regulate cellular responses to DNA double-strand breaks. Cell 123:1213–1226, doi:S0092-8674(05)01165-7 [pii]10.1016/j.cell.2005.09.038

Toledo LI et al (2011) A cell-based screen identifies ATR inhibitors with synthetic lethal properties for cancer-associated mutations. Nat Struct Mol Biol 18:721–727. doi:10.1038/nsmb.2076

Vaupel P, Kallinowski F, Okunieff P (1989) Blood flow, oxygen and nutrient supply, and metabolic microenvironment of human tumors: a review. Cancer Res 49:6449–6465

Vousden KH, Lane DP (2007) p53 in health and disease. Nat Rev Mol Cell Biol 8:275–283. doi:nrm2147 [pii]10.1038/nrm2147

Wang Y et al (2000) BASC, a super complex of BRCA1-associated proteins involved in the recognition and repair of aberrant DNA structures. Genes Dev 14:927–939

Watters D et al (1997) Cellular localisation of the ataxia-telangiectasia (ATM) gene product and discrimination between mutated and normal forms. Oncogene 14:1911–1921

Yang J et al (2009) Small-molecule activation of p53 blocks hypoxia-inducible factor 1alpha and vascular endothelial growth factor expression in vivo and leads to tumor cell apoptosis in normoxia and hypoxia. Mol Cell Biol 29:2243–2253. doi:MCB.00959-08 [pii]10.1128/MCB.00959-08

Ye C et al (2007) Expression patterns of the ATM gene in mammary tissues and their associations with breast cancer survival. Cancer 109:1729–1735. doi:10.1002/cncr.22592

Yee KS, Vousden KH (2005) Complicating the complexity of p53. Carcinogenesis 26:1317–1322. doi:bgi122 [pii]10.1093/carcin/bgi122

Young SD, Marshall RS, Hill RP (1988) Hypoxia induces DNA overreplication and enhances metastatic potential of murine tumor cells. Proc Natl Acad Sci U S A 85:9533–9537

Yu G et al (2004) Expression of ATM protein and its relationship with p53 in pancreatic carcinoma with tissue array. Pancreas 28:421–426. doi:00006676-200405000-00011 [pii]

Yuan J, Narayanan L, Rockwell S, Glazer PM (2000) Diminished DNA repair and elevated mutagenesis in mammalian cells exposed to hypoxia and low pH. Cancer Res 60:4372–4376

Zou L, Elledge SJ (2003) Sensing DNA damage through ATRIP recognition of RPA-ssDNA complexes. Science 300:1542–1548

Chapter 3
Emerging Roles of Non-Coding RNAs in the Hypoxic Response

Mircea Ivan, Xiaoling Zhong, Simona Greco and Fabio Martelli

Abstract Hypoxia is a key component of the tumor microenvironment and represents a well-documented source of therapeutic failure in clinical oncology. Recent work has provided support to the idea that noncoding RNAs (ncRNAs), and in particular microRNAs (miRNAs), may play important roles in the adaptive response to low oxygen in tumors. Specifically, all published studies agree that the induction of microRNA-210 (miR-210) is a consistent feature of the hypoxic response in both normal and transformed cells. miR-210 is a robust target of hypoxia-inducible factor (HIF), and its overexpression has been detected in a variety of diseases with a hypoxic component, including most solid tumors. High levels of miR-210 have been linked to an in vivo hypoxic signature and to adverse prognosis in breast and pancreatic cancer patients. A wide variety of miR-210 targets have been identified, pointing to roles in mitochondrial metabolism, angiogenesis, differentiation, DNA damage response, and cell survival. Such targets are suspected to affect the development of tumors in multiple ways; therefore, an increased knowledge about miR-210's actions may lead to novel diagnostic and therapeutic approaches in the cancer field.

Keywords Hypoxia · MicroRNA · Cancer · Biomarker · MiR-210 · Mitochondria · Apoptosis · Metabolism

M. Ivan (✉) · X. Zhong
Department of Medicine, Indiana University, Indianapolis, IN, USA
e-mail: mivan@iupui.edu

M. Ivan
Department of Microbiology and Immunology, Indiana University, Indianapolis, IN, USA

F. Martelli (✉) · S. Greco
Molecular Cardiology Laboratory, IRCCS-Policlinico San Donato, Via Morandi 30, 20097 Milan, Italy
e-mail: fabio.martelli@grupposandonato.it

G. Melillo (ed.), *Hypoxia and Cancer,* Cancer Drug Discovery and Development, DOI 10.1007/978-1-4614-9167-5_3,

3.1 Hypoxia

Tissue hypoxia is a key feature of virtually all solid tumors (Semenza 2010) and regulates diverse processes such as energy metabolism, cell survival and proliferation, angiogenesis, adhesion, and motility. The hypoxic adaptation is critical for tumor expansion, invasion, and metastasis, and has been well documented as a source of therapeutic failure in clinical oncology. There is solid evidence from both clinical research and animal models supporting the paradigm that tumors with extensive low oxygen tension are more likely to exhibit poor prognosis and resistance to anticancer therapy. Therefore, a more complete understanding of cellular adaptation to oxygen deprivation is paramount for developing therapeutic strategies of increased efficacy (Wilson and Hay 2011).

Tumor cells react to hypoxia in part via a transcriptional program that is largely orchestrated by the core component of the hypoxia-sensing machinery, the hypoxia-inducible factors (HIFs) (Wang and Semenza 1993;Wang and Semenza 1995; Wang et al. 1995). As a transcription factor, in hypoxic cells HIF binds specifically to the promoter of hundreds of genes, thought to be instrumental for tumor growth (Levy et al. 1995; Semenza et al. 1994; Firth et al. 1994; Firth et al. 1995), aspect which is discussed in detail in other chapters of the book. Although historically gene induction by low oxygen has dominated hypoxia research, more recently we have become increasingly aware of the significance of gene repression promoted by low oxygen tension (Bindra et al. 2004; Nakamura et al. 2008). In a timely fashion, during the past 5 years a multitude of reports have demonstrated that specific microRNAs (miRNAs) are involved in the hypoxic response and contribute to the repression of biologically important genes by low oxygen tension (Donker et al. 2007; Camps et al. 2008; Giannakakis et al. 2008; Pulkkinen et al. 2008; Gee et al. 2010; Fasanaro et al. 2008).

3.2 Noncoding RNA—Wide Roles in Physiology and Disease

During the past decade, tens of thousands of noncoding RNAs (ncRNAs) have been found to be transcribed from the mammalian genome and, despite failing to be translated into proteins, many of them are known or suspected to exhibit biological functions. However, given their high—and still growing—number, the vast majority lack any known function. Among them, one particular family—miRNAs—has received special attention, and, as a result, we have accumulated disproportionate knowledge about their role (Amaral et al. 2008) in physiology and disease. miRNAs are endogenous short RNAs ($\sim$ 22 nucleotides in length), which were first discovered in 1993 through genetic screens in Caenorhabditis elegans (Lee et al. 1993) and are now unanimously recognized as key posttranscriptional gene expression regulators. To date more than 2,000 miRNAs have been identified in humans and predicted to regulate at least one-third of all protein-encoding genes (Lewis et al. 2005).

miRNA biogenesis is initiated mainly by RNA polymerase II as long primary transcripts (pri-miRNA), typically containing the cap structure and the poly(A) tails. This feature predicts the presence of a wealth of pri-miRNAs alongside mRNA in most full transcriptome analyses databases. In the second step, pri-miRNAs are processed by the nuclear RNase III Drosha, leading to 60–70nt hairpin-shaped intermediates, termed precursor miRNAs (pre-miRNAs). These are subsequently exported out of the nucleus and cleaved by the cytoplasmic RNase III Dicer into a short miRNA duplex. One strand of this short-lived duplex is degraded by an unknown nuclease, while the other strand is retained as mature miRNA (Lee et al. 2004), which is incorporated into the RNA-induced silencing complex (RISC).

The mature miRNA leads the RISC to recognize the target mRNA by sequence complementarity, usually found in the 3′-untranslated region (3′UTR). This generally leads to translational inhibition and/or degradation of the target mRNAs (Djuranovic et al. 2011, 2012). It is currently thought that thousands of classic (protein-coding) genes are regulated via such mechanisms (Bartel 2009). Given that a single miRNA can target multiple mRNAs, often hundreds, and a single mRNA might contain several sequences in the 3′UTR sequence for miRNA recognition, miRNAs are involved in virtually all known biological processes, such as differentiation, proliferation, death, and metabolism (Kloosterman and Plasterk 2006; Leung and Sharp 2010). This relative lack of specificity poses significant challenges for the miRNA research field, in particular when it comes to the identification of biologically meaningful targets.

While being highly relevant for physiology, miRNAs have been at least as deeply investigated in most pathological settings, and arguably in no disease more than cancer. Genome-wide miRNA profiling analyses demonstrated that different cancer types tend to exhibit specific signatures of deregulated miRNAs (Lu et al. 2005; Calin and Croce 2006), including cancers of colon (Michael et al. 2003), breast (Iorio et al. 2005), brain (Ciafre et al. 2005), liver (Murakami et al. 2006), and lung (Yanaihara et al. 2006). While mechanisms behind the specific shifts of profiles in tumors have long remained elusive, recent understanding of miRNA responses to microenvironmental stresses and oncogenic alterations has provided useful clues. Given the central role of low oxygen tension in shaping tumor biology, there is no surprise that hypoxia has been closely investigated as a contributor to the altered neoplastic miRNA-ome.

In the following sections, we attempt to summarize the current knowledge about the involvement of miRNAs in the hypoxic response and anticipate opportunities for cancer diagnosis and treatment derived from this rapidly expanding information.

3.3 Hypoxia-Regulated miRNAs, a New Arm of the Response to Oxygen Deprivation

During the past 5 years, groups from diverse research fields have performed screens for hypoxia-regulated miRNAs in very wide cellular contexts. In contrast to coding genes, the emerging picture is that miRNAs exhibit significantly more variability in their response to low oxygen. Drawing the line under more than a dozen of independent screens, performed with various technologies, miR-210 stands out as the only miRNA consistently upregulated in virtually all experimental systems tested so far (Gee et al. 2010; Kulshreshtha et al. 2007; Camps et al. 2008; Giannakakis et al. 2008; Foekens et al. 2008; Porkka et al. 2007; Lawrie et al. 2008).

miR-210 is generally recognized as one of the most robust HIF-1 targets (Fasanaro et al. 2008; Kulshreshtha et al. 2007; Crosby et al. 2009a, b), and its level may accurately mirror HIF-1 activity in vivo and in vitro. Interestingly, there seems to be a striking discrepancy between the effects of HIF-1 (dramatic) and HIF-2 (weak or absent) on miR-210 expression (Crosby et al. 2009a, b; Huang et al. 2009).

3.3.1 Mir-210: A Small Mirror of HIF Activity with Clinical Implications

miR-210 is upregulated in most solid tumors, and its levels correlate with a negative clinical outcome (Camps et al. 2008; Gee et al. 2010; Giannakakis et al. 2008; Foekens et al. 2008; Porkka et al. 2007; Lawrie et al. 2008; Iorio and Croce 2012). Moreover, miR-210 levels are correlated with a genetic signature of hypoxia, suggesting that its overexpression in tumors is the direct consequence of decreased oxygen tension in the microenvironment.

Given its close relationship with HIF, it is hardly surprising that miR-210 is particularly abundant in renal clear cell carcinomas (RCCs) (Juan et al. 2010), which express abnormally high levels of HIF, due to the genetic inactivation of the von Hippel–Lindau tumor suppressor, the well-recognized E3-ligase for HIF (Ohh et al. 2000).

While the subsequent discussions concentrate on tumor-related miR-210 functions, the significance of HIF regulation of miR-210 seems to extend well beyond tumor biology. This should not come as a surprise, as hypoxia/HIF represents central components of other clinical conditions, which have major impacts on morbidity and mortality, including cardiac and peripheral ischemia as well as cerebrovascular diseases (Semenza et al. 2010), cardiac hypertrophy and failure (van Rooij et al. 2006; Thum et al. 2007; Greco et al. 2012; Jeyaseelan et al. 2008), limb ischemia (Pulkinnen et al. 2008; Jayaseelan et al. 2008) and in, ischemic wounds (Biswas et al. 2010), acute myocardial infarction (Bostjancic et al. 2009), atherosclerosis obliterans (Li et al. 2011), and preeclampsia (Pineles et al. 2007; Zhu et al. 2009; Enquobahrie et al. 2011).

These statements are not meant to imply that the hypoxic response is limited to the upregulation of miR-210. Indeed, other mature miRNAs with proven importance in tumor biology have been reported to be upregulated by at least two or three studies, such as miR-21, miR-181a, miR-23, miR-24, miR-373, miR-103, and miR-107 (Kulshreshtha et al. 2007; Lawrie et al. 2008; Crosby et al. 2009a, b). This variability continues to puzzle the field and may be at the root of the differences in hypoxia responses between tissues, arguably more so than the “classic” protein-coding genes.

3.3.2 Mir-210 Targets in Cancer: How Many is too Many?

A comprehensive understanding of miRNA roles in cancer (or in any disease for that matter) requires a comprehensive knowledge of their biologically relevant targets; however, the search for such targets remains the bottleneck of miRNA research.

For miR-210, in silico searches for targets employing prediction programs, such as TargetScan, PicTar, MicroCosm, and Dianalab, reveal a very complex spectrum of candidates, including genes involved in proliferation, DNA repair, chromatin remodeling, metabolism, and cell migration (Pulkkinen et al. 2008; Fasanaro et al. 2008; Camps et al. 2008; Giannakakis et al. 2008; Crosby et al. 2009a, b; Chan et al. 2009; Chen et al. 2010; Mizuno et al. 2009; Qin et al. 2010). These programs generally search for complementarity between the miRNA "seed" sequences and the 3′UTR of all known genes, and the resulting lists of candidates may contain hundreds of members. When the search is extended to the 5′UTR and the coding region, the yield is even higher. It is also becoming increasingly apparent that "seed" binding is not always sufficient, as other features of the surrounding sequences can affect binding efficacy. No specific algorithm is generally acknowledged as the most sensitive or accurate and, to further complicate matters, the lists of candidates generated by these different algorithms usually exhibit very limited overlap. Finally, relying on one particular program may well lead to missing most targets. For example, PicTar and TargetScan predict relatively few targets for human miR-210, and most of the experimentally validated targets are not predicted by any of these programs. Such difficulties prompted the use of integrated bioinformatic and experimental approaches in order to identify relevant miR-210 targets. One widely used approach takes advantage of the ability of many miRNAs to induce target mRNA destabilization. Gene expression is measured in the whole transcriptome or in a subset of candidates in cells where a specific miRNA is either overexpressed or knocked down. Then, only those transcripts modulated in the expected direction are further analyzed with target prediction software programs to distinguish between direct and indirect targets. For instance, both Zhang et al. (Zhang et al. 2009) and Puissegur et al. (Puissegur et al. 2011) began their search by identifying transcripts that were down-modulated upon forced miR-210 expression in colorectal and lung adenocarcinoma cancer cell lines, respectively. The downregulated genes which also contained a predicted miR-210 binding site were analyzed further, to confirm the identity of MAX dimerization protein (MNT) (Zhang et al. 2009), NDUFA4, and SDHD (Puissegur et al. 2011) as bona fide targets.

Alternative approaches were based on a direct analysis of the mRNAs associated with the RISC complex. Our groups used a combination of proteomic and transcriptomic techniques in order to identify miR-210-modulated genes (Fasanaro et al. 2009). Specifically, proteomic profiling of human primary endothelial cells identified at least ten down-modulated proteins in miR-210-expressing cells. In additional experiments, 52 transcripts were found to be both induced upon miR-210 knockdown and down-modulated by miR-210 expression. Very few, if any, of these genes were found by target identification algorithms. However, a low-stringency search revealed that these genes were enriched in miR-210 seed-complementary sequences.

The analysis of the mRNAs, associated with the RISC loaded with miR-210 and purified by immunoprecipitation, revealed that the complex was significantly enriched for 16 candidate targets. Intriguingly, the seed matches for nine of these genes were localized in the 5′UTR or in the coding region. Surprisingly, a ncRNA involved in the epigenetic inactivation of the X chromosome (Xist) was identified in the RISC immunoprecipitate. The significance of this interaction is still elusive, but it may reveal a previously unsuspected layer of interaction between ncRNAs. Finally, 15 targets predicted by PicTar and TargetScan were also validated in our study, underlining the usefulness of such programs, especially in conjunction with additional experimental approaches.

Very recently, the complexity of the rules regulating miRNA/mRNA interaction was further made evident by our identification of ROD1 as a miR-210 "seedless" target (Fasanaro et al. 2010). Indeed, while the miR-210 seed region displayed very limited complementarity to ROD1, this targeting was mediated by the central region of miR-210 binding within the coding sequence of ROD1.

Using a similar strategy, Huang (Huang et al. 2009) and colleagues compared RISC immunoprecipitates of normoxic and hypoxic breast cancer cells, an effort that led to the identification of more than 200 mRNAs that were enriched following cell exposure to hypoxia. An in silico search revealed that 50 of these were recognized as direct miR-210 targets by a prediction software, and a subset of these genes was further validated by independent techniques. Notably, there are very few targets in common between the above-mentioned reports. miR-210 targets, which have been experimentally validated, and their potential regulatory functions are summarized in Table 3.1 and Fig. 3.1.

The jury is still out on whether this will turn out to be the norm for miRNAs in general, and we may need to include a significant number of future analyses before reaching a conclusion. One needs to bear in mind that the number of identified targets is still relatively limited, especially when compared to computational predictions. A very general question related to miR-210-published targets, and very valid for any miRNA, is whether a significant but low-fold (e.g., 1.2–1.5) change in target expression is biologically relevant. The answer may depend on the gene in question and additional contributions from other pathways that may offset or enhance the effect of miR-210. Another caveat is that the validation of targets is often performed by manipulating the expression of the miRNAs well outside the physiologic window of expression, using transfected miRNA mimics and antagonists.

With all these limitations in mind, we review some of the better validated miR-210 targets and discuss their relevance for the hypoxic response and/or cancer biology.

3.3.3 Mir-210, Mitochondrial Metabolism, and Oxidative Stress

Following exposure to hypoxia, cell metabolism shifts from mitochondrial oxidative phosphorylation to glycolysis (often termed the "Pasteur effect"). This effect is generally thought to be a consequence of HIF activation, which induces most glycolytic enzymes, pyruvate dehydrogenase kinase, and represses mitochondrial biogenesis

Table 3.1 mir-210-validated argets and physio-pathological pathways

Validated target	Relevant to biological process	References
ABCB9	Positive regulation of T-cell-mediated cytotoxicity; protein transport	Fasanaro et al. 2009
AcvR1b	Induction of apoptosis; negative regulation of cell growth; receptor signaling protein serine/threonine kinase activity; transforming growth factor beta-activated receptor activity	Fasanaro et al. 2009; Mizuno et al. 2009
AIFM3	Activation of cysteine-type endopeptidase activity involved in apoptotic process by cytochrome c; cell redox homeostasis; electron transport chain	Yang et al. 2012
APC	Canonical Wnt receptor signaling pathway; cell adhesion; cell cycle arrest; positive regulation of apoptotic process; response to DNA damage stimulus	Fasanaro et al. 2009
ATP11C	Cation transport; ion transmembrane transport; phospholipid translocation	Fasanaro et al. 2009
BDNF	Nervous system development; positive regulation of synapse assembly	Fasanaro et al. 2009; Giannakakis et al. 2008
CASP8AP2	Cellular response to mechanical stimulus; induction of apoptosis via death domain receptors; signal transduction	Kim et al. 2009
CBX1	Negative regulation of transcription, DNA-dependent	Fasanaro et al. 2009
CDK10	Negative regulation of cell proliferation; traversing start control point of mitotic cell cycle	Fasanaro et al. 2009
CHD9	Transcription, DNA dependent; small molecule metabolic process; chromatin modification	Fasanaro et al. 2009
CLASP2	Establishment or maintenance of cell polarity; cell division; axon guidance	Fasanaro et al. 2009
COX10	Mitochondrial electron transport, cytochrome c to oxygen	Chen et al. 2010
CPEB2	Regulation of translation	Fasanaro et al. 2009
CTGF	Rregulation of cell growth	Hu et al. 2010
DAPK	Apoptotic process; intracellular protein kinase cascade	Hu et al. 2010
DDAH1	Nitric oxide mediated signal transduction; positive regulation of angiogenesis; arginine catabolic process; citrulline metabolic process	Fasanaro et al. 2009
E2F3	Positive regulation of transcription, DNA-dependent	Fasanaro et al. 2009; Giannakakis et al. 2008

Table 3.1 (continued)

Validated target	Relevant to biological process	References
EFNA3	Cell–cell signaling	Fasanaro et al. 2009
EGR3	Cell migration involved in sprouting angiogenesis; cellular response to fibroblast growth factor stimulus; cellular response to vascular endothelial growth factor stimulus	Chen et al. 2012
ELK3	Negative regulation of transcription, DNA dependent; angiogenesis; wound healing	Fasanaro et al. 2009
FGFRL1	Ventricular septum morphogenesis; skeletal system development; negative regulation of cell proliferation; heart valve morphogenesis	Huang et al. 2009; Tsuchiya et al. 2011
GIT2	Protein binding, regulation of ARF GTPase activity, ARF GTPase activator activity, regulation of G-protein-coupled receptor protein signaling pathway, nucleoplasm, zinc ion binding	Chen et al. 2012
GPD1L	Carbohydrate metabolic process; triglyceride biosynthetic process	Fasanaro et al. 2009
HECTD1	Protein ubiquitination involved in ubiquitin-dependent protein catabolic process	Fasanaro et al. 2009
HOXA1	Artery morphogenesis; central nervous system neuron differentiation; neuromuscular process; positive regulation of transcription from RNA polymerase II promoter	Huang et al. 2009
HOXA3	Angiogenesis; organ formation; transcription, DNA dependent	Fasanaro et al. 2009; Giannakakis et al. 2008
HOXA9	Transcription, DNA dependent; multicellular organismal development; endothelial cell activation	Huang et al. 2009
HYPB	Angiogenesis; coronary vasculature morphogenesis; pericardium development	Giannakakis et al. 2008
ISCU	Iron-sulfur cluster assembly; nitrogen fixation	Chan et al. 2009; Chen et al. 2010, Fasanaro et al. 2009; Favaro et al. 2010
KCMF1	Ligase activity; zinc ion binding	Giannakakis et al. 2008
KIAA1161	Positive regulation of insulin-like growth factor receptor signaling pathway; positive regulation of protein kinase B signaling cascade; skeletal muscle fiber development	Fasanaro et al. 2009
MDGA1	Spinal cord association neuron differentiation; neuron migration; brain development	Fasanaro et al. 2009
MIB1	Notch signaling pathway; heart and blood development; negative regulation of neuron differentiation	Fasanaro et al. 2009
MID1IP1	Lipid biosynthetic process; negative regulation of microtubule depolymerization	Fasanaro et al. 2009

Table 3.1 (continued)

Validated target	Relevant to biological process	References
MNT	Regulation of cell cycle; regulation of apoptotic process; cell aging	Zhang et al. 2009
NCAM1	Cell adhesion; axon guidance	Fasanaro et al. 2009
NDUFA4	Mitochondrial electron transport	Giannakakis et al. 2008; Puissegur et al. 2011
NIPBL	Response to DNA damage stimulus; positive regulation of histone deacetylation; cellular protein localization; developmental growth	Fasanaro et al. 2009
NP1	Axonogenesis involved in innervation; central nervous system development; synaptic transmission	Pulkkinen et al. 2008
NPTX1	Central nervous system development; synaptic transmission	Fasanaro et al. 2009, Pulkkinen et al. 2008
P4HB	Cell redox homeostasis; lipid metabolic process	Fasanaro et al. 2009
PTPN1	Negative regulation of insulin receptor signaling pathway; actin cytoskeleton reorganization; JAK-STAT cascade involved in growth hormone signaling pathway	Fasanaro et al. 2009
RAD52	Mitotic recombination	Crosby et al. 2009a, b; Fasanaro et al. 2009
SDHD	Respiratory electron transport chain; tricarboxylic acid cycle	Merlo et al. 2012; Puissegur et al. 2011
SEH1L	Carbohydrate metabolic process; cell division; attachment of spindle microtubules	Fasanaro et al. 2009
SERTAD2	Negative regulation of cell growth	Fasanaro et al. 2009
SHIP-1	Apoptotic process; blood coagulation; determination of adult lifespan	Lee et al. 2012
SMCHD1	Inactivation of X chromosome by DNA methylation	Fasanaro et al. 2009
TCF7L2	Transcription regulation; Wnt signaling pathway	Qin et al. 2010
TFRC	Cellular iron ion homeostasis	Yoshioka et al. 2012
TNPO1	Protein translocation; mRNA metabolic process	Fasanaro et al. 2009
UBQLN1	Apoptotic process; response to hypoxia; regulation of protein ubiquitination	Fasanaro et al. 2009
VMP1	Autophagy; cell junction assembly; cell–cell adhesion	Ying et al. 2011
XIST	Chromosome X inactivation	Fasanaro et al. 2009
ZNF462	Negative regulation of DNA binding; positive regulation of transcription from RNA polymerase II promoter; transcription, DNA dependent	Chen et al. 2012

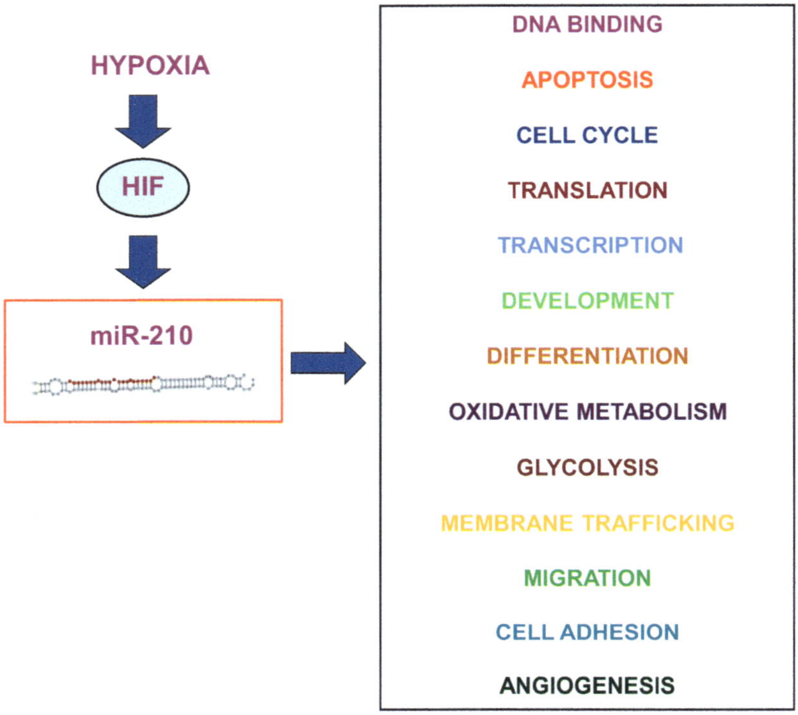

Fig. 3.1 Hypoxia regulates miR-210 via the hypoxia-inducible factor (HIF), which in turn mediates the expression of factors that are implicated in various physiopathological pathways

(Zhang et al. 2007). Consistent data from several groups have demonstrated that miR-210 significantly contributes to this metabolic shift by downregulating the function of the electron transport chain (ETC). One component of this mechanism is direct repression by miR-210 of the iron–sulfur (Fe–S) cluster scaffold proteins, ISCU1 and ISCU2 (Chan et al. 2009; Chen et al. 2010; Fasanaro et al. 2009; Favaro et al. 2010). ISCU1/2 is part of an ancient machinery that facilitates the assembly of Fe–S clusters that are then incorporated into enzymes involved in the tricarboxylic acid cycle (TCA), such as aconitase, as well as mitochondrial respiratory complexes I, II, and III (Tong and Rouault 2006). ISCU is arguably the most consistent target reported to date, and its importance is underscored by the data obtained from human tumors. Moreover, the role of ISCU as a metabolic regulator has been well established, as its mutations are associated with hereditary lactic acidosis, characterized by myopathy and exercise intolerance (Mochel et al. 2008). To further strengthen the case for ISCU as a biologically and pathologically relevant target, its levels negatively correlate with miR-210 in tumor data sets (Favaro et al. 2010). Moreover, the expression of ISCU is highly predictive of a favorable prognosis in breast cancer, the exact opposite association compared to miR-210. Therefore, the miR-210–ISCU relationship is detectable even in the high complexity of the tumor regulatory mechanisms.

Another mitochondrial miR-210 target that has been rather consistently reported is nicotinamide adenine dinucleotide (NADH) dehydrogenase (ubiquinone) 1 alpha subcomplex 4 (NDUFA4) (Giannakakis et al. 2008). Until very recently, this was thought to be an intrinsic part of Complex I, but very recent work has demonstrated its interaction with Complex IV (Balsa et al. 2012). Additionally, evidence has been presented for the succinate dehydrogenase complex, subunit D (SDHD), a subunit of Complex II, as a direct target for miR-210 (Puissegur et al. 2011). As SDHD is also a tumor suppressor (Pasini and Stratakis 2009), this link lends additional support to the protumorigenic role of miR-210.

An intriguing target that is predicted by several programs and has been experimentally confirmed is glycerol-3-phosphate dehydrogenase 1-like gene (GPD1-L) (Fasanaro et al. 2009), which may be involved in the Pasteur effect by modulating NAD + /NADH ratios (Liu et al. 2010). Although the biochemical function of GPD1-L remains to be elucidated, it is highly homologous to glycerol-3-phosphate dehydrogenase (GPD), the catalyst of the glycerol phosphate shuttle, which transfers electrons from cytoplasmic NADH to the mitochondrial ETC. Thus, the forced expression of miR-210 in normoxia represses mitochondrial respiration and enhances glycolysis, which leads to decreased levels of adenosine triphosphate (ATP), potentially explaining the observation that high amounts of miR-210 are not beneficial when oxygen supply is optimal. In contrast, upon hypoxia, repression of the ETC decreases the mismatch with reduced oxygen tension, thus increasing ATP levels (Chan et al. 2009), and minimizing the impact of hypoxia on cell energy production. The work of Kelly et al. (2011) has indicated that miR-210 via the destabilization of GPD1-L contributes to the inactivation of HIF prolyl hydroxylase activity and therefore increases HIF expression. This would indicate that miR-210 is both downstream and upstream of HIF signaling. Puissegur et al. (2011) also provided the first evidence that high miR-210 also contributes to the maintenance of high HIF during hypoxia; therefore, miR-210 and HIF form a feed-forward loop. This mechanism may be dependent on the suppression of ISCU, SDHD, NDUFA4, and the resulting generation of ROS and TCA metabolites, such as succinate, which are discussed below.

It has long been accepted that tumors largely rely on glycolysis even when the oxygen supply is normal (Warburg effect). With regard to the possible involvement of miR-210 in this effect, the best candidate tumors are the RCCs. miR-210 expression has been shown to be very high despite high oxygen supply compared to other solid tumors. Whether miR-210 is more generally involved in the Warburg effect is currently unclear.

Collectively, these data support the notion that miR-210 contributes to a metabolic shift in tumors toward a high glycolytic phenotype, which may benefit tumor growth.

The effect of miR-210 on ETC should also impact the production of mitochondrial reactive oxygen species (ROS), an expected consequence of electron leakage. Indeed, miR-210 expression increases oxidative stress in normoxic conditions and this is, at least in part, mediated by ISCU (Chan et al. 2009; Favaro et al. 2010). However, conflicting results were reported in hypoxia. Favaro et al. (2010) saw hypoxic induction of ROS in cancer lines and reversal with the miR-210 antagonist. Conversely, Chan et al. (2009), working on endothelial cells, did not detect any change in ROS production after exposing the cells to hypoxia; however, they saw an increase when miR-210 was blocked. This discrepancy warrants additional investigation and it may be a reflection of the underlying differences between normal versus cancer cells. Additionally, miR-210 may exhibit differential effects on various species of ROS, which should be addressed using more specific reagents.

The status of miR-210 expression in tumors may have therapeutic implications. While high miR-210 is generally linked to a more adverse prognosis in breast, head and neck, and pancreatic tumors (Gee et al. 2010; Camps et al. 2008; Greither et al. 2010), it may also sensitize them to various types of experimental drugs, such as antiglycolytic agents. The first in vitro evidence was provided by Chen et al. (2010), who reported that overexpressing miR-210 rendered cells significantly more susceptible to killing by 3-bromo-pyruvate, an inhibitor of the glycolytic pathway. Molecules of this class, such as 2-deoxyglucose or dichloroacetate, have been considered as promising therapeutic agents; however, they are yet to fulfill their promise in clinical settings. Identification of gene signatures that include miR-210 may help identify subsets of patients who could benefit from such agents.

3.3.4 Angiogenesis

Angiogenesis is a complex multistep process involving extravasation of plasma proteins, degradation of extracellular matrix, endothelial cell migration and proliferation, and capillary tube formation (Detmar 2000). This process normally occurs during embryonic development and rarely in the adult with exceptions like wound healing and pathological processes such as tumor growth and ischemic response. Tumor growth is highly dependent on angiogenesis to establish a nutrient supply compatible with cell survival and proliferation. Hypoxia, due to an imbalance between oxygen consumption by tumor cells with high metabolic activities (Gatenby and Gillies 2004) or oxygen delivery by poorly formed vasculature (Brown and Giaccia 1998), stimulates compensatory angiogenesis mediated by potent angiogenic factors, such as the vascular endothelial growth factor (VEGF) (Sitohy et al. 2012).

Specific miRNAs are documented to be involved in different aspects of the angiogenic response, as either positive or negative regulators of this complex process (Wu et al. 2009; Wang et al. 2009). It is not surprising that miR-210, as the prototypical hypoxia-miR, has been investigated for its potential role in hypoxia-induced angiogenesis. Forced miR-210 expression in normoxic endothelial cells stimulates the formation of capillary-like structures, as well as VEGF-induced cell migration (Fasanaro et al. 2008; Lou et al. 2012; Liu et al. 2012). One target relevant for this function appears to be Ephrin-A3 (EFNA3), which is repressed by miR-210 and hypoxia (Pulkkinen et al. 2008; Fasanaro et al. 2008; Fasanaro et al. 2009). Consistently, override of repression antagonizes both tubulogenesis and VEGF-mediated endothelial chemotaxis.

The regulation of EFNA3 during hypoxia seems more complex when ischemic responses are examined (Pulkkinen et al. 2008). Contrary to the expectation of miRNA-mediated repression, EFNA3 was expressed at high levels in postischemic mouse hippocampus, exhibiting a similar trend to miR-210 (Pulkkinen et al. 2008).

Although a series of possible explanations may be hypothesized, it is worth noting that EFNA3 transcription is induced by hypoxia (Fasanaro et al. 2008). Thus, EFNA3 protein levels are determined by the balance of mRNA induction and miRNA repression, and hence the outcome may change in different pathological contexts.

Information gleaned from the noncancer systems may well turn out to be very relevant for miR-210 in tumors. For example, umbilical cord blood CD34+ cells expanded in VEGF-containing medium had a significant upregulation in miR-210 expression, which, when transplanted into mouse ischemic hindlimb, significantly improved the tissue perfusion/capillary density (Alaiti et al. 2012). Thus, it will be of great interest to investigate whether the stimulation of miR-210 by VEGF also occurs in tumor cells that express VEGF receptors. This hypothetical positive feedback between VEGF and miR-210 may promote the autocrine tumor growth and paracrine stimulation of angiogenesis. Furthermore, the existence of this feedback may partly explain why anti-VEGF therapy has limited efficacy, and a combinatorial use of VEGF and miR-210 inhibitors might enhance antitumor efficacy.

3.3.5 miRNAs in DNA Damage Response

Genome integrity is under constant threat by a variety of intrinsic and extrinsic genotoxic stresses, including ionizing radiation, environmental mutagens, ROS, ultraviolet light, and chemo- or radiotherapeutic agents. Cellular responses to DNA damage involve a complex network of processes that detect and repair genomic lesions. miRNAs have been demonstrated to be involved in these processes (Landau and Slack 2011; Wan et al. 2011). For instance, treatment with different types of DNA-damaging agents has been shown to result in the differential activation of miRNAs (Simone et al. 2009). Zhang et al. provided direct evidence that as many as a quarter of the tested miRNAs are significantly induced upon DNA damage in an ataxia telangiectasia mutated (ATM)-dependent manner (Zhang et al. 2011). Recent data indicate that miR-210 targets RAD52 (Crosby et al. 2009a, b; Fasanaro et al. 2009), which participates in the homologous recombination (HR)-mediated repair of double-stranded breaks (Sung and Robberson 1995; Benson et al. 1998; Shinohara and Ogawa 1998; Chapman et al. 2012), by assisting the loading of RAD51 onto DNA. As it is well established that HR repair activity is compromised in hypoxic cells (Bindra et al. 2004, 2005), miR-210-mediated suppression of RAD52 may provide an additional mechanism by which HR repair is suppressed under low oxygen tension. Whether miR-210's action on RAD52 is enough to measurably affect DNA damage is not entirely clear; however, recent indications to support this hypothesis have been provided (Faraonio et al. 2012).

3.3.6 Regulation of Apoptosis

Programmed cell death (apoptosis) is essential for tissue homeostasis in multicellular organisms. Cellular stresses, including hypoxia, are main triggers of the apoptotic program, and a multitude of miRNAs have been shown to regulate this process. In cancer, evasion from apoptotic responses is critical for tumor progression, as transformed cells need to overcome the adverse conditions present in their microenvironment. In severe hypoxia, the survival of cells is enhanced by a functional miR-210 (Fasanaro et al. 2008; Favaro et al. 2010; Kulshreshtha et al. 2007), although the relevant targets for this effect remain unclear. The antiapoptotic effects of miR-210 are arguably better understood in noncancer models, such as myocardial infarction (Hu et al. 2010). It was recently shown that ischemic preconditioning enhances stem cell survival via miR-210 targeting a putative gene encoding for the apoptotic protein caspase-8-associated protein-2 (CASP8AP2) (Kim et al. 2009). Whether this gene is relevant for cancer cells is still unknown. More recently, the same research group identified another miRNA, miR-107, as an apoptotic inhibitor via a putative target gene, PDCD10 (programmed cell death-10). While the effect is independent of miR-210, the combination of miR-107 and miR-210 has an additive effect (Kim et al. 2012), which is significant, as miR-107 is hypoxia inducible in several cell types (Kulshreshtha et al. 2007).

Hypoxia-induced miRNAs seem to be only a part of the story, as low oxygen is also known to repress both coding genes and miRNAs (Kulshreshtha et al. 2007). Yan et al. provided evidence that the oncogenic *miR*-17–92 cluster is repressed by anoxia-induced p53, and the repression is due to p53 binding to a site in the proximal region of the *miR*-17–92 promoter. This process is likely to have functional consequences, as it is involved in the sensitization to hypoxia-induced apoptosis (Yan et al. 2009). To further substantiate the link, they also showed that pri-miR-17–92 expression was well correlated with the p53 status in colorectal carcinomas. However, it is still unclear whether in clinical samples the expression of *miR*-17–92 is inversely correlated with hypoxia-inducible genes.

To summarize, evidence for miRNA-mediated regulation of apoptosis in hypoxia is emerging for both normal and cancer cells, and the effect seems to involve, at least in some cell types, direct effects on the caspase-dependent apoptotic pathway.

3.3.7 Cell Cycle

Hypoxia has a profound impact on cell proliferation, via a multitude of pathways. In some cell types, extended exposure to hypoxia leads to downregulation of a large number of cell cycle genes, including cyclins and positive regulators of cell cycle transition (Hammer et al. 2007). The relationship is much more complex, as many cells tend to proliferate better under low oxygen conditions (Krick et al. 2005); therefore, it is conceivable that both pro-proliferative and anti-proliferative miRNAs may be relevant for cell cycle responses to decreased oxygen tension. Our work has established that one of the cell cycle-regulating targets of miR-210 is E2F3 (Fasanaro et al. 2009), one of the promoters of G1/S transition (Lees et al. 1993; Leone et al. 1998). Again, the relative contribution of this link to the cell cycle extension in hypoxia is still unclear. More recently, independent work has suggested that the miR-210 effect on the cell cycle is much broader, as its expression was correlated with the downregulation of a group of mitosis-related genes, including Plk1, Cdc25B, Cyclin F, Bub1B, and Fam83D (He et al. 2012).

Tsuchiya et al. found that overexpressed miR-210 induced cell cycle arrest in G(1)/G(0) and G(2)/M. They also identified the fibroblast growth factor receptor-like 1 (FGFRL1) as a target of miR-210 and showed that decreased miR-210 expression in esophageal squamous cell carcinomas leads to the derepression of FGFRL1 and accelerated cell cycle progression (Tsuchiya et al. 2011). However, one cannot generally state that miR-210 induction in hypoxia negatively regulates cell cycle progression, based on the evidence that miR-210 may also activate the myc pathway, via downregulation of the c-Myc antagonist MNT (Zhang et al. 2009).

Finally, miR-210 expression at physiological levels did not affect the proliferation of normoxic primary endothelial cells (Fasanaro et al. 2008). Therefore, the net impact of miR-210 on cell proliferation very likely depends on a multitude of factors, and caution should be exercised before making very general statements on this subject.

3.4 Hypoxia-Regulated miRNAs as Biomarkers

In 2008, miRNAs were discovered to be present in body fluids (Gilad et al. 2008), opening the way for the utilization of circulating miRNAs as non- or minimally invasive biomarkers for the diagnosis of cancer or treatment-response predictors (Cortez et al. 2011; Allegra et al. 2012). To date, at least 79 miRNAs have been reported as plasma or serum miRNA biomarker candidates for solid and hematologic tumors (Allegra et al. 2012). miR-210 was reported to be increased in the serum from patients with diffuse large B-cell lymphoma (Lawrie et al. 2008), clear cell renal carcinoma (Zhao et al. 2012), or pancreatic cancer (Wang et al. 2009; Ho et al. 2010). Circulating miR-210 levels were also correlated with sensitivity to trastuzumab (a human epidermal growth factor receptor 2 monoclonal antibody), tumor presence, and lymph node metastases in breast cancer patients (Jung et al. 2012), suggesting that plasma miR-210 may be used to monitor the response to trastuzumab-containing therapies (Cortez et al. 2011).

One of the miRNA characteristics that increase their attractiveness as biomarker candidates is their high stability under conditions (increased temperature, pH change, or extended storage) that lead to the degradation of most mRNAs. It has been suggested that the stable feature of miRNAs may be partially attributed to the exosomal packaging. The release of exosomes into the extracellular environment provides an opportunity to detect exosome components in body fluids, which may enable the description of molecular events occurring in the cells (Iguchi et al. 2010). Interestingly, hypoxia has been demonstrated to promote the release of exosomes by cultured breast cancer cells, as evaluated by the real-time polymerase chain reaction (PCR) assay of miR-210 (King et al. 2012); therefore, one can speculate that the elevated levels of circulating miR-210 directly reflect the hypoxic state of tumor cells.

Besides the body fluids, surgically resected tumor tissues have been used to analyze miRNAs, and the detected expression levels or patterns of miRNAs can be potential biomarkers. A very recent work performed on 161 tumor samples from Japanese breast cancer patients revealed that the degree of miR-210 expression is inversely correlated with disease-free survival, and a high expression level of miR-210 is therefore an indication of poor prognosis (Toyama et al. 2012).

The recent development of anti-miRNA agents such as locked nucleic acid (LNA) represents a significant step in the therapeutic targeting of miRNAs (Iorio and Croce 2012; Stenvang et al. 2008; Lanford et al. 2010). LNA is a conformational RNA analog that binds complementary RNA with high affinity and specificity. The LNA-mediated silencing of miRNA function in vitro and in vivo supports the potential of LNA in the therapeutic intervention of cancer-associated miRNAs (Stenvang et al. 2008).

A newer tool for the inhibition of miRNA activities is based on the peptide nucleic acid (PNA) chemistry (Gambari et al. 2011; Fabbri et al. 2011a, b). PNAs are DNA analogs in which the sugar–phosphate backbone is replaced by N-(2-aminoethyl) glycine units. PNAs targeting several miRNAs have already been developed and their biological effects studied both in vitro and in vivo. For example, polyarginine-PNAs directed against miR-210 (as well as miR-221) showed a very high affinity for their target and efficient cellular uptake without the need for transfection reagents, leading to the inhibition of miRNA activity and altered expression of miR-210-regulated targets in K562 human chronic myelogenous leukemia cells. It is conceivable that the inactivation of miRNAs involved in hypoxic adaptation, in combination with other anticancer agents, may be a viable strategy to target a tumor compartment that poses significant therapeutic challenges.

Finally, more powerful experimental models are anticipated to answer many of the outstanding questions regarding hypoxia miRNAs in general, and miR-210 in particular. To date, there are relatively few genetic miRNA inactivation animal models, but their availability is anticipated to grow dramatically in the near future.

Acknowledgments This work was supported by NIH R01 funding and the American Cancer Society (MI, USA), the Italian Ministry of Health (Ministero della Salute), and the Italian Association for Cancer Research (AIRC).

References

Alaiti MA, Ishikawa M, Masuda H, Simon DI, Jain MK, Asahara T, Costa MA (2012) Up-regulation of miR-210 by vascular endothelial growth factor in ex vivo expanded CD34 +cells enhances cell-mediated angiogenesis. J Cell Mol Med 16:2413–2421

Allegra A, Alonci A, Campo S, Penna G, Petrungaro A, Gerace D, Musolino C (2012) Circulating microRNAs: New biomarkers in diagnosis, prognosis and treatment of cancer (Review). Int J Oncol 41:1897–1912

Amaral PP, Dinger ME, Mercer TR, Mattick JS (2008) The eukaryotic genome as an RNA machine. Science 319:1787–1789

Balsa E, Marco R, Perales-Clemente E, Szklarczyk R, Calvo E, Landazuri MO, Enriquez JA (2012) NDUFA4 is a subunit of complex IV of the mammalian electron transport chain. Cell Metab 16:378–386

Bartel DP (2009) MicroRNAs: target recognition and regulatory functions. Cell 136:215–233

Benson FE, Baumann P, West SC (1998) Synergistic actions of Rad51 and Rad52 in recombination and DNA repair. Nature 391:401–404

Bindra RS, Schaffer PJ, Meng A, Woo J, Maseide K, Roth ME, Lizardi P, Hedley DW, Bristow RG, Glazer PM (2004) Down-regulation of Rad51 and decreased homologous recombination in hypoxic cancer cells. Mol Cell Biol 24:8504–8518

Bindra RS, Gibson SL, Meng A, Westermark U, Jasin M, Pierce AJ, Bristow RG, Classon MK, Glazer PM (2005) Hypoxia-induced down-regulation of BRCA1 expression by E2Fs. Cancer Res 65:11597–11604

Biswas S, Roy S, Banerjee J, Hussain SR, Khanna S, Meenakshisundaram G, Kuppusamy P, Friedman A, Sen CK (2010) Hypoxia inducible microRNA 210 attenuates keratinocyte proliferation and impairs closure in a murine model of ischemic wounds. Proc Natl Acad Sci U S A 107:6976–6981

Bostjancic E, Zidar N, Glavac D (2009) MicroRNA microarray expression profiling in human myocardial infarction. Dis Markers 27:255–268

Brown JM, Giaccia AJ (1998) The unique physiology of solid tumors: opportunities (and problems) for cancer therapy. Cancer Res 58:1408–1416

Calin GA, Croce CM (2006) MicroRNA signatures in human cancers. Nature reviews. Cancer 6:857–866

Camps C, Buffa FM, Colella S, Moore J, Sotiriou C, Sheldon H, Harris AL, Gleadle JM, Ragoussis J (2008) hsa-miR-210 Is induced by hypoxia and is an independent prognostic factor in breast cancer. Clin Cancer Res 14:1340–1348

Chan SY, Zhang YY, Hemann C, Mahoney CE, Zweier JL, Loscalzo J (2009) MicroRNA-210 controls mitochondrial metabolism during hypoxia by repressing the iron-sulfur cluster assembly proteins ISCU1/2. Cell Metab 10:273–284

Chapman JR, Taylor MR, Boulton SJ (2012) Playing the end game: DNA double-strand break repair pathway choice. Mol Cell 47:497–510

Chen Z, Li Y, Zhang H, Huang P, Luthra R (2010) Hypoxia-regulated microRNA-210 modulates mitochondrial function and decreases ISCU and COX10 expression. Oncogene 29:4362–4368

Chen WY, Liu WJ, Zhao YP, Zhou L, Zhang TP, Chen G, Shu H (2012) Induction, modulation and potential targets of miR-210 in pancreatic cancer cells. Hepatobiliary Pancreat Dis Int 11:319–324

Ciafre SA, Galardi S, Mangiola A, Ferracin M, Liu CG, Sabatino G, Negrini M, Maira G, Croce CM, Farace MG (2005) Extensive modulation of a set of microRNAs in primary glioblastoma. Biochem Biophys Res Commun 334:1351–1358

Cortez MA, Bueso-Ramos C, Ferdin J, Lopez-Berestein G, Sood AK, Calin GA (2011) MicroRNAs in body fluids–the mix of hormones and biomarkers. Nat Rev Clin Oncol 8:467–477

Crosby ME, Devlin CM, Glazer PM, Calin GA, Ivan M (2009a) Emerging roles of microRNAs in the molecular responses to hypoxia. Curr Pharm Des 15:3861–3866

Crosby ME, Kulshreshtha R, Ivan M, Glazer PM (2009b) MicroRNA regulation of DNA repair gene expression in hypoxic stress. Cancer Res 69:1221–1229

Detmar M (2000) Tumor angiogenesis. J Investig Dermatol Symp Proc 5:20–23

Djuranovic S, Nahvi A, Green R (2011) A parsimonious model for gene regulation by miRNAs. Science 331:550–553

Djuranovic S, Nahvi A, Green R (2012) miRNA-mediated gene silencing by translational repression followed by mRNA deadenylation and decay. Science 336:237–240

Donker RB, Mouillet JF, Nelson DM, Sadovsky Y (2007) The expression of Argonaute2 and related microRNA biogenesis proteins in normal and hypoxic trophoblasts. Mol Hum Reprod 13:273–279

Enquobahrie DA, Abetew DF, Sorensen TK, Willoughby D, Chidambaram K, Williams MA (2011) Placental microRNA expression in pregnancies complicated by preeclampsia. Am J Obstet Gynecol 204:e112–121

Fabbri E, Brognara E, Borgatti M, Lampronti I, Finotti A, Bianchi N, Sforza S, Tedeschi T, Manicardi A, Marchelli R, Corradini R, Gambari R (2011a) miRNA therapeutics: delivery and biological activity of peptide nucleic acids targeting miRNAs. Epigenomics 3:733–745

Fabbri E, Manicardi A, Tedeschi T, Sforza S, Bianchi N, Brognara E, Finotti A, Breveglieri G, Borgatti M, Corradini R, Marchelli R, Gambari R (2011b) Modulation of the biological activity of microRNA-210 with peptide nucleic acids (PNAs). ChemMedChem 6:2192–2202

Faraonio R, Salerno P, Passaro F, Sedia C, Iaccio A, Bellelli R, Nappi TC, Comegna M, Romano S, Salvatore G, Santoro M, Cimino F (2012) A set of miRNAs participates in the cellular senescence program in human diploid fibroblasts. Cell Death Differ 19:713–721

Fasanaro P, D'Alessandra Y, Di Stefano V, Melchionna R, Romani S, Pompilio G, Capogrossi MC, Martelli F (2008) MicroRNA-210 modulates endothelial cell response to hypoxia and inhibits the receptor tyrosine kinase ligand Ephrin-A3. J Biol Chem 283:15878–15883

Fasanaro P, Greco S, Lorenzi M, Pescatori M, Brioschi M, Kulshreshtha R, Banfi C, Stubbs A, Calin GA, Ivan M, Capogrossi MC, Martelli F (2009) An integrated approach for experimental target identification of hypoxia-induced miR-210. J Biol Chem 284:35134–35143

Favaro E, Ramachandran A, McCormick R, Gee H, Blancher C, Crosby M, Devlin C, Blick C, Buffa F, Li JL, Vojnovic B, Pires das Neves R, Glazer P, Iborra F, Ivan M, Ragoussis J, Harris AL (2010) MicroRNA-210 regulates mitochondrial free radical response to hypoxia and krebs cycle in cancer cells by targeting iron sulfur cluster protein ISCU. PLoS One 5:e10345

Fasanaro P, Romani S, Voellenkle C, Maimone B, Capogrossi MC, Martelli F (2012) ROD1 Is a Seedless Target Gene of Hypoxia-Induced miR-210. PLoS One 7:e44651

Firth JD, Ebert BL, Pugh CW, Ratcliffe PJ (1994) Oxygen-regulated control elements in the phosphoglycerate kinase 1 and lactate dehydrogenase A genes: similarities with the erythropoietin 3′ enhancer. Proc Natl Acad Sci U S A 91:6496–6500

Firth JD, Ebert BL, Ratcliffe PJ (1995) Hypoxic regulation of lactate dehydrogenase A. Interaction between hypoxia-inducible factor 1 and cAMP response elements. J Biol Chem 270:21021–21027

Foekens JA, Sieuwerts AM, Smid M, Look MP, Weerd V de, Boersma AW, Klijn JG, Wiemer EA, Martens JW (2008) Four miRNAs associated with aggressiveness of lymph node-negative, estrogen receptor-positive human breast cancer. Proc Natl Acad Sci U S A 105:13021–13026

Gambari R, Fabbri E, Borgatti M, Lampronti I, Finotti A, Brognara E, Bianchi N, Manicardi A, Marchelli R, Corradini R (2011) Targeting microRNAs involved in human diseases: a novel approach for modification of gene expression and drug development. Biochem Pharmacol 82:1416–1429

Gatenby RA, Gillies RJ (2004) Why do cancers have high aerobic glycolysis? Nat Rev Cancer 4:891–899

Gee HE, Camps C, Buffa FM, Patiar S, Winter SC, Betts G, Homer J, Corbridge R, Cox G, West CM, Ragoussis J, Harris AL (2010) hsa-mir-210 is a marker of tumor hypoxia and a prognostic factor in head and neck cancer. Cancer 116:2148–2158

Giannakakis A, Sandaltzopoulos R, Greshock J, Liang S, Huang J, Hasegawa K, Li C, O'Brien-Jenkins A, Katsaros D, Weber BL, Simon C, Coukos G, Zhang L (2008) miR-210 links hypoxia with cell cycle regulation and is deleted in human epithelial ovarian cancer. Cancer Biol Ther 7:255–264

Gilad S, Meiri E, Yogev Y, Benjamin S, Lebanony D, Yerushalmi N, Benjamin H, Kushnir M, Cholakh H, Melamed N, Bentwich Z, Hod M, Goren Y, Chajut A (2008) Serum microRNAs are promising novel biomarkers. PLoS One 3:e3148

Greco S, Fasanaro P, Castelvecchio S, D'Alessandra Y, Arcelli D, Di Donato M, Malavazos A, Capogrossi MC, Menicanti L, Martelli F (2012) MicroRNA dysregulation in diabetic ischemic heart failure patients. Diabetes 61:1633–1641

Greither T, Grochola LF, Udelnow A, Lautenschlager C, Wurl P, Taubert H (2010) Elevated expression of microRNAs 155, 203, 210 and 222 in pancreatic tumors is associated with poorer survival. Int J Cancer 126:73–80

Hammer S, To KK, Yoo YG, Koshiji M, Huang LE (2007) Hypoxic suppression of the cell cycle gene CDC25A in tumor cells. Cell Cycle 6:1919–1926

He J, Wu J, Xu N, Xie W, Li M, Li J, Jiang Y, Yang BB, Zhang Y (2012) MiR-210 disturbs mitotic progression through regulating a group of mitosis-related genes. Nucleic Acids Res 41:498–508
Ho AS, Huang X, Cao H, Christman-Skieller C, Bennewith K, Le QT, Koong AC (2010) Circulating miR-210 as a novel hypoxia marker in pancreatic cancer. Transl Oncol 3:109–113
Hu S, Huang M, Li Z, Jia F, Ghosh Z, Lijkwan MA, Fasanaro P, Sun N, Wang X, Martelli F, Robbins RC, Wu JC (2010) MicroRNA-210 as a novel therapy for treatment of ischemic heart disease. Circulation 122:124–131
Huang X, Ding L, Bennewith KL, Tong RT, Welford SM, Ang KK, Story M, Le QT, Giaccia AJ (2009) Hypoxia-inducible mir-210 regulates normoxic gene expression involved in tumor initiation. Mol Cell 35:856–867
Iguchi H, Kosaka N, Ochiya T (2010) Versatile applications of microRNA in anti-cancer drug discovery: from therapeutics to biomarkers. Curr Drug Discov Technol 7:95–105
Iorio MV, Croce CM (2012) MicroRNA dysregulation in cancer: diagnostics, monitoring and therapeutics. A comprehensive review. EMBO Mol Med 4:143–159
Iorio MV, Ferracin M, Liu CG, Veronese A, Spizzo R, Sabbioni S, Magri E, Pedriali M, Fabbri M, Campiglio M, Menard S, Palazzo JP, Rosenberg A, Musiani P, Volinia S, Nenci I, Calin GA, Querzoli P, Negrini M, Croce CM (2005) MicroRNA gene expression deregulation in human breast cancer. Cancer Res 65:7065–7070
Jeyaseelan K, Lim KY, Armugam A (2008) MicroRNA expression in the blood and brain of rats subjected to transient focal ischemia by middle cerebral artery occlusion. Stroke 39:959–966
Juan D, Alexe G, Antes T, Liu H, Madabhushi A, Delisi C, Ganesan S, Bhanot G, Liou LS (2010) Identification of a microRNA panel for clear-cell kidney cancer. Urology 75:835–841
Jung EJ, Santarpia L, Kim J, Esteva FJ, Moretti E, Buzdar AU, Di Leo A, Le XF, Bast RC Jr, Park ST, Pusztai L, Calin GA (2012) Plasma microRNA 210 levels correlate with sensitivity to trastuzumab and tumor presence in breast cancer patients. Cancer 118:2603–2614
Kelly TJ, Souza AL, Clish CB, Puigserver P (2011) A hypoxia-induced positive feedback loop promotes hypoxia-inducible factor 1alpha stability through miR-210 suppression of glycerol-3-phosphate dehydrogenase 1-like. Mol Cell Biol 31:2696–2706
Kim HW, Haider HK, Jiang S, Ashraf M (2009) Ischemic preconditioning augments survival of stem cells via miR-210 expression by targeting caspase-8-associated protein 2. J Biol Chem 284:33161–33168
Kim HW, Mallick F, Durrani S, Ashraf M, Jiang S, Haider KH (2012) Concomitant activation of miR-107/PDCD10 and Hypoxamir-210/Casp8ap2 and their role in cytoprotection during ischemic preconditioning of stem cells. Antioxid Redox Signal 17:1053–1065
King HW, Michael MZ, Gleadle JM (2012) Hypoxic enhancement of exosome release by breast cancer cells. BMC Cancer 12:421
Kloosterman WP, Plasterk RH (2006) The diverse functions of microRNAs in animal development and disease. Dev Cell 11:441–450
Krick S, Hanze J, Eul B, Savai R, Seay U, Grimminger F, Lohmeyer J, Klepetko W, Seeger W, Rose F (2005) Hypoxia-driven proliferation of human pulmonary artery fibroblasts: cross-talk between HIF-1alpha and an autocrine angiotensin system. FASEB J 19:857–859
Kulshreshtha R, Ferracin M, Negrini M, Calin GA, Davuluri RV, Ivan M (2007) Regulation of microRNA expression: the hypoxic component. Cell Cycle 6:1426–1431
Kulshreshtha R, Ferracin M, Wojcik SE, Garzon R, Alder H, Agosto-Perez FJ, Davuluri R, Liu CG, Croce CM, Negrini M, Calin GA, Ivan M (2007) A microRNA signature of hypoxia. Mol Cell Biol 27:1859–1867
Landau DA, Slack FJ (2011) MicroRNAs in mutagenesis, genomic instability, and DNA repair. Semin Oncol 38:743–751
Lanford RE, Hildebrandt-Eriksen ES, Petri A, Persson R, Lindow M, Munk ME, Kauppinen S, Orum H (2010) Therapeutic silencing of microRNA-122 in primates with chronic hepatitis C virus infection. Science 327:198–201
Lawrie CH, Gal S, Dunlop HM, Pushkaran B, Liggins AP, Pulford K, Banham AH, Pezzella F, Boultwood J, Wainscoat JS, Hatton CS, Harris AL (2008) Detection of elevated levels of

tumour-associated microRNAs in serum of patients with diffuse large B-cell lymphoma. Br J Haematol 141:672–675
Lee DW, Futami M, Carroll M, Feng Y, Wang Z, Fernandez M, Whichard Z, Chen Y, Kornblau S, Shpall EJ, Bueso-Ramos CE, Corey SJ (2012) Loss of SHIP-1 protein expression in high-risk myelodysplastic syndromes is associated with miR-210 and miR-155. Oncogene 31:4085–4094
Lee RC, Feinbaum RL, Ambros V (1993) The C. elegans heterochronic gene lin-4 encodes small RNAs with antisense complementarity to lin-14. Cell 75:843–854
Lee Y, Kim M, Han J, Yeom KH, Lee S, Baek SH, Kim VN (2004) MicroRNA genes are transcribed by RNA polymerase II. EMBO J 23:4051–4060
Lees JA, Saito M, Vidal M, Valentine M, Look T, Harlow E, Dyson N, Helin K (1993) The retinoblastoma protein binds to a family of E2F transcription factors. Mol Cell Biol 13:7813–7825
Leone G, DeGregori J, Yan Z, Jakoi L, Ishida S, Williams RS, Nevins JR (1998) E2F3 activity is regulated during the cell cycle and is required for the induction of S phase. Genes Dev 12:2120–2130
Leung AK, Sharp PA (2010) MicroRNA functions in stress responses. Mol Cell 40:205–215
Levy AP, Levy NS, Wegner S, Goldberg MA (1995) Transcriptional regulation of the rat vascular endothelial growth factor gene by hypoxia. J Biol Chem 270:13333–13340
Lewis BP, Burge CB, Bartel DP (2005) Conserved seed pairing, often flanked by adenosines, indicates that thousands of human genes are microRNA targets. Cell 120:15–20
Li T, Cao H, Zhuang J, Wan J, Guan M, Yu B, Li X, Zhang W (2011) Identification of miR-130a, miR-27b and miR-210 as serum biomarkers for atherosclerosis obliterans. Clin Chim Acta 412:66–70
Liu F, Lou YL, Wu J, Ruan QF, Xie A, Guo F, Cui SP, Deng ZF, Wang Y (2012) Upregulation of MicroRNA-210 regulates renal angiogenesis mediated by activation of VEGF signaling pathway under ischemia/perfusion injury in vivo and in vitro. Kidney Blood Press Res 35:182–191
Liu M, Liu H, Dudley SC Jr (2010) Reactive oxygen species originating from mitochondria regulate the cardiac sodium channel. Circ Res 107:967–974
Lou YL, Guo F, Liu F, Gao FL, Zhang PQ, Niu X, Guo SC, Yin JH, Wang Y, Deng ZF (2012) miR-210 activates notch signaling pathway in angiogenesis induced by cerebral ischemia. Mol Cell Biochem 370:45–51
Lu J, Getz G, Miska EA, Alvarez-Saavedra E, Lamb J, Peck D, Sweet-Cordero A, Ebert BL, Mak RH, Ferrando AA, Downing JR, Jacks T, Horvitz HR, Golub TR (2005) MicroRNA expression profiles classify human cancers. Nature 435:834–838
Merlo A, Quiros SB de, Secades P, Zambrano I, Balbin M, Astudillo A, Scola B, Aristegui M, Suarez C, Chiara MD (2012) Identification of a Signaling Axis HIF-1alpha/MicroRNA-210/ISCU independent of sdh mutation that defines a subgroup of head and neck paragangliomas. J Clin Endocrinol Metab 97:E2194–2200
Michael MZ SMOC, Holst Pellekaan NG van, Young GP, James RJ (2003) Reduced accumulation of specific microRNAs in colorectal neoplasia. Mol Cancer Res 1:882–891
Mizuno Y, Tokuzawa Y, Ninomiya Y, Yagi K, Yatsuka-Kanesaki Y, Suda T, Fukuda T, Katagiri T, Kondoh Y, Amemiya T, Tashiro H, Okazaki Y (2009) miR-210 promotes osteoblastic differentiation through inhibition of AcvR1b. FEBS lett 583:2263–2268
Mochel F, Knight MA, Tong WH, Hernandez D, Ayyad K, Taivassalo T, Andersen PM, Singleton A, Rouault TA, Fischbeck KH, Haller RG (2008) Splice mutation in the iron-sulfur cluster scaffold protein ISCU causes myopathy with exercise intolerance. Am J Hum Genet 82:652–660
Murakami Y, Yasuda T, Saigo K, Urashima T, Toyoda H, Okanoue T, Shimotohno K (2006) Comprehensive analysis of microRNA expression patterns in hepatocellular carcinoma and non-tumorous tissues. Oncogene 25:2537–2545
Nakamura H, Tanimoto K, Hiyama K, Yunokawa M, Kawamoto T, Kato Y, Yoshiga K, Poellinger L, Hiyama E, Nishiyama M (2008) Human mismatch repair gene, MLH1, is transcriptionally repressed by the hypoxia-inducible transcription factors, DEC1 and DEC2. Oncogene 27: 4200–4209
Ohh M, Park CW, Ivan M, Hoffman MA, Kim TY, Huang LE, Pavletich N, Chau V, Kaelin WG (2000) Ubiquitination of hypoxia-inducible factor requires direct binding to the beta-domain of the von Hippel-Lindau protein. Nat Cell Biol 2:423–427
Pasini B, Stratakis CA (2009) SDH mutations in tumorigenesis and inherited endocrine tumours: lesson from the phaeochromocytoma-paraganglioma syndromes. J Intern Med 266:19–42

Pineles BL, Romero R, Montenegro D, Tarca AL, Han YM, Kim YM, Draghici S, Espinoza J, Kusanovic JP, Mittal P, Hassan SS, Kim CJ (2007) Distinct subsets of microRNAs are expressed differentially in the human placentas of patients with preeclampsia. Am J Obstet Gynecol 196:e261–266

Porkka KP, Pfeiffer MJ, Waltering KK, Vessella RL, Tammela TL, Visakorpi T (2007) MicroRNA expression profiling in prostate cancer. Cancer Res 67:6130–6135

Puissegur MP, Mazure NM, Bertero T, Pradelli L, Grosso S, Robbe-Sermesant K, Maurin T, Lebrigand K, Cardinaud B, Hofman V, Fourre S, Magnone V, Ricci JE, Pouyssegur J, Gounon P, Hofman P, Barbry P, Mari B (2011) miR-210 is overexpressed in late stages of lung cancer and mediates mitochondrial alterations associated with modulation of HIF-1 activity. Cell Death Differ 18:465–478

Pulkkinen K, Malm T, Turunen M, Koistinaho J, Yla-Herttuala S (2008) Hypoxia induces microRNA miR-210 in vitro and in vivo ephrin-A3 and neuronal pentraxin 1 are potentially regulated by miR-210. FEBS Lett 582:2397–2401

Qin L, Chen Y, Niu Y, Chen W, Wang Q, Xiao S, Li A, Xie Y, Li J, Zhao X, He Z, Mo D (2010) A deep investigation into the adipogenesis mechanism: profile of microRNAs regulating adipogenesis by modulating the canonical Wnt/beta-catenin signaling pathway. BMC Genomics 11:320

Semenza GL (2010) Defining the role of hypoxia-inducible factor 1 in cancer biology and therapeutics. Oncogene 29:625–634

Semenza GL (2010) Vascular responses to hypoxia and ischemia. Arterioscler Thromb Vasc Biol 30:648–652

Semenza GL, Roth PH, Fang HM, Wang GL (1994) Transcriptional regulation of genes encoding glycolytic enzymes by hypoxia-inducible factor 1. J Biol Chem 269:23757–23763

Shinohara A, Ogawa T (1998) Stimulation by Rad52 of yeast Rad51-mediated recombination. Nature 391:404–407

Simone NL, Soule BP, Ly D, Saleh AD, Savage JE, Degraff W, Cook J, Harris CC, Gius D, Mitchell JB (2009) Ionizing radiation-induced oxidative stress alters miRNA expression. PLoS One 4:e6377

Sitohy B, Nagy JA, Dvorak HF (2012) Anti-VEGF/VEGFR therapy for cancer: reassessing the target. Cancer Res 72:1909–1914

Stenvang J, Silahtaroglu AN, Lindow M, Elmen J, Kauppinen S (2008) The utility of LNA in microRNA-based cancer diagnostics and therapeutics. Semin Cancer Biol 18:89–102

Sung P, Robberson DL (1995) DNA strand exchange mediated by a RAD51-ssDNA nucleoprotein filament with polarity opposite to that of RecA. Cell 82:453–461

Thum T, Galuppo P, Wolf C, Fiedler J, Kneitz S, Laake LW van, Doevendans PA, Mummery CL, Borlak J, Haverich A, Gross C, Engelhardt S, Ertl G, Bauersachs J (2007) MicroRNAs in the human heart: a clue to fetal gene reprogramming in heart failure. Circulation 116:258–267

Tong WH, Rouault TA (2006) Functions of mitochondrial ISCU and cytosolic ISCU in mammalian iron-sulfur cluster biogenesis and iron homeostasis. Cell Metab 3:199–210

Toyama T, Kondo N, Endo Y, Sugiura H, Yoshimoto N, Iwasa M, Takahashi S, Fujii Y, Yamashita H (2012) High expression of microRNA-210 is an independent factor indicating a poor prognosis in Japanese triple-negative breast cancer patients. Jpn J Clin Oncol 42:256–263

Tsuchiya S, Fujiwara T, Sato F, Shimada Y, Tanaka E, Sakai Y, Shimizu K, Tsujimoto G (2011) MicroRNA-210 regulates cancer cell proliferation through targeting fibroblast growth factor receptor-like 1 (FGFRL1). J Biol Chem 286:420–428

Rooij E van, Sutherland LB, Liu N, Williams AH, McAnally J, Gerard RD, Richardson JA, Olson EN (2006) A signature pattern of stress-responsive microRNAs that can evoke cardiac hypertrophy and heart failure. Proc Natl Acad Sci the U S A 103:18255–18260

Wan G, Mathur R, Hu X, Zhang X, Lu X (2011) miRNA response to DNA damage. Trends Biochem Sci 36:478–484

Wang GL, Semenza GL (1993) General involvement of hypoxia-inducible factor 1 in transcriptional response to hypoxia. Proc Natl Acad Sci U S A 90:4304–4308

Wang GL, Semenza GL (1995) Purification and characterization of hypoxia-inducible factor 1. J Biol Chem 270:1230–1237

Wang GL, Jiang BH, Rue EA, Semenza GL (1995) Hypoxia-inducible factor 1 is a basic-helix-loop-helix-PAS heterodimer regulated by cellular O2 tension. Proc Natl Acad Sci U S A 92:5510–5514

Wang J, Chen J, Chang P, LeBlanc A, Li D, Abbruzzesse JL, Frazier ML, Killary AM, Sen S (2009) MicroRNAs in plasma of pancreatic ductal adenocarcinoma patients as novel blood-based biomarkers of disease. Cancer Prev Res (Phila) 2:807–813

Wang S, Olson EN (2009) AngiomiRs–key regulators of angiogenesis. Curr Opin Genet Dev 19:205–211

Wilson WR, Hay MP (2011) Targeting hypoxia in cancer therapy. Nat Rev Cancer 11:393–410

Wu F, Yang Z, Li G (2009) Role of specific microRNAs for endothelial function and angiogenesis. Biochem Biophys Res Commun 386:549–553

Yan HL, Xue G, Mei Q, Wang YZ, Ding FX, Liu MF, Lu MH, Tang Y, Yu HY, Sun SH (2009) Repression of the miR-929217–92 cluster by p53 has an important function in hypoxia-induced apoptosis. EMBO J 28:2719–2732

Yanaihara N, Caplen N, Bowman E, Seike M, Kumamoto K, Yi M, Stephens RM, Okamoto A, Yokota J, Tanaka T, Calin GA, Liu CG, Croce CM, Harris CC (2006) Unique microRNA molecular profiles in lung cancer diagnosis and prognosis. Cancer Cell 9:189–198

Yang W, Sun T, Cao J, Liu F, Tian Y, Zhu W (2012) Downregulation of miR-210 expression inhibits proliferation, induces apoptosis and enhances radiosensitivity in hypoxic human hepatoma cells in vitro. Exp Cell Res 318:944–954

Ying Q, Liang L, Guo W, Zha R, Tian Q, Huang S, Yao J, Ding J, Bao M, Ge C, Yao M, Li J, He X (2011) Hypoxia-inducible microRNA-210 augments the metastatic potential of tumor cells by targeting vacuole membrane protein 1 in hepatocellular carcinoma. Hepatology 54:2064–2075

Yoshioka Y, Kosaka N, Ochiya T, Kato T (2012) Micromanaging Iron Homeostasis: Hypoxia-Inducible Micro-RNA-210 suppresses iron homeostasis-related proteins. J Biol Chem 287:34110–34119

Zhang H, Gao P, Fukuda R, Kumar G, Krishnamachary B, Zeller KI, Dang CV, Semenza GL (2007) HIF-1 inhibits mitochondrial biogenesis and cellular respiration in VHL-deficient renal cell carcinoma by repression of C-MYC activity. Cancer Cell 11:407–420

Zhang X, Wan G, Berger FG, He X, Lu X (2011) The ATM kinase induces microRNA biogenesis in the DNA damage response. Mol Cell 41:371–383

Zhang Z, Sun H, Dai H, Walsh RM, Imakura M, Schelter J, Burchard J, Dai X, Chang AN, Diaz RL, Marszalek JR, Bartz SR, Carleton M, Cleary MA, Linsley PS, Grandori C (2009) MicroRNA miR-210 modulates cellular response to hypoxia through the MYC antagonist MNT. Cell Cycle 8:2756–2768

Zhao A, Li G, Peoc'h M, Genin C, Gigante M (2012) Serum miR-210 as a novel biomarker for molecular diagnosis of clear cell renal cell carcinoma. Exp Mol Pathol 94:115–120

Zhu XM, Han T, Sargent IL, Yin GW, Yao YQ (2009) Differential expression profile of microRNAs in human placentas from preeclamptic pregnancies vs normal pregnancies. Am J Obstet Gynecol 200:e661–667

Chapter 4
Hypoxia and Reactive Oxygen Species

Agnes Görlach

Abstract In recent years, superoxide and other reactive oxygen species (ROS) have been acknowledged to act not only as unwanted and even toxic byproducts of aerobic metabolism but also as important signaling molecules in various physiological and pathophysiological conditions. This has broadened the field of oxygen signaling in a substantial way given the fact that superoxide is derived from molecular oxygen. In this regard ROS and ROS-dependent signaling pathways appear to be connected in different ways to the pathways involved in the adaptation towards a low-oxygen environment.

One of the major pathways regulated by oxygen availability relies on the activity of hypoxia-inducible transcription factors (HIFs). Originally described to be only induced and activated under hypoxia, accumulating evidence suggests that HIFs play a more general role in response to diverse cellular activators and stressors, many of which use ROS as signal transducers. On the other hand, the HIF pathway has also been implicated in controlling some important ROS-generating systems. Thus, an important cross talk exists between ROS signaling systems and the HIF pathway which may have substantial consequences for the pathogenesis of various disorders including cancer.

Keywords NADPH oxidase · HIF · Hypoxia · Reactive oxygen species · NFkB · Tumor · Signaling

4.1 Introduction: Oxygen Signaling in the Tumor Environment

An adequate supply of oxygen is required for the proper functioning of all aerobic organisms. Thus, inadequate oxygen supply to cells, tissues, and organs requires efficient mechanisms to counteract these potentially detrimental conditions. Inadequate

A. Görlach (✉)
Experimental and Molecular Pediatric Cardiology, German Heart Center Munich
at the TU Munich, Lazarettstraße 36, 80636 Munich, Germany
e-mail: goerlach@dhm.mhn.de

G. Melillo (ed.), *Hypoxia and Cancer,* Cancer Drug Discovery
and Development, DOI 10.1007/978-1-4614-9167-5_4,

oxygen availability including hypoxia or hypoxemia has been considered to be an essential element in the pathogenesis of a variety of diseases, including cardiovascular and pulmonary diseases as well as cancer.

In many tumors, a significant fraction of hypoxic cells has been identified. Fast tumor growth and inadequate tumor vascularization have been considered as important factors causing tumor hypoxia. The development of hypoxic fractions in a tumor has been related to resistance to therapy and often correlates with poor prognosis. However, this state is not fixed and hypoxic areas in tumors are frequently supplied with oxygen again when vascularization physically or functionally improves leading to reoxygenation events.

A key element in the cellular adaptation to hypoxia is a family of transcription factors termed hypoxia-inducible factors (HIFs). These transcription factors are sensitive to oxygen availability and activate complex gene expression patterns directed to adapt to low-oxygen conditions. HIF transcription factors also have been found to be upregulated in diseases associated with hypoxia/hypoxemia including pulmonary hypertension and cancer.

An adequate supply of O_2 is beneficial for cell and organ function, allowing production of energy in the mitochondrial respiratory chain as well as ordered functioning of many cellular enzymes and processes. However, aerobic respiratory activity in the mitochondria has also been linked to the generation of highly reactive O_2 derivates termed reactive oxygen species (ROS). When ROS are formed in high quantities overwhelming endogenous and/or exogenous antioxidant capacity damage to cellular proteins, RNA, DNA, and lipids can occur. Oxidative stress has been linked to the pathology of different diseases including tumor initiation and progression. However, ROS at low levels can also be generated in a controlled way by different enzyme systems, of which NADPH oxidases are the most prominent ones. This low-output ROS generation has been acknowledged to contribute to cellular signaling events rather than to cell-destroying pathways.

In this regard the generation and function of ROS have been discussed also in the context of signaling under hypoxic conditions. On the other hand, although initially counterintuitive, the HIF pathway has been found to be also sensitive to ROS. These observations suggest various interactions and cross talks between ROS and hypoxia signaling pathways. Intriguingly, both hypoxia and ROS signaling pathways have been associated with initiation and progression of a substantial number of diseases, of which cancer has been of particular importance.

4.2 Biochemistry of ROS and Antioxidants

Superoxide anion radicals ($O_2^-\bullet$) are formed from molecular oxygen by acquisition of an electron and can further react to other ROS such as hydrogen peroxide (H_2O_2), hydroxyl radicals ($OH\bullet$), peroxynitrite ($ONOO^-$), hypochlorous acid (HOCl), and singlet oxygen ($^1O_2$) (Fig. 4.1). $O_2^-\bullet$ is not freely diffusible, but can cross membranes via ion channels. Extracellular $O_2^-\bullet$ has been shown to enter the cell via the

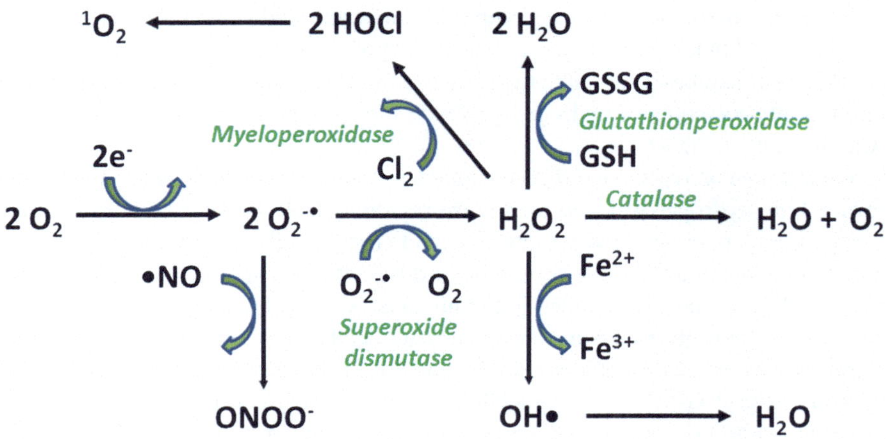

Fig. 4.1 Generation of reactive oxygen species. Electron transfer on molecular oxygen results in the formation of superoxide anion radical which can, with the help of superoxide dismutases, be transformed to hydrogen peroxide. In the presence of nitric oxide $O_2^-\bullet$ can readily react to peroxynitrite. H_2O_2 can in the presence of Fe(II) promote hydroxyl radical formation. In the presence of chloride, hypochlorous acid is formed by the help of myeloperoxidase which can then further react to singlet oxygen. Hydrogen peroxide can also be transformed to water by catalase or by glutathione peroxidases

anion blocker-sensitive chloride channel-3 (ClC-3) (Hawkins et al. 2007), while mitochondrial outer membrane voltage-dependent anion channels (VDACs) can direct superoxide flux from mitochondria to cytosol (Han et al. 2003). H_2O_2 on the other hand, which is not a radical, is diffusible and has therefore been frequently considered to act as a second messenger. In the presence of iron, superoxide and hydrogen peroxide can lead to the formation of highly reactive hydroxyl radicals which can damage cellular proteins, RNA, DNA, and lipids. Interaction of ROS with nitric oxide (NO) or fatty acids can lead to the formation of peroxynitrite or peroxyl radicals, respectively, which are also highly reactive. In the presence of chloride, peroxidases can catalyze the generation of hypochlorous acid (HOCl) and singlet oxygen ($^1O_2$) from hydrogen peroxide (Görlach and Kietzmann 2007).

The generation of ROS is not unopposed in the cell since antioxidant enzymes and antioxidant scavengers contribute to control the levels of ROS and to prevent oxidative stress reactions. Within this system, the nuclear transcription factor Nrf2 has been considered to play an important role in regulating gene expression of antioxidant enzymes (Hybertson et al. 2011).

The reaction of superoxide into oxygen and hydrogen peroxide is catalyzed by three superoxide dismutases (SODs). They differ in their subcellular location and catalytic metal moiety. While the dimeric Cu/Zn SOD resides in the cytosol, the extracellular SOD (ECSOD) is a tetramer complexed with Cu and Zn as well. A tetramer complexed with Mn (MnSOD) is found in the mitochondria (Maier and Chan 2002).

Hydrogen peroxide can be decomposed into water and oxygen by catalase, an enzyme found in nearly all living organisms exposed to oxygen. Catalase resides in peroxysomes and has a very high turnover rate: one molecule of catalase can convert millions of molecules of hydrogen peroxide into water and oxygen every second (Chelikani et al. 2004).

Glutathione peroxidases (GPXs) belong to a family of selenoproteins which can reduce lipid hydroperoxides to their corresponding alcohols. They can also reduce free hydrogen peroxide to water, thereby oxidizing the reduced monomeric GSH to glutathione disulfide (GSSG). So far, eight different isoforms of glutathione peroxidase (GPX1–8) have been identified in humans. Among them, cytoplasmic GPX1, whose preferred substrate is hydrogen peroxide, is the most common one. GPX4 is commonly expressed but at lower levels than GPX1 and preferentially reduces lipid hydroperoxides; GPX2 is an extracellular enzyme which is primarily found in the intestine, while extracellular GPX3 is especially abundant in plasma (Muller et al. 2007).

Thioredoxins (TRXs) are small homodimeric selenoenzymes containing an FAD molecule and a COOH-terminal selenocysteine residue in their Gly-Cys-SeCys-Gly active site. TRXs enable the reduction of oxidized proteins by cysteine thiol–disulfide exchange. They can also act as electron donors to peroxidases and ribonucleotide reductases. TRXs are kept in a reduced state by the flavoenzymes thioredoxin reductase (TRXR) in an NADPH-dependent reaction. Mammalian cells contain two TRX systems: the cytosolic TRX1/TRXR1 and the mitochondrial TRX2/TRXR2. Loss-of-function mutation of either of the two human thioredoxin genes is lethal. The related glutaredoxins share many of the functions of thioredoxins, but are reduced by glutathione rather than a specific reductase (Berndt et al. 2007).

Peroxiredoxins (Prxs) belong to a family of thiol-specific antioxidant proteins, also termed as the thioredoxin peroxidases or alkyl-hydroperoxide-reductase-C22 proteins. Prxs exert their protective antioxidant role in cells through their peroxidase activity whereby hydrogen peroxide, peroxynitrite, and a wide range of organic hydroperoxides (ROOH) are reduced and detoxified (Wood et al. 2003). Antioxidant scavengers are predominantly of dietary origin. These biomolecules include tocopherol, ascorbic acid, carotenoids, uric acid, and polyphenols.

4.3 Sources of ROS Generation

Various enzymatic systems have been shown to be able to release ROS. Thereby, NADPH oxidases are unique in that their sole function is to generate ROS whereas in the mitochondrial respiratory chain, ROS generation is considered as a byproduct of oxidative metabolism. Other systems able to generate ROS, mostly $O_2^{-}\bullet$ or H_2O_2, include the arachidonic acid pathway, the cytochrome P450 family, glucose oxidase, amino acid oxidases, xanthine oxidase, or, under certain conditions, even NO synthases.

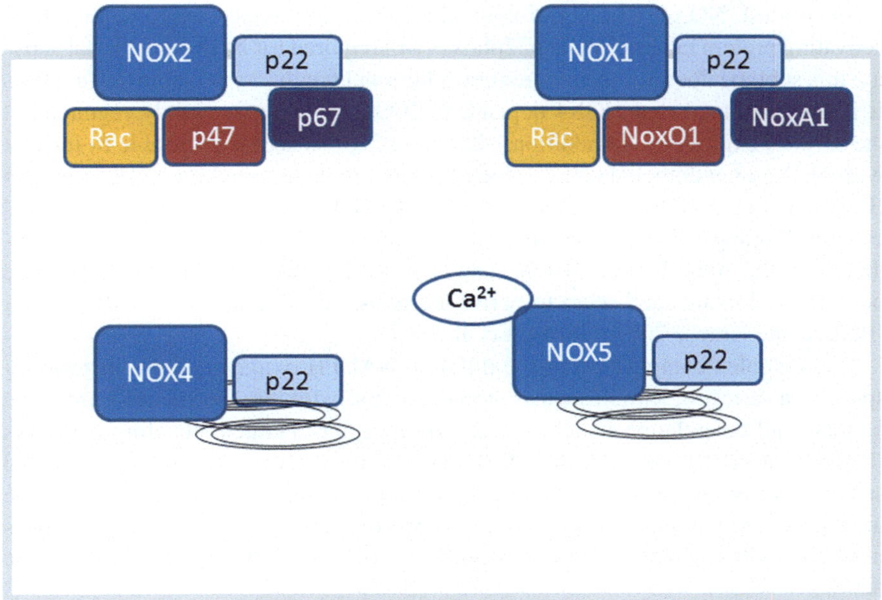

Fig. 4.2 Different NADPH oxidases are expressed in the cell. NADPH oxidases consist of a central catalytic core, the NOX protein, which forms a cytochrome b558 with p22phox. Additional cofactors dependent on the NOX type are required for full activity. NOX2 (formerly known as gp91phox) requires the GTPase Rac, as well as p47phox and p67phox. NOX1 which is also present at the plasma membrane contains instead of p47phox NoxO1, and instead of p67phox NoxA1. The intracellular NADPH oxidases, NOX4 and NOX5, do not require additional cofactors

4.3.1 NADPH Oxidases

NADPH oxidases have been initially identified in leukocytes where they are responsible for the respiratory burst as part of the innate immune response. Subsequently, NADPH oxidases have been identified in many nonphagocytic cells including vascular cells and tumor cells (for review, see Görlach et al. 2002; Bedard and Krause 2007; Babior et al. 2004, Brown and Griendling 2009; Lambeth et al. 2007).

NADPH oxidases are multiprotein enzymes which transfer an electron from NADPH to oxygen to generate $O_2^-\bullet$. The single-electron reduction further requires FAD and two hemes which are localized in a transmembrane NOX protein which forms together with the protein p22phox, a cyctochrome b558. To date five homologous NOX proteins termed NOX1–NOX5 have been identified (Fig. 4.2). Dependent on the specific NOX protein, additional cofactors are required for proper activation of some of these enzymes, in particular for NOX1, NOX2, and NOX3. The main function of these regulatory proteins is to bring NADPH in close proximity to FAD to facilitate electron flow which is accompanied by an internal refolding in order to transfer the electron from FAD to heme and later on to oxygen.

In contrast, NOX4- and NOX5-containing enzymes do not seem to require these activating protein factors. While p22phox is still required for NOX4 functional activity, this subtype does not seem to require additional regulatory proteins for function. It is thus assumed that NOX4-dependent ROS generation is mainly regulated at the level of expression. In addition, although p22phox seems to be able to interact with NOX5 in endothelial cells, it does not seem to be required for NOX5 function (BelAiba et al. 2007b). However, several of the NOX5 variants contain N-terminal calcium-binding EF hands which confer calcium sensitivity of this enzyme. Furthermore, the more distinct DUOX1 and DUOX2 enzymes have additional putative peroxidase domains and seem to perform a two-electron reduction yielding H_2O_2 (Bedard and Krause 2007; Lambeth et al. 2007).

The complex composition of the different NADPH oxidases allows them to respond to a variety of stimuli such as growth factors, cytokines, hormones, vasoactive factors, and coagulation factors mainly via receptor-operated signaling pathways (Petry et al. 2010). In addition, NADPH oxidase activity can also be regulated at the level of expression, whereby transcriptional mechanisms seem to be the most prevalent pathways. This allows NADPH oxidases to also respond to changes in microenvironmental conditions including variations in oxygen availability (see below).

4.3.2 Mitochondrial ROS Generation

Early evidence, mainly from studies using isolated mitochondria, suggested that a small percentage of oxygen used by the electron transport chain is not completely reduced to water but is instead converted to $O_2^-\bullet$. ROS generation by mitochondria has been related to respiratory activity during which a series of redox reactions occurs along which electrons are transferred from a donor molecule (NADH or QH2) to O_2, concluding at complex IV (cytochrome oxidase), where molecular O_2 is reduced to water. While initially the rate of ROS production in isolated mitochondria was considered to be 1–2 % of the oxygen consumed in the respiratory chain (Chance et al. 1979), the rate of mitochondrial $O_2^-\bullet$ production in vivo is still unclear, but may be in fact below these levels (Aon et al. 2010). Mitochondrial generation of $O_2^-\bullet$ is dependent not only on the oxygen concentration but also on the concentration of potential electron donors and the second-order rate constants for the reactions between them (Murphy 2009). At complex I, significant $O_2^-\bullet$ production can still occur even when mitochondria are not making ATP, which leads to a high proton-motive force and a reduced coenzyme Q pool, or when there is a high NADH/NAD$^+$ ratio in the mitochondrial matrix. $O_2^-\bullet$ production also has been described to occur at complex III where superoxide is formed via the Q-cycle and can enter either the mitochondrial intermembrane space or the matrix (Rigoulet et al. 2011; Turrens 2003). Once in the matrix, $O_2^-\bullet$ is converted first to H_2O_2 by MnSOD, and then to water by GPX. In the intermembrane space, $O_2^-\bullet$ can be converted to H_2O_2 by CuZnSOD, or can be scavenged by cytochrome c or can be transferred to the cytosol via VDACs (Han et al. 2003). There is some controversy in the literature regarding the relative contribution of complexes I and III to $O_2^{-}\bullet$ generation and it appears that it depends on the cell or tissue type and/or on the respiratory status.

In contrast to NADPH oxidases, mitochondrial ROS generation has only recently been related to receptor–ligand interactions. However, due to the dependency of mitochondrial activity on oxygen availability, mitochondrial ROS generation has been linked to oxygen signaling pathways (see below).

4.4 Regulation of HIFs by Oxygen

Cellular adaptation to hypoxia is primarily mediated by transcription factors of the HIF family. These heterodimers belonging to the basic helix–loop–helix and Per-ARNT-Sim (PAS) domain-containing proteins consist of an α-subunit and a β-subunit. The latter one is also known as arylhydrocarbon receptor-nuclear translocator ARNT (Hoffman et al. 1991). To date, three family members are known, HIF-1 (Semenza and Wang 1992; Beck et al. 1993), HIF-2 (also termed EPAS-1) (Tian et al. 1997), and HIF-3 (Gu et al. 1998), whereby oxygen sensitivity is mediated by the α-subunits, while the β-subunit is constitutively expressed (reviewed by Wenger 2002).

4.4.1 Regulation of HIF-α Protein Stability by Oxygen

Under normoxic conditions, HIF-α proteins are unstable. This is brought about by the HIF-α oxygen-dependent degradation domain (ODD) (Huang et al. 1998) and two transactivation domains (TADs) referred to as amino-terminal TADN and carboxy-terminal TADC (Jiang et al. 1997; Pugh et al. 1997). Under normoxic conditions, HIF-α proteins are hydroxylated on at least two proline residues within the ODD (Jaakkola et al. 2001; Ivan et al. 2001) which leads to binding of the von Hippel–Lindau tumor suppressor protein (pVHL) and subsequent ubiquitinylation and proteasomal degradation. HIF-α hydroxylation at the ODD is mediated by prolylhydroxylase domain-containing proteins (PHD-1, -2, -3), also known as HIF-prolyl 4-hydroxylases (HIF-P4H). These enzymes vary in their subcellular localization and target affinity (for review, see Kaelin and Ratcliffe 2008). A fourth HIF-1α hydroxylating enzyme has been identified as PHD4 or P4H-TM. It is located in the endoplasmic reticulum, and seems to be more closely related to procollagen prolylhydroxylases than the other PHDs (Koivunen et al. 2007).

At the TADC, HIF-1α and HIF-2α are additionally hydroxylated at an asparagine residue by factor inhibiting HIF (FIH) under normoxic conditions. This prevents the recruitment of the coactivator CREB binding protein (CBP/p300) and thus transcriptional activity (Mahon et al. 2001). Since the TADC is lacking in HIF-3α this protein may differentially act compared to the other HIF-α subunits. In fact, there is increasing evidence that, in particular, murine isoforms of this protein counteract the function of the other HIFs which may help to fine-tune the adaptation towards hypoxia (Makino et al. 2001).

Since the activity of the HIF hydroxylating enzymes has been considered to be strictly dependent on oxygen availability, PHDs appear to be attractive elements in the oxygen sensing cascade. However, in addition to oxygen, hydroxylase activity is also dependent on the presence of its cofactors Fe(II), 2-oxoglutarate, and ascorbate (for review, see Kaelin and Ratcliffe 2008), suggesting a certain redox sensitivity of these enzymes.

In addition to hydroxylation and subsequent ubiquitinylation, other hydroxylase-independent posttranslational modifications of HIF-α, aimed to increase HIF-α levels under hypoxia, have been recently described, which include sumoylation (Bae et al. 2004), neddylation (Ryu et al. 2011), and acetylation/deacetylation by Sirt1 (Dioum et al. 2009). In addition, HIF-1α stability has also been reported to be enhanced by S-nitrosylation at Cys533 in the ODD independently of its hydroxylation status (Li et al. 2007).

4.4.2 Regulation of HIF-α Translation and Transcription by Oxygen

In contrast to many other translation processes, HIF-1α mRNA translation has been shown to be efficient also under hypoxic conditions (Görlach et al. 2000b), and subsequent studies demonstrated that HIF-1α protein expression can persist during hypoxia, owing to an internal ribosome entry site (IRES) in the 5′-untranslated region of its mRNA (Lang et al. 2002) where the polypyrimidine tract binding protein (PTB) can specifically interact with the HIF-1α IRES (Schepens et al. 2005). Additional studies provided evidence that interaction of RNA-binding proteins including HuR, CPEB1, and CPEB2 with HIF-1α mRNA sequences is also involved in the translational response to hypoxia (Galban and Gorospe 2009). More recently, also HIF-2α mRNA has been shown to escape the translational block under hypoxia and to be enriched in polysomes from hypoxic cells (Thomas and Johannes 2007). An iron-responsive element in the 5′UTR has been shown to promote translation of HIF-2α mRNA not only in response to iron, but also in response to hypoxia (Sanchez et al. 2007; Zimmer et al. 2008).

While the synthesis of many proteins is reduced under hypoxic conditions, HIF-1α and HIF-2α are able to escape this control, thus allowing efficient translation under normoxic and hypoxic conditions (Thomas and Johannes 2007; Gorlach et al. 2000; Hui et al. 2006). Both, hypoxic translation of HIF-1α and of HIF-2α, have been shown to require influx of extracellular calcium, stimulation of classical protein kinase C-alpha (cPKC-alpha), and the activity of mammalian target of rapamycin, mTOR. The translational component has been estimated to contribute to approximately 40–50 % of HIF-α protein accumulation after 3 h at 1 % O_2 (Hui et al. 2006).

The HIF-1α and HIF-2α genes have been shown to contain an IRES, which seems to be crucially involved in their translational maintenance under hypoxic conditions (Lang et al. 2002; Sanchez et al. 2007; Zimmer et al. 2008; Young et al. 2008). Although only limited data exist describing translational control of HIF-α, it has been shown that PTB can specifically interact with the HIF-1α IRES, and that this interaction is enhanced in hypoxic conditions (Schepens et al. 2005).

However, prolonged hypoxia has been shown to result in decreased levels of HIF-1α mRNA in several cell lines including endothelial cells (Chamboredon et al. 2011 and unpublished own observations). While interaction with a transcriptional repressor mechanism cannot be excluded, other data suggest that the presence of many AU-rich instability elements (ARE) indicative of short-lived mRNAs in the 3′-UTR may also contribute to the regulation of HIF-1α under hypoxia. In fact, several proteins have been shown to bind to the 3′UTR and to control mRNA translation, including PTB, cytoplasmic polyadenylation element-binding proteins (CPEB-1 and -2), and HuR (Schepens et al. 2005; Galban et al. 2008; Hagele et al. 2009). The HIF-1α mRNA was also recently identified as a target for the miR-17–92 microRNA (miRNA) cluster (Hagele et al. 2009). The mRNA-destabilizing protein tristetraprolin (TTP) may contribute to decreased HIF-1α mRNA levels under prolonged hypoxia by binding to the most distal ARE present in the HIF-1α 3′-UTR (Chamboredon et al. 2011).

Finally, in addition to the regulation by protein stabilization and translation, HIF-1α has been shown to be regulated at the transcriptional level. While heat shock factors have been suggested to limit basal HIF-1α transcription (Chen et al. 2011), acute hypoxia has been shown to increase binding of the redox-sensitive transcription factor NFκB to the HIF-1α promoter (Belaiba et al. 2007a, Rius et al. 2008).

4.5 ROS Modulate HIFs

Early evidence indicating redox sensitivity of HIF-1 was provided by studies demonstrating that treatment of purified HIF-1 with H_2O_2, diamide, or N-ethyl-maleimide prevented the ability to bind DNA under hypoxic conditions, while prior application of dithiothreitol could preserve DNA binding suggesting that HIF-1 DNA binding requires reducing conditions (Wang et al. 1995). In addition, Trx has been shown to enhance HIF-1α protein levels, and this effect was then related to the interaction of the Trx effector Ref-1 with the HIF-1α TADN and TADC. The latter reaction seemed to be dependent on the redox state of cysteine 800 in HIF-1α and cysteine 848 in HIF-2α (Huang et al. 1996; Welsh et al. 2002; Ema et al. 1999; Carrero et al. 2000; Lando et al. 2000). In addition, mutation of cysteine 800 prevented the decrease in HIF-1α TADC activity in response to hydroxyl radicals (OH•) (Liu et al. 2004) supporting the notion that a reducing environment is preferential for stabilization and functioning of HIF-α proteins.

In contrast, while administration of external H_2O_2 or expression of MnSOD seemed to prevent hypoxic induction of HIF-1α in tumor cells (Huang et al. 1996;

Wang et al. 2005), application of low concentrations of H_2O_2 under normoxic conditions or overexpression of Cu/ZnSOD or MnSOD was able to increase the levels of HIF-1α in particular in vascular cells (BelAiba et al. 2004; Bonello et al. 2007; Diebold et al. 2010a; Gorlach et al. 2001) but also in Hep3B cells (Chandel et al. 2000). Variations in the sensitivity of the HIF pathway towards H_2O_2 have been observed and related to cell type-specific properties such as the levels and activities of antioxidant enzymes present in a given cell.

Subsequent evidence was provided that HIF-α proteins are responsive to a variety of nonhypoxic stimuli in an ROS-dependent manner. Examples for such signaling pathways include insulin (Kietzmann et al. 2003; Treins et al. 2002), PDGF, TGF-β, IGF-1, EGF and HGF (Gorlach et al. 2001; Tacchini et al. 2004; Richard et al. 2000; Liu et al. 2006b; Fukuda et al. 2003), thrombin (Gorlach et al. 2001), angiotensin-II (Richard et al. 2000), urotensin-II (Diebold et al. 2012), TNF-alpha and cytokines (Stiehl et al. 2002; Haddad and Land 2001), carbachol (Hirota et al. 2004), oxLDL (Shatrov et al. 2003), mechanical stress (Kim et al. 2002), the HIV-1 accessory protein Vpr (Deshmane et al. 2009) as well as "hypoxic mimetic" $CoCl_2$ (BelAiba et al. 2004; Chandel et al. 2000). Several studies identified NADPH oxidases as important sources of ROS in the regulation of HIF-α in vascular cells (Bonello et al. 2007; Diebold et al. 2010a; Gorlach et al. 2001; Diebold et al. 2008, 2012; Petry et al. 2012) The findings that NADPH oxidases regulate HIF-α were also substantiated by the in vivo observation that HIF-1α levels were elevated in carotid lesions of mice overexpressing p22phox in smooth muscle cells (Khatri et al. 2004) while knockdown of p22phox diminished thrombin-induced HIF-1α levels in these cells (Görlach et al. 2001). Similarly, NOX4 was shown to control HIF-1α and HIF-2α levels in smooth muscle cells (Bonello et al. 2007; Diebold et al. 2010a). Recent data indicate that NOX2 and NOX5 which are important for endothelial ROS generation (BelAiba et al. 2007b; Gorlach et al. 2000a) also play a role in the regulation of HIF-1α (Diebold et al. 2012; Petry et al. 2012) in these cells in response to thrombin and vasoactive peptides. NOX2 was also involved in the regulation of HIF-2α by lipopolysaccharides (LPS) in microglia (Oh et al. 2008).

Increasing evidence suggests that NADPH oxidases are also expressed in tumor cells and are important regulators of tumor growth and progression (Block and Gorin 2012). Since HIF transcription factors play a central role in tumor biology, a cross talk between NADPH oxidases and HIFs may be important also in tumor biology. In line with this assumption, tumor treatment with hyperthermia was shown to enhance NOX1 and subsequently HIF-1α levels in tumor cells (Moon et al. 2010). NOX1 was also shown to increase HIF-1α levels in lung tumor cells (Goyal et al. 2007) while NOX4 knockdown decreased HIF-1α levels in ovarian cancer cells (Xia et al. 2007). In line with this, depletion of NOX4 or NOX1 reduced HIF-2α levels in VHL-deficient 786-O or RCC4 renal carcinoma cells suggesting that ROS may act via a VHL-dependent pathway (Block et al. 2007; Maranchie and Zhan 2005). Interestingly p22phox was shown to maintain HIF-2α protein levels through inactivation of tuberin and downstream activation of ribosomal protein S6 kinase 1/4E-BP1 pathway (Block et al. 2010). In addition to NOX1 and NOX4, NOX3, which has been considered to be only expressed in fetal tissues and in the inner ear, has been shown to be expressed in HepG2 cells, where it contributes to HIF-1α regulation by insulin (Carnesecchi et al. 2006). These data indicate that NADPH oxidases are important sources of ROS in a variety of non-hypoxic pathways in vascular and tumor cells.

In addition to NADPH oxidases, recent evidence suggested that mitochondrial ROS generation may also contribute to stimulus-activated ROS levels under non-hypoxic conditions since induction of HIF-1α by angiotensin-II was repressed in smooth muscle cells upon application of a complex III inhibitor, by Rieske Fe–S protein siRNA, or by the mitochondrial-targeted antioxidant SkQ1 (Patten et al. 2010). Also, normoxic stabilization of HIF-1α by thrombopoietin required ROS production and mitochondrial function as was determined by using inhibitors of mitochondrial respiratory chain as well as the unspecific flavin inhibitor DPI which also inhibits NADPH oxidases (Yoshida et al. 2008). Further studies have to dissect the relative role of NADPH oxidases versus mitochondrial ROS generation in the control of the HIF pathway under non-hypoxic conditions.

4.6 Mechanisms of HIF Regulation by ROS

Similar to the situation under hypoxic conditions there is now increasing evidence that ROS can regulate HIF-α levels by different mechanisms.

4.6.1 Regulation of HIF-α stability by ROS

Although inhibition of HIF-α hydroxylation has been considered to be specifically related to hypoxic conditions, increasing evidence suggests that oxygen availability is not the sole regulator of PHD activity. Evidence that ROS can affect HIF-α stability via interference with the PHD/pVHL pathway was provided by findings that H_2O_2 application or NOX4 overexpression increased HIF-2α TADN and TADC activity and this response was abolished upon mutation of the target prolines or asparagines, respectively (Diebold et al. 2010a). In line with this, H_2O_2 decreased pVHL binding of HIF-2α (Diebold et al. 2010a). Similar observations were made with HIF-1α indicating that ROS may mediate HIF-α stability by affecting hydroxylation and pVHL binding (unpublished observation).

Since ROS are known to be able to oxidize Fe(II) to Fe(III) it was reasonable to assume that the availability of Fe(II) may be disturbed under these conditions. In *junD*-deficient cells that exhibit chronic oxidative stress, it was shown that ROS interfered with Fe(II) availability in the HIF prolylhydroxylase catalytic site possibly by a Fenton-type reaction, thus diminishing HIF-1*α* hydroxylation and allowing its accumulation (Gerald et al. 2004). Interestingly, in thrombin-treated cells with NOX4-dependent increase in HIF-α levels, the availability of Fe(II) was also decreased indicating that ROS generated by NADPH oxidases are able to diminish the pool of available Fe(II) (Diebold et al. 2010a). Conversely, it was shown that the addition of iron to cultured human prostate adenocarcinoma cells stimulated HIF-1α degradation (Knowles et al. 2003) indicating that the balance between Fe(II) and Fe(III) is of major importance in controlling HIF-α levels.

In this context, it has been assumed that the major role of ascorbate, another cofactor of PHD/FIH hydroxylase activity, is to maintain Fe(II) levels in the cell by providing a radical cycling system for regeneration of Fe(II) (Epstein et al. 2001). Although the mechanism is not entirely known yet, it is proposed that ascorbate reduces Fe(III) to Fe(II) within the enzyme active site, thus rendering the enzyme active. Additionally, ascorbate might enhance the provision of Fe(II) from an intracellular pool such as ferritin by conversion of Fe(III) into Fe(II). In this case, it is proposed that superoxide ion, generated during the iron-promoted oxidation of ascorbate, acts as a reductant of ferritin iron (Boyer and McCleary 1987).

In the cellular system, there is now ample evidence that provision with ascorbate is able to regulate HIF-α levels under non-hypoxic conditions (BelAiba et al. 2004; Diebold et al. 2010a; Gorlach et al. 2001; Knowles et al. 2003; Page et al. 2008). Ascorbate was able to decrease induction of HIF-α TADN and TADC activity by thrombin and H_2O_2, while it restored pVHL binding to HIF-2α in the presence of thrombin and H_2O_2 suggesting that in vivo activity of PHD and FIH is strongly dependent on ascorbate (Diebold et al. 2010a). Interestingly, thrombin and angiotensin-II decreased cellular ascorbate levels (Diebold et al. 2010a; Page et al. 2008) further suggesting that ascorbate availability may provide an important mechanism in the regulation of HIF-α under non-hypoxic conditions. Interestingly, recent in vitro studies suggested that also other reducing agents such as glutathione and dithiothreitol promote HIF hydroxylase activity further indicating that the cellular redox state is important in controlling PHD activity (Flashman et al. 2010). In accordance with this, it was observed that the redox cycler DMNQ increased HIF-1α accumulation via reduction of pVHL binding suggesting that DMNQ was interfering with PHD activity (Kohl et al. 2006) while the same substance prevented accumulation of HIF-1α induced by TNF-α (Sandau et al. 2001) indicating that the current redox state is an essential determinant in regulation of HIF-α stability via PHD activity. Such a notion may also explain the heterogeneous results observed on the role of ascorbate in regulating hypoxic HIF-α levels. In line with this, it was recently shown that the systemic response to hypoxia was preserved in vitamin C-deprived Gulo$^{-/-}$ knockout mice. Furthermore, glutathione was found to fully substitute for vitamin C requirement of all three PHD isoforms in vitro. Mutation of a previously recognized redox-sensitive cysteine in PHD2 (Mecinovic et al. 2009) increased basal hydroxylation rates and conferred resistance to oxidative damage in vitro, suggesting that this surface-accessible PHD2 cysteine residue is a target of antioxidative protection by vitamin C and glutathione (Nytko et al. 2011).

Recent data indicated that HIF-1α may also be regulated by neddylation involving NEDD8 (Ryu et al. 2010). This ubiquitin-like protein covalently binds to its substrate proteins and, thus, regulates their stabilities and functions. Similar to the situation with ubiquitinylation, ROS prevented degradation of HIF-1α in the presence of NEDD8. It was also shown that the Sentrin/SUMO-specific protease 3 (SENP3) was stabilized by ROS resulting in enhanced HIF-1 transactivation. This was brought about by SENP3-dependent de-SUMOylation of the HIFα coactivator p300 (Huang et al. 2009). Although the relative importance of these pathways has not been clarified yet, these findings further support the importance of ROS to act as versatile regulators of HIF-α under non-hypoxic conditions.

4.6.2 Regulation of HIF-α Transcription and Translation by Oxygen

In addition to regulation of HIF-α at the level of protein stability, non-hypoxic induction of HIF-1α has been described to be regulated by a transcriptional mechanism in an ROS-dependent manner (for review, see Gorlach 2009). Several studies reported that cytokines, proinflammatory and prothrombotic factors, growth factors, and vasoactive peptides including interleukin-1, HGF, angiotensin-II, or LPS which are all known to increase ROS levels can induce HIF-1α mRNA levels in several cell types (Tacchini et al. 2004; Diebold et al. 2008; Frede et al. 2006; Page et al. 2002). Subsequently, it was shown that HIF-1α is a direct target gene of NFκB (Belaiba et al. 2007a; Rius et al. 2008; Bonello et al. 2007; van Uden et al. 2008; Görlach and Bonello 2008), and that ROS derived from NADPH oxidases or direct application of H_2O_2 regulate NFκB-dependent HIF-1α transcription (Bonello et al. 2007; Petry et al. 2012; Diebold et al. 2010b). These findings indicate that transcriptional regulation of HIF-1α by ROS-sensitive activation of NFκB may represent an important mechanism how agonists can induce HIF-1α under non-hypoxic conditions and provide a pathophysiologically interesting link between these two important redox-sensitive transcription factors with various implications not only for inflammatory conditions, but also for cardiovascular diseases and cancer.

A role of ROS in increasing HIF-1α transcription was also shown in cells with a mutation in mitochondrial NADH dehydrogenase subunit 6 (ND6), and this pathway involved phosphatidylinositol 3-kinase (PI3K)-Akt, protein kinase C (PKC), and histone deacetylase (HDAC), although the specific transcription factor involved in this response has not been identified (Koshikawa et al. 2009). In contrast to HIF-1α, only limited data are available on the regulation of HIF-2α mRNA and the contribution of ROS. It has been shown that NOX4 depletion decreased HIF-2α mRNA levels in RCC4 cells (Maranchie and Zhan 2005) although the underlying mechanisms have not been resolved. Since the NFκB binding site which mediates transcriptional regulation of HIF-1α does not seem to be conserved in the HIF-2α promoter, this may be an important factor determining nonredundant functions of HIF-1α and HIF-2α in hypoxic and non-hypoxic conditions. In addition, it was recently shown that in hematopoietic stem cells stimulated with erythropoietin, HIF-2α was a direct Stat5 target gene. Although not explicitly studied, this mechanism may also involve ROS (Fatrai et al. 2011).

During normoxia, HIF-1α protein synthesis has been considered to be regulated by cap-dependent signaling processes, mediated mainly through PI3K/Akt and mTOR, in particular, in response to tyrosine kinase signaling (Zhong et al. 2000; Laughner et al. 2001). ROS derived from NADPH oxidases and mitochondria have been shown to be able to activate PI3K/Akt signaling in normal and malignant cells (Görlach and Kietzmann 2007; Görlach et al. 2001; Djordjevic et al. 2005b; Wenner 2012) and have been implicated in translational regulation of HIF-1α in smooth muscle cells in response to angiotensin-II (Page et al. 2002). In addition, it was proposed that the PI3K pathway in conjunction with NFκB may be involved in the translational regulation of HIF-1α in response to TNF-α (Zhou et al. 2004). In addition, ROS-dependent, but NFκB-independent, translation of HIF-1α was observed in human myeloid cells stimulated with LPS (Nishi et al. 2008). Recently, it was shown that HIF-2α mRNA translation was controlled by p22phox by a mechanism involving stabilization of Rictor-associated mTORC2 complex in the absence of VHL through an eIF4E-dependent mRNA translational mechanism (Nayak et al. 2012). However, the relative importance of ROS-dependent HIF-α translational mechanisms compared to transcriptional mechanisms and protein stabilization has not been clarified, yet.

4.7 ROS and HIF: the Mitochondrial Case

While evidence that ROS are regulating HIF-α levels are compelling now under non-hypoxic conditions, the role of ROS in the control of HIF-α under hypoxic conditions has been a matter of controversy for many years. Initial studies using antioxidants observed a protective effect on HIF-1α stability under hypoxia–reoxygenation conditions while the effect on hypoxic HIF-1α was not clearly evident (Haddad et al. 2000) Similarly, ascorbate treatment was unable to prevent hypoxic induction of HIF-1α in different cellular models (Knowles et al. 2003; Brown and Nurse 2008).

Conversely, the use of mitochondrial inhibitors and also subsequent molecular approaches and mitochondria-depleted cells suggested that mitochondria due to their ability to generate ROS were required to enhance HIF-1α induction under hypoxic conditions (Chandel et al. 2000). These latter findings found support by studies reporting that ROS levels increase under hypoxic conditions, while this response was decreased upon inhibition of mitochondrial complex III (Bell et al. 2007), complex II (Guzy et al. 2008), or complex I (Agani et al. 2000). Further studies suggested that a functional mitochondrial chain is required for enhanced ROS generation under hypoxia and thus induction of HIF-1α, and that the source of the ROS is possibly a ubisemiquinone at the Q_o site of complex III (Bell et al. 2007; Hamanaka and Chandel 2009).

In contrast, it was also reported that hypoxic induction of HIF-1α was not affected in mitochondria-depleted cells or cells treated with mitochondrial inhibitors (Srinivas et al. 2001; Vaux et al. 2001) while other studies reported that induction of HIF-1α by hypoxia was independent of ROS generation. For example, overexpression of thioredoxin reductase blocked hypoxic generation of ROS but had no effect on hypoxia-dependent HIF-1α stabilization (Naranjo-Suarez et al. 2012).

However, it was also suggested that the activity of the mitochondrial respiratory chain itself was responsible for decreased hypoxic induction of HIF-1α in the presence of mitochondrial respiratory chain inhibitors since mitochondria are the major cellular sink for oxygen (Wenger 2006; Doege et al. 2005; Hagen et al. 2003). Accordingly, it was proposed that decreased electron transport chain activity of mitochondria induced by various inhibitors results in increased cytoplasmic oxygen concentration which in turn leads to PHD reactivation and destabilization of HIF-1α or HIF-2α (Chua et al. 2010). In addition, overexpression of an alternative oxidase which transfers electrons from coenzyme Q to oxygen to form water, thus bypassing mitochondrial complex III, decreased superoxide production from complex III while oxygen consumption was maintained. Interestingly, these manipulations did not affect HIF-1α stabilization in hypoxia suggesting that complex III-derived ROS do not contribute to hypoxic induction of HIF-1α. Finally, it has been shown that inhibition of the mitochondrial F0F1-ATPase increases the mitochondrial membrane potential and slows down electron transport concomitant with increased superoxide production while it prevents hypoxic stabilization of HIF-1α (Gong and Agani 2005). In support of this, CHCHD4, which stands for Coiled-Coil-Helix-Coiled-Coil-Helix Domain Containing 4, has been recently identified to regulate mitochondrial oxygen consumption (Yang et al. 2012). Thus, overexpression of CHDHD4 increased mitochondrial respiratory rates which results in decreased intracellular oxygen concentrations, and increased HIF-1α stabilization in hypoxia independent of the presence of antioxidants.

Under aerobic conditions, cells have been reported to switch to COX4-1 regulatory subunit of cytochrome c oxidase (complex IV) but to the COX4-2 subunit under hypoxic conditions, and this switch is transcriptionally regulated by HIF-1. Interestingly, COX4-2 was able to optimize electron flux under hypoxic conditions and did not promote ROS generation. Thus, it was suggested that under conditions of chronic hypoxia, complex IV can become a source of increased ROS production only if the COX4 subunit switch does not occur (Fukuda et al. 2007).

In support of this, when isolated mitochondria were maintained at low O_2 concentrations, ROS production decreased as the oxygen concentration was lowered from approx. 5 μM O_2 to anoxia (Hoffman et al. 2007). This finding suggests that the decrease in oxygen concentration itself does not affect $O_2^{-}\bullet$ production directly, for example, by altering the stability of the ubisemiquinone radical at complex III, the rate of degradation or release of H_2O_2 from mitochondria or the sidedness of $O_2^{-}\bullet$ release from complex III. In addition, mitochondria isolated from the thorax of hypoxia-adapted flies showed decreased respiratory rates during state 3 and state 4 with a sharp decrease in complex II activity and modest increases in complexes I and III activity and a 60 % reduction in superoxide leakage from mitochondria (Ali et al. 2012).

Subsequently, various studies reported that under hypoxia ROS concentrations possibly derived from mitochondria were decreased (Hoffman et al. 2007; Hoffman and Brookes 2009; Wang et al. 2008; Jung et al. 2008; Mehta et al. 2008; Tuttle et al. 2007). Concomitant with decreased mitochondrial ROS production HIF-1α was stabilized under hypoxia while upon reoxygenation, ROS levels increased in contrast to HIF-1α levels (Wang et al. 2008; Archer and Michelakis 2002).

4.8 NADPH Oxidases in the Hypoxic Environment

In addition to mitochondrial ROS generation, NADPH oxidases have been also investigated with regard to their role in hypoxic ROS and HIF signaling using mostly in vivo models.

A variety of studies was performed investigating the situation under conditions of hypoxia/ischemia and reoxygenation or intermittent hypoxia, subsuming that increased ROS generation under reoxygenated conditions would derive from NADPH oxidases.

Interestingly, a variety of ischemia-reoxygenation animal models associated, for example, with stroke or cardiac infarction, and diverse protocols of intermittent hypoxia-simulating sleep apnea were able to show increased ROS levels in various tissues of relevance such as brain, heart, and lungs. Using primarily NOX2-/- or p47phox-/- mouse models, ROS levels and tissue damage were shown to be diminished in the majority of cases (Doerries et al. 2007; Murdoch et al. 2006; Kleikers et al. 2012; Nair et al. 2011). While in many of those models, HIF-α protein levels have not been determined, in lung tumor cells NOX1 was shown to contribute to upregulation of HIF-1α in response to intermittent hypoxia although to a much lesser extent than to hypoxic upregulation of HIF-1α (Malec et al. 2010). NOX4 was shown to contribute not only to ROS generation in response to cyclic hypoxia in different brain tumors, but also to tumor progression and radioresistance under these conditions, similar to HIF-1α (Hsieh et al. 2012). In hindlimb ischemia, NOX2 mediated HIF-1α regulation in the bone marrow (Urao et al. 2012). In addition, NOX2 was also shown to contribute to enhanced HIF-1α levels in the carotid body and in PC cells in response to intermittent hypoxia (for review, see Prabhakar and Semenza 2012).

In many cases, these observations were accompanied by increased levels of NADPH oxidase subunits including p22phox, p47phox, and different NOX proteins. Since it has been shown that several NADPH oxidase subunits can be induced by ROS, including p22phox, NOX2, NOX4, and Rac1 (Diebold et al. 2009, 2012; Djordjevic et al. 2005a), increased ROS levels in the reoxygenation/reperfusion periods may be responsible for such an upregulation thereby promoting sustained ROS generation under these conditions.

Interestingly, however, it was also shown that NADPH oxidases play a role in the response to acute or chronic hypoxia in different animal models. For example, compared to wild-type animals exposed to chronic hypoxia, ROS production was decreased in pulmonary arteries from NOX2–/– or gp91phox–/– mice concomitant with decreased right ventricular hypertrophy and pulmonary vascular remodeling (Liu et al. 2006a). Similarly, hypoxia-induced ROS levels were reduced in p47phox-deficient carotid body cells (He et al. 2005). However, in NOX2–/– mice, the hypoxic response in glomus cells was not altered compared to wild-type mice (He et al. 2002; Roy et al. 2000; Archer et al. 1999).

On the other hand, in perfused lungs, $O_2^-\bullet$ release was decreased upon hypoxia in wild-type animals as determined by EPR. Interestingly, p47phox deficiency already decreased pulmonary $O_2^{-}\bullet$ levels under normoxia and this decrease was not further changed under hypoxic conditions (Weissmann et al. 2006). While no measurements of HIF-α levels have been performed in these studies, the discrepant findings in the regulation of ROS levels in response to hypoxia, in particular, in the in vivo situation might be explained by our observations that vessels exposed to hypoxia for longer time periods show decreased ROS generation when ROS measurements were performed under complete hypoxic conditions (unpublished observation). However, when ROS measurements were performed under normoxic conditions after exposure to hypoxia, which mimics the situation under hypoxia/reoxygenation conditions, ROS levels were increased. Similarly, ROS levels were elevated in vessels and organs derived from mice exposed to 2 weeks of hypoxia where sample preparation had to be performed under normoxic conditions (unpublished observation), suggesting that even short normoxic periods following exposure to hypoxia can exacerbate ROS release similar to reoxygenation.

In this regard, it has been reported that hypoxia can increase the levels of several NADPH oxidase subunits, including NOX4 (Diebold et al. 2010; Mittal et al. 2007). When ROS levels were determined under normoxia following a hypoxic incubation period maximizing NOX4 levels, ROS levels were elevated compared to normoxic conditions. This response was diminished when NOX4 was depleted by shRNA. However, when ROS levels were determined by the same method in cells exposed to short-term hypoxia not sufficient to induce NOX4, ROS levels were lower than in normoxic cells. Overexpression of NOX4 restored ROS levels to normoxic values suggesting that induction of NOX4 and possibly other NADPH oxidase subunits under hypoxic conditions may be of importance to maintain ROS levels under hypoxia and to allow enhanced ROS generation under reoxygenation conditions (Diebold et al. 2010). In addition, hypoxic induction of NOX4 has been shown to also contribute to hypoxic HIF1α upregulation in different cell types similar to the situation under normoxia (Diebold et al. 2010; Li et al. 2010). Strikingly, depletion of this transcription factor prevented adaptive ROS production after hypoxia–reoxygenation similar to the situation with NOX4 depletion. Subsequent studies including site-directed mutagenesis of a putative HIF-binding site in the NOX4 promoter and chromatin immunoprecipitation revealed that NOX4 is a HIF-α target gene (Diebold et al. 2010). Subsequently, functional hypoxia response elements were identified in the human NOX2 and Rac1 promoters indicating that NADPH oxidases as genuine HIF target genes are involved in the adaptive response to hypoxia (Diebold et al. 2008, 2012). Experiments in mice deficient in endothelial HIF-1α confirmed the relevance of this transcription factor in the regulation of NADPH oxidases (Diebold et al. 2012). Although the exact importance of hypoxia and HIF-dependent upregulation of NADPH oxidases is still not clarified, one may speculate that a certain level of ROS is important for maintaining basal cellular functions under oxygen-limited conditions and may thus help to protect against apoptosis or cell death at least at the cellular level. Furthermore, in leukocytes and other immune cells, induction of NADPH oxidases under hypoxic conditions may contribute to initiation and propagation of inflammatory conditions frequently associated with conditions of oxygen

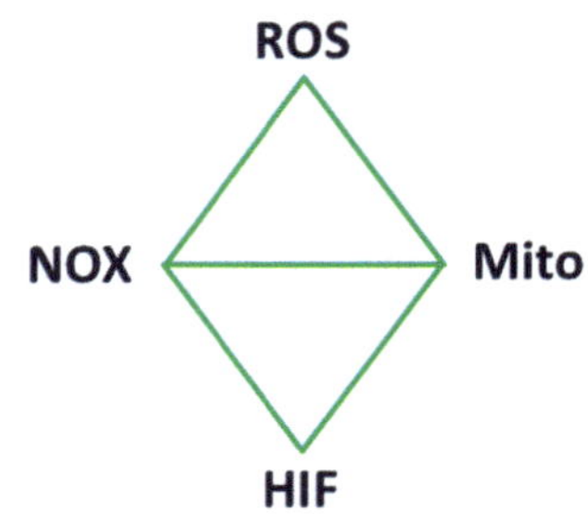

Fig. 4.3 Cross talk between HIF and ROS involving NOX and mitochondria. Reactive oxygen species (ROS) derived from NADPH oxidases (NOX) or mitochondria respiratory chain (Mito) can activate HIF transcription factors which on the other hand can regulate NOX and mitochondrial respiratory chain products

deficiency. This would also explain the protective effects of NADPH oxidase deficiency seen in different animal models of intermittent and sustained hypoxia. Novel findings indicating a cross talk of NADPH oxidases with mitochondrial activity or even suggesting mitochondrial NADPH oxidase activity may further complicate the picture (Schulz et al. 2012).

In essence, the increasingly tightening link between hypoxia, HIF, ROS, mitochondria, and NADPH oxidases provides intriguing insights into the complex mechanisms of redox regulation of the HIF pathway under different conditions (Fig. 4.3). Clearly, the HIF pathway is a central adaptive mechanism taking place in response to hypoxia, and HIF itself also acts an effector of ROS-regulated signaling pathways under various stress conditions. Thus, redox-sensitive HIF activation may play an important role in many disorders characterized not only by hypoxia but also by hypoxia/reoxygenation events. The increasing awareness that hypoxia is not a stable status in tumors, but that fluctuating conditions between hypoxia and reoxygenation most probably are more prevalent in vivo, further draw attention to the relevance of the ROS–HIF cross in cancer.

Acknowledgments This work was supported by DFG GO709/4-5 and the Seventh European Framework Programme (Metoxia).

References

Agani FH, Pichiule P, Chavez JC, LaManna JC (2000) The role of mitochondria in the regulation of hypoxia-inducible factor 1 expression during hypoxia. J Biol Chem 275:35863–35867

Ali SS, Hsiao M, Zhao HW, Dugan LL, Haddad GG et al (2012) Hypoxia-adaptation involves mitochondrial metabolic depression and decreased ROS leakage. PLoS One 7:e36801

Aon MA, Cortassa S, O'Rourke B (2010) Redox-optimized ROS balance: a unifying hypothesis. Biochim Biophys Acta 1797:865–877

Archer S, Michelakis E (2002) The mechanism(s) of hypoxic pulmonary vasoconstriction: potassium channels, redox O(2) sensors, and controversies. News Physiol Sci 17:131–137

Archer SL, Reeve HL, Michelakis E, Puttagunta L, Waite R et al (1999) O2 sensing is preserved in mice lacking the gp91 phox subunit of NADPH oxidase. Proc Natl Acad Sci USA 96:7944–7949

Babior BM (2004) NADPH oxidase. Curr Opin Immunol 16:42–47

Bae SH, Jeong JW, Park JA, Kim SH, Bae MK et al (2004) Sumoylation increases HIF-1alpha stability and its transcriptional activity. Biochem Biophys Res Commun 324:394–400

Beck I, Weinmann R, Caro J (1993) Characterization of hypoxia-responsive enhancer in the human erythropoietin gene shows presence of hypoxia-inducible 120-Kd nuclear DNA-binding protein in erythropoietin-producing and nonproducing cells. Blood 82:704–711

Bedard K, Krause KH (2007) The NOX family of ROS-generating NADPH oxidases: physiology and pathophysiology. Physiol Rev 87:245–313

BelAiba RS, Djordjevic T, Bonello S, Flugel D, Hess J et al (2004) Redox-sensitive regulation of the HIF pathway under non-hypoxic conditions in pulmonary artery smooth muscle cells. Biol Chem 385:249–257

Belaiba RS, Bonello S, Zahringer C, Schmidt S, Hess J et al (2007a) Hypoxia up-regulates hypoxia-inducible factor-1alpha transcription by involving phosphatidylinositol 3-kinase and nuclear factor kappaB in pulmonary artery smooth muscle cells. Mol Biol Cell 18:4691–4697

BelAiba RS, Djordjevic T, Petry A, Diemer K, Bonello S et al (2007b) NOX5 variants are functionally active in endothelial cells. Free Radic Biol Med 42:446–459

Bell EL, Klimova TA, Eisenbart J, Moraes CT, Murphy MP et al (2007) The Qo site of the mitochondrial complex III is required for the transduction of hypoxic signaling via reactive oxygen species production. J Cell Biol 177:1029–1036

Berndt C, Lillig CH, Holmgren A (2007) Thiol-based mechanisms of the thioredoxin and glutaredoxin systems: implications for diseases in the cardiovascular system. Am J Physiol Heart Circ Physiol 292:H1227–H1236

Block K, Gorin Y (2012) Aiding and abetting roles of NOX oxidases in cellular transformation. Nat Rev Cancer 12:627–637

Block K, Gorin Y, New DD, Eid A, Chelmicki T et al (2010) The NADPH oxidase subunit p22phox inhibits the function of the tumor suppressor protein tuberin. Am J Pathol 176:2447–2455

Bonello S, Zahringer C, BelAiba RS, Djordjevic T, Hess J et al (2007) Reactive oxygen species activate the HIF-1alpha promoter via a functional NFkappaB site. Arterioscler Thromb Vasc Biol 27:755–761

Boyer RF, McCleary CJ (1987) Superoxide ion as a primary reductant in ascorbate-mediated ferritin iron release. Free Radic Biol Med 3:389–395

Brown DI, Griendling KK (2009) Nox proteins in signal transduction. Free Radic Biol Med 47:1239–1253

Brown ST, Nurse CA (2008) Induction of HIF-2alpha is dependent on mitochondrial O2 consumption in an O2-sensitive adrenomedullary chromaffin cell line. Am J Physiol Cell Physiol 294:C1305–C1312

Carnesecchi S, Carpentier JL, Foti M, Szanto I (2006) Insulin-induced vascular endothelial growth factor expression is mediated by the NADPH oxidase NOX3. Exp Cell Res 312:3413–3424

Carrero P, Okamoto K, Coumailleau P, O'Brien S, Tanaka H et al (2000) Redox-regulated recruitment of the transcriptional coactivators CREB-binding protein and SRC-1 to hypoxia-inducible factor 1alpha. Mol Cell Biol 20:402–415

Chamboredon S, Ciais D, Desroches-Castan A, Savi P, Bono F et al (2011) Hypoxia-inducible factor-1alpha mRNA: a new target for destabilization by tristetraprolin in endothelial cells. Mol Biol Cell 22:3366–3378

Chance B, Sies H, Boveris A (1979) Hydroperoxide metabolism in mammalian organs. Physiol Rev 59:527–605

Chandel NS, McClintock DS, Feliciano CE, Wood TM, Melendez JA et al (2000) Reactive oxygen species generated at mitochondrial complex III stabilize hypoxia-inducible factor-1alpha during hypoxia: a mechanism of O2 sensing. J Biol Chem 275:25130–25138

Chelikani P, Fita I, Loewen PC (2004) Diversity of structures and properties among catalases. Cell Mol Life Sci 61:192–208

Chen R, Liliental JE, Kowalski PE, Lu Q, Cohen SN (2011) Regulation of transcription of hypoxia-inducible factor-1alpha (HIF-1alpha) by heat shock factors HSF2 and HSF4. Oncogene 30:2570–2580

Chua YL, Dufour E, Dassa EP, Rustin P, Jacobs HT et al (2010) Stabilization of hypoxia-inducible factor-1alpha protein in hypoxia occurs independently of mitochondrial reactive oxygen species production. J Biol Chem 285:31277–31284
Deshmane SL, Mukerjee R, Fan S, Del Valle L, Michiels C et al (2009) Activation of the oxidative stress pathway by HIV-1 Vpr leads to induction of hypoxia-inducible factor 1alpha expression. J Biol Chem 284:11364–11373
Diebold I, Djordjevic T, Hess J, Görlach A (2008) Rac-1 promotes pulmonary artery smooth muscle cell proliferation by upregulation of plasminogen activator inhibitor-1: role of NFkappaB-dependent hypoxia-inducible factor-1alpha transcription. Thromb Haemost 100:1021–1028
Diebold I, Djordjevic T, Petry A, Hatzelmann A, Tenor H et al (2009) Phosphodiesterase 2 mediates redox-sensitive endothelial cell proliferation and angiogenesis by thrombin via Rac1 and NADPH oxidase 2. Circ Res 104:1169–1177
Diebold I, Flugel D, Becht S, Belaiba RS, Bonello S et al (2010a) The hypoxia-inducible factor-2alpha is stabilized by oxidative stress involving NOX4. Antioxid Redox Signal 13:425–436
Diebold I, Petry A, Djordjevic T, Belaiba RS, Fineman J et al (2010b) Reciprocal regulation of Rac1 and PAK-1 by HIF-1alpha: a positive-feedback loop promoting pulmonary vascular remodeling. Antioxid Redox Signal 13:399–412
Diebold I, Petry A, Hess J, Gorlach A (2010c) The NADPH oxidase subunit NOX4 is a new target gene of the hypoxia-inducible factor-1. Mol Biol Cell 21:2087–2096
Diebold I, Petry A, Sabrane K, Djordjevic T, Hess J et al (2012) The HIF1 target gene NOX2 promotes angiogenesis through urotensin-II. J Cell Sci 125:956–964
Dioum EM, Chen R, Alexander MS, Zhang Q, Hogg RT et al (2009) Regulation of hypoxia-inducible factor 2alpha signaling by the stress-responsive deacetylase sirtuin 1. Science 324:1289–1293
Djordjevic T, BelAiba RS, Bonello S, Pfeilschifter J, Hess J et al (2005a) Human urotensin II is a novel activator of NADPH oxidase in human pulmonary artery smooth muscle cells. Arterioscler Thromb Vasc Biol 25:519–525
Djordjevic T, Pogrebniak A, BelAiba RS, Bonello S, Wotzlaw C et al (2005b) The expression of the NADPH oxidase subunit p22phox is regulated by a redox-sensitive pathway in endothelial cells. Free Radic Biol Med 38:616–630
Doege K, Heine S, Jensen I, Jelkmann W, Metzen E (2005) Inhibition of mitochondrial respiration elevates oxygen concentration but leaves regulation of hypoxia-inducible factor (HIF) intact. Blood 106:2311–2317
Doerries C, Grote K, Hilfiker-Kleiner D, Luchtefeld M, Schaefer A et al (2007) Critical role of the NAD(P)H oxidase subunit p47phox for left ventricular remodeling/dysfunction and survival after myocardial infarction. Circ Res 100:894–903
Ema M, Hirota K, Mimura J, Abe H, Yodoi J et al (1999) Molecular mechanisms of transcription activation by HLF and HIF1alpha in response to hypoxia: their stabilization and redox signal-induced interaction with CBP/p300. EMBO J 18:1905–1914
Epstein AC, Gleadle JM, McNeill LA, Hewitson KS, O'Rourke J et al (2001) C. elegans EGL-9 and mammalian homologs define a family of dioxygenases that regulate HIF by prolyl hydroxylation. Cell 107:43–54
Fatrai S, Wierenga AT, Daenen SM, Vellenga E, Schuringa JJ (2011) Identification of HIF2alpha as an important STAT5 target gene in human hematopoietic stem cells. Blood 117:3320–3330
Flashman E, Davies SL, Yeoh KK, Schofield CJ (2010) Investigating the dependence of the hypoxia-inducible factor hydroxylases (factor inhibiting HIF and prolyl hydroxylase domain 2) on ascorbate and other reducing agents. Biochem J 427:135–142
Frede S, Stockmann C, Freitag P, Fandrey J (2006) Bacterial lipopolysaccharide induces HIF-1 activation in human monocytes via p44/42 MAPK and NF-kappaB. Biochem J 396:517–527
Fukuda R, Kelly B, Semenza GL (2003) Vascular endothelial growth factor gene expression in colon cancer cells exposed to prostaglandin E2 is mediated by hypoxia-inducible factor 1. Cancer Res 63:2330–2334
Fukuda R, Zhang H, Kim JW, Shimoda L, Dang CV et al (2007) HIF-1 regulates cytochrome oxidase subunits to optimize efficiency of respiration in hypoxic cells. Cell 129:111–122

Galban S, Gorospe M (2009) Factors interacting with HIF-1alpha mRNA: novel therapeutic targets. Curr Pharm Des 15:3853–3860

Galban S, Kuwano Y, Pullmann R Jr, Martindale JL, Kim HH et al (2008) RNA-binding proteins HuR and PTB promote the translation of hypoxia-inducible factor 1alpha. Mol Cell Biol 28:93–107

Gerald D, Berra E, Frapart YM, Chan DA, Giaccia AJ et al (2004) JunD reduces tumor angiogenesis by protecting cells from oxidative stress. Cell 118:781–794

Gong Y, Agani FH (2005) Oligomycin inhibits HIF-1alpha expression in hypoxic tumor cells. Am J Physiol Cell Physiol 288:C1023–C1029

Gorlach A (2009) Regulation of HIF-1 at the Transcriptional Level. Curr Pharm Des 15(33):3844–3852

Görlach A, Bonello S (2008) The cross-talk between NF-kappaB and HIF-1: further evidence for a significant liaison. Biochem J 412:e17–19

Görlach A, Kietzmann T (2007) Superoxide and derived reactive oxygen species in the regulation of hypoxia-inducible factors. Methods Enzymol 435:421–446

Gorlach A, Brandes RP, Nguyen K, Amidi M, Dehghani F et al (2000a) A gp91phox containing NADPH oxidase selectively expressed in endothelial cells is a major source of oxygen radical generation in the arterial wall. Circ Res 87:26–32

Görlach A, Camenisch G, Kvietikova I, Vogt L, Wenger RH et al (2000b) Efficient translation of mouse hypoxia-inducible factor-1alpha under normoxic and hypoxic conditions. Biochim Biophys Acta 1493:125–134

Görlach A, Diebold I, Schini-Kerth VB, Berchner-Pfannschmidt U, Roth U et al (2001) Thrombin activates the hypoxia-inducible factor-1 signaling pathway in vascular smooth muscle cells: Role of the p22(phox)-containing NADPH oxidase. Circ Res 89:47–54

Görlach A, Kietzmann T, Hess J (2002) Redox signaling through NADPH oxidases: involvement in vascular proliferation and coagulation. Ann NY Acad Sci 973:505–507

Goyal P, Weissmann N, Grimminger F, Hegel C, Bader L et al (2004) Upregulation of NAD(P)H oxidase 1 in hypoxia activates hypoxia-inducible factor 1 via increase in reactive oxygen species. Free Radic Biol Med 36:1279–1288

Gu YZ, Moran SM, Hogenesch JB, Wartman L, Bradfield CA (1998) Molecular characterization and chromosomal localization of a third alpha-class hypoxia inducible factor subunit, HIF3alpha. Gene Expr 7:205–213

Guzy RD, Sharma B, Bell E, Chandel NS, Schumacker PT (2008) Loss of the SdhB, but Not the SdhA, subunit of complex II triggers reactive oxygen species-dependent hypoxia-inducible factor activation and tumorigenesis. Mol Cell Biol 28:718–731

Haddad JJ, Land SC (2001) A non-hypoxic, ROS-sensitive pathway mediates TNF-alpha-dependent regulation of HIF-1alpha. FEBS Lett 505:269–274

Haddad JJ, Olver RE, Land SC (2000) Antioxidant/pro-oxidant equilibrium regulates HIF-1alpha and NF-kappa B redox sensitivity. Evidence for inhibition by glutathione oxidation in alveolar epithelial cells. J Biol Chem 275:21130–21139

Hagele S, Kuhn U, Boning M, Katschinski DM (2009) Cytoplasmic polyadenylation-element-binding protein (CPEB) 1 and 2 bind to the HIF-1alpha mRNA 3'-UTR and modulate HIF-1alpha protein expression. Biochem J 417:235–246

Hagen T, Taylor CT, Lam F, Moncada S (2003) Redistribution of intracellular oxygen in hypoxia by nitric oxide: effect on HIF1alpha. Science 302:1975–1978

Hamanaka RB, Chandel NS (2009) Mitochondrial reactive oxygen species regulate hypoxic signaling. Curr Opin Cell Biol 21:894–899

Han D, Antunes F, Canali R, Rettori D, Cadenas E (2003) Voltage-dependent anion channels control the release of the superoxide anion from mitochondria to cytosol. J Biol Chem 278:5557–5563

Hawkins BJ, Madesh M, Kirkpatrick CJ, Fisher AB (2007) Superoxide flux in endothelial cells via the chloride channel-3 mediates intracellular signaling. Mol Biol Cell 18:2002–2012

He L, Chen J, Dinger B, Sanders K, Sundar K et al (2002) Characteristics of carotid body chemosensitivity in NADPH oxidase-deficient mice. Am J Physiol Cell Physiol 282:C27–C33

He L, Dinger B, Sanders K, Hoidal J, Obeso A et al (2005) Effect of p47phox gene deletion on ROS production and oxygen sensing in mouse carotid body chemoreceptor cells. Am J Physiol Lung Cell Mol Physiol 289:L916–L924
Hirota K, Fukuda R, Takabuchi S, Kizaka-Kondoh S, Adachi T et al (2004) Induction of hypoxia-inducible factor 1 activity by muscarinic acetylcholine receptor signaling. J Biol Chem 279:41521–41528
Hoffman DL, Brookes PS (2009) Oxygen sensitivity of mitochondrial reactive oxygen species generation depends on metabolic conditions. J Biol Chem 284:16236–16245
Hoffman EC, Reyes H, Chu FF, Sander F, Conley LH et al (1991) Cloning of a factor required for activity of the Ah (dioxin) receptor. Science 252:954–958
Hoffman DL, Salter JD, Brookes PS (2007) Response of mitochondrial reactive oxygen species generation to steady-state oxygen tension: implications for hypoxic cell signaling. Am J Physiol Heart Circ Physiol 292:H101–H108
Hsieh CH, Wu CP, Lee HT, Liang JA, Yu CY et al (2012) NADPH oxidase subunit 4 mediates cycling hypoxia-promoted radiation resistance in glioblastoma multiforme. Free Radic Biol Med 53:649–658
Huang LE, Arany Z, Livingston DM, Bunn HF (1996) Activation of hypoxia-inducible transcription factor depends primarily upon redox-sensitive stabilization of its alpha subunit. J Biol Chem 271:32253–32259
Huang LE, Gu J, Schau M, Bunn HF (1998) Regulation of hypoxia-inducible factor 1alpha is mediated by an O2-dependent degradation domain via the ubiquitin-proteasome pathway. Proc Natl Acad Sci USA 95:7987–7992
Huang C, Han Y, Wang Y, Sun X, Yan S et al (2009) SENP3 is responsible for HIF-1 transactivation under mild oxidative stress via p300 de-SUMOylation. EMBO J 28:2748–2762
Hui AS, Bauer AL, Striet JB, Schnell PO, Czyzyk-Krzeska MF (2006) Calcium signaling stimulates translation of HIF-alpha during hypoxia. FASEB J 20:466–475
Hybertson BM, Gao B, Bose SK, McCord JM (2011) Oxidative stress in health and disease: the therapeutic potential of Nrf2 activation. Mol Aspects Med 32:234–246
Ivan M, Kondo K, Yang H, Kim W, Valiando J et al (2001) HIFalpha targeted for VHL-mediated destruction by proline hydroxylation: implications for O2 sensing. Science 292:464–468
Jaakkola P, Mole DR, Tian YM, Wilson MI, Gielbert J et al (2001) Targeting of HIF-alpha to the von Hippel-Lindau ubiquitylation complex by O2-regulated prolyl hydroxylation. Science 292:468–472
Jiang BH, Zheng JZ, Leung SW, Roe R, Semenza GL (1997) Transactivation and inhibitory domains of hypoxia-inducible factor 1alpha. Modulation of transcriptional activity by oxygen tension. J Biol Chem 272:19253–19260
Jung SN, Yang WK, Kim J, Kim HS, Kim EJ et al (2008) Reactive oxygen species stabilize hypoxia-inducible factor-1 alpha protein and stimulate transcriptional activity via AMP-activated protein kinase in DU145 human prostate cancer cells. Carcinogenesis 29:713–721
Kaelin WG Jr, Ratcliffe PJ (2008) Oxygen sensing by metazoans: the central role of the HIF hydroxylase pathway. Mol Cell 30:393–402
Khatri JJ, Johnson C, Magid R, Lessner SM, Laude KM et al (2004) Vascular oxidant stress enhances progression and angiogenesis of experimental atheroma. Circulation 109:520–525
Kietzmann T, Samoylenko A, Roth U, Jungermann K (2003) Hypoxia-inducible factor-1 and hypoxia response elements mediate the induction of plasminogen activator inhibitor-1 gene expression by insulin in primary rat hepatocytes. Blood 101:907–914
Kim HH, Lee SE, Chung WJ, Choi Y, Kwack K et al (2002) Stabilization of hypoxia-inducible factor-1alpha is involved in the hypoxic stimuli-induced expression of vascular endothelial growth factor in osteoblastic cells. Cytokine 17:14–27
Kleikers PW, Wingler K, Hermans JJ, Diebold I, Altenhofer S et al (2012) NADPH oxidases as a source of oxidative stress and molecular target in ischemia/reperfusion injury. J Mol Med (Berl) 90:1391–1406

Knowles HJ, Raval RR, Harris AL, Ratcliffe PJ (2003) Effect of ascorbate on the activity of hypoxia-inducible factor in cancer cells. Cancer Res 63:1764–1768

Kohl R, Zhou J, Brune B (2006) Reactive oxygen species attenuate nitric-oxide-mediated hypoxia-inducible factor-1alpha stabilization. Free Radic Biol Med 40:1430–1442

Koivunen P, Tiainen P, Hyvarinen J, Williams KE, Sormunen R et al (2007) An endoplasmic reticulum transmembrane prolyl 4-hydroxylase is induced by hypoxia and acts on hypoxia-inducible factor alpha. J Biol Chem 282:30544–30552

Koshikawa N, Hayashi J, Nakagawara A, Takenaga K (2009) Reactive oxygen species-generating mitochondrial DNA mutation up-regulates hypoxia-inducible factor-1alpha gene transcription via phosphatidylinositol 3-kinase-Akt/protein kinase C/histone deacetylase pathway. J Biol Chem 284:33185–33194

Lambeth JD, Kawahara T, Diebold B (2007) Regulation of Nox and Duox enzymatic activity and expression. Free Radic Biol Med 43:319–331

Lando D, Pongratz I, Poellinger L, Whitelaw ML (2000) A redox mechanism controls differential DNA binding activities of hypoxia-inducible factor (HIF) 1alpha and the HIF-like factor. J Biol Chem 275:4618–4627

Lang KJ, Kappel A, Goodall GJ (2002) Hypoxia-inducible factor-1alpha mRNA contains an internal ribosome entry site that allows efficient translation during normoxia and hypoxia. Mol Biol Cell 13:1792–1801

Laughner E, Taghavi P, Chiles K, Mahon PC, Semenza GL (2001) HER2 (neu) signaling increases the rate of hypoxia-inducible factor 1alpha (HIF-1alpha) synthesis: novel mechanism for HIF-1-mediated vascular endothelial growth factor expression. Mol Cell Biol 21:3995–4004

Li F, Sonveaux P, Rabbani ZN, Liu S, Yan B et al (2007) Regulation of HIF-1alpha stability through S-nitrosylation. Mol Cell 26:63–74

Li J, Wang JJ, Yu Q, Chen K, Mahadev K et al (2010) Inhibition of reactive oxygen species by Lovastatin downregulates vascular endothelial growth factor expression and ameliorates blood-retinal barrier breakdown in db/db mice: role of NADPH oxidase 4. Diabetes 59:1528–1538

Liu Q, Berchner-Pfannschmidt U, Moller U, Brecht M, Wotzlaw C et al (2004) A Fenton reaction at the endoplasmic reticulum is involved in the redox control of hypoxia-inducible gene expression. Proc Natl Acad Sci USA 101:4302–4307

Liu JQ, Zelko IN, Erbynn EM, Sham JS, Folz RJ (2006a) Hypoxic pulmonary hypertension: role of superoxide and NADPH oxidase (gp91phox). Am J Physiol Lung Cell Mol Physiol 290:L2–L10

Liu LZ, Hu XW, Xia C, He J, Zhou Q et al (2006b) Reactive oxygen species regulate epidermal growth factor-induced vascular endothelial growth factor and hypoxia-inducible factor-1alpha expression through activation of AKT and P70S6K1 in human ovarian cancer cells. Free Radic Biol Med 41:1521–1533

Mahon PC, Hirota K, Semenza GL (2001) FIH-1: a novel protein that interacts with HIF-1alpha and VHL to mediate repression of HIF-1 transcriptional activity. Genes Dev 15:2675–2686

Maier CM, Chan PH (2002) Role of superoxide dismutases in oxidative damage and neurodegenerative disorders. Neuroscientist 8:323–334

Makino Y, Cao R, Svensson K, Bertilsson G, Asman M et al (2001) Inhibitory PAS domain protein is a negative regulator of hypoxia-inducible gene expression. Nature 414:550–554

Malec V, Gottschald OR, Li S, Rose F, Seeger W et al (2010) HIF-1 alpha signaling is augmented during intermittent hypoxia by induction of the Nrf2 pathway in NOX1-expressing adenocarcinoma A549 cells. Free Radic Biol Med 48:1626–1635

Maranchie JK, Zhan Y (2005) Nox4 is critical for hypoxia-inducible factor 2-alpha transcriptional activity in von Hippel-Lindau-deficient renal cell carcinoma. Cancer Res 65:9190–9193

Mecinovic J, Chowdhury R, Flashman E, Schofield CJ (2009) Use of mass spectrometry to probe the nucleophilicity of cysteinyl residues of prolyl hydroxylase domain 2. Anal Biochem 393:215–221

Mehta JP, Campian JL, Guardiola J, Cabrera JA, Weir EK et al (2008) Generation of oxidants by hypoxic human pulmonary and coronary smooth-muscle cells. Chest 133:1410–1414

Mittal M, Roth M, Konig P, Hofmann S, Dony E et al (2007) Hypoxia-dependent regulation of nonphagocytic NADPH oxidase subunit NOX4 in the pulmonary vasculature. Circ Res 101:258–267

Moon EJ, Sonveaux P, Porporato PE, Danhier P, Gallez B et al (2010) NADPH oxidase-mediated reactive oxygen species production activates hypoxia-inducible factor-1 (HIF-1) via the ERK pathway after hyperthermia treatment. Proc Natl Acad Sci USA 107:20477–20482

Muller FL, Lustgarten MS, Jang Y, Richardson A, Van Remmen H (2007) Trends in oxidative aging theories. Free Radic Biol Med 43:477–503

Murdoch CE, Grieve DJ, Cave AC, Looi YH, Shah AM (2006) NADPH oxidase and heart failure. Curr Opin Pharmacol 6:148–153

Murphy MP (2009) How mitochondria produce reactive oxygen species. Biochem J 417:1–13

Nair D, Dayyat EA, Zhang SX, Wang Y, Gozal D (2011) Intermittent hypoxia-induced cognitive deficits are mediated by NADPH oxidase activity in a murine model of sleep apnea. PLoS One 6:e19847

Naranjo-Suarez S, Carlson BA, Tsuji PA, Yoo MH, Gladyshev VN et al (2012) HIF-independent regulation of thioredoxin reductase 1 contributes to the high levels of reactive oxygen species induced by hypoxia. PLoS One 7:e30470

Nayak BK, Feliers D, Sudarshan S, Friedrichs WE, Day RT et al (2012) Stabilization of HIF-2alpha through redox regulation of mTORC2 activation and initiation of mRNA translation. Oncogene 32(25):3147–3155

Nishi K, Oda T, Takabuchi S, Oda S, Fukuda K et al (2008) LPS induces hypoxia-inducible factor 1 activation in macrophage-differentiated cells in a reactive oxygen species-dependent manner. Antioxid Redox Signal 10:983–995

Nytko KJ, Maeda N, Schlafli P, Spielmann P, Wenger RH et al (2011) Vitamin C is dispensable for oxygen sensing in vivo. Blood 117:5485–5493

Oh YT, Lee JY, Yoon H, Lee EH, Baik HH et al (2008) Lipopolysaccharide induces hypoxia-inducible factor-1 alpha mRNA expression and activation via NADPH oxidase and Sp1-dependent pathway in BV2 murine microglial cells. Neurosci Lett 431:155–160

Page EL, Robitaille GA, Pouyssegur J, Richard DE (2002) Induction of hypoxia-inducible factor-1alpha by transcriptional and translational mechanisms. J Biol Chem 277:48403–48409

Page EL, Chan DA, Giaccia AJ, Levine M, Richard DE (2008) Hypoxia-inducible factor-1alpha stabilization in nonhypoxic conditions: role of oxidation and intracellular ascorbate depletion. Mol Biol Cell 19:86–94

Patten DA, Lafleur VN, Robitaille GA, Chan DA, Giaccia AJ et al (2010) Hypoxia-inducible factor-1 activation in nonhypoxic conditions: the essential role of mitochondrial-derived reactive oxygen species. Mol Biol Cell 21:3247–3257

Petry A, Weitnauer M, Gorlach A (2010) Receptor activation of NADPH oxidases. Antioxid Redox Signal 13:467–487

Petry A, Belaiba RS, Weitnauer M, Görlach A (2012) Inhibition of endothelial nitric oxyde synthase increases capillary formation via Rac1-dependent induction of hypoxia-inducible factor-1alpha and plasminogen activator inhibitor-1. Thromb Haemost 108:849–862

Prabhakar NR, Semenza GL (2012) Adaptive and maladaptive cardiorespiratory responses to continuous and intermittent hypoxia mediated by hypoxia-inducible factors 1 and 2. Physiol Rev 92:967–1003

Pugh CW, O'Rourke JF, Nagao M, Gleadle JM, Ratcliffe PJ (1997) Activation of hypoxia-inducible factor-1; definition of regulatory domains within the alpha subunit. J Biol Chem 272:11205–11214

Richard DE, Berra E, Pouyssegur J (2000) Nonhypoxic pathway mediates the induction of hypoxia-inducible factor 1alpha in vascular smooth muscle cells. J Biol Chem 275:26765–26771

Rigoulet M, Yoboue ED, Devin A (2011) Mitochondrial ROS generation and its regulation: mechanisms involved in H(2)O(2) signaling. Antioxid Redox Signal 14:459–468

Rius J, Guma M, Schachtrup C, Akassoglou K, Zinkernagel AS et al (2008) NF-kappaB links innate immunity to the hypoxic response through transcriptional regulation of HIF-1alpha. Nature 453:807–811

Roy A, Rozanov C, Mokashi A, Daudu P, Al-mehdi AB et al (2000) Mice lacking in gp91 phox subunit of NAD(P)H oxidase showed glomus cell [Ca(2+)](i) and respiratory responses to hypoxia. Brain Res 872:188–193

Ryu JH, Li SH, Park HS, Park JW, Lee B et al (2010) NEDD8 conjugation stabilizes HIF-alpha subunits reactive oxygen species-dependently. J Biol Chem 286(9):6963–6970

Ryu JH, Li SH, Park HS, Park JW, Lee B et al (2011) Hypoxia-inducible factor alpha subunit stabilization by NEDD8 conjugation is reactive oxygen species-dependent. J Biol Chem 286:6963–6970

Sanchez M, Galy B, Muckenthaler MU, Hentze MW (2007) Iron-regulatory proteins limit hypoxia-inducible factor-2alpha expression in iron deficiency. Nat Struct Mol Biol 14:420–426

Sandau KB, Zhou J, Kietzmann T, Brune B (2001) Regulation of the hypoxia-inducible factor 1alpha by the inflammatory mediators nitric oxide and tumor necrosis factor-alpha in contrast to desferroxamine and phenylarsine oxide. J Biol Chem 276:39805–39811

Schepens B, Tinton SA, Bruynooghe Y, Beyaert R, Cornelis S (2005) The polypyrimidine tract-binding protein stimulates HIF-1alpha IRES-mediated translation during hypoxia. Nucleic Acids Res 33:6884–6894

Schulz E, Wenzel P, Munzel T, Daiber A (2012) Mitochondrial redox signaling: interaction of mitochondrial reactive oxygen species with other sources of oxidative stress. Antioxid Redox Signal

Semenza GL, Wang GL (1992) A nuclear factor induced by hypoxia via de novo protein synthesis binds to the human erythropoietin gene enhancer at a site required for transcriptional activation. Mol Cell Biol 12:5447–5454

Shatrov VA, Sumbayev VV, Zhou J, Brune B (2003) Oxidized low-density lipoprotein (oxLDL) triggers hypoxia-inducible factor-1alpha (HIF-1alpha) accumulation via redox-dependent mechanisms. Blood 101:4847–4849

Srinivas V, Leshchinsky I, Sang N, King MP, Minchenko A et al (2001) Oxygen sensing and HIF-1 activation does not require an active mitochondrial respiratory chain electron-transfer pathway. J Biol Chem 276:21995–21998

Stiehl DP, Jelkmann W, Wenger RH, Hellwig-Burgel T (2002) Normoxic induction of the hypoxia-inducible factor 1alpha by insulin and interleukin-1beta involves the phosphatidylinositol 3-kinase pathway. FEBS Lett 512:157–162

Tacchini L, De Ponti C, Matteucci E, Follis R, Desiderio MA (2004) Hepatocyte growth factor-activated NF-kappaB regulates HIF-1 activity and ODC expression, implicated in survival, differently in different carcinoma cell lines. Carcinogenesis 25:2089–2100

Thomas JD, Johannes GJ (2007) Identification of mRNAs that continue to associate with polysomes during hypoxia. RNA 13:1116–1131

Tian H, McKnight SL, Russell DW (1997) Endothelial PAS domain protein 1 (EPAS1), a transcription factor selectively expressed in endothelial cells. Genes Dev 11:72–82

Treins C, Giorgetti-Peraldi S, Murdaca J, Semenza GL, Van Obberghen E (2002) Insulin stimulates hypoxia-inducible factor 1 through a phosphatidylinositol 3-kinase/target of rapamycin-dependent signaling pathway. J Biol Chem 277:27975–27981

Turrens JF (2003) Mitochondrial formation of reactive oxygen species. J Physiol 552:335–344

Tuttle SW, Maity A, Oprysko PR, Kachur AV, Ayene IS et al (2007) Detection of reactive oxygen species via endogenous oxidative pentose phosphate cycle activity in response to oxygen concentration: implications for the mechanism of HIF-1alpha stabilization under moderate hypoxia. J Biol Chem 282:36790–36796

Urao N, McKinney RD, Fukai T, Ushio-Fukai M (2012) NADPH oxidase 2 regulates bone marrow microenvironment following hindlimb ischemia: role in reparative mobilization of progenitor cells. Stem Cells 30:923–934

van Uden P, Kenneth NS, Rocha S (2008) Regulation of hypoxia-inducible factor-1alpha by NF-kappaB. Biochem J 412:477–484
Vaux EC, Metzen E, Yeates KM, Ratcliffe PJ (2001) Regulation of hypoxia-inducible factor is preserved in the absence of a functioning mitochondrial respiratory chain. Blood 98:296–302
Wang GL, Jiang BH, Semenza GL (1995) Effect of altered redox states on expression and DNA-binding activity of hypoxia-inducible factor 1. Biochem Biophys Res Commun 212(2):550–556
Wang M, Kirk JS, Venkataraman S, Domann FE, Zhang HJ et al (2005) Manganese superoxide dismutase suppresses hypoxic induction of hypoxia-inducible factor-1alpha and vascular endothelial growth factor. Oncogene 24:8154–8166
Wang W, Fang H, Groom L, Cheng A, Zhang W et al (2008) Superoxide flashes in single mitochondria. Cell 134:279–290
Weissmann N, Zeller S, Schafer RU, Turowski C, Ay M et al (2006) Impact of mitochondria and NADPH oxidases on acute and sustained hypoxic pulmonary vasoconstriction. Am J Respir Cell Mol Biol 34:505–513
Welsh SJ, Bellamy WT, Briehl MM, Powis G (2002) The redox protein thioredoxin-1 (Trx-1) increases hypoxia-inducible factor 1alpha protein expression: Trx-1 overexpression results in increased vascular endothelial growth factor production and enhanced tumor angiogenesis. Cancer Res 62:5089–5095
Wenger RH (2002) Cellular adaptation to hypoxia: O2-sensing protein hydroxylases, hypoxia-inducible transcription factors, and O2-regulated gene expression. FASEB J 16:1151–1162
Wenger RH (2006) Mitochondria: oxygen sinks rather than sensors? Med Hypotheses 66:380–383
Wenner CE (2012) Targeting mitochondria as a therapeutic target in cancer. J Cell Physiol 227:450–456
Wood ZA, Poole LB, Karplus PA (2003) Peroxiredoxin evolution and the regulation of hydrogen peroxide signaling. Science 300:650–653
Xia C, Meng Q, Liu LZ, Rojanasakul Y, Wang XR et al (2007) Reactive oxygen species regulate angiogenesis and tumor growth through vascular endothelial growth factor. Cancer Res 67:10823–10830
Yang J, Staples O, Thomas LW, Briston T, Robson M et al (2012) Human CHCHD4 mitochondrial proteins regulate cellular oxygen consumption rate and metabolism and provide a critical role in hypoxia signaling and tumor progression. J Clin Invest 122:600–611
Yoshida K, Kirito K, Yongzhen H, Ozawa K, Kaushansky K et al (2008) Thrombopoietin (TPO) regulates HIF-1alpha levels through generation of mitochondrial reactive oxygen species. Int J Hematol 88:43–51
Young RM, Wang SJ, Gordan JD, Ji X, Liebhaber SA et al (2008) Hypoxia-mediated selective mRNA translation by an internal ribosome entry site-independent mechanism. J Biol Chem 283:16309–16319
Zhong H, Chiles K, Feldser D, Laughner E, Hanrahan C et al (2000) Modulation of hypoxia-inducible factor 1alpha expression by the epidermal growth factor/phosphatidylinositol 3-kinase/PTEN/AKT/FRAP pathway in human prostate cancer cells: implications for tumor angiogenesis and therapeutics. Cancer Res 60:1541–1545
Zhou J, Callapina M, Goodall GJ, Brune B (2004) Functional integrity of nuclear factor kappaB, phosphatidylinositol 3'-kinase, and mitogen-activated protein kinase signaling allows tumor necrosis factor alpha-evoked Bcl-2 expression to provoke internal ribosome entry site-dependent translation of hypoxia-inducible factor 1alpha. Cancer Res 64:9041–9048
Zimmer M, Ebert BL, Neil C, Brenner K, Papaioannou I et al (2008) Small-molecule inhibitors of HIF-2a translation link its 5'UTR iron-responsive element to oxygen sensing. Mol Cell 32:838–848

Chapter 5
Hypoxia and Gene Expression

Maria Carla Bosco and Luigi Varesio

Abstract Analysis of gene expression is a promising approach to understand tumor hypoxia and to relate biology to prognosis and response to therapy. Understanding the response to hypoxia of every component of the tumor mass is indispensable to understand the biology of the system. However, the plasticity and distribution of hypoxic areas make it difficult to gauge tumor hypoxia by combining the information of individual components. Global information on tumor gene expression revealed by the analysis of the transcriptome of the tumor mass provides a global view of the hypoxic status of the tissue. The global and analytic assessment of tumor hypoxia, revealed by the whole tumor mass or by its individual components, will be discussed with particular emphasis on tumor-infiltrating leukocytes.

Keywords Hypoxia · Gene expression · Signature · Tumor-infiltrating leukocytes · Prognosis · Biomarkers

5.1 Introduction

5.1.1 Tumor Hypoxia

The tumor mass is highly heterogeneous with regard to vessel density and organization, resulting in a patchwork of anoxic, hypoxic, and well-oxygenated areas. Within the tumor mass there is a mosaic of biologically different cells comprising neoplastic, stromal, and infiltrating immune cells, all of which manifest a typical pattern of

M. C. Bosco (✉) · L. Varesio
Laboratory of Molecular Biology, Gaslini Institute,
Largo Gaslini 5, 16147 Genoa, Italy
e-mail: mariacarlabosco@ospedale-gaslini.ge.it

L. Varesio
e-mail: luigivaresio@ospedale-gaslini.ge.it

G. Melillo (ed.), *Hypoxia and Cancer,* Cancer Drug Discovery and Development, DOI 10.1007/978-1-4614-9167-5_5,

response to hypoxia (de Visser and Coussens 2005; Schreiber et al. 2011). Oxygen tension (pO_2) decrease in the tumor microenvironment impacts on cancer cells as well as on leukocyte biology, on tumor proliferation and metastatic potential, and on inflammatory responses, thereby conditioning the outcome of the disease through complex and multifaceted effects on every tumor component (Harris 2002; Imtiyaz and Simon 2010; Knowles and Harris 2007; Semenza 2011; Sitkovsky and Lukashev 2008).

Cellular adaptation to hypoxia is mediated by transcriptional changes quantified by the 10,924 publications identified by the Medline query "Hypoxia and gene expression." Hypoxia-inducible factors (HIFs) are often referred to as the master regulators of the transcriptional response to hypoxia, and HIF-dependent events are often taken as a paradigm of hypoxia. A recent review (Semenza 2012) and other chapters in this book will expand the issue of HIF-mediated responses. However, HIF can be induced by several signals under nonhypoxic conditions, and the hypoxic status impacts on gene expression through transcription factors other than HIF (Battaglia et al. 2008; Cummins and Taylor 2005; Rius et al. 2008).

Hypoxia elicits the expression of several microRNAs , which can modify the transcriptional/translational profile of the cell (Pocock 2011). The presence and function of microRNAs in the cellular response and transcriptional profile are addressed elsewhere in this book. The net result of transcriptional activity and transcript stability is a steady-state level of messenger RNA (mRNA) expression that provides a global picture of the hypoxic phenotype and that will be discussed in more detail here.

Analysis of gene expression is a promising way to understand tumor hypoxia and to relate biology to prognosis and response to therapy. Comprehensive analysis of the response to hypoxia of every component of the tumor mass is one way to identify its role in tumor progression, regression, and response to therapy. This analytic approach requires the ability to merge the knowledge acquired on individual cell types in a more general model accounting for the overall status of the tumor. However, the plasticity and distribution of hypoxic areas make it difficult to gauge tumor hypoxia by combining the information of individual components. Another way to address the question of tumor hypoxia definition is to utilize global information on tumor gene expression revealed by the analysis of the transcriptome of the tumor mass. This strategy is made possible by high-throughput microarray technologies and provides a global description of the pattern of gene expression in the tumor mass and in the hypoxic component. The global and analytic assessment of tumor hypoxia, revealed by the whole tumor mass or by its individual components, will be discussed in the following sections.

5.1.2 Hypoxia Core Genes

The gene expression profile of individual cells exposed to hypoxia is composed of thousands of genes modulated by the activation of the master hypoxia regulators, the HIF pathway, HIF-independent transcription, microRNA activity, secondary response to hypoxia-inducible genes, etc. Hypoxia response is evolutionarily conserved in all mammalian cells because life on earth evolved under hypoxic conditions and common key processes needed to sustain life under conditions of limited oxygen availability exist. Hypoxic cells need alternative energy sources to compensate for the inhibition of oxidative metabolism and the malfunctioning of O_2-dependent enzymes (Harris 2002; Semenza 2011; Wenger et al. 2005) and they activate transcriptional pathways essential to cope with reduced oxygenation, upregulating genes involved in glycolytic metabolism and glucose transport or associated with nonglycolytic metabolism and ion transport (Bosco et al. 2006; Bosco et al. 2008b; Bosco et al. 2011; Fang et al. 2009a; Gaber et al. 2009; Ricciardi et al. 2008). Genes belonging to the gluconeogenesis, glycolisis, or glucose metabolism pathways (GLUT1, GAPDH, PGK1, ALDOC, ALDOA, HK1, HK2, ENO1, LDHA), pH regulation (carbonic anhydrase IX—CAIX), and angiogenesis (vascular endothelial growth factor—VEGF) are induced under hypoxic conditions in most cells/tissues examined. We may envision the existence of a "core cluster of hypoxia genes" defining a constant feature of the hypoxic status of cells of different type, origin, and functional state. The existence of such a group of genes is further suggested by the growing body of evidence demonstrating that the transcriptional response to HIF is a common denominator of many hypoxic cells.

However, when comparing different cell types profiled by microarray analysis in search of common hypoxia-responsive genes, we observe that differences are predominant and common genes are hard to find. One reason is that not every cell responds to hypoxia in an identical fashion during physiological and pathological adaptation. Different cells of the human body have diverse energy requirements, operate in different microenvironments, and are physiologically exposed to different ranges of O_2 concentrations. Furthermore, cell transformation impacts on the response to hypoxia and hypoxia-induced transcriptional pathways may be modified as well. The heterogeneity of the transcriptional response to hypoxia among human primary cells from several anatomical locations including coronary, artery, and kidney epithelial or mesenchymal cells was shown by Chi et al. (2006). We have analyzed the gene expression profile of nine neuroblastoma cell lines cultured under normoxic or hypoxic conditions. Neuroblastoma cell lines were heterogeneous, like the primary tumors, with regard to differentiation, chromosomal alteration, and MYCN amplification. Each cell line showed more than 200 genes inducible by hypoxia belonging to the expected pathways such as glucose metabolism related. However, none of the genes was concomitantly modulated by hypoxia in every cell line, demonstrating a complete lack of core hypoxia cluster under these circumstances. It was further shown that the strong transcriptional response of some of the cell lines to MYCN amplification was responsible for masking the response to hypoxia and blurring the clustering approach (Fardin et al. 2009). Benita et al. (2009) reported that only 17 genes were commonly shared by 6 cell lines representing different cell lineages exposed to hypoxia, including classic HIF target genes (ENO1, BHLHB2, BNIP3). Part of this selectivity in the response could also be attributed to the fact that HIF-independent pathways could play a variable, albeit important, role. In fact, only 23 % of the genes inducible by hypoxia in each cell line showed an HIF-binding site in the promoter region. Analysis of the HIF-binding sites in the promoter of hypoxia-inducible genes in macrophages and dendritic cells indicated a much higher correlation between gene upregulation and the presence of a Hypoxia Responsive Element (HRE) in the promoter (Blengio et al. 2013; Bosco et al. 2011; Pierobon et al. 2013; Ricciardi et al. 2008), presumably reflecting the fact that these cells are nonproliferating and cell cycle transcriptome is not modulated by hypoxia.

In conclusion, the concept of hypoxia core genes is becoming fuzzy. A large number of genes are expressed in response to hypoxia and, among them, we can find significant representation of energy-related biochemical pathways, HIF-dependent genes, demonstrating a core "functional response." The latter however, does not easily translate into a cluster of genes universally expressed in response to hypoxia, given the variability of the individual response to hypoxia. Conversely, even the genes that are the most constant representative of the response to hypoxia (e.g., VEGF) are also susceptible to modulation by nonhypoxic stimuli preventing the establishment of a biunivocal relationship between hypoxia and expression of specific genes.

5.2 Tumor Hypoxia Gene Expression

The importance of hypoxia in conditioning tumor aggressiveness is documented by an extensive literature (Brown and William 2004; Harris 2002; Lin and Yun 2010; Lu and Kang 2010; Semenza 2009). Hypoxia activates specific genes encoding angiogenic, metabolic, and metastatic factors (Carmeliet et al. 1998; Chan and Giaccia 2007), thus contributing to the acquisition of the tumor aggressive phenotype (Carmeliet et al. 1998; Harris 2002; Rankin and Giaccia 2008). The link between hypoxia and poor prognosis is supported by a lot of data in humans and in animal models that associate hypoxic areas of the tumor with leukocyte infiltrate, radiotherapy and drug resistance, and metastatic spread. However, the translation of this concept to the clinical practice is rather complex due to the difficulties in quantifying tumor hypoxia, patchy in nature, in the tumor mass (Wilson and Hay 2011). The variability in the levels of hypoxia among individual tumors, even within a single disease subtype, calls for tools that can be used to quantify tumor hypoxia in a clinical setting. To this end, different approaches exist relying on physicochemical measurements, biomarkers, or the use of gene expression signatures.

5.2.1 Measurement of Tissue Hypoxia

Efforts to measure tissue O_2 levels have been made for long time. Vaupel's group pioneered the use of O_2-sensing electrodes (Vaupel and Thews 1974) that are still being used (Vaupel and Kelleher 2013). This approach generated relevant information but it is highly impractical for clinical use, because it is invasive, may have a restricted accessibility of many tumors, and cannot be applied to archival clinical material.

An alternative approach is to rely on exogenous markers of hypoxia derived from the infusion of tracers in the organism: pimonidazole (Raleigh et al. 1999), fluorine-18 fluoromisonidazole (18F-MISO) (Gross et al. 1995), flourine 18-fluoroazomycin arabinoside (F-FAZA) (Busk et al. 2008), 2-(2-nitro-1H-imidazol-1-yl)-*N*-(2,2,3,3,3-pentafluoropropyl) acetamide (EF5) (Evans et al. 2000), and copper(II) complex of diacetyl-2,3-bis(*N*(4)-methyl-3-thiosemicarbazonato) ligand (Cu-ATSM) (Holland et al. 2009). These tracers are bound at pO_2 levels below 10 mm Hg and revealed by appropriate antibodies, thereby imaging the distribution of hypoxia. However, the use of these tracers is implemented mainly in animal models because they would have to be administrated systemically to the patient. Hypoxia markers are qualitative rather than quantitative and it is difficult to derive an objective, reproducible measure of the hypoxic status of the tumor as a whole. Additional limitations of these methods are that they are linked to the analysis of biopsies, which may not be easily available and/or may not be representative of the tumor mass (Jogi et al. 2004).

Positron emission tomography (PET) methods are undergoing active development in measuring hypoxia (Mees et al. 2009). One strategy depends on radiolabeled antibodies against carbonic anhydrase 9 (CA9) (Hoeben et al. 2010). This approach has potential for monitoring hypoxia to the extent that CA9 can be considered a specific hypoxia-inducible factor-1 (HIF-1) reporter and that HIF-1 activity is regulated by hypoxia. Several derivatives of nitroimidazole are now being studied to identify the best tracer with high uptake and low toxicity (Chadwick et al. 2011; Synnott et al. 2010). Among these, 18F-MISO is the most extensively studied. Although the 18F-MISO-PET technique is noninvasive and allows the serial imaging of hypoxia, the accumulation of 18F-MISO in hypoxic tumors is relatively low. This results in a low signal-to-noise ratio and hence a poor contrast between hypoxic tumors and surrounding normal tissues. Cu-ATSM holds a good promise as an agent for delineating hypoxia within tumor with PET also because it is well tolerated by patients. Halle et al. (2012) utilized magnetic resonance (MR) images and dynamic contrast-enhanced imaging (DCE-MRI) to identify the prognostic parameter A(Brix) that was inversely related to HIF1α protein expression, consistent with the possible application in defining tumor hypoxia.

5.2.2 *Biomarkers*

A widely used approach is the evaluation of endogenous hypoxia markers expressed by the cells/tissues as a result of low pO_2. Immunohistochemistry is commonly used in patients' biopsies to detect both endogenous and exogenous genes generally associated with the hypoxic status such as HIF-1α and some of its target genes, including GLUT1, CAIX, and VEGF. This approach stems from the concept of hypoxia core genes consistently expressed in hypoxic areas and it provides a qualitative appraisal of hypoxia distribution within the tumor. However, the relevant biomarkers to use have to be evaluated in every instance depending on the target cell, given the aforementioned variability of the gene expression response to hypoxia among different cell types. Furthermore, one limitation associated with the use of these markers is their potential regulation by non-hypoxia-related factors (for example, pH or the concentrations of metabolites such as glucose and glutamine).

A recent review of the correlation of hypoxia-related protein expression with prognosis in oral squamous cell carcinoma patients identified unique roles of hypoxia pathways in tumor progression. However, it was difficult to transfer the association between markers and hypoxia established experimentally to the tumor contest in vivo (Eckert et al. 2012).

The significance of hypoxia biomarkers in predicting cancer patients' outcome has been evaluated in more than 200 studies on carcinoma, sarcoma, brain tumor leukemia, and others. In general, one or two biomarkers, detected by immunohistochemistry or enzyme-linked immunosorbent assay (ELISA), were utilized to generate survival curves categorizing the patient's outcome. Though most of the studies showed significance of these genes in predicting prognosis, there was no definitive prognostic individual marker (reviewed in Moon et al. 2007), suggesting the need for further studies with standardized methods to examine gene expression and the use of multiple genes as biomarkers.

5.2.3 Hypoxia Signatures

The gene expression signature is a cluster of genes whose expression in cells or tissues is related to a specific physiological status of the cell (e.g., hypoxia) or to a pathological situation (e.g., tumor). The variability of tumor cell transcriptional response to hypoxia coupled with the high number of genes measurable by microarray technology prompted the organization of this information into gene expression signatures. Currently available hypoxia signatures are driven by two aims: (1) define a hypoxia signature as general as possible to be applied to most cell types and malignancies as in the case of the core hypoxic gene cluster previously discussed and (2) define tumor or cell-type-specific signatures that provide the greatest ability to differentiate the normoxic from the hypoxic state.

Hypoxia signatures may contain a few or hundreds of genes. The higher the number of genes, the greater is the information contained as well as the background noise of the signature. Hence, careful bioinformatic approaches have to be used to maximize the discriminatory capacity of the signature while reducing the noise to a minimum. Signatures may be composed of biochemically unrelated genes that, together, generate a robust classifier of the hypoxic state. One strategy to define a hypoxia signature relies on experimental data on gene expression of normoxic and hypoxic cells/tissues. Appropriate feature selection algorithms are applied (Saeys et al. 2007) to identify the sets of genes that can better discriminate between the normoxic and hypoxic statuses of the cell. Alternatively, hypoxia signatures can be obtained in silico by bioinformatic approaches based on genes known from the literature to be linked to the manifestation of hypoxia (Buffa et al. 2010). A collection of "seeds," known hypoxia-inducible genes, are used as starting point from which to build an association network. The approach defines a "neighborhood of co-expression" around each seed gene, and then connects seeds that have a significant degree of overlap between their neighborhoods. This analysis leads to hypoxia signatures, or metagenes, highly enriched for hypoxia-regulated pathways. Genes with the highest connectivity are also the most prognostic, and the metagene consists of a small number of top-ranked genes, including VEGFA, SLC2A1, and PGAM1.

One of the first hypoxia signatures was reported by Chi et al. (2006) who analyzed gene expression of primary cells from several anatomic locations, including coronary artery and renal tubules, cultured in vitro under hypoxia or normoxia. They derived a signature called "epithelial hypoxia signature." This signature was developed through complementary DNA microarray technology and cluster analysis. Seigneuric et al. (2007) used the previously mentioned data by Chi et al. (2006) to derive gene signatures associated selectively with the early hypoxic response of the cell. Fardin et al. (2010a) described the hypoxic signature selective for neuroblastoma tumors (NB-hypo). This signature was derived from the analysis of the gene expression profile of several neuroblastoma cell lines taking advantage of the strong discriminating power of the l1-l2 feature selection technique combined with the analysis of differential gene expression (Fardin et al. 2009; Fardin et al. 2010a; Fardin et al. 2010b). Definition of hypoxia signature through a metagene (Winter et al. 2007) was described in head and neck squamous cell carcinomas (HNSCCs) by clustering genes whose expression pattern was similar to well-known hypoxia-regulated genes like CAIX or GLUT 1. The metagene contained 99 genes, some of which were also previously described to be induced in vitro by hypoxia. Extension of the metagene approach led to the definition of a signature common to hypoxic HNSCC and breast cancer (Buffa et al. 2010). A seven-gene set associated with chronic hypoxia was derived from the analysis of gene expression of the human HepG2 liver cells cultured for 72 hours under mild hypoxic conditions (2 %) and applied to the prognosis of hepatocellular carcinoma (HCC) (Malenstein et al. 2010). A 15-gene hypoxia signature was derived from in vitro studies of five human squamous cell carcinoma cell lines (Toustrup et al. 2011) and the derived classifier was tested for its impact on hypoxic modification of radiotherapy in head and neck cancer (Toustrup et al. 2012a).

An interesting collection of gene clusters related to hypoxia can be found among the 8,513 gene sets in the Molecular Signatures Database (MSigDB, current release v3.1; http://www.broadinstitute.org/gsea/msigdb/index.jsp) (Subramanian et al. 2005). MSigDB places emphasis on a genomic, unbiased approach to the definition of gene sets; therefore an important component of MSigDB is the collection of gene sets from published expression profiles. In addition to gene sets curated from prior knowledge (such as GO, BioCarta), experimental sets provide an unbiased readout of a biological state. MSigDB contains gene sets easily used for gene set enrichment analysis (GSEA), a computational method that determines whether an a priori defined set of genes shows statistically significant, concordant differences between two biological states (e.g., normoxic and hypoxic). This analysis may prove to be instrumental to validate prognostic signatures with respect to their relationship to hypoxia and was used to define the hypoxia metagene (Buffa et al. 2010). Searching for "hypoxia" in the database outputs 43 sets representing gene clusters associated with different cell lines and primary cells exposed to hypoxia (see Table 5.1). Interestingly, none of the genes of the 43 hypoxia sets is common to all of them, underlying the trend of poor overlapping among hypoxia signatures (Varesio unpublished observation). Many publications on hypoxia-induced gene expression list hundreds of modulated genes that are interesting from the biological standpoint but inappropriate to serve as signature because of the high associated background noise. The genes included in this collection represent a further elaboration of the comparison between normoxic and hypoxia state and include some bioinformatic filters to select clusters based on discriminatory capacity between normoxic and hypoxic state.

The existing hypoxia signatures and gene sets tend to be selective for the cellular system in which they were derived, although the information on the generality of the hypoxia signature application is quite limited. This property of the hypoxia signatures can be an asset in trying to dissect the hypoxia response of the tumor mass in the individual cellular components. Assessment of hypoxia through signatures does not generate an absolute measure of tissue pO_2, but it can rank a series of tumors on the bases of the expression of several hypoxia genes resulting in a clinically relevant prognostic indicator .

5.3 Hypoxia-Driven Gene Expression in Tumor-Infiltrating Leukocytes

Within the tumor microenvironment, there is a close interaction between cancer cells and infiltrating leukocytes which influence each other's biology and gene expression patterns. Clinical and experimental evidence suggests that chronic inflammation is causally linked to tumorigenesis and that the degree of inflammation and the type of immune cells present at tumor sites are responsible for tilting the balance between tumor progression and regression (Ben Baruch 2006; de Visser and Coussens 2005; Li et al. 2005; Palazon et al. 2012; Vicari et al. 2002). Immune cells can not only suppress tumor growth by destroying cancer cells or inhibiting their proliferation but also promote tumor progression by conditioning the microenvironment to facilitate tumor proliferation (Lamagna et al. 2006; Palazon et al. 2012; Qian and Pollard 2010; Schreiber et al. 2011; Sica and Bronte 2007). Reduced tumor oxygenation causes divergent regulation of innate and adaptive immunity (Palazon et al. 2012; Sica et al. 2011), promoting innate immune cell recruitment, survival, differentiation, and proangiogenic and proinflammatory responses (Bosco and Varesio 2010; Bosco et al. 2008b; Cramer et al. 2003; Imtiyaz and Simon 2010; Knowles and Harris 2007; Murdoch et al. 2004; Nizet and Johnson 2009; Peyssonnaux et al. 2005), while decreasing adaptive immune cell proliferation, effector

Table 5.1 Hypoxia-related gene sets. (The 43 sets of sequences related to hypoxia were obtained from the Molecular Signatures Database (MSigDB), which is a collection of annotated gene sets (Subramanian et al. 2005) that can be used in conjunction with GSEA. GSEA and MSigDB are maintained by the GSEA team with the support of the MSigDB Scientific Advisory Board and funded by the National Cancer Institute, National Institutes of Health, and National Institute of General Medical Sciences, USA)

Gene set	No. of genes	Contributor	Keyword
LU_TUMOR_ANGIOGENESIS_UP	25	Robertson	Angiogenesis
HELLEBREKERS_SILENCE_DURING_TUM_ANG	81	Robertson	Angiogenesis
HU_ANGIOGENESIS_UP	23	Saunders	Angiogenesis
HU_ANGIOGENESIS_DN	41	Saunders	Angiogenesis
KEGG_GLYCOLYSIS_GLUCONEOGENESIS	62	KEGG	Glycolysis
MOOTHA_GLYCOLYSIS	31	Mootha	Glycolysis
ELVIDGE_HIF1A_TARGETS_UP	72	Liberzon	HIF
ELVIDGE_HIF2A_TARGETS_UP	6	Liberzon	HIF
ELVIDGE_HIF1A_AND_HIF2A_TARGETS_UP	42	Liberzon	HIF
ELVIDGE_HIF1A_AND_HIF2A_TARGETS_DN	140	Liberzon	HIF
SEMENZA_HIF1_TARGETS	36	Newman	HIF
MANALO_HYPOXIA_UP	146	Stafford	Hypoxia
MENSE_HYPOXIA_UP	129	Vogelsang	Hypoxia
MIZUKAMI_HYPOXIA_UP	12	Robertson	Hypoxia
MIZUKAMI_HYPOXIA_DN	6	Robertson	Hypoxia
JIANG_HYPOXIA_VIA_VHL	36	Liberzon	Hypoxia
ELVIDGE_HYPOXIA_BY_DMOG_DN	67	Liberzon	Hypoxia
JIANG_HYPOXIA_CANCER	87	Newman	Hypoxia
LEONARD_HYPOXIA	48	Newman	Hypoxia
HARRIS_HYPOXIA	81	Newman	Hypoxia
WINTER_HYPOXIA_UP	126	Liberzon	Hypoxia
WINTER_HYPOXIA_DN	67	Liberzon	Hypoxia
WEINMANN_ADAPTATION_TO_HYPOXIA_UP	30	Liberzon	Hypoxia
WEINMANN_ADAPTATION_TO_HYPOXIA_DN	42	Liberzon	Hypoxia
MAINA_HYPOXIA_VHL_TARGETS_UP	6	Saunders	Hypoxia
KONDO_HYPOXIA	10	Liberzon	Hypoxia

Table 5.1 (continued)

Gene set	No. of genes	Contributor	Keyword
FARDIN_HYPOXIA_9	9	Liberzon	Hypoxia
FARDIN_HYPOXIA_11	32	Liberzon	Hypoxia
CHUANG_OXIDATIVE_STRESS_RESPONSE_DN	11	Newman	Oxidative
CHUANG_OXIDATIVE_STRESS_RESPONSE_UP	29	Newman	Oxidative
WEIGEL_OXIDATIVE_STRESS_RESPONSE	35	Newman	Oxidative
WEIGEL_OXIDATIVE_STRESS_BY_HNE_AND_H2O2	39	Newman	Oxidative
WEIGEL_OXIDATIVE_STRESS_BY_TBH_AND_H2O2	36	Newman	Oxidative
WEIGEL_OXIDATIVE_STRESS_BY_HNE_AND_TBH	60	Newman	Oxidative
KEGG_OXIDATIVE_PHOSPHORYLATION	135	KEGG	Oxidative
ABE_VEGFA_TARGETS_2HR	35	Newman	VEGF
ABE_VEGFA_TARGETS_30MIN	29	Newman	VEGF
ABE_VEGFA_TARGETS	20	Newman	VEGF
KEGG_VEGF_SIGNALING_PATHWAY	76	KEGG	VEGF
JIANG_VHL_TARGETS	145	Newman	VHL
ABDULRAHMAN_KIDNEY_CANCER_VHL_DN	14	Liberzon	VHL
MAINA_VHL_TARGETS_UP	10	Saunders	VHL
MAINA_VHL_TARGETS_DN	20	Saunders	VHL

functions, and survival (Caldwell et al. 2001; Raber et al. 2012; Sica et al. 2011; Sitkovsky and Lukashev 2008), eventually resulting in the amplification of innate inflammatory reactions and inhibition of adaptive immunity (Palazon et al. 2012). Furthermore, hypoxia can have both pro- and antiapoptotic consequences depending on the cellular context, inducing cell death (Sitkovsky and Lukashev 2008; Sun et al. 2010) or survival (Roiniotis et al. 2009) of distinct immune cell populations. In general, it appears that hypoxic immune cells are genomically and functionally reprogrammed toward a tumor-promoting direction (Fig. 5.1).

Analysis of hypoxia-driven transcriptional profile of tumor-infiltrating immune cells cannot be easily carried out because the patchy nature of tumor hypoxia prevents the selective purification of hypoxic cells. Furthermore, reoxygenation of the samples may occur during the preparation of single-cell suspensions, introducing a spurious perturbation of gene expression (Bosco et al. 2008a). Analysis of the gene expression profile of leukocyte subsets purified from peripheral blood of normal donors and exposed to in vitro hypoxia is a way to assess the hypoxic response in these cells (Bosco et al. 2008b; Bosco and Varesio 2012). Results of these studies provide the basic information to be integrated into a more comprehensive model of the hypoxic tumor mass components. Leukocyte population presents a typical functional response to hypoxia linked to cell lineage, differentiation state, and biology, mirrored by large clusters of differentially expressed genes. Furthermore, clusters of genes shared by different hypoxic leukocyte subpopulations can be identified representing a limited core hypoxia response linked, in part, to similar functional activities.

5.3.1 Mononuclear Phagocytes

Mononuclear phagocytes (MPs) represent a prominent component of the inflammatory infiltrate in most malignant tumors (Knowles and Harris 2007; Mantovani and Sica 2010; Qian and Pollard 2010) where they are recruited by tumor-derived factors as circulating monocytes (Mn), subsequently accumulating within poorly vascularized and hypoxic tumor areas and differentiating into macrophages (Mf) (Bennaceur et al. 2008; Bosco et al. 2008b; Crowther et al. 2001; Knowles and Harris 2007; Mantovani and Sica 2010; Murdoch et al. 2004; Palucka et al. 1998; Schmieder et al. 2012). Mn and Mf have the potential to exert antitumor responses, through the expression of effector and immunostimulatory functions, and to contribute to tumor progression, by exerting immunosuppressive, proangiogenic, and tissue-disrupting activities (Burke et al. 2003; Knowles and Harris 2007; Lamagna et al. 2006; Lewis and Pollard 2006; Murdoch et al. 2004; Qian and Pollard 2010; Sica and Bronte 2007). Hypoxia regulates the balance between expression of anti- and protumoral activities (Crowther et al. 2001; Knowles and Harris 2007; Lamagna et al. 2006; Lewis and Murdoch 2005). The observation of a correlation between the extent of hypoxia and tumor-associated Mf (TAM) infiltration, tumor progression, and poor patient prognosis supports the role of hypoxia in the promotion of TAM protumoral functions (Grimshaw and Balkwill 2001; Imtiyaz and Simon 2010; Lamagna et al.

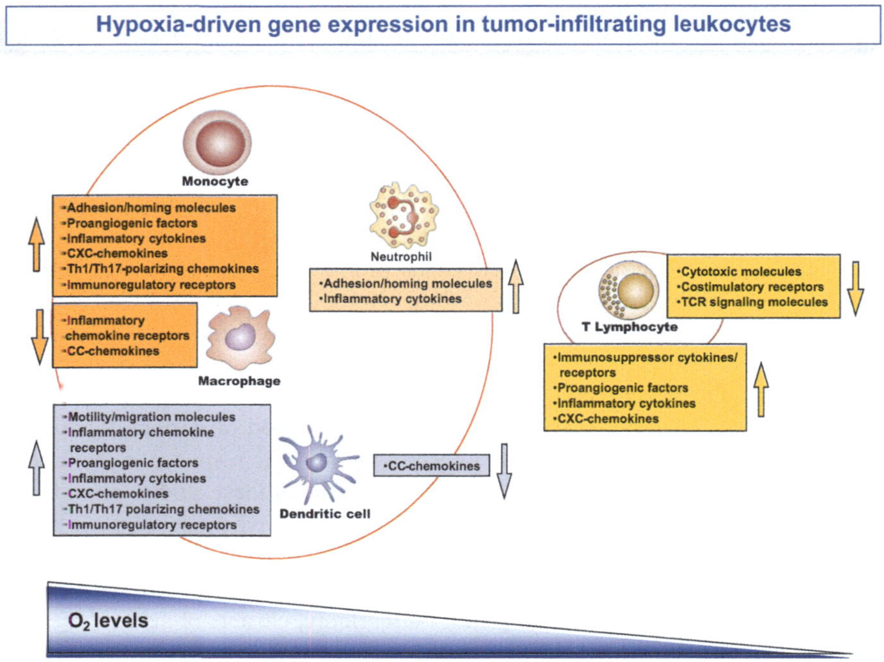

Fig. 5.1 Hypoxia differentially regulates gene expression in leukocyte subsets. Some genes are shared among subsets although the majority are unique of a specific lineage and differentiation/maturation stage. Hypoxia promotes innate immune cell recruitment and extravasation, proangiogenic activity, and proinflammatory responses, but it decreases adaptive immune cell activation and effector functions

2006; Murdoch et al. 2004; Pollard 2004; Schmieder et al. 2012; Sica and Bronte 2007). Myeloid dendritic cells (DCs) are also represented in human tumors (Lin et al. 2010; Lin et al. 2006; Vicari et al. 2002) and their immunogenic capacity may be conditioned by the tumor microenvironment. (Lin et al. 2010; Steinman and Banchereau 2007). Tumor-associated DCs (TADCs) display mostly an immature phenotype, characterized by defective activation and migration to draining lymph nodes and tolerance to tumor antigens, and promote tumor progression and dissemination (Bennaceur et al. 2008; Gabrilovich 2004; Lin et al. 2010; Vicari et al. 2002). Hypoxia is one of the signals contributing to DC dysfunction (Bosco and Varesio 2012; Rama et al. 2008; Yang et al. 2009).

Microarray analyses have defined the gene expression profiles of different hypoxic MP populations, demonstrating their functional reprogramming through the differential expression of clusters of genes critical for their adaptation to the hypoxic environment. Hypoxia modulates the expression of both genes common to cells belonging to distinct differentiation/maturation stages, which are probably representative of the transcriptome of Mn lineage cells (Blengio et al. 2013; Bosco et al. 2006; Bosco et al. 2011; Bosco and Varesio 2012; Fang et al. 2009a; Ricciardi et al. 2008;

Yang et al. 2009), and genes specific for a certain stage of differentiation, potentially reflecting the cell genetic makeup (Bosco et al. 2008b; Bosco and Varesio 2012).

An example of the common and unique gene expression changes that characterize distinct MP populations is represented by the modulation of a cluster of genes implicated in cell motility/migration and tissue remodeling. MP migratory activity depends on a defined repertoire of adhesion molecules and chemokine receptors (Imhof and Aurrand-Lions 2004; Mantovani et al. 2004). As documented by microarray analysis, hypoxia tightly regulates the expression of several of these genes, contributing to both Mn recruitment and Mn/Mf retainment in tumor lesions (Bosco et al. 2008b). On the one hand, it upregulates the integrin family members, CD11b and CD18, the fractalkine receptor, CX3CR1 (Bosco et al. 2006; Bosco et al. 2008b), the CXCL12/SDF-1 receptor, CXCR4 (Bosco et al. 2006; Schioppa et al. 2003), the angiopoietin 2 (Ang2) receptor, tie-2 (Lewis et al. 2007), and the VEGF receptor, FLT1 (Bosco et al. 2006), thus increasing circulating human Mn adhesion to the endothelium, extravasation, and recruitment to O_2-deprived tumor regions (Bosco et al. 2008b; Imhof and Aurrand-Lions 2004; Knowles and Harris 2007; Kong et al. 2004; Lewis et al. 2007; Mantovani et al. 2004; Murdoch et al. 2004). On the other hand, it differentially modulates the expression of genes coding for specific inflammatory chemokine receptors and pro/antichemotactic genes impairing the chemotaxis of both primary Mn and differentiated Mf after recruitment to the tumor, thus preventing their migration away from the hypoxic areas and promoting their retainment/concentration (Bosco et al. 2008b; Imtiyaz and Simon 2010). Downregulation of the inflammatory chemokine receptors, CCR2, CCR5, and CCR1 (Bosco et al. 2006), and consequent impairment of chemotaxis to their ligands were demonstrated in human Mn (Turner et al. 1999). Decreased CCR5 expression by hypoxia was also reported in mouse Mf (Bosco et al. 2004b) and human TAM isolated from ovarian carcinoma (Sica et al. 2000). Hypoxia also inhibits chemoattractant-induced migration in both Mn and Mf by upregulating the MAPK phosphatase 1 (MKP-1), which decreases MAPK phosphorylation required for cell migration (Grimshaw and Balkwill 2001), and the regulator of G-protein signaling 1, a member of a new class of G-protein-signaling deactivators which inactivates several chemotactic receptors (Bosco et al. 2006). Furthermore, hypoxia decreases the expression of the cell motility-promoting factor, autotaxin, while upregulating the Mn-arrest GRO family chemokines, CXCL2 and CXCL3, and the Mf migration inhibitory factor (MIF) both in Mn and Mf (Bosco et al. 2006; Fang et al. 2009a; Knowles and Harris 2007). Modulation of genes involved in cell motility and migration, such as those controlling actin cytoskeleton, adherence junction, focal adhesion, and transendothelial migration, is also an important feature of the transcriptome of Mn-derived immature DCs generated under hypoxia (Bosco et al. 2008b; Ricciardi et al. 2008; Yang et al. 2009). However, as opposed to Mn and Mf, hypoxia stimulates in DCs the upregulation of the inflammatory chemokine receptors, CCR2, CCR3, and CCR5, in addition to CXCR4 and CX3CR1, associated with increased responsiveness to specific ligands (Bosco et al. 2008b; Ricciardi et al. 2008; Yang et al. 2009), suggesting that the hypoxic environment stimulates a migratory program in these cells increasing their redistribution within pathologic tissues (Cavanagh and Von Andrian 2002) and migration away from hypoxic areas to secondary lymphoid organs where they can fulfill their antigen-presenting functions and trigger T-cell responses (Lanzavecchia and Sallusto 2001; Rossi and Young 2005; Steinman and Banchereau 2007). Modulation of genes coding for molecules implicated in tissue remodeling, which is also a common denominator of Mn, Mf, and DC response to hypoxia, may contribute to the regulation of their migratory behavior. Distinct members of the matrix metalloproteinases (MMPs), a family of proteolytic enzymes that degrade the extracellular matrix (ECM) and allow leukocyte migration, are differentially modulated by hypoxia in MP depending on the cell differentiation stage (Bosco et al. 2006; Burke et al. 2003; Knowles and Harris 2007; White et al. 2004). Interestingly, however, the gene coding for MMP9 was similarly downregulated in both human Mn and DCs, resulting in reduced migration (Bosco et al. 2006; Rahat et al. 2006), whereas genes encoding tissue inhibitor of metalloproteinase 1 (TIMP1) and tissue factor pathway inhibitors 1 and 2 (TFPI1,2), which inhibit cell migratory activity by reducing ECM degradation, were upregulated in hypoxic Mn, Mn-derived Mf (MDMs), and DCs (Bosco et al. 2006; Fang et al. 2009a), thus representing common hypoxia gene targets in MP.

MP hypoxic profiles also reflect a profound influence of low pO_2 on the expression of specific functions (Harris 2002; Imtiyaz and Simon 2010; Semenza 2011; Sitkovsky and Lukashev 2008). A common proinflammatory, proangiogenic, and protumoral response to hypoxia is expressed throughout the differentiation of Mn into TAM and DCs (Blengio et al. 2013; Bosco et al. 2006; Bosco et al. 2008b; Burke et al. 2003; Fang et al. 2009a; Ricciardi et al. 2008; Schioppa et al. 2003; White et al. 2004). Upregulation of over 30 genes coding for proteins with a primary role on endothelial cell survival, proliferation, adhesion, chemotaxis, and carcinogenesis, such as VEGF, MIF, adrenomedullin, coagulation factor III, acid fibroblast growth factor 1, and osteopontin (OPN) (Rodrigues et al. 2007; Xu et al. 2010), is detectable in primary human Mn exposed to short-term hypoxia (Bosco et al. 2006), human Mn-derived Mf (MDMs) (Burke et al. 2003; Fang et al. 2009a; Harris 2002; Knowles and Harris 2007; White et al. 2004), and DCs (Blengio et al. 2013; Bosco et al. 2008b; Bosco et al. 2011; Ricciardi et al. 2008; Yang et al. 2009) generated under conditions of chronic hypoxia or isolated *ex vivo*. Another set of hypoxia-inducible genes shared by human Mn, MDMs, mouse Mf, Mf cell lines, and human DCs (3068; Fang et al. 2009b) is that coding for proinflammatory cytokines/receptors, such as various components of the IL-1 and TNF receptor/ligand superfamily, IL-6, and IL-23A. Mn and Mf exposed to hypoxia also share the inducibility of the tie-2 receptor, which leads to the downregulation of the antiangiogenic and T-cell-activating cytokine IL-12 in Ang2-treated hypoxic cells (Murdoch et al. 2007) and of arginase-1 (Arg1), which impairs T-cell signal transduction and function acting through the depletion of arginine (Bosco et al. 2006; Louis et al. 1998), as a mechanism of T-cell suppression (Coffelt et al. 2011; Raber et al. 2012; Sica and Bronte 2007). However, hypoxia can also activate a cytokine gene expression pattern unique for a specific cell differentiation stage, as indicated by the finding that a few genes upregulated in hypoxic Mn, MDMs, or DCs were not increased in any of the other Mn lineage populations or were even downregulated (Blengio et al. 2013; Bosco et al. 2006; Fang et al. 2009a; White et al. 2004).

Deregulated chemokine network appears a major culprit in the development, growth, angiogenesis, and organ-specific invasion of several malignancies (Ben Baruch 2006; Kulbe et al. 2004). Interestingly, modulation of the chemokine gene expression profile is a feature shared by the hypoxic transcriptome of different MP populations, although different combinations of molecules are selectively expressed depending on their stage of differentiation. Human Mn exposure to hypoxia profoundly modulates the expression of genes coding for chemokines, with a characteristic dichotomy resulting in the upregulation of ELR + CXC chemokines active on neutrophils, such as CXCL2, CXCL3, CXCL5, CXCL6, and CXCL8 (Bosco et al. 2006), and in the inhibition of chemokines predominantly active on Mn/Mf, T lymphocytes, natural killer (NK) cells, basophils, and/or DCs, such as CCL2, CCL8, CCL15, CCL18, CCL19, CCL23, and CXCL11 (Bosco et al. 2006; Bosco et al. 2008a; Bosco et al. 2008b). Similar upregulation of CXCL8 and CXCL2 and downregulation of CCL2 were demonstrated in hypoxic human MDMs and mouse Mf (Bosco et al. 2004a; Fang et al. 2009a; White et al. 2004), whereas overexpression of CXCL1, CCL3, and CCL5 was reported in Mf and MDM (Fang et al. 2009a; Zampetaki et al. 2004), but not in Mn (Bosco et al. 2006). Bosco et al. (2006, 2008a) and Battaglia et al. (2008) demonstrated that the induction of CCL20, a selective chemoattractant for iDCs, effector/memory T lymphocytes, and naive B cells (Schutyser et al. 2003), is shared by hypoxic Mn and Mf in vitro and in vivo by synovial Mn lineage cells infiltrating the hypoxic joints of juvenile idiopathic arthritis patients. Collectively, these data suggest that hypoxic Mn and Mf exert a similar chemotactic activity on specific immune cell subsets but utilize different combinations of chemokines in relationship to their differentiation stage. Profound changes in the expression of genes coding for chemokines also characterize the gene profile of immature and mature DCs generated under hypoxia, which share the upregulation of those active on neutrophils and activated/memory Th1 lymphocytes concomitantly with the downmodulation of those predominantly chemotactic for naive/resting T cells and Mn or associated with type II polarization, although with some differences in relationship to their maturation stage. Specifically, CXCL8, CXCL1, CXCL10, and CCL20 are induced and CCL18, CCL26 CCL23, CCL24, CCL14, CCL13, and CCL2 are inhibited in immature DCs (Elia et al. 2008; Ricciardi et al. 2008), whereas CXCL2, CXCL3, CXCL5, CXCL6, CXCL8, CCL20, CCL3, and CCL5 are upregulated and CCL18 and CCL23 downregulated in mature DCs (Blengio et al. 2013; Yang et al. 2009). Because some of the hypoxia-inducible CXC chemokines are also potent inducers of angiogenesis and play a critical role in the progression and metastatic spread of certain tumors (Ghadjar et al. 2009; Hwang et al. 2004; Strieter et al. 2006), their increased production by different hypoxic MP subtypes represents another common mechanism accounting for their proangiogenic/protumoral phenotype. Furthermore, the concomitant upregulation by hypoxia of CC chemokines involved in the pathway leading to Th1 immunity and Th17 chemotaxis and differentiation, such as CCL3, CCL5, and CCL20 (Brand 2009; Nistala et al. 2008; Schutyser et al. 2003; Wedderburn et al. 2000), and inhibition of those playing a role in Th2 polarization, Treg generation, and in the maintenance of tolerance, such as CCL18 and CCL23 (Vulcano et al. 2003), shared by all the MP subsets, is compatible with a shift of MP-driven responses toward a Th1/Th17-polarized proinflammatory direction upon adaptation to the hypoxic environment (Kapsenberg 2003; Lanzavecchia and Sallusto 2001; Moser and Murphy 2000).

Hypoxic human Mn, MDMs, and DCs also share the differential modulation of a cluster of genes coding for immunoregulatory cell surface (IRS) receptors (Bosco et al. 2006; Bosco et al. 2011; Bosco and Varesio 2012; Fang et al. 2009a; Pierobon et al. 2013; Yang et al. 2009). These molecules sense pathogen-associated molecular pattern (PAMP)- and damage-associated molecular pattern (DAMP)-mediated inhibitory and activating signals in the environment and are implicated in the regulation of cell maturation, immunogenicity, pro/antiinflammatory and effector functions (Colonna et al. 2000; Kawai and Akira 2011). Interestingly, both common and distinct members of these receptor families were modulated in the different MP populations (Bosco et al. 2006; Bosco et al. 2011; Fang et al. 2009a; Yang et al. 2009), suggesting that hypoxia can selectively influence the IRS receptor profile of MP depending on their differentiation/maturation stage and functional orientation, thus affecting their response toward an antitumor or a tumor-promoting direction (Bosco et al. 2006). Among the hypoxia-inducible receptors shared by all MP populations, of particular relevance is the triggering receptor expressed on myeloid cells (TREM)-1, that is up-regulated on hypoxic Mn and MDMs and induced *ex novo* on immature and mature DCs generated under long-term hypoxia both in vitro and in vivo (Bosco et al. 2006, 2011; Fang et al. 2009a; Pierobon et al. 2013). TREM-1 engagement triggers further secretion of proinflammatory and Th1-priming cytokines and chemokines, such as IL-1β, TNFα, IL-6, CXCL8, CCL4, CCL5, CCL2, CCL17, OPN, and IL-12 (Bosco et al. 2006; Bosco et al. 2011; Fang et al. 2009a; Pierobon et al. 2013). TREM-1 sustained expression may thus represent another mechanism by which the hypoxic tumor environment can promote the Th1/Th17-polarizing inflammatory responses of MP (Bosco et al. 2011; Bosco and Varesio 2012; Bouchon et al. 2000; Sharif and Knapp 2008). TREM-1 was recently shown to have a role in amplifying MP inflammatory responses in tumors. Murat et al. (2009) showed that TREM-1 is induced by hypoxia in TAM infiltrating glioblastoma, and its expression in TAM, Kupffer cells, and hepatic stellate cells infiltrating non-small cell lung cancer and HCC appears to correlate with cancer progression, recurrence, and poor prognosis (Ho et al. 2008; Liao et al. 2012; Wu et al. 2012).

5.3.2 T Lymphocytes

Adaptive immune cells (T lymphocytes and B lymphocytes) mediate inflammation-associated carcinogenesis and neoplastic progression (de Visser and Coussens 2005). The few data currently available on the gene expression regulation by hypoxia in these cells point toward an inhibitory activity of hypoxia on adaptive immune cell function.

Both CD4+ and CD8+ T cells infiltrate the tumor stroma, although they tend to remain associated with vessels in less-hypoxic areas, and only few can be found around necrotic areas.

Hypoxia can affect T-cell development, proliferation, and effector functions (Caldwell et al. 2001). The hypoxic tumor microenvironment contributes to T-cell immunosuppression by impairing their tumoricidal activity through the inhibition of T-cell receptor (TCR) signaling and costimulation, skewing their cytokine profile, and modifying the expression of costimulatory receptors (Caldwell et al. 2001; Gaber et al. 2009; Lukashev et al. 2006; Neumann et al. 2005; Sitkovsky and Lukashev 2008; Thiel et al. 2007). Gaber et al. (2009) showed by functional genome analysis that exposure to hypoxia causes profound alterations in the transcriptome of both quiescent and phytohemagglutinin (PHA)-stimulated CD4+ T cells. Substantial effects are exerted on genes involved in immune response, transcription and transcriptional regulation, protein folding, translation, and modification, cell proliferation, apoptosis, angiogenesis and vasomotor regulation, migration, and ECM remodeling (Gaber et al. 2009). Microarray analysis also demonstrated increased expression by hypoxia of adenosine inhibitory receptors A2A/A2B (Gaber et al. 2009), whose activation by extracellular adenosine released under hypoxic conditions leads to accumulation of immunosuppressive intracellular cyclic AMP repressing T-cell functions (Palazon et al. 2012; Sitkovsky and Lukashev 2008) and inducing T-cell apoptosis (Sun et al. 2010), suggestive of a positive feedback mechanism of regulation of hypoxia-mediated immunosuppressive metabolic pathway. Hypoxia also mediates suppression of several molecules involved in redox metabolisms. However, similarly to what was shown for MP, several genes coding for proinflammatory and proangiogenic cytokines and growth factors (including VEGF, IL1β, TNFα, IL-6, MIF, IL-23A, chemokines and their receptors, such as CXCL8, CXCL1, CXCL2, CCL5, CCL3, CCL4, CXCR4, CCR7, CCR6, CCR5, and CXCR6, immunoregulatory receptors, and costimulatory molecules, such as CD69, CD47, CD96, CD28, TLR2, and CD137) are upregulated by hypoxia in CD4+ cells (Gaber et al. 2009; Palazon et al. 2012). Therefore, hypoxia induces a set of genes involved in angiogenesis, inflammation, and T-cell activation, common to cells belonging to innate and adaptive immunity.

5.3.3 Other Tumor-Infiltrating Leukocytes

Tumor infiltrate includes other biologically relevant cell types which have a dual role on tumor progression. Neutrophils can promote tumor destruction, but they may also enhance the growth, invasiveness, and metastatic potential of tumor cells (Sica et al. 2011). Similarly to what shown in Mn, hypoxia mediates neutrophils' adherence to the endothelium and extravasation by inducing β2 integrin expression (Sica et al. 2011), as well as recruitment/accumulation into the tumors in response to members of the CXC chemokine family and inflammatory mediators released by endothelial and tumor-infiltrating cells (Mantovani et al. 2004). Walmsley et al. (2005) demonstrated that O_2 deprivation promotes neutrophil survival, phagocytosis, and release of proinflammatory cytokines including the neutrophil survival factor, CCL4, resulting in the promotion of neutrophil inflammatory functions (Imtiyaz and Simon 2010; Sica et al. 2011). However, no data are currently available for what concerns the modulation of neutrophil gene expression profile in the hypoxic microenvironment.

NK cells can kill a wide range of tumor cells (Moretta and Moretta 2004), but tumors have developed different escape mechanisms to avoid NK-cell immuno-surveillance (Pietra et al. 2012; Platonova et al. 2011). Limited information is currently available on the impact of hypoxia on NK cell functions. It was shown that low pO_2 inhibits NK-cell differentiation and cytotoxic activity (Fink et al. 2003; Sceneay et al. 2012; Yun et al. 2011; Balsamo et al. 2013). However, no evidence has been provided up to now on the regulation by hypoxia of the gene expression profile in these cells.

5.4 Prognostic Potential of Hypoxia Gene Expression

Clues to the prognosis of cancer might be present at the time of surgical removal in the pattern of hypoxic gene expression of the primary tumor. Every cell type is bound to respond to hypoxia and the gene expression profile of the tumor will have, to some extent, the footprint of hypoxia which, in turn, might be a prognostic indicator. Gene expression studies are of importance to delineate the pathways and the key functional responses that characterize the functional status of different cells under hypoxic conditions, as we have seen for immune cells. Analysis of differential gene expression between normoxic and hypoxic state pinpoints the major molecular events responsible for the hypoxic phenotype. A second major accomplishment of hypoxia gene expression profile studies is to identify gene clusters that, by representing the hypoxic state of a cell or a tissue, have prognostic implication in predicting the disease outcome.

Most of the studies aimed at generating prognostic gene expression classifiers for cancer patients utilized signatures obtained by supervised mathematical computation of the entire gene expression profile of primary tumors, followed by selection of the signature that better partitions two cohorts of patients (e.g., good and poor outcome). This approach, also referred to as top-down, generates good classifiers, but the assembly of the signature is purely mathematical and composed of a very heterogeneous cluster of genes that do not shed light on underlying biology. The use of a biological feature, rather than an algorithm-based classifier, links tumor selection to a molecular program that can be associated with the progress and the management of the tumor. This approach, referred to as bottom-up or biology-driven, is the leading rational to test the potential of hypoxia signatures as prognostic indicators. This approach relies on two premises: (1) hypoxia signatures are a measure of tumor hypoxia in vivo (see Sect. 2.3) and (2) tumor hypoxia is associated with poor prognosis. In this contest, hypoxia signature would be not only a prognostic factor but also the indicator for hypoxia as a potential cause of poor outcome and, hence, the indicator of a therapeutic strategy. For example, hypoxia is associated with radioresistance, and definition of tumor hypoxia could be instrumental to guide patients' selection for radiotherapy.

Most of the hypoxia gene signatures previously described were tested on cohorts of patients for which the gene expression profile of the tumor was available. Chi et al. (2006) reported that the epithelial hypoxia signature was associated with poorer prognosis in breast and ovarian cancer. Furthermore, they found that a set of renal tumors could be stratified into two groups, one with high and one with low expression of the hypoxia-response genes. The high-hypoxia-response group included clear-cell renal cell carcinomas, which frequently present high levels of HIF-1α and/or HIF-2α because of the loss of functional pVHL. They also showed that the generated signature was an independent predictor of poor prognosis, proving its potential in clinical decision making.

The same data set was exploited by Seigneuric et al. (2007) who developed an "early hypoxia" and a "late hypoxia" gene expression signature. Applying these signatures to a validation set of patients with primary breast cancer treated with surgery plus radiotherapy and systemic therapy when needed, the "early hypoxia" signature had a significant prognostic and superior impact compared with the "late hypoxia" signature. This suggested that genes responding early to hypoxia might be the most informative as hypoxia markers.

The 99 gene signature by Winter et al. (2007) proved to be an independent prognostic factor for recurrence-free survival when validated in a publicly available head and neck cancer data set and was furthermore a significant prognostic factor in a published breast cancer series. Further studies from the same group described the "common hypoxia metagene" (Buffa et al. 2010). The prognostic relevance of this signature was assessed in four independent data sets (one breast, two lung, and one head and neck) and, in general, the common hypoxia metagene outperformed the original 99 gene signature.

Using computational methods, Malenstein et al. (2010) identified seven genes responding to two characteristics: differentially expressed under chronic hypoxia and correlation with poor prognostic indicators in the training sets. The source of genes under investigation derives from the hypoxia response; the final classifier is defined on the basis of the prognostic power. This approach is strongly biased toward class prediction and possibly less representative of the hypoxic status of the tumor. They demonstrated that the gene signature was associated with poor survival and early recurrence in 135 patients.

Fardin et al. (2010a) profiled gene expression of the tumors of neuroblastoma patients and divided them according to the 62 probeset signature expression values by K-means clustering. The signature successfully stratified the neuroblastoma patients into good and poor prognosis groups. Multivariate Cox analysis revealed that the NB-hypo is a significant independent predictor after controlling for commonly used risk factors including the amplification of MYCN oncogene. NB-hypo increases the resolution of the MYCN stratification by dividing patients with MYCN not-amplified tumors into good and poor outcome, suggesting that hypoxia is associated with the aggressiveness of neuroblastoma tumor independently from MYCN amplification.

Halle et al. (2012) utilizing MR images identified the prognostic DCE-MRI parameter A(Brix). On the basis of 78 patients with cervical cancer subjected to curative chemoradiotherapy, it was determined that a 31-gene-expression signature characteristic of tumors with low A(Brix) was associated with elevated tumor hypoxia.

This DCE-MRI hypoxia gene signature showed prognostic impact in an independent validation cohort of 109 patients. These findings suggest the use of DCE-MRI to noninvasively identify patients with hypoxia-related chemioradioresistance.

Hypoxia classifiers were utilized for predicting the impact of radiotherapy in the treatment of HNSCC. The fact that hypoxia increases tumor resistance to the effects of radiotherapy has been known for over 70 years. In 1953 Gray et al. showed that hypoxia exists in human tumors and that necrosis occurs about 100 μm from the nearest blood vessel which is also the diffusion distance of soluble O_2 (Gray et al. 1953). The relationship between tumor hypoxia and radiation resistance poses a recurrent problem, reducing the response to this treatment in some patients. The heterogeneity of tumors with respect to hypoxia indicates the possibility to stratify patients on the bases of the potential responsiveness to this treatment. Toustrup et al. (2012a) evaluated a 15-gene hypoxia classifier in relation to its prognostic and predictive impact for hypoxic modification of radiotherapy with nimorazole in a cohort of patients with HNSCC who had been randomized to either nimorazole or placebo in conjunction with radiotherapy (Toustrup et al. 2012b). The "less" hypoxic tumors had a better loco-regional tumor response to radiotherapy and, by adding nimorazole to the "more" hypoxic tumors, the response of these tumors could be improved to a level equal to the "less" hypoxic tumors. Thus, the classifier was capable of predicting which tumors benefited from the addition of hypoxic modification of radiotherapy and also which tumors did not benefit from the additive treatment. These studies also confirm that a gene signature focusing on the cumulative molecular information expressed by the tumor cells under hypoxia will produce a promising strategy for hypoxic classification of tumors. The characterization of the tumor at diagnosis is a valuable tool for deciding the treatment, and poor outcome prediction by hypoxia classifiers may identify the tumors that, as a result of the hypoxic status, express high genetic instability (Vaupel 2004), contain undifferentiated or cancer stem cells (Edsjo et al. 2007; Lin and Yun 2010), or have a higher metastatic potential (Lu and Kang 2010). These characteristics of the primary tumor may be those that initiated the aggressiveness of the disease and could be targeted by individualized treatment. Many therapeutic agents are being developed to target hypoxia (for review see Wilson and Hay 2011) or cells in a hypoxic environment with gene therapy (Carta et al. 2001; Greco and Scott 2007; Muthana et al. 2011) and are being tested in the clinic.

In conclusion, each tumor type is characterized by risk factors derived from the association of clinical parameters with biological and molecular determinations. Hypoxia signature in primary tumors is a good predictor of patients' outcome and/or disease progression. However, the prediction is not totally accurate because the correlation between hypoxia and poor prognosis is not absolute and several factors will overcome the negative effects of hypoxia including the host response to therapy. Therefore, hypoxia signature could be successfully used to improve patient stratification in conjunction with other risk factors and will find a primary role steering the choice of therapy toward hypoxia-related targets or identifying the patients refractory to hypoxia-sensitive therapies (Fig. 5.2).

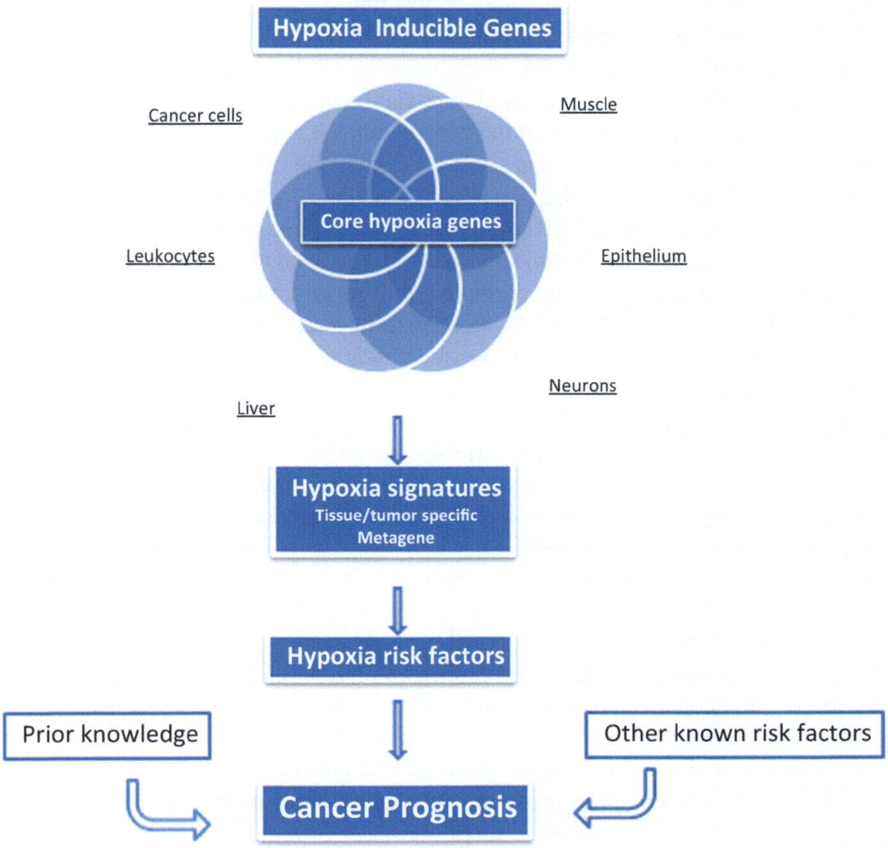

Fig. 5.2 Hypoxia-inducible genes are heterogeneous and differ depending on the cell type. However, core hypoxia genes, shared by many tissues, can be identified. Bioinformatic analysis allows the organization of the hypoxia-modulated genes into signatures and related risk factors for stratification of cancer patients. Robust indicators for cancer prognosis can be obtained by combining hypoxia risk factors and other information from prior knowledge on the disease and its course

5.5 Final Remarks

mRNA assessment by microarray analysis is becoming an affordable and reliable method to characterize hypoxia response. However, microRNAs, noncoding RNA, protein patterns, and transcription factor analysis promise to generate equally important information to define the biology of hypoxia as well as hypoxia differential features. Integration is the keyword for an effective use of these rapidly growing sets of information, and functional genomics is instrumental for translating the existing information into knowledge. Next-generation sequencing, and specifically RNA sequencing, coupled with epigenomics and epigenetics (Perez-Perri et al. 2011; Watson et al. 2010), will bring in a plethora of valuable information. However, the full exploitation of this wealth of data may be lost unless the bioinformatic component of biological research is ready to deal with it.

The diversity of the cancer prognostic signatures raises the question of whether improvement of the predictive power of hypoxia signature could be achieved by combining the discriminating power of the existing ones (Cornero et al. 2012; Fan et al. 2011). Prognostic gene expression signatures have often similar performances despite the lack of gene overlapping, suggesting that they relate to a common biological feature but derive from a highly variable environment (Fan et al. 2006; Haibe-Kains et al. 2008). The combination of the information contained in these signatures should improve the accuracy and/or the predictive power. Ensemble methods were originally developed to enhance classification performance (Tan and Gilbert 2003) and have been recently applied to biomarker identification and feature selection (Abeel et al. 2010). The problem of merging signatures or data sets was recently addressed in breast cancer where it was shown that multiple signatures can lead to robust prognostic when combined with clinical variables and large databases of gene expression (Xu et al. 2008). Furthermore, Nuyten et al. (2008) demonstrated the relevance of combining biological gene expression signatures with an independent predictor for outcome in breast cancer patients. Recently, Fang et al. (2011) reported the generation of a prognostic model combining hundreds of gene expression signatures to clinical–pathological factors utilizing the least absolute shrinkage and selection operator method and a Cox proportional hazards approach. Cornero et al. (2012) described the design, generation, and properties of a classifier based on an ensemble approach that merges 20 heterogeneous, neuroblastoma-related gene signatures, including hypoxia signatures, to blend their discriminating power, rather than numeric values, into a single, highly accurate patients' outcome predictor.

Clinicians have recognized for some time that a combination of clinical variables provides better prediction tools than single variables. Several algorithms can perform this task, but the translation of the computational results to the clinic requires the use of explicit statements coupled with the capability of blending prior knowledge on the disease with newly acquired information from high-throughput technology. Communication problems often hinder the translation of the basic findings to the clinical world, and attention should be paid also to the diffusion of the message hidden in algorithms. Classification methods capable of constructing models described by a set of intelligible rules (if <premise> then <consequence>) were developed to overcome this drawback. Rule generation techniques produce not only the subset of variables actually correlated with the pathology of interest, but also explicit intelligible conditions that determine a specific diagnosis/prognosis. Traditionally, this task was assigned to algorithms generating decision-tree structures that can be expressed as multiple rules. Recent results have indicated the shadow clustering (SC) technique, which leads to final models called logic learning machines (LLMs), as a good paradigm for cancer patients' stratification on the bases of clinical, biochemical risk factors and prognostic signatures (Muselli and Ferrari 2011).

References

Abeel T, Helleputte T, Van de PY et al (2010) Robust biomarker identification for cancer diagnosis with ensemble feature selection methods. Bioinformatics 26:392–398

Balsamo M, Manzini C, Pietra G, Raggi F, Blengio F, Mingari MC, Varesio l, Moretta l, Bosco MC, Vitale M (2013) Hypoxia downregulates the expression of activating receptors involved in NK cell-mediated target cell killing without affecting ADCC. Eur J Immunol. doi: 10.1002/eji.201343448 [Epub ahead of print]

Battaglia F, Delfino S, Merello E et al (2008) Hypoxia transcriptionally induces macrophage-inflammatory protein-3{alpha}/CCL-20 in primary human mononuclear phagocytes through nuclear factor (NF)-{kappa}B. J Leukoc Biol 83:648–662

Ben Baruch A (2006) Inflammation-associated immune suppression in cancer: the roles played by cytokines, chemokines and additional mediators. Semin Cancer Biol 16:38–52

Benita Y, Kikuchi H, Smith AD et al (2009) An integrative genomics approach identifies Hypoxia Inducible Factor-1 (HIF-1)-target genes that form the core response to hypoxia. Nucleic Acids Res 37:4587–4602

Bennaceur K, Chapman J, Brikci-Nigassa L et al (2008) Dendritic cells dysfunction in tumour environment. Cancer Lett 272:186–196

Blengio F, Raggi F, Pierobon D et al (2013) The hypoxic environment reprograms the cytokine/chemokine expression profile of human mature dendritic cells. Immunobiology 218(1):76–89

Bosco MC, Varesio L (2010) Monocytic cell gene regulation by the hypoxic synovial environment in juvenile idiopathic arthritis: implications for disease pathogenesis. J Clin Rheumatol Musculoskel Med 1:47–55

Bosco MC, Varesio L (2012) Dendritic cell reprogramming by the hypoxic environment. Immunobiology 217:1241–1249

Bosco MC, Puppo M, Pastorino S et al (2004a) Hypoxia selectively inhibits monocyte chemoattractant protein-1 production by macrophages. J Immunol 172:1681–1690

Bosco MC, Reffo G, Puppo M et al (2004b) Hypoxia inhibits the expression of the CCR5 chemokine receptor in macrophages. Cell Immunol 228:1–7

Bosco MC, Puppo M, Santangelo C et al (2006) Hypoxia modifies the transcriptome of primary human monocytes: modulation of novel immune-related genes and identification of CC-chemokine ligand 20 as a new hypoxia-inducible gene. J Immunol 177:1941–1955

Bosco MC, Delfino S, Ferlito F et al (2008a) Hypoxic synovial environment and expression of macrophage inflammatory protein MIP-3a/CCL20 in Juvenile Idiopathic Arthritis. Arthritis Rheum 58:1833–1838

Bosco MC, Puppo M, Blengio F et al (2008b) Monocytes and dendritic cells in a hypoxic environment: Spotlights on chemotaxis and migration. Immunobiology 213:733–749

Bosco MC, Pierobon D, Blengio F et al (2011) Hypoxia modulates the gene expression profile of immunoregulatory receptors in human mature dendritic cells: identification of TREM-1 as a novel hypoxic marker in vitro and in vivo. Blood 117:2625–2639

Bouchon A, Dietrich J, Colonna M (2000) Cutting edge: inflammatory responses can be triggered by TREM-1, a novel receptor expressed on neutrophils and monocytes. J Immunol 164:4991–4995

Brand S (2009) Crohn's disease: Th1, Th17 or both? The change of a paradigm: new immunological and genetic insights implicate Th17 cells in the pathogenesis of Crohn's disease. Gut 58: 1152–1167

Brown JM, William WR (2004) Exploiting tumour hypoxia in cancer treatment. Nat Rev Cancer 4:437–447

Buffa FM, Harris AL, West CM et al (2010) Large meta-analysis of multiple cancers reveals a common, compact and highly prognostic hypoxia metagene. Br J Cancer 102:428–435

Burke B, Giannoudis A, Corke KP et al (2003) Hypoxia-induced gene expression in human macrophages: implications for ischemic tissues and hypoxia-regulated gene therapy. Am J Pathol 163:1233–1243

Busk M, Horsman MR, Jakobsen S et al (2008) Cellular uptake of PET tracers of glucose metabolism and hypoxia and their linkage. Eur J Nucl Med Mol Imaging 35:2294–2303
Caldwell CC, Kojima H, Lukashev D et al (2001) Differential effects of physiologically relevant hypoxic conditions on T lymphocyte development and effector functions. J Immunol 167: 6140–6149
Carmeliet P, Dor Y, Herbert JM et al (1998) Role of HIF-1alpha in hypoxia-mediated apoptosis, cell proliferation and tumour angiogenesis. Nature 394:485–490
Carta L, Pastorino S, Melillo G et al (2001) Engineering of macrophages to produce IFN-gamma in response to hypoxia. J Immunol 166:5374–5380
Cavanagh LL, Von Andrian UH (2002) Travellers in many guises:The origins and destinations of dendritic cells. Immun Cell Biol 80:448–462
Chadwick W, Boyle JP, Zhou Y et al (2011) Multiple oxygen tension environments reveal diverse patterns of transcriptional regulation in primary astrocytes. PLoS One 6:e21638
Chan DA, Giaccia AJ (2007) Hypoxia, gene expression, and metastasis. Cancer Metastasis Rev 26:333–339
Chi JT Wang Z, Nuyten DS et al (2006) Gene expression programs in response to hypoxia: cell type specificity and prognostic significance in human cancers. PLoS Med 3:e47
Coffelt SB, Chen YY, Muthana M et al (2011) Angiopoietin 2 stimulates TIE2-expressing monocytes to suppress T cell activation and to promote regulatory T cell expansion. J Immunol 186: 4183–4190
Colonna M, Nakajima H, Cella M (2000) A family of inhibitory and activating Ig-like receptors that modulate function of lymphoid and myeloid cells. Semin Immunol 12:121–127
Cornero A, Acquaviva M, Fardin P et al (2012) Design of a multi-signature ensemble classifier predicting neuroblastoma patients' outcome. BMC Bioinformatics 13(Suppl 4):S13
Cramer T, Yamanishi Y, Clausen BE et al (2003) HIF-1a is essential for myeloid cell-mediated inflammation. Cell 112:645–657
Crowther M, Brown NJ, Bishop ET et al (2001) Microenvironmental influence on macrophage regulation of angiogenesis in wounds and malignant tumors. J Leukoc Biol 70:478–490
Cummins EP, Taylor CT (2005) Hypoxia-responsive transcription factors. Pflugers Arch 450: 363–371
de Visser KE, Coussens LM (2005) The interplay between innate and adaptive immunity regulates cancer development. Cancer Immunol Immunother 54:1143–1152
Eckert AW, Kappler M, Schubert J et al (2012) Correlation of expression of hypoxia-related proteins with prognosis in oral squamous cell carcinoma patients. Oral Maxillofac Surg 16:189–196
Edsjo A, Holmquist L, Pahlman S (2007) Neuroblastoma as an experimental model for neuronal differentiation and hypoxia-induced tumor cell dedifferentiation. Semin Cancer Biol 17: 248–256
Elia AR, Cappello P, Puppo M et al (2008) Human dendritic cells differentiated in hypoxia downmodulate antigen uptake and change their chemokine expression profile. J Leukoc Biol 84: 1472–1482
Evans SM, Hahn S, Pook DR et al (2000) Detection of hypoxia in human squamous cell carcinoma by EF5 binding. Cancer Res 60:2018–2024
Fan C, Oh DS, Wessels L et al (2006) Concordance among gene-expression-based predictors for breast cancer. N Engl J Med 355:560–569
Fan C, Prat A, Parker JS et al (2011) Building prognostic models for breast cancer patients using clinical variables and hundreds of gene expression signatures. BMC Med Genomics 4:3
Fang HY, Hughes R, Murdoch C et al (2009a) Hypoxia-inducible factors 1 and 2 are important transcriptional effectors in primary macrophages experiencing hypoxia. Blood 114:844–859
Fang HY, Hughes R, Murdoch C et al (2009b) Hypoxia-inducible factors 1 and 2 are important transcriptional effectors in primary macrophages experiencing hypoxia. Blood 114:844–859
Fardin P, Barla A, Mosci S et al (2009) The l1-l2 regularization framework unmasks the hypoxia signature hidden in the transcriptome of a set of heterogeneous neuroblastoma cell lines. BMC Genomics 10:474

Fardin P, Barla A, Mosci S et al (2010a) A biology-driven approach identifies the hypoxia gene signature as a predictor of the outcome of neuroblastoma patients. Mol Cancer 9:185

Fardin P, Cornero A, Barla A et al (2010b) Identification of multiple hypoxia signatures in neuroblastoma cell lines by l(1)-l(2) regularization and data reduction. J Biomed Biotechnol 2010:878709

Fink T, Ebbesen P, Koppelhus U et al (2003) Natural killer cell-mediated basal and interferon-enhanced cytotoxicity against liver cancer cells is significantly impaired under in vivo oxygen conditions. Scand J Immunol 58:607–612

Gaber T, Haupl T, Sandig G et al (2009) Adaptation of human CD4+ T cells to pathophysiological hypoxia: a transcriptome analysis. J Rheumatol 36:2655–2669

Gabrilovich D (2004) Mechanisms and functional significance of tumour-induced dendritic-cell defects. Nat Rev Immunol 4:941–952

Ghadjar P, Rubie C, Aebersold DM et al (2009) The chemokine CCL20 and its receptor CCR6 in human malignancy with focus on colorectal cancer. Int J Cancer 125:741–745

Gray LH, Conger AD, Ebert M et al (1953) The concentration of oxygen dissolved in tissues at the time of irradiation as a factor in radiotherapy. Br J Radiol 26:638–648

Greco O, Scott S (2007) Tumor hypoxia and targeted gene therapy. Int Rev Cytol 257:181–212

Grimshaw MJ, Balkwill FR (2001) Inhibition of monocyte and macrophage chemotaxis by hypoxia and inflammation—a potential mechanism. Eur J Immunol 31:480–489

Gross MW, Karbach U, Groebe K et al (1995) Calibration of misonidazole labeling by simultaneous measurement of oxygen tension and labeling density in multicellular spheroids. Int J Cancer 61:567–573

Haibe-Kains B, Desmedt C, Piette F et al (2008) Comparison of prognostic gene expression signatures for breast cancer. BMC Genomics 9:394

Halle C, Andersen E, Lando M et al (2012) Hypoxia-induced gene expression in chemoradioresistant cervical cancer revealed by dynamic contrast-enhanced MRI. Cancer Res 72:5285–5295

Harris AL (2002) Hypoxia—a key regulatory factor in tumour growth. Nat Rev Cancer 2:38–47

Ho CC, Liao WY, Wang CY et al (2008) TREM-1 expression in tumor-associated macrophages and clinical outcome in lung cancer. Am J Respir Crit Care Med 177:763–770

Hoeben BA, Kaanders JH, Franssen GM et al (2010) PET of hypoxia with 89Zr-labeled cG250-F(ab')2 in head and neck tumors. J Nucl Med 51:1076–1083

Holland JP, Lewis JS, Dehdashti F (2009) Assessing tumor hypoxia by positron emission tomography with Cu-ATSM. Q J Nucl Med Mol Imaging 53:193–200

Hwang J, Kim CW, Son KN et al (2004) Angiogenic activity of human CC chemokine CCL15 in vitro and in vivo. FEBS Lett 570:47–51

Imhof BA, Aurrand-Lions M (2004) Adhesion mechanisms regulating the migration of monocytes. Nat Rev Immunol 4:432–444

Imtiyaz HZ, Simon MC (2010) Hypoxia-inducible factors as essential regulators of inflammation. Curr Top Microbiol Immunol 345:105–120

Jogi A, Vallon-Christersson J, Holmquist L et al (2004) Human neuroblastoma cells exposed to hypoxia: induction of genes associated with growth, survival, and aggressive behavior. Exp Cell Res 295:469–487

Kapsenberg ML (2003) Dendritic-cell control of pathogen-driven T-cell polarization. Nat Rev Immunol 3:984–993

Kawai T, Akira S (2011) Toll-like receptors and their crosstalk with other innate receptors in infection and immunity. Immunity 34:637–650

Knowles HJ, Harris AL (2007) Macrophages and the hypoxic tumour microenvironment. Front Biosci 12:4298–4314

Kong T, Eltzschig HK, Karhausen J et al (2004) Leukocyte adhesion during hypoxia is mediated by HIF-1-dependent induction of beta2 integrin gene expression. Proc Natl Acad Sci U S A 101:10440–10445

Kulbe H, Levinson NR, Balkwill F et al (2004) The chemokine network in cancer–much more than directing cell movement. Int J Dev Biol 48:489–496

Lamagna C Aurrand-Lions M, Imhof BA (2006) Dual role of macrophages in tumor growth and angiogenesis. J Leukoc Biol 80:705–713
Lanzavecchia A, Sallusto F (2001) Regulation of T cell immunity by dendritic cells. Cell 106: 263–266
Lewis C, Murdoch C (2005) Macrophage responses to hypoxia: implications for tumor progression and anti-cancer therapies. Am J Pathol 167:627–635
Lewis CE, Pollard JW (2006) Distinct role of macrophages in different tumor microenvironments. Cancer Res 66:605–612
Lewis CE, De Palma M, Naldini L (2007) Tie2-expressing monocytes and tumor angiogenesis: regulation by hypoxia and angiopoietin-2. Cancer Res 67:8429–8432
Li Q, Withoff S, Verma IM (2005) Inflammation-associated cancer: NF-kappaB is the lynchpin. Trends Immunol 26:318–325
Liao R, Sun TW, Yi Y et al (2012) Expression of TREM-1 in hepatic stellate cells and prognostic value in hepatitis B-related hepatocellular carcinoma. Cancer Sci 103:984–992
Lin A, Schildknecht A, Nguyen LT et al (2010) Dendritic cells integrate signals from the tumor microenvironment to modulate immunity and tumor growth. Immunol Lett 127:77–84
Lin KW, Jacek T, Jacket R (2006) Dendritic cells heterogeneity and its role in cancer immunity. Cancer Res Ther 2:35–40
Lin Q, Yun Z (2010) Impact of the hypoxic tumor microenvironment on the regulation of cancer stem cell characteristics. Cancer Biol Ther 9:949–956
Louis CA, Reichner JS, Henry WL, Jr. et al (1998) Distinct arginase isoforms expressed in primary and transformed macrophages: regulation by oxygen tension. Am J Physiol 274:R775–R782
Lu X, Kang Y (2010) Hypoxia and hypoxia-inducible factors: master regulators of metastasis. Clin Cancer Res 16:5928–5935
Lukashev D, Klebanov B, Kojima H et al (2006) Cutting edge: hypoxia-inducible factor 1 and its activation-inducible short isoform I.1 negatively regulate functions of CD4+ and CD8+ T lymphocytes. J Immunol 177:4962–4965
Malenstein H, Gevaert O, Libbrecht L et al (2010) A seven-gene set associated with chronic hypoxia of prognostic importance in hepatocellular carcinoma. Clin Cancer Res 16:4278–4288
Mantovani A, Sica A (2010) Macrophages, innate immunity and cancer: balance, tolerance, and diversity. Curr Opin Immunol 22:231–237
Mantovani A, Allavena P, Sozzani S et al (2004) Chemokines in the recruitment and shaping of the leukocyte infiltrate of tumors. Semin Cancer Biol 14:155–160
Mees G, Dierckx R, Vangestel C et al (2009) Molecular imaging of hypoxia with radiolabelled agents. Eur J Nucl Med Mol Imaging 36:1674–1686
Moon EJ, Brizel DM, Chi JT et al (2007) The potential role of intrinsic hypoxia markers as prognostic variables in cancer. Antioxid Redox Signal 9:1237–1294
Moretta L, Moretta A (2004) Unravelling natural killer cell function: triggering and inhibitory human NK receptors. EMBO J 23:255–259
Moser M, Murphy KM (2000) Dendritic cell regulation of TH1-TH2 development. Nat Immunol 1:199–205
Murat A, Migliavacca E, Hussain SF et al (2009) Modulation of angiogenic and inflammatory response in glioblastoma by hypoxia. PLoS One 4:e5947
Murdoch C, Giannoudis A, Lewis CE (2004) Mechanisms regulating the recruitment of macrophages into hypoxic areas of tumors and other ischemic tissues. Blood 104:2224–2234
Murdoch C, Tazzyman S, Webster S et al (2007) Expression of Tie-2 by human monocytes and their responses to angiopoietin-2. J Immunol 178:7405–7411
Muselli M, Ferrari E (2011) Coupling Logical Analysis of Data and Shadow Clustering for Partially Defined Positive Boolean Function Reconstruction. IEEE Transactions on Knowledge Data Engineering 23:37–50
Muthana M, Giannoudis A, Scott SD et al (2011) Use of macrophages to target therapeutic adenovirus to human prostate tumors. Cancer Res 71:1805–1815

Neumann AK, Yang J, Biju MP et al (2005) Hypoxia inducible factor 1 alpha regulates T cell receptor signal transduction. Proc Natl Acad Sci U S A 102:17071–17076

Nistala K, Moncrieffe H, Newton KR et al (2008) Interleukin-17-producing T cells are enriched in the joints of children with arthritis, but have a reciprocal relationship to regulatory T cell numbers. Arthritis Rheum 58:875–887

Nizet V, Johnson RS (2009) Interdependence of hypoxic and innate immune responses. Nat Rev Immunol 9:609–617

Nuyten DS, Hastie T, Chi JT et al (2008) Combining biological gene expression signatures in predicting outcome in breast cancer: An alternative to supervised classification. Eur J Cancer 44:2319–2329

Palazon A, Aragones J, Morales-Kastresana A et al (2012) Molecular pathways: hypoxia response in immune cells fighting or promoting cancer. Clin Cancer Res 18:1207–1213

Palucka KA, Taquet N, SanchezChapuis F et al (1998) Dendritic cells as the terminal stage of monocyte differentiation. J Immunol 160:4587–4595

Perez-Perri JI, Acevedo JM, Wappner P (2011) Epigenetics: new questions on the response to hypoxia. Int J Mol Sci 12:4705–4721

Peyssonnaux C, Datta V, Cramer T et al (2005) HIF-1alpha expression regulates the bactericidal capacity of phagocytes. J Clin Invest 115:1806–1815

Pierobon D, Bosco MC, Blengio F et al (2013) Chronic hypoxia reprograms human immature dendritic cells by inducing a proinflammatory phenotype and TREM-1 expression. Eur J Immunol 43:949–966

Pietra G, Manzini C, Rivara S et al (2012) Melanoma cells inhibit natural killer cell function by modulating the expression of activating receptors and cytolytic activity. Cancer Res 72: 1407–1415

Platonova S, Cherfils-Vicini J, Damotte D et al (2011) Profound coordinated alterations of intratumoral NK cell phenotype and function in lung carcinoma. Cancer Res 71:5412–5422

Pocock R (2011) Invited review: decoding the microRNA response to hypoxia. Pflugers Arch 461:307–315

Pollard JW (2004) Tumour-educated macrophages promote tumour progression and metastasis. Nat Rev Cancer 4:71–78

Qian BZ, Pollard JW (2010) Macrophage diversity enhances tumor progression and metastasis. Cell 141:39–51

Raber P, Ochoa AC, Rodriguez PC (2012) Metabolism of L-Arginine by Myeloid-Derived Suppressor Cells in Cancer: Mechanisms of T cell suppression and Therapeutic Perspectives. Immunol Invest 41:614–634

Rahat MA, Marom B, Bitterman H et al (2006) Hypoxia reduces the output of matrix metalloproteinase-9 (MMP-9) in monocytes by inhibiting its secretion and elevating membranal association. J Leukoc Biol 79:706–718

Raleigh JA, Chou SC, Arteel GE et al (1999) Comparisons among pimonidazole binding, oxygen electrode measurements, and radiation response in C3H mouse tumors. Radiat Res 151:580–589

Rama I, Brune B, Torras J et al (2008) Hypoxia stimulus: An adaptive immune response during dendritic cell maturation. Kidney Int 73:816–825

Rankin EB, Giaccia AJ (2008) The role of hypoxia-inducible factors in tumorigenesis. Cell Death Differ 15:678–685

Ricciardi A, Elia AR, Cappello P et al (2008) Transcriptome of hypoxic immature dendritic cells: modulation of chemokine/receptor expression. Mol Cancer Res 6:175–185

Rius J, Guma M, Schachtrup C et al (2008) NF-kappaB links innate immunity to the hypoxic response through transcriptional regulation of HIF-1alpha. Nature 453:807–811

Rodrigues LR, Teixeira JA, Schmitt FL et al (2007) The role of osteopontin in tumor progression and metastasis in breast cancer. Cancer Epidemiol Biomarkers Prev 16:1087–1097

Roiniotis J, Dinh H, Masendycz P et al (2009) Hypoxia prolongs monocyte/macrophage survival and enhanced glycolysis is associated with their maturation under aerobic conditions. J Immunol 182:7974–7981

Rossi M, Young JW (2005) Human dendritic cells: potent antigen-presenting cells at the crossroads of innate and adaptive immunity. J Immunol 175:1373–1381
Saeys Y, Inza I, Larranaga P (2007) A review of feature selection techniques in bioinformatics. Bioinformatics 23:2507–2517
Sceneay J, Chow MT, Chen A et al (2012) Primary tumor hypoxia recruits CD11b+/Ly6Cmed/Ly6G+ immune suppressor cells and compromises NK cell cytotoxicity in the premetastatic niche. Cancer Res 72:3906–3911
Schioppa T, Uranchimeg B, Saccani A et al (2003) Regulation of the Chemokine Receptor CXCR4 by Hypoxia. J Exp Med 198:1391–1402
Schmieder A, Michel J, Schonhaar K et al (2012) Differentiation and gene expression profile of tumor-associated macrophages. Semin Cancer Biol 22:289–297
Schreiber RD, Old LJ, Smyth MJ (2011) Cancer immunoediting: integrating immunity's roles in cancer suppression and promotion. Science 331:1565–1570
Schutyser E, Struyf S, Van Damme J (2003) The CC chemokine CCL20 and its receptor CCR6. Cytokine Growth Factor Rev 14:409–426
Seigneuric R, Starmans MH, Fung G et al (2007) Impact of supervised gene signatures of early hypoxia on patient survival. Radiother Oncol 83:374–382
Semenza GL (2009) Regulation of cancer cell metabolism by hypoxia-inducible factor 1. Semin Cancer Biol 19:12–16
Semenza GL (2011) Oxygen sensing, homeostasis, and disease. N Engl J Med 365:537–547
Semenza GL (2012) Hypoxia-inducible factors: mediators of cancer progression and targets for cancer therapy. Trends Pharmacol Sci 33:207–214
Sharif O, Knapp S (2008) From expression to signaling: roles of TREM-1 and TREM-2 in innate immunity and bacterial infection. Immunobiology 213:701–713
Sica A, Bronte V (2007) Altered macrophage differentiation and immune dysfunction in tumor development. J Clin Invest 117:1155–1166
Sica A, Saccani A, Bottazzi B et al (2000) Defective expression of the Monocyte Chemotactic Protein-1 Receptor CCR2 in macrophages associated with human ovarian carcinoma. J Immunol 164:733–738
Sica A, Melillo G, Varesio L (2011) Hypoxia: a double-edged sword of immunity. J Mol Med 89:657–665
Sitkovsky M, Lukashev D (2008) Regulation of immune cells by local-tissue oxygen tension: HIF1 alpha and adenosine receptors. Nat Rev Immunol 5:712–721
Steinman RM, Banchereau J (2007) Taking dendritic cells into medicine. Nature 449:419–426
Strieter RM, Burdick MD, Mestas J et al (2006) Cancer CXC chemokine networks and tumour angiogenesis. Eur J Cancer 42:768–778
Subramanian A, Tamayo P, Mootha VK et al (2005) Gene set enrichment analysis: a knowledge-based approach for interpreting genome-wide expression profiles. PNAS 102:15545–15550
Sun J, Zhang Y, Yang M et al (2010) Hypoxia induces T-cell apoptosis by inhibiting chemokine C receptor 7 expression: the role of adenosine receptor A(2). Cell Mol Immunol 7:77–82
Synnott JM, Guida A, Mulhern-Haughey S et al (2010) Regulation of the hypoxic response in Candida albicans. Eukaryot Cell 9:1734–1746
Tan AC, Gilbert D (2003) Ensemble machine learning on gene expression data for cancer classification. Appl Bioinformatics 2:S75–S83
Thiel M, Caldwell CC, Kreth S et al (2007) Targeted deletion of HIF-1alpha gene in T cells prevents their inhibition in hypoxic inflamed tissues and improves septic mice survival. PLoS One 2:e853
Toustrup K, Sorensen BS, Nordsmark M et al (2011) Development of a hypoxia gene expression classifier with predictive impact for hypoxic modification of radiotherapy in head and neck cancer. Cancer Res 71:5923–5931
Toustrup K, Sorensen BS, Alsner J et al (2012a) Hypoxia gene expression signatures as prognostic and predictive markers in head and neck radiotherapy. Semin Radiat Oncol 22:119–127

Toustrup K, Sorensen BS, Lassen P et al (2012b) Gene expression classifier predicts for hypoxic modification of radiotherapy with nimorazole in squamous cell carcinomas of the head and neck. Radiother Oncol 102:122–129
Turner L, Scotton C, Negus R et al (1999) Hypoxia inhibits macrophage migration. Eur J Immunol 29:2280–2287
Vaupel P (2004) The role of hypoxia-induced factors in tumor progression. Oncologist 9 (Suppl 5):10–17
Vaupel P, Kelleher DK (2013) Blood flow and oxygenation status of prostate cancers. Adv Exp Med Biol 765:299–305
Vaupel P, Thews G (1974) PO2 distribution in tumor tissue of DS-carcinosarcoma. Oncology 30:475–484
Vicari AP, Caux C, Trinchieri G (2002) Tumour escape from immune surveillance through dendritic cell inactivation. Semin Cancer Biol 12:33–42
Vulcano M, Struyf S, Scapini P et al (2003) Unique regulation of CCL18 production by maturing dendritic cells. J Immunol 170:3843–3849
Walmsley SR, Print C, Farahi N et al (2005) Hypoxia-induced neutrophil survival is mediated by HIF-1alpha-dependent NF-kappaB activity. J Exp Med 201:105–115
Watson JA, Watson CJ, McCann A et al (2010) Epigenetics, the epicenter of the hypoxic response. Epigenetics 5:293–296
Wedderburn LR, Robinson N, Patel A et al (2000) Selective recruitment of polarized T cells expressing CCR5 and CXCR3 to the inflamed joints of children with juvenile idiopathic arthritis. Arthritis Rheum 43:765–774
Wenger RH, Stiehl DP, Camenisch G (2005) Integration of oxygen signaling at the consensus HRE. Sci STKE 2005:re12
White JR, Harris RA, Lee SR et al (2004) Genetic amplification of the transcriptional response to hypoxia as a novel means of identifying regulators of angiogenesis. Genomics 83:1–8
Wilson WR, Hay MP (2011) Targeting hypoxia in cancer therapy. Nat Rev Cancer 11:393–410
Winter SC, Buffa FM, Silva P et al (2007) Relation of a hypoxia metagene derived from head and neck cancer to prognosis of multiple cancers. Cancer Res 67:3441–3449
Wu J, Li J, Salcedo R et al (2012) The proinflammatory myeloid cell receptor TREM-1 controls Kupffer cell activation and development of hepatocellular carcinoma. Cancer Res 72:3977–3986
Xu L, Tan AC, Winslow RL et al (2008) Merging microarray data from separate breast cancer studies provides a robust prognostic test. BMC Bioinformatics 9:125
Xu M, Mizoguchi I, Morishima N et al (2010) Regulation of antitumor immune responses by the IL-12 family cytokines, IL-12, IL-23, and IL-27. Clin Dev Immunol 2010:832454
Yang M, Ma C, Liu S et al (2009) Hypoxia skews dendritic cells to a T helper type 2-stimulating phenotype and promotes tumour cell migration by dendritic cell-derived osteopontin. Immunology 128:e237–e249
Yun S, Lee SH, Yoon SR et al (2011) Oxygen tension regulates NK cells differentiation from hematopoietic stem cells in vitro. Immunol Lett 137:70–77
Zampetaki A, Mitsialis SA, Pfeilschifter J et al (2004) Hypoxia induces macrophage inflammatory protein-2 (MIP-2) gene expression in murine macrophages via NF-kappaB: the prominent role of p42/p44 and PI3 kinase pathways. FASEB J 18:1090–1092

Chapter 6
Carbonic Anhydrase IX: From Biology to Therapy

Silvia Pastorekova and Claudiu T. Supuran

Abstract Growing tumor tissues develop a stressful microenvironment characterized by hypoxia and acidosis. Tumor cells can survive these stresses via induction of adaptive transcriptional changes mediated primarily by the hypoxia-inducible factor (HIF), and via stimulation of ion transport machinery maintaining normal intracellular pH. In addition, through these adaptive responses tumor cells acquire new features endowing them with selective advantage in migration, invasion, metastasis, and resistance to therapy. Carbonic anhydrase IX (CA IX), a highly active cancer-related carbonic anhydrase isoform, is linked to both hypoxia, as a direct transcriptional target of HIF, and acidosis, as a component of mechanisms that facilitate ion transport across the plasma membrane and thereby counteract the intracellular accumulation of acidic metabolic products. Expression pattern of CA IX in human tumors reflects the activation of the HIF pathway by physiologic hypoxia, genetic defects, and/or oncogenic events. Moreover, CA IX plays an active role not only in pH regulation but also in cell migration and invasion. Thus, it is often exploited and/or investigated as an intrinsic marker of hypoxia, a prognostic indicator, and a therapeutic target for antibodies or inhibitors of the enzyme activity. It is believed that these CA IX-targeted therapeutic approaches can mediate the selective killing of CA IX-positive cells or sensitize tumor cells to conventional treatment modalities. In addition, both CA IX-specific antibodies and CA IX-selective inhibitors can serve as imaging tools allowing for selection of patients potentially benefiting from CA IX-directed therapy. Recent advances in understanding CA IX regulation and functional involvement in tumor progression as well as development of CA IX-binding drugs provide novel opportunities for treatment of hypoxic tumors.

S. Pastorekova (✉)
Department of Molecular Medicine, Institute of Virology, Slovak Academy of Sciences, Dubravska cesta 9, 845 05 Bratislava, Slovak Republic
e-mail: silvia.pastorekova@savba.sk

C. T. Supuran
Polo Scientifico, NEUROFARBA Department, Section of Pharmaceutical and Nutriceutical Sciences, Università degli Studi di Firenze, Via Ugo Schiff 6, Sesto Fiorentino, 50019 Florence, Italy
e-mail: claudiu.supuran@unifi.it

G. Melillo (ed.), *Hypoxia and Cancer*, Cancer Drug Discovery and Development, DOI 10.1007/978-1-4614-9167-5_6,

Keywords Carbonic anhydrase · Hypoxia · Cancer · pH regulation · Bicarbonate transport · Inhibitor · Sulfonamide · Anticancer therapy

Abbreviations

CA IX	carbonic anhydrase IX protein
CA9	carbonic anhydrase 9 gene/promoter
CAI	carbonic anhydrase inhibitor
Car	mouse CA gene
FIH	factor inhibiting HIF
GLUT	glucose transporter
HIF	hypoxia-inducible factor
HRE	hypoxia-response element
LDH	lactate dehydrogenase
MAPK	mitogen activated protein kinase
MCT	monocarboxylate transporter
NBC	Na^+/bicarbonate co-transporter
NHE	Na^+/H^+ exchanger
PDK	pyruvate dehydrogenase kinase
PG	proteoglycan-like domain
PHD	prolyl hydroxylase
pHe	extracellular pH
pHi	intracellular pH
PI3K	phosphatidyl inositol-3 kinase
RCC	renal cell carcinoma
VEGF	vascular endothelial growth factor
VHL	von Hippel Lindau

6.1 Introduction

Cancer develops through accumulation of mutations and epigenetic changes in tumor cells leading to acquisition of selective advantage in proliferation and survival (Hanahan and Weinberg 2000). Moreover, in the growing tumor tissue, subpopulations of cells have to overcome different constraints, such as physiological barriers caused by increased interstitial pressure and abnormal vasculature, leading to the lack of nutrients and oxygen (hypoxia). To be able to expand, tumor cells induce remodeling of their expression program and gain new phenotypic features, including the capability to stimulate angiogenesis, rebuild metabolism, decrease adhesion, adjust proliferation, increase migration and invasion, etc. This adaptation to physiological stresses in the tumor tissue is a fundamental aspect of tumor progression that supports metastasis and resistance to anticancer therapy (Harris 2002).

One of the principal responses to hypoxia occurs in metabolism, which is shifted from oxidative phosphorylation towards glycolysis that usually persists also in reoxygenated tumor cells. Thisis because the glycolytic metabolism can generate energy independently of oxygen supply and also support biosynthetic pathways, and is therefore advantageous for tumor growth (Kim and Dang 2006; Vander Heiden 2009). However, intracellular accumulation of the acidic products of such oncogenic metabolism is incompatible with proliferation and survival as many biosynthetic reactions depend on neutral or slightly alkaline intracellular pH. Therefore, tumor cells activate mechanisms of ion transport and acid–base control that can accomplish efficient extrusion of acidic metabolites and maintain intracellular pH within the optimal values. On the other hand, the increased extrusion of the metabolic waste linked with its reduced removal by aberrant tumor vasculature leads to extracellular acidosis that supports tumor progression and resistance to therapy (Stubbs et al. 2000; Brahimi-Horn and Pouyssegur 2007). Carbonic anhydrase IX (CA IX) is an important component of these pH-regulating mechanisms in tumor cells exposed to stresses from hypoxia and metabolic acidosis (Pastorekova et al. 2008).

CA IX belongs to the α family of carbonic anhydrases, the zinc metalloenzymes that catalyze the reversible conversion of carbon dioxide to bicarbonate ions and protons (Pastorekova et al. 2004). Fifteen human CA isoforms display variable degrees of activity, molecular features, subcellular localization (cytosolic, mitochondrial, transmembrane, and secreted), distribution in tissues (broad or confined to certain tissues), expression levels, kinetic properties, and sensitivity to inhibitors (mostly sulfonamides and their derivatives). Twelve active isoenzymes (CA I–IV, VA, VB, VI, VII, IX, XII–XIV) possess a conserved active site including three histidine residues that participate in the coordination of a zinc ion and the fourth histidine residue that functions as a proton shuttle (Supuran 2007, 2008). Three CA isoforms (VIII, X, and XI) lacking one of the three important histidines are catalytically inactive. There are also two CA-related proteins that are known as receptor protein tyrosine phosphatases (RPTPβ/γ) and have an incompletely conserved, inactive CA domain pocket, which works as a receptor site binding the neuronal adhesion molecule, contactin, and thereby mediates adhesive communication between astroglial cells and gonadotrophin-releasing hormone neurones (Peles et al. 1995; Parent et al. 2007).

Carbonic anhydrases play fundamental roles in various physiological processes (including respiration, digestion, and production of body fluids). Abnormal levels of various isoenzymes were found in conjunction with different human pathologies, such as glaucoma, osteopetrosis, renal failure, neurological disorders, etc. (Supuran 2008). Decreased expression of the cytosolic isoenzymes CA I, II, III, and XIII was detected in advanced colorectal, hepatocellular, and lung tumors compared to corresponding normal tissues (Mori et al. 1993; Kivela et al. 2001; Chiang et al. 2002; Kuo et al. 2003; Kummola et al. 2005) suggesting their relationship to cell differentiation. Nevertheless, CA II is strongly expressed in GIST, in tumor endothelium of melanomas and gliomas, and is a part of the whole genome sequencing-derived metastatic gene signature of colorectal cancer suggesting its involvement in tumor biology in certain tissues and under certain situations (Yoshiura et al. 2005; Haapasalo et al. 2007; Ki et al. 2007). The mitochondrial CA V isoform represents another potential player in cancer, because its catalytic activity supporting the biosythesis of metabolic intermediates might be required for highly proliferating glycolytic tumor cells, although no relevant data are available so far.

On the other hand, increasing number of studies demonstrate the expression of catalytically inactive isoforms CA VIII and XI in cancer (namely colorectal and lung carcinomas) and indicate that they can be involved in tumor growth and invasion, apparently via enzyme-independent functions (Morimoto et al. 2005; Aspatwar et al. 2010). Actually, among the isoforms broadly distributed in differentiated normal tissues, CA XII is the one that can be obviously designated as cancer-related, since it is abnormally expressed in some tumor types. Tureci et al. (1998) detected CA XII in human renal cell carcinoma (RCC) by serological expression screening with autologous antibodies and showed that its messenger RNA (mRNA) is overexpressed in about 10 % of RCC patients. Ivanov et al. (1998) cloned CA XII as a novel von Hippel Lindau tumor suppressor protein (pVHL) target using RNA differential display. Afterwards, expression of CA XII was also demonstrated in breast, ovarian, colorectal, cervical, and lung carcinoma and was linked to differentiated phenotype, good prognosis, and/or favorable response to therapy (Kivela et al. 2000; Hynninen et al. 2006; Wykoff et al. 2001; Yoo et al. 2010; Ilie et al. 2011). The only link between CA XII and poor prognosis was made in oral squamous carcinoma (Chien et al. 2012). Interestingly, CA XII was also proposed as a sero-diagnostic marker for lung cancer and as a marker of breast cancer lymph node metastasis (Kobayashi et al. 2012; Tafreshi et al. 2012). The latest study showed the expression of CA XII in T-cell acute lymphoblastic leukemia/lymphoma (Lounnas et al. 2013).

CA IX occupies a special position among the other CA isoforms because of its limited expression in normal tissues, but very broad distribution in many solid tumors, often in connection with advanced cancer. Thus, CA IX is the only "genuine" cancer-associated CA isoenzyme that presumably evolved to operate in the context of abnormal tumor phenotype and physiology. The molecular features, upstream regulatory mechanisms, and downstream functional consequences described below support this view.

6.2 Molecular Characteristics of CA IX

Carbonic anhydrase IX, CA IX, originally named MN protein, was first detected in the human HeLa cell line (derived from cervical carcinoma) by the monoclonal antibody M75, which showed that it is a plasma membrane antigen regulated by cell density and associated with a tumorigenic phenotype (Pastorekova et al. 1992; Zavada et al. 1993). Expression of the MN antigen was found in various tumor cell lines and surgical tumor specimens, but not in the corresponding normal tissues (Zavada et al. 1993; Liao et al. 1994).

Cloning and sequencing of the MN complementary DNA (cDNA) and gene showed that the encoded protein contains a well-conserved CA domain located in the large extracellular part of the molecule (Pastorek et al. 1994; Opavsky et al. 1996). It shows a significant amino acid sequence identity to the secreted isozyme CA VI (40.8 %) and the cytosolic isozyme CA II (35.8 %)and contains all amino acids required for the catalytic activity. Exon–intron distribution in the genomic region

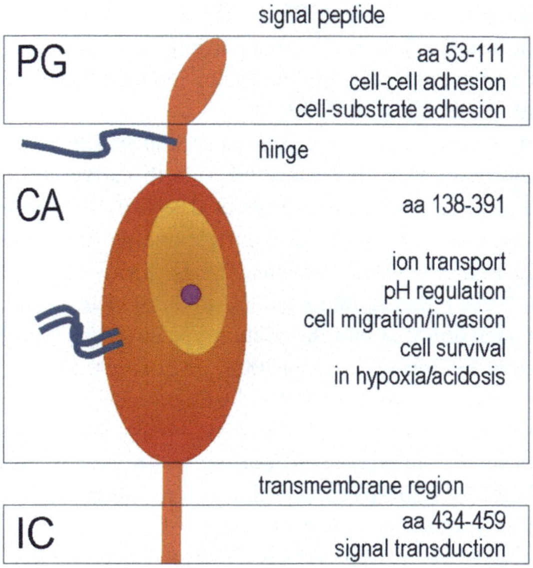

Fig. 6.1 Scheme of the domain composition of the mature CA IX protein, which is composed of a proteoglycan-like region (*PG*) connected by hinge to a carbonic anhydrase domain (*CA*), followed by a transmembrane region (TM) and an intracellular tail (*IC*). Blue side chains represent modifications by O- and N-glycosylation. Numbers besides the drawing designate amino acids (*aa*) encompassed within the individual domains (numbering corresponds to the cDNA nucleotide sequence including the portion coding for the signal peptide). Biological functions attributed to the domains based on the experimental evidence are listed on the right side

coding for the CA domain is similar to that in other α-CA genes. Because at the time of MN protein discovery, it was the ninth mammalian CA identified, it was renamed CA IX (Hewett-Emmett and Tashian 1996). Interestingly, CA IX appears to belong to the most evolutionarily ancient human carbonic anhydrases.

The CA domain of CA IX is C-terminally extended with a single transmembrane region and a short intracellular tail. CA IX also contains an N-terminal extension (see Fig. 6.1), which was initially incorrectly sequenced, but was then corrected by genomic sequencing and recognized as a region homologous to the keratan sulfate-attachment domain of a large proteoglycan aggrecan (Opavsky et al. 1996). This PG-like region appears to be modified by a keratan sulfate glucosaminoglycan chain of about 3.5 kDa (Hilvo et al. 2008). This might have important implications for the biological function of CA IX, because sulfated sugar chains can mediate interactions with the components of the extracellular matrix as it was shown for the CD-44 metastasis-related protein that contains a similar modification (Takahashi et al. 1996). Interestingly, the PG region is absent from the other CA isozymes. On the other hand, a similar PG–CA domain combination is found in extracellular parts of human receptor protein tyrosine phosphatases (RPTPβ/γ) and a related rat phosphacan, which contain inactive CA pockets and are expressed in the form of PGs (Barnea et al. 1994). An additional N-glycosylation site is present in the CA domain and binds a high-mannose sugar (Pastorek et al. 1994). The cytoplasmic tail of CA IX contains three phosphorylation sites (T, S, and Y) that participate in modulation of the catalytic activity of CA IX and in epidermal growth factor-induced signal transduction to Akt kinase (Dorai et al. 2005; Ditte et al. 2011).

The PG region contains a repetitive sequence that serves as a linear epitope for the monoclonal antibody M75 that can bind to both native and denatured CA IX without cross-reactivity to other CAs(Zavada et al. 2000). In immunoblotting, CA IX is represented by two bands of 58/54 kDa corresponding to the monomeric form of the protein. Under nonreducing conditions, CA IX forms an oligomer of 153 kDa (Pastorekova et al. 1992). Initial data from the study of virus pseudotypes suggested that native CA IX is assembled into the pseudotype envelope membrane as a trimer, similarly to viral glycoproteins (Harrison 2008). In contrast, a recent crystallographic study supports a dimeric constitution of the CA IX protein (Alterio et al. 2009). Since this latter study was accomplished with the recombinant catalytic domain in absence of the other parts of the molecule and correct posttranslational modifications, it cannot be considered for definitive resolution of the nature of the CA IX oligomer, and thus this issue still remains controversial.

The enzyme activity of CA IX is very high, even exceeding that of CA II (Wingo et al. 2001; Hilvo et al. 2008). Moreover, it is an inhibitor-avid and targetable enzyme that can be inhibited by different classes of compounds, such as sulfonamides, sulfamates, and their derivatives, as well as by physiologically relevant ions, natural products (coumarins), inhibitors of unrelated enzymes, some chemotherapeutic drugs, etc. (Supuran 2008; Neri and Supuran 2011; Buller et al. 2011). Some compounds can act even at subnanomolar concentrations towards a purified recombinant CA domain and work much better with CA IX than with other CA isozymes; see in more detail later in this chapter.

6.3 Expression Pattern and Regulation of CA IX

6.3.1 CA IX Protein Distribution in Human Tissues

The CA IX expression in normal human tissues is limited to the epithelia of the gastrointestinal tract. Variable levels of CA IX were also detected in lining cells of the body cavity, in the female and male reproductive tract, in ventricular linings of the central nervous system (CNS), and in the choroid plexus, but majority of normal tissues are CA IX negative (Ivanov et al. 2001; Karhumaa et al. 2001; Hynninen et al. 2004). The highest CA IX levels are found in all cell types of the glandular gastric mucosa. CA IX is also present in the epithelium of the small intestine and the proximal colon, but its level decreases towards the rectum (Pastorekova et al. 1997). Intestinal CA IX is expressed in the epithelial cells of the crypts, which possess the greatest proliferative activity (Saarnio et al. 1998a). Moreover, CA IX is expressed in the gallbladder mucosa and in the pancreatic ducts. In all gastrointestinal epithelial cells, CA IX is localized in the basolateral membranes suggesting its possible involvement in intercellular communication, maintenance of tissue integrity, and regulation of basolateral ion transport. In accord, CA IX-deficient mice display gastric hyperplasia, numerous cysts, and aberrant cell lineage development with an increased number of surface pit cells and glandular atrophy. However, the knockout mice do not show any significant change in gastric pH, hydrochloric acid production, and systemic electrolyte status (Gut et al. 2002). Excess dietary salt has no significant effect on the severity of pit cell hyperplasia (Leppilampi et al. 2005). Microarray analysis revealed that CA IX deficiency is connected to differential expression of genes involved in developmental processes, cell differentiation, and immune response and downregulation of genes for several digestive enzymes (Kallio et al. 2010). Interestingly, CA IX-deficient mice also show age-dependent behavioral changes (abnormal locomotion activity and poor performance in a memory test) and a morphological disruption of brain histology with valuolar degeneration (Pan et al. 2012). These results suggest that CA IX may, directly or indirectly, play novel functions in the brain tissue.

On the other hand, CA IX is frequently found in diverse human tumors, including carcinomas of the uterine cervix, kidney, brain, head and neck, esophagus, lung, colon, breast, ovaria, endometrium, vulva, bladder, thyroid gland, etc. (Liao et al. 1994, 1997; McKiernan et al. 1997; Turner et al. 1997; Saarnio et al. 1998; Vermylen et al. 1999; Ivanov et al. 2001; Haapasalo et al. 2005; Kowalewska et al. 2005; Niemela et al. 2007; Jarvela et al. 2008; Koperek et al. 2011). For an unknown reason, CA IX is generally absent from the normal prostate tissues as well as from the prostate carcinomas (Smyth et al. 2010).

6.3.2 Transcriptional Regulation of the *CA9* Gene

This expression pattern of CA IX principally reflects the strong transcriptional regulation of the *CA9* gene by the hypoxia-inducible factor 1 (HIF-1) in response to hypoxia. Molecular mechanisms underlying HIF-1 assembly from the oxygen-regulated α and constitutive β subunits and control of its transactivation capacity are described in detail elsewhere in this book and briefly summarized in Fig. 6.2. Importantly, HIF-1 regulates the *CA9* gene expression through the hypoxia response element (HRE) localized immediately upstream of the transcription initiation site, which makes it a part of the basal *CA9* promoter (Wykoff et al. 2000). In addition, SP-1 transcription factor binding just upstream of HRE is required for the full transcriptional activation of the *CA9* gene (Kaluz et al. 2002, 2003). The highest expression of CA9 is achieved at high density combined with low oxygen and under both conditions, mitogen activated protein kinase (MAPK) and phosphatidyl-inositol-3-kinase (PI3K) pathways mediate signaling to HIF-1 and SP1 and cross-talk in activation of CA9 transcription (Kaluz et al. 2002; Kopacek et al. 2005). In conditions of severe hypoxia, the transcription of *CA9* is regulated through both the HIF-1 and unfolded protein response (UPR) pathways. The UPR response is mediated by the ATF4 transcription factor that binds directly to the *CA9* promoter and is associated with loss of the transcriptional

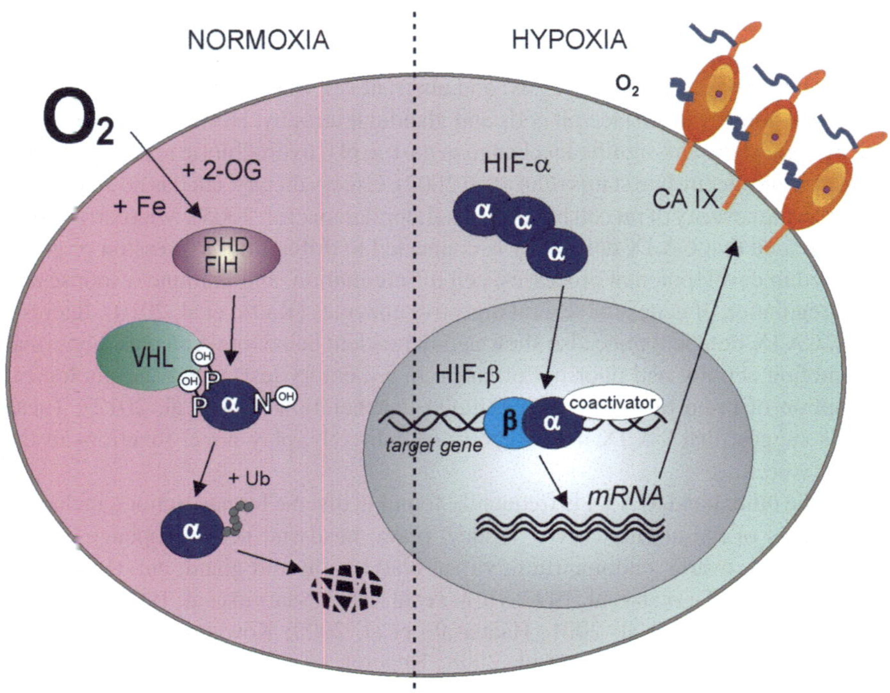

Fig. 6.2 Mechanism of the HIF-mediated transcriptional activation of target genes, including the gene coding for *CA IX*, in response to hypoxia. The *left part* shows the situation in normoxia. In the presence of oxygen, prolyl hydroxylases (*PHD*) and factor inhibiting HIF (*FIH*) modify α subunit of hypoxia-inducible factor (*HIF-α*). Hydroxylation of the conserved prolines in *HIF-α* molecule leads to recognition of *HIF-α* by the von Hippel Lindau (*VHL*) tumor suppressor protein, which mediates *HIF-α* ubiquitilation and degradation in the proteasome. Hydroxylation of the N-terminal asparagine prevents binding of coactivators. As a result, normoxic cells principally do not contain active *HIF-α*. The *right part* illustrates the situation in hypoxia, when the lack of oxygen maintains hydroxylases inactive so that they are unable to modify *HIF-α*. This allows *HIF-α* to escape recognition by *VHL* and prevents degradation. As a result, *HIF-α* accumulates in the cytoplasm, enters the nucleus, interacts with the coactivator, dimerizes with the constitutive *HIF-β* subunit and forms an active transcription complex, which induces transcription of genes containing a hypoxia response element (HRE)

repressive histone 3 lysine 27 tri-methylation mark (van den Beucken et al. 2009). Expression of the *CA9* gene can be also positively modulated by extracellular acidosis in a cell type-specific manner (Ihnatko et al. 2006).

Additional pathways regulating CA9 expression were disclosed by the group of Bonafè in the investigations of the mammary stem cells grown in mammospheres. They found that p66Shc/Notch-3 interplay modulates self-renewal (by inducing Notch-ligand Jagged-1) and hypoxia survival (by inducing CA9) in mammary gland stem/progenitor cells (Sansone et al. 2007a, b). They also recently demonstrated that the peroxisome proliferator-activated receptor-α/hypoxia inducible factor-1α (PPAR-α/HIF-1α) interplay induces the expression of CA9 along with interleukin-6 (IL-6) and apolipoprotein E (Papi et al. 2013).

On the other hand, CA9 expression can be subjected to negative regulation by the pVHL, in accord with the role of pVHL in the negative control of HIF-α protein stability (see Fig. 6.2). This is particularly important in the RCCs that often contain *VHL* gene mutations leading to the loss of pVHL function. Moreover, CA9 transcription can be repressed by the protein complex consisting of Microrchidia 2 (MORC2) and histone deacetylase 4 (HDAC4). MORC2 binds to the promoter element upstream of the SP1–HIF region and HDAC4 deacetylases Histone 3 and thereby reduces the CA9 transactivation (Shao et al. 2010). CA9 is also regulated by epigenetic mechanisms. Hypomethylation of CpG sites at positions of -74 and -6 is required for CA9 expression, particularly in pVHL-defective RCCs (Cho et al. 2000, Ashida et al. 2002). In non-RCC tumor cells, demethylation of -74 CpG site represents an adverse factor modifying the effect of high cell density (Jakubickova et al. 2005).

6.3.3 Other Regulatory Mechanisms Involved in the Control of CA IX Expression

Regulation of the CA9 expression occurs also at posttranscriptional and posttranslational levels and involves diverse regulatory mechanisms that affect splicing, mRNA stability, translation, posttranslational modifications, ECD shedding, endocytosis, etc.

Alternative splicing of the CA9 transcript generates either a full-length mRNA composed of all 11 exons that encodes a plasma membrane-localized and functionally competent CA IX protein, or a much less abundant truncated mRNA lacking the exons 8 and 9, which encodes the cytoplasmic/secreted form of CA IX protein with a decreased enzyme activity (Barathova et al. 2008). The longer transcript is increasingly produced in response to hypoxia and is linked to the tumor phenotype, whereas the shorter one is expressed independently of hypoxia and is therefore detectable also under normoxic conditions and in the normal tissues. This is very important for the correct design of primers used in the clinical studies of cancer-and/or hypoxia-related expression of CA9, where the presence of an alternatively spliced variant in absence of the full-length transcript can provide a false-positive signal. Recent studies support the view that only the full-length CA9 transcript is valuable for prognotic purposes (Malentacchi et al. 2009, 2012).

The CA IX protein is very stable, with its half-life corresponding to about 38 h as determined in reoxygenated HeLa cells and thus respresents a marker of both present and past hypoxia (Rafajova et al. 2004). Hypoxia can also drive posttranlational modification of the CA IX protein by phophorylation of threonine in its intracellular tail through the PKA-related pathway (see also below) (Ditte et al. 2011). Moreover, the cell surfcae abundance of the CA IX protein is regulated by the shedding of its ectodomain (ECD) into the culture medium and into body fluids of tumor patients (Kobayashi et al 2012; Kock et al 2011). Constitutive shedding of CA IX ECD affects about 10 % of the cell-associated protein and is sensitive to metalloprotease-inhibitor batimastat. Interestingly, hypoxia maintains the normal rate of the basal ECD release, thus leading to a concomitant increase in cell-associated and extracellular levels of CA IX. The CA IX ECD shedding can be elevated by treatment with different compounds that affect phosphorylation signaling and cell viability. This activated shedding was shown to be mediated by a tumor necrosis factor-α (TNF-α) converting enzyme (TACE/ADAM17) (Zatovicova et al. 2005). Thus, cleavage of the CA IX ECD is a regulated process that responds to signal transduction-related stimuli and may contribute to adaptive changes in the protein composition of tumor cells and of their microenvironment. The shed CA IX ECD can be detected in the body fluids of cancer patients and can potentially serve as a circulating cancer biomarker for patient selection, monitoring, and management in lung, vulvar, breast, and renal cancer (Ilie et al. 2010; Kock et al. 2011; Muller et al. 2011; Takacova et al. 2013).

6.4 Clinical Importance of CA IX Expression in Tumor Tissues

6.4.1 CA IX as an Intrinsic Marker of Hypoxia

Tumor hypoxia is a clinically relevant phenomenon influencing response to therapy, cancer progression, and survival of patients. The detection of hypoxic regions within tumors is therefore very important for the stratification of patients for suitable treatments and for prognosis of the disease development. Currently available approaches to hypoxia detection are based on different methods, which are mostly invasive and restrict measurements to easily accessible tumor sites. Further limitations include low resolution of measurements, inaccurate interpretation of oxygenation profiles, and inability to use them retrospectively. On the other hand, the use of an intrinsic hypoxic marker is simple, reproducible, and can be performed on a routine basis, without the need of special equipment, using both prospective and retrospective samples. However, various hypoxia-induced molecules exhibit different time and magnitude of induction, stability, and many other parameters indicating that there is no ideal marker of hypoxia, because each molecule reflects a different aspect of tumor cell response to hypoxia. However, the comparison of more markers can provide a better assessment of the tumor phenotype (Bache et al. 2006; Kappler et al. 2008; Rademakers et al. 2011).

CA IX is one of the most strongly hypoxia-induced proteins that appear to be the best available indicator of chronic hypoxia (Potter and Harris 2004). Its clinical relevance is supported by numerous studies based on immunohistochemical or reverse transcriptase polymerase chain reaction (RT PCR) detection of CA IX in different tumor types, where its expression is associated with different histopathological parameters, disease progression, treatment outcomes, and patient survival. CA IX also correlates with the level of hypoxia measured by needle electrodes or by the chemical marker of hypoxia, pimonidazole (Loncaster et al. 2001; Olive et al. 2001; Airley et al. 2003; Hutchison et al. 2004; Iakovlev et al. 2007), and with the expression of HIF-1α in several tumor types (Hui et al. 2002; Swinon et al. 2004; Velugel et al. 2005; Kim et al. 2005; Brennan et al. 2006; Korkolopoulou et al. 2007). Pimonidazole binds to severely hypoxic tissues as a function of drug availability and accumulation and usually detects chronic hypoxia present in the narrow perinecrotic regions, whereas CA IX is found in a relatively wide perinecrotic area including modestly hypoxic and/or reoxygenated cells. It is therefore not surprising that comparison of their spatial distribution allows for the discrimination between actually hypoxic and reoxygenated areas (Shin et al. 2007). On the other hand, HIF-1α levels rise immediately in response to reduced oxygen and instantaneously decline upon reoxygenation. Thus, the expression of HIF-1α peaks at shorter distances to blood vessels, declines in the most hypoxic regions that border necrosis, and detects regions of hypoxia at the time of biopsy or tumor removal (Sobhanifar et al. 2005). Median distance between a blood vessel and a beginning of CA IX expression is 80 μm (range 40–140) in head and neck carcinoma and bladder carcinoma (Beasley et al. 2001; Turner et al. 2002) and is about 90 μm in nonsmall-lung carcinoma (Swinon et al. 2004). Thus, CA IX is found between the borders of the HIF-1α zone and the zone of EF5, a chemical marker of hypoxia suggesting that CA IX induction requires lower oxygen levels than HIF-1α, but higher levels than EF5 (or pimonidazole). This corresponds with the finding of Olive et al. (2001) that the cells expressing CA IX belong to a broader perinecrotic area and are viable, clonogenic, and resistant to killing by ionizing radiation. Furthermore, CA IX expression shows only a partial overlap with VEGF and glucose transporter (GLUT-1) indicating their different upstream regulation and posttranslational stability in reoxygenated cells (Giatromanolaki et al. 2001; Tomes et al. 2003; Vordermark et al. 2005; Kim et al. 2005; Beasley et al. 2001; Sorensen et al. 2005).

6.4.2 CA IX Correlation With Clinical Parameters and Other Tumor Biomarkers

This expression pattern affects the relationship of CA IX with clinical variables. CA IX significantly correlates with high tumor grade, necrosis, treatment outcome, and poor prognosis in patients with the breast and lung carcinomas (Brennan et al. 2006; Wykoff et al. 2001; Kon-no et al. 2006; Hussain et al. 2007; Generali et al. 2006; Chia et al. 2001), with necrosis, advanced stage, and treatment outcome in head and neck cancer (Beasley et al. 2001; Le et al. 2007), with necrosis, histological grade, and survival of patients with brain tumors (Haapasalo et al. 2005; Korkolopoulou et al. 2005), with poor prognosis in esophageal and gastric cancers (Chen et al. 2005; Driesen et al. 2006), with metastases in primary cervical cancer (Kim et al. 2006), etc. In the cDNA microarray study of the hypoxia transcriptome in human bladder cancer, CA IX was used as a surrogate marker of hypoxia to which array mRNA changes were correlated in order to define the in vivo hypoxia profile and identify new hypoxia-regulated genes (Ord et al. 2005). Similarly, CA IX was used as one of ten hypoxia-regulated genes to cluster a hypoxia metagene in head and neck squamous cell cancers, which was a significant prognostic factor in published head and neck as well as in breast cancer data sets (Winter et al. 2007). Thus, CA IX seems to be valuable for both research purposes and clinical practice. Its detection by the monoclonal antibody M75 does not require an invasive approach or metabolic incorporation before biopsy removal and can be standardized for routine use.

CA IX expression is very high and frequent in clear cell RCCs, due to an inactivating mutation in the VHL gene that affects the majority of RCCs and leads to loss of pVHL function (Gnarra et al. 1994), activation of the HIF pathway, and constitutive upregulation of HIF targets (including CA IX) independently of physiological hypoxia (Wiesener et al. 2001; Mandriota et al. 2002). In contrast to other tumor types, CA IX expression in RCC decreases with increasing tumor grade and stage (with a very high cut-off value at 85 % of CA IX-positive cells) (Uemura et al. 1999; Bui et al. 2003, 2004; Sandlund et al. 2007). This is a consequence of HIF-1α decline in advanced RCC and its replacement by a HIF-2α tumor growth-promoting isoform that induces targets other than CA IX (Raval et al. 2005). Nevertheless, CA IX can be upregulated by interferons (IFN-α and IFN-γ), which are known to improve the therapeutic responses in 5–25 % of patients with metastatic RCC (Brouwers et al. 2003). Furthermore, high CA IX expression was suggested as an important predictor of better outcome in RCC patients receiving IL-2-based therapy (Atkins et al. 2005).

In non-RCC tumors, where CA IX distribution reflects physiological hypoxia or oncogenic activation of the HIF pathway, CA IX indicates poor prognosis and correlates with various prognostic factors, including c-ErbB2 (Bartosova et al. 2002; Schoppmann et al. 2012); epidermal growth factor receptor (EGFR), c-ErbB2, and MUC-1 (Giatromanolaki et al. 2001); EGFR and matrix metalloproteinase (MMP-9) (Swinson et al. 2004); osteopontin and CD44 (Le et al. 2006); p53 and p300 (Vleugel et al. 2006); estrogen receptor, progesterone receptor, bcl-2, p53, c-ErbB2, and Ki-67 (Generali et al. 2006); cyclin E, cyclin A2, and Ki-67 (Brennan et al. 2006); lysyl oxidase, ephrin A1, and galectin-1 (Le et al. 2007); Slug (Storci et al. 2008); and manganese superoxide dismutase (Jarvela et al. 2008). The association of CA IX with these components of oncogenic and other regulatory pathways suggests their possible cross talk. Some of its pieces recently emerged through studies of the regulatory and mainly functional apects of CA IX.

6.5 CA IX as a Functional Component of Tumor Biology

6.5.1 CA IX Role in Adaptation to Hypoxia and Acidosis

Association with tumors, relationship to hypoxia, and high catalytic activity suggest a role for CA IX in the adaptation of tumor cells to their microenvironment, particularly to hypoxia-promoted pericellular acidosis that requires intensified pH regulation. Acidosis can result from the accumulation of acidic metabolic waste of the hypoxia-induced anaerobic glycolysis (dominating over oxidative phosphorylation) and/or aerobic glycolysis typical of oncogenic metabolism (Fukumura and Jain 2007). In hypoxia, HIF-1 coordinates the upregulation of virtually all glycolytic enzymes and glucose transporters via HRE elements in the promoters of their genes (Dang and Semenza 1999; Brahimi-Horn and Pouyssegur 2007; Vander Heiden et al. 2009). HIF-1 also strongly induces lactate dehydrogenase isoforms, LDH-A and LHD-5, which convert pyruvate to lactate, as well as pyruvate dehydrogenase kinase 1, which prevents the entry of pyruvate into the tricarboxylic acid (TCA) cycle, thereby supporting the switch to glycolytic metabolism. In addition, the MYC transcription factor coordinates the expression of many glycolytic enzymes independently of hypoxia (Vander Heiden et al. 2009).

Such a predominantly glycolytic metabolism leads to the excessive production of lactate, protons, and also CO_2, which cannot be efficiently removed from the pericellular microenvironment because of defective vasculature and elevated interstitial pressure in solid tumors (Helmlinger et al. 2002; Fukumura and Jain 2007). Glycolysis-deficient cells can also generate acidic tumors, with CO_2 as the main source of acidity. Extracellular acidosis then causes transient intracellular acidification, which is incompatible with cell growth and survival. Therefore, tumor cells have to actively regulate their cytoplasmic pH to maintain neutral or slightly alkaline values favorable to various intracellular processes, see Fig. 6.3. They can do it via increased levels and/or activities of pumps, ion transporters, and exchangers that extrude lactate and protons, including the H^+/monocarboxylate transporter (MCT), the Na^+/H^+ exchanger (NHE), and the vacuolar H^+/ATP pump. On the other hand, anion exchangers (AEs) and Na + /bicarbonate co-transporters (NBCs) import bicarbonate for buffering intracellular pH (Fang et al. 2008; Parks et al. 2011).

Extracellular acidosis is typical of hypoxic tumors, where the mean profiles of partial oxygen pressure correlate with intratumoral pH values (although this correlation is not always present in individual tumor regions). Moreover, the VHL/HIF pathway controls the expression of several components of the pH-regulating molecular machinery including anion exchanger 2 (AE2), NHE1, and MCT4 (Karumanchi et al. 2001; Shimoda et al. 2006; Ullah et al. 2006). Interestingly, experiments with three-dimensional (3D) tumor spheroids show that pHi recovery from intracellular acidification in the hypoxic core cells is slower upon inhibition of bicarbonate transporters than upon inhibition of NHE, supporting the idea that bicarbonate transport is of principal importance in coping with intracellular acidosis (Hulikova et al. 2011).

Acidic pHe increases the expression of angiogenic factors and proteases, stimulates migration and invasion, impairs immune functions, and thereby contributes to tumor progression. In addition, it enhances resistance to therapy (Raghunand et al. 2003; Wojtkowiak et al. 2011; Thews et al. 2006; Vaupel 2010; Swietach et al. 2012). Experimental data show that CA IX is implicated in the development of extracellular acidosis in tumors via its catalytic activity as described below. In accord with this concept, CA IX knockdown enhances the effect of bevacizumab treatment, reducing tumor growth rate in vivo (McIntyre et al. 2012).

6.5.2 CA IX and pH Regulation Through the Bicarbonate Transport Metabolon

Extracellular position of the active site and high catalytic activity predispose CA IX to play a role in the pericellular metabolism of CO_2 . In contrast to the other CA isoforms, CA IX is inhibited by bicarbonate but not by lactate and therefore can

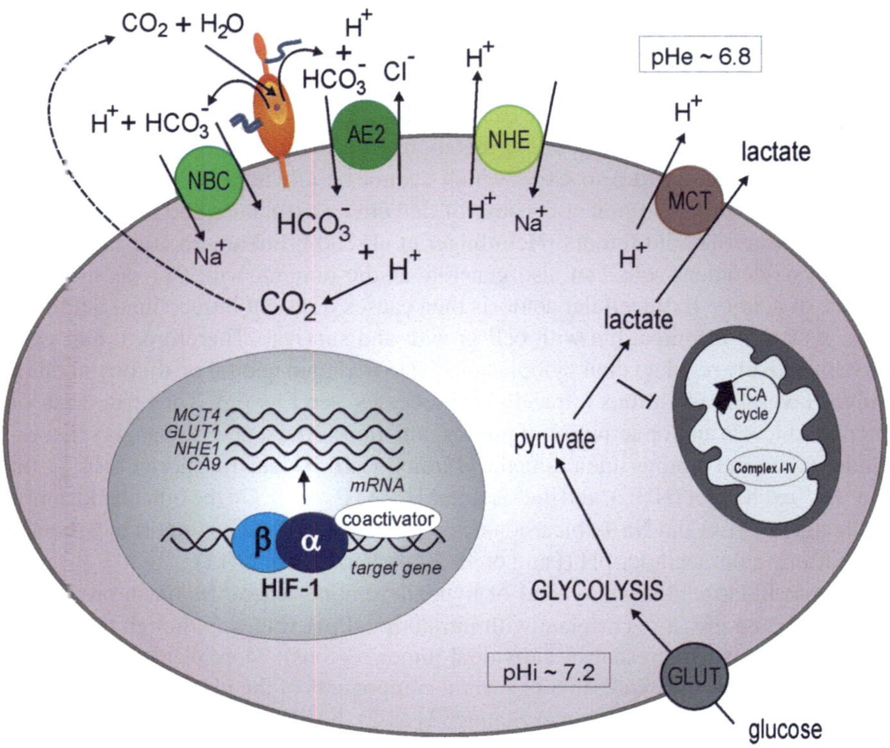

Fig. 6.3 pH regulation in hypoxic tumor cells. *HIF-1* induces expression of glucose transporters (*GLUT1*) and glycolytic enzymes thereby shifting metabolism to glycolysis that produces excess of lactate and protons. To maintain neutral intracellular pH, these acidic products are extruded by the HIF-regulated monocarboxylate transporter (*MCT4*) and the Na+/H+ exchanger (*NHE1*) and acidify the extracellular microenvironment. Tumor cells also produce high amount of CO_2 that diffuses through the plasma membrane and contributes to the extracellular acidosis. Pericellular CO_2 is hydrated to bicarbonate ions and protons in a reaction catalyzed by hypoxia-activated carbonic anhydrase IX (*CA IX*). *CA IX*-facilitated production of bicarbonate ions is coupled to inward transport to the cytoplasm by bicarbonate transporters (*NBC1* and *AE2*) where these ions buffer protons and neutralize intracellular pH. This reaction produces CO_2 that leaves the cell by diffusion and may enter a new round of hydration

perform its enzyme function also in the tumor microenvironment, which is rich in lactate produced by glycolysis and is deprived of bicarbonate (Innocenti et al. 2005). It seems plausible that under these conditions CA IX primarily catalyzes the hydration of CO_2 to bicarbonate ions and protons. Protons remain outsidecells and further acidify their microenvironment, whereas bicarbonate ions are actively transported to the cell cytoplasm by bicarbonate transporters. This process requires spatial and functional cooperation between bicarbonate transporter(s) and CA IX on the surface of tumor cells, in the form of a "bicarbonate transport metabolon" complex. Such a metabolon is based on the locally concentrated production of bicarbonate ions directly coupled to their transport (McMurtrie et al. 2004). This prevents the

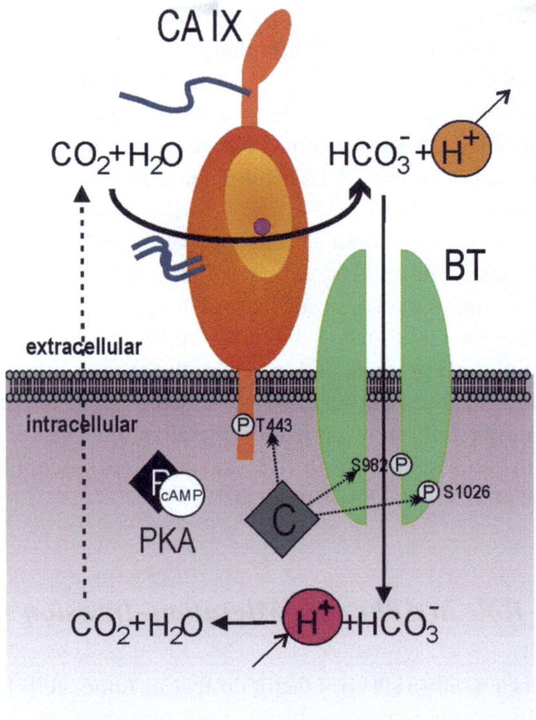

Fig. 6.4 The proposed *CA IX* role in the bicarbonate transport metabolon acting in tumor cells. *CA IX* catalyzes hydration of pericellular CO_2 to bicarbonate and proton. Protons remain outside of the cell and contribute to extracellular acidification. Bicarbonate ions are taken into bicarbonate transporters (*BT*) and transported to cytoplasm, where they are converted back to CO_2 in a reaction that consumes intracellular protons. This leads to neutralization of intracellular pH. Based on the experimental data, the metabolon is induced in hypoxia through the catalytic subunit (*C*) of the hypoxia-activated protein kinase A (*PKA*). *PKA* phosphorylates *CA IX* and also NBC1 and this results in a spatially coordinated and functionally coupled activation of the *CA IX*-catalyzed production of bicarbonate ions of the NBC1-mediated bicarbonate transport across the membrane

dissipation of bicarbonate ions in the acidic pericellular milieu and accelerates their flux across the plasma membrane, which is very important for pH regulation and intense ion movement in tumor cells, especially in hypoxia. In summary, CA IX produces bicarbonate ions which are imported to cytoplasm where they are dehydrated back to CO_2 and then leave the cells by diffusion. The dehydration reaction consumes intracellular protons and helps to neutralize the cell interior. On the other hand, the extracellular hydration of CO_2 results in the net production of protons, which remain outsidethe cells and feed acidosis as illustrated in Fig. 6.4. This model is supported by the experimental evidence for the in vitro interaction between the extracellular catalytic domain of CA IX and AE2 as well as that between CA IX and NBCe1 (namely its extracellular loop 4) and also by the demonstration that CA IX accelerates bicarbonate flux through these transporters (Morgan et al. 2007; Orlowski et al. 2012).

Experiments performed with the cells constitutively expressing CA IX suggest that the catalytic performance of CA IX is stimulated by hypoxia (Svastova et al. 2004). This occurs through hypoxia-elevated cyclic adenosine monophosphate (cAMP) levels that activate protein kinase A, which in turn phosphorylates the threonine residue in the intracellular tail of CA IX. Thereby, PKA stimulates inside-out signaling to the extracellular catalytic domain of CA IX resulting in activation of the enzyme-coupled pH regulation (Ditte et al. 2011).

Thus, CA IX contributes to both acidification of the extracellular milieu and alkalinization of the intracellular pH and thereby supports the survival of tumor cells exposed to hypoxia and/or excess of acidic products of the oncogenic metabolism. This is demonstrated in studies using 3D tumor cell spheroids, where CA IX minimizes the intracellular pH gradient and increases the extracellular pH gradient from the spheroid surface to the spheroid core (Swietach et al. 2009, 2010). Moreover, CA IX-deficient tumor cells show an inability to adapt to acidosis in vitro, reduced capacity to grow in tumor xenografts in vivo, and increased susceptibility to radiation treatment (Chiche et al. 2009; Doyen et al. 2012).

6.5.3 CA IX Role in Adhesion–Migration–Invasion

However, CA IX is not only a survival factor protecting tumor cells from hypoxia and acidosis but also functionally involved in the processes connected with cell adhesion, migration, and invasion as documented by several studies exploiting cellular models either with overexpression or suppression of CA IX or with expression of enzyme-dead mutants or following inhibition with chemical compounds or antibodies. To briefly summarize what we know at present: (1) CA IX reduces E-cadherin-mediated cell–cell adhesion via the mechanism that involves a competitive interaction with β-catenin (Svastova et al. 2003); (2) CA IX mediates adhesion of cells to solid support (Zavada et al. 2000); (3) CA IX is involved in cell spreading and formation of adhesion contacts (Radvak et al. 2013); (4) CA IX relocalizes to lamellipodia and facilitates cell migration (Svastova et al. 2012); (5) CA9 induces cell motility through aberrant Rho-GTPase signal transduction (Shin et al. 2011); and (6) CA IX increases cell invasion (Svastova et al. 2012). Certain aspects of these processes clearly depend on pH regulation and thus participation of CA IX can be at least in part explained by its catalytic activity (Fig. 6.4). Nevertheless, experimental data suggest also an enzyme-independent role for CA IX, presumably mediated by the N-terminal PG-like domain, although further investigations towards elucidation of this role are needed.

Catalytic activity of CA IX is clearly important for its role in cell migration, which requires the establishment of correct pH gradients along the longitudinal cell axis, with acidic pHe and alkaline pHi at the cell front and alkaline pHe and acidic pHi at the rear end (Stock et al. 2007; Stock and Schwab 2009; Martin et al. 2011). The formation of this gradient is accompanied by relocalization of ion transporters (including NHE1, NBCe1, and AE2) to lamellipodia, where they carry out intense ion transport (Klein et al. 2000; Schwab 2001). CA IX is relocalized to the same areas and cooperates with this pH-regulating machinery. In the lamellipodia of hypoxic tumor cells stimulated to migration, CA IX interacts with AE2 and NBCe1 as proven by a proximity ligation assay. This first, direct, in situ evidence supports the existence of the spatially and functionally coordinated bicarbonate transport metabolon (Svastova et al. 2012; Svastova and Pastorekova 2013).

6.5.4 Other Emerging Functional Aspects of CA IX

Additional potential functions of CA IX are still emerging. Kim et al. (2012) showed that CA IX induces mTOR phosphorylation and angiogenesis and that these effects are inhibited through CA IX interaction with DKK1 protein. Characterization of the CA IX interactome reveals that CA IX interacts with proteins involved in nuclear/cytoplasmic transport, gene transcription, and protein stability and suggests the existence of nuclear CA IX protein subpopulations with a potential intracellular function, distinct from the crucial CA IX role at the cell surface (Buanne et al. 2013).

6.6 Therapeutic Targeting of CA IX

6.6.1 CA IX Targeting Through Monoclonal Antibodies

CA IX is principally targetable by specific monoclonal antibodies and by inhibitors of the catalytic activity. The first, immunotherapeutic option takes the advantage of the strong association of CA IX in tumors and rare expression in normal tissues. This is expected to allow for selective antibody-mediated killing of tumor cells that express CA IX. The antibodies can either attract cytotoxic cellular components of the immune system or deliver cytotoxic drugs or toxins. The second approach, exploiting the inhibitors CA IX, is based on the fact that the enzyme activity of CA IX is important for the survival of tumor cells in stressful tumor microenvironment and for their ability to migrate and invade.

The immunotherapeutic approach was consistently studied in whole series of preclinical and clinical experiments in RCC models and patients with the CA IX-specific monoclonal antibody G250 and its humanized, chimeric, and bispecific variants (Oosterwijk et al. 1986; Luiten et al. 1996, 1997). This antibody was generated against an RCC-associated antigen G250 that was later identified as CA IX (Oosterwijk et al. 1986; Grabmaier et al. 2004). Chimeric G250 (composed of the variable regions of murine G250 and constant regions derived from human immunoglobulin G, IgG) utilizes antibody-dependent cell cytotoxicity (ADCC) as an effector mechanism. This antibody, known under commercial names RENCAREX® or GIRENTUXIMAB®, showed good safety, tolerability, and a promising efficacy profile in the phase I and II clinical trials with more than 100 patients with metastatic RCC (Davis et al. 2007; Siebels et al. 2011). Unfortunately, despite initial expectations (Reichert 2011), the recently finished phase III clinical trial (ARISER), targeted at patients with nonmetastatic RCC, showed no significant improvement of disease-free survival among patients treated with RENCAREX® compared to placebo, possibly due to a study design based on an adjuvant setting after the surgical removal of the primary tumor cells expressing the CA IX target. This approach led to removal of the target before the treatment, and thus it questions the setting of the trial rather than the suitability of the antibody itself.

Based on the evidence that determination of CA IX expression in nephrectomy specimens of RCC patients improves prognostic accuracy, CA IX expression might be used to stratify metastatic ccRCC patients for systemic treatment. Clinical studies with the CA IX-specific monoclonal antibody cG250 have provided unequivocal evidence that ccRCC lesions can be imaged with radiolabeled cG250, known under commercial name REDECTANE® (Divgi et al. 2007; Stillebroer et al. 2010).

In addition, development and preclinical characterization of new CA IX-specific antibodies in different tumor types is still ongoing. Zatovicova et al. (2003) generated a series of mouse monoclonal antibodies directed to the catalytic domain of CA IX. One of them, monoclonal antibody VII/20, induces efficient receptor-mediated internalization and elicits a marked anticancer effect in the mouse xenograft model of colorectal carcinoma (Zatovicova et al. 2010). High-affinity human monoclonal antibodies (A3 and CC7) specific to human CA IX were produced by phage technology and tested in a colorectal model (Ahlskog et al. 2009). Xu et al. (2010) constructed and used paramagnetic proteoliposomes of CA IX to screen 27 billion human single-chain antibody phage display libraries and identified a panel of 13 human CA IX-specific antibodies that are diverse in their functions including induction of surface CA IX internalization into endosomes and inhibition of the CA activity. Another antibody, named MSC8, inhibiting CA IX enzyme activity by up to 57 and 76 % was generated by Murri-Plesko et al. (2011).

Moreover, a human 3ee9 monoclonal antibody conjugated to monomethyl auristatin E through a self-cleavable linker (BAY 79-4620) showed potent antitumor efficacy in human xenograft models inducing partial-to-complete tumor shrinkage even following a single dose depending on the CA IX expression levels (Petrul et al. 2012). These preclinical data support the development of the described antibodies for the treatment of cancer patients with CA IX overexpressing tumors. Finally, rabbit polyclonal antibody specific for CA IX together with the mouse monoclonal antibody specific for CA XII were used as the imaging probes for detection of breast cancer lymph-node metastasis in nude mice bearing mammary fat pad tumors (Tafreshi et al. 2012). Thus, antibody-based imaging probes have potential for noninvasive staging of cancer in the clinic and elimination of unneeded surgery, which is costly and associated with morbidities.

CA IX can be also potentially targeted through specific peptides identified by phage display technology, including CaIX-P1 peptide binding to the extracellular domain of CA IX (Askoxylakis et al. 2010; 2012) and peptide binding to the epitope for the M75 monoclonal antibody in the N-terminal PG domain of CA IX (Zavada et al. 2000).

6.6.2 CA IX Targeting Through Inhibitors of Catalytic Activity

Inhibitors of CA enzyme activity represent a new class of emerging anticancer drugs (Supuran and Scozzafava 2000; Supuran et al. 2001). Initially, some sulfonamide derivatives showed antiproliferative effects in tumor cell lines cultured under standard normoxic conditions. Because sulfonamides generally show non-selective inhibitory activity towards different CA isoenzymes, it was difficult to determine their target. As we know currently, at least in some cell lines it could be CA IX. Indeed, different classes of sulfonamides and related compounds can efficiently inhibit CA IX. In addition to classical sufonamides, such as acetazolamide (AZM), methazolamide, ethoxzolamide, and dichlorophenamide, good-to-excellent CA IX inhibitory properties were proven for aromatic and heterocyclic sulfonamides (Vullo et al. 2003; Jaiswal et al. 2004; Ozensoy et al. 2005), halogenosulfanilamide and halogenophenylaminobenzolamide derivatives (Ilies et al. 2003), lipophilic sulfonamides (Franchi et al. 2003), sulfamates (Winum et al. 2003a), aliphatic sulfamates (Winum et al. 2003b), fluorine-containing sulfonamides (Vullo et al. 2004), sulfonamides incorporating 1,3,5-triazine and 1,2,4-triazine moieties (Garaj et al. 2004, 2005), sulfonamides derived from 4-isothiocyanato-benzolamide (Cecchi et al. 2004), E7070 sulfonamide developed originally as an anticancer agent blocking the cell cycle (Abbate et al. 2004), sulfonamides incorporating thioureido-sulfanilyl scaffolds (Pucetti et al. 2005), novel sulfanilamide/AZM derivatives obtained by the tail approach (Turkmen et al. 2005), bis-sulfamates (Winum et al. 2005a), sulfonamides incorporating hydrazine moieties (Winum et al. 2005b), 1,3,4-thiadiazole- and 1,2,4-triazole-thiols (Almajan et al. 2005), N-hydroxysulfamides (Winum et al. 2005c), polyfluorinated aromatic/heterocyclic sulfonamides (Pastorekova et al. 2005), benzo[b]thiophene 1,1 dioxide sulfonamides (Innocenti et al. 2005), substituted difluoromethanesulfonamides (Cecchi et al. 2005), indanesulfonamides (Thiry et al. 2006), and copper(II) complexes of polyamino-polycarboxylamido aromatic/heterocyclic sulfonamides (Rami et al. 2008). Interestingly, CA IX can be also efficiently inhibited by nanomolar concentrations of celecoxib and valdecoxib sulfonamide inhibitors of cyclooxygenase 2 (COX-2), which is a key enzyme of arachidonic acid metabolism involved in colorectal carcinoma (Weber et al. 2004; Dogne et al. 2007). This may suggest that the mode of action of these COX-2 inhibitors could involve targetingCAs. Moreover, CA IX can be inhibited by sulfamates, inhibitors of steroid sulfatase, which plays a role in the production of active steroids and is implicated in breast cancer (Winum et al. 2003). Further sulthiame, a clinically used antiepileptic, is a potent inhibitor of CA IX (Temperini et al. 2007). Recent studies showed that clinically used drugs, such as the tyrosine kinase inhibitors imanitib and nilotinib as well as statins, inhibit the catalytic activity of CAs, albeit without selectivity for CA IX (Parkkila et al. 2009, 2012).

CA IX selectivity of the inhibitors can be successfully achieved by modulating the physical and chemical properties of the compounds via attachment of different side chains and other modifications (Scozzafava et al. 2004; Pastorekova et al. 2004; Supuran 2008). Certain alterations can introduce or improve the membrane impermeability so that the inhibitor can bind only or predominantly to extracellularly exposed

CAs (Casey et al. 2004; Pastorekova et al. 2004). The other changes can modify the size or surface topology to fit better into the active site cavity of CA IX than into other isoforms. Some types of modifications enhanced the efficiency of inhibition so that they work even at subnanomolar concentrations (Vullo et al. 2004) when analyzed against the recombinant catalytic domain of CA IX. These extensive studies revealed several compounds with a reasonable selectivity ratio favoring inhibition activities against CA IX compared to other isoforms, in particular CA XII and CA II. For example, selectivity ratios for the inhibition of CA IX over the cytosolic isozymes CA I and II were in the range of 107–955 for glycosyl-thioureido-sulfonamides (Smaine et al. 2007). Another sophisticated strategy was used to generate hypoxia-activatable inhibitors. Different 2-mercapto-substituted-benzenesulfonamides and their disulfides/sulfones showed consistently increased inhibitory power (52.8- to 243-fold) over the corresponding oxidized (S-S type) derivatives (Saczewski et al. 2006). The best representatives out of these differentially acting derivatives can serve as lead compounds for further development of CA IX-specific inhibitors with therapeutic potential against cancer.

Initial preclinical data obtained with several types of compounds are very promising. As demonstrated by Svastova et al. (2004) and Cecchi et al. (2005), fluorescein-conjugated carbonic anhydrase inhibitor thioureido-homosulfanilamide (FITC-CAI) can reduce the extracellular acidification and bind only to hypoxic cells expressing CA IX, but neither to CA IX-positive normoxic cells nor to the CA IX-negative controls. The FITC-CAI binding to hypoxic cells is perturbed by reoxygenation (Dubois et al. 2007). This binding pattern may be determined by the accessibility of the active site to inhibitors during hypoxia presumably related to folding changes consequent to phosphorylation of the intracellular domain by the hypoxia-activated PKA as described above (Ditte et al. 2011). Of course, we cannot exclude other mechanisms such as additional modification and/or hypoxia-induced protein–protein interactions. Importantly, selective binding of FITC-CAI to CA IX-expressing hypoxic cells suggests that labeled CA IX inhibitors can be potentially used as tools for in vivo imaging of hypoxic tumors. Data from animal experiments imply that this is a feasible approach (Dubois et al. 2007, 2010) that differs from imaging with antibodies, which bind to CA IX independently of the physiological circumstances and therefore can image both hypoxic and post-hypoxic CA IX-expressing tumors.

CA inhibitors can be also used in anticancer therapy either alone, with the purposeof compromising the survival of tumor cells due to perturbed pH control, or combined with conventional chemotherapeutics to improve their uptake and efficacy due to a disturbed pH gradient (Neri and Supuran 2011).

An anticancer and/or antimetastatic effect was demonstrated for several compounds inhibiting CA activity. Treatment of mice harboring CA IX-positive 4T1 mammary tumors with CAIX-specific small molecule inhibitors resulted in significant inhibition of tumor growth and metastasis formation in both spontaneous and experimental models of metastasis, without inhibitory effects on CAIX-negative tumors (Lou et al. 2011). A similar effect was observed for an ureido-substituted benzenesulfonamide in the same metastatic breast cancer model (Pacchiano et al. 2011).

Finally, an S4 compound of the sulfamate class of CA IX inhibitors in nanomolar concentrations can decrease tumor cell migration and spreading in vitro and reduce metastatic tumor burden in the lung of mice with orthotopic breast xenografts (Gieling et al. 2012).

Changes in extracellular pH can modulate therapeutic responses to conventional chemotherapy by influencing the uptake of weakly electrolytic anticancer drugs and the efficiency of radiotherapy (Vukovic and Tannock 1997; Kozin et al. 2001; Raghunand et al. 2003; Stubbs et al. 2000). In particular, reduction of extracellular acidosis can increase the uptake and cytotoxic effects of weakly basic drugs, including doxorubicin. Interestingly, chronic ingestion of sodium bicarbonate solution was shown to enhance the capacity of doxorubicin to decrease the tumor size (Raghunand et al. 2003). AAM, a classical CA inhibitor, also reduced in vivo growth of tumor when given alone and produced additive tumor growth delays when administered in combination with various chemotherapeutic compounds (Teicher et al. 1993). Specific inhibition of CA IX activity by indanesulfonamide enhanced the effect of tumor irradiation (Dubois et al. 2011). Gieling et al. (2013) showed that the CA inhibitor AZM enhanced doxorubicin toxicity but reduced melphalan toxicity in cell lines with high CA IX expression under anoxic conditions and that the toxicity changes reflected modification of passive drug uptake.

Recent developments have brought new compounds with novel modes of inhibition, more selective effects towards CA IX, and/or considerable antitumor and antimetastatic effects in preclinical in vivo experiments. These include coumarins (natural compounds from plants) and their derivatives, some of which can significantly inhibit the growth of primary breast tumors in mice (Maresca and Supuran 2010; Maresca et al. 2010; Touisni et al. 2011; Carta et al. 2012). Phenol-based inhibitors differ from all other classes of inhibitors known to date as they bind between the phenol-binding site and the coumarin-binding site, filling thus the middle of the enzyme cavity. They exploit different interactions with amino acid residues and water molecules from the CA active site compared to other classes of inhibitors, offering the possibility of designing CAIs with an interesting inhibition profile compared to the clinically used sulfonamides/sulfamates (Durdagi et al. 2011).

CA IX inhibitors could be also potentially used for selective delivery of therapeutic moieties such as isotopes and agents cytotoxic to tumor cells. Winum et al. (2005d) designed and synthesized boron-containing inhibitors with high affinity for CA IX, which may have potential application in boron neutron capture therapy (BNTC) for tumors that are nonresponsive to classical therapeutic modalities.

6.7 Conclusion

Accumulating evidence suggests that the hypoxia-induced CA IX is an integral component of the cancer physiology, which acts as both the survival factor and the facilitator of tumor progression. This functional involvement in tumor phenotype together with the cancer/hypoxia-related expression pattern make CA IX a clinically exploitable molecule for diagnostic as well as therapeutic purposes. Different targeting tools available at present, including specific monoclonal antibodies, peptides, and inhibitors, offer many opportunities for the clinical utilization of CA IX. Although some of them have already shown promising effects either in the preclinical experiments or even in the clinical studies, there is still need for further investigations, with a refined mode of application, dosage, tumor type, group of patients, and other important parameters. Only such a consistent attitude can prove the usefulness of these promising drugs as anticancer therapeutics.

Acknowledgment The authors' research is supported by grants from the EU 7th Framework program (Collaborative project METOXIA), from the Research and Development Support Agency (DO7RP-0017-09 and APVV-0658-11), and from the Research & Development Operational Program funded by the ERDF (project ITMS 26240120027).

References

Abbate F, Casini A, Owa T et al (2004) Carbonic anhydrase inhibitors: E7070, a sulfonamide anticancer agent, potently inhibits cytosolic isozymes I and II, and transmembrane, tumor-associated isozyme IX. Bioorg Med Chem Lett 14:225–229

Ahlskog JK, Schliemann C, Mårlind J et al (2009) Human monoclonal antibodies targeting carbonic anhydrase IX for the molecular imaging of hypoxic regions in solid tumours. Br J Cancer 101:645–657

Airley RE, Loncaster J, Raleigh JA et al (2003) GLUT-1 and CA IX as intrinsic markers of hypoxia in carcinoma of the cervix: relationship to pimonidazole binding. Int J Cancer 104:85–91

Almajan GL, Innocenti A, Pucetti L et al (2005) Carbonic anhydrase inibitors: inhibition of the cytosolic and tumor-associated carbonic anhydrase isozymes I, II and IX with a series of 1,3,4-thiadiazole- and 1,2,4-triazole-thiols. Bioorg Med Chem Lett 15:2347–2352

Alterio V, Hilvo M, Di FA (2009) Crystal structure of the catalytic domain of the tumor-associated human carbonic anhydrase IX. Proc Natl Acad Sci U S A 106:16233–16238

Ashida S, Nishimori I, Tanimura M et al (2002) Effects of von Hippel-Lindau gene mutation and methylation status on expression of transmembrane carbonic anhydrases in renal cell carcinoma. J Cancer Res Clin Oncol 128:561–568

Askoxylakis V, Garcia-Boy R, Rana S et al (2010) A new peptide ligand for targeting human carbonic anhydrase IX, identified through the phage display technology. PLoS One 5(12):e15962. doi 10.1371/journal.pone.0015962.

Askoxylakis V, Ehemann V, Rana S et al (2012) Binding of the phage display derived peptide CaIX - P1 on human colorectal carcinoma cells correlates with the expression of carbonic anhydrase IX. Int J Mol Sci 13:13030–13048

Aspatwar A, Tolvanen ME, Ortutay C et al (2010) Carbonic anhydrase related protein VIII and its role in neurodegeneration and cancer. Curr Pharm Des 16:3264–3276

Atkins M, Regan M, McDermott D et al (2005) Carbonic anhydrase IX expression predicts outcome of interleukin 2 therapy for renal cancer. Clin Cancer Res 11:3714–3721

Bache M, Reddemann R, Said HM (2006) Immunohistochemical detection of osteopontin in advanced head-and-neck cancer: prognostic role and correlation with oxygen electrode measurements, hypoxia-inducible-factor-1alpha-related markers, and hemoglobin levels. Int J Radiat Oncol Biol Phys 66:1481–1487

Barathova M, Takacova M, Holotnakova T et al (2008) Alternative splicing variant of the hypoxia marker carbonic anhydrase IX expressed independently of hypoxia and tumour phenotype. Br J Cancer 98:129–136

Barnea G, Grumet M, Milev P et al (1994) Receptor tyrosine phosphatase beta is expressed in the form of proteoglycan and binds to the extracellular matrix protein tenascin. J Biol Chem 269:14349–14352
Bartosova M, Parkkila S, Pohlodek K et al (2002) Expression of carbonic anhydrase IX in breast is associated with malignant tissues and related to overexpression of c-erbB2. J Pathol 197:1–8
Beasley NJP, Wykoff CC, Watson PH et al (2001) Carbonic anhydrase IX, an endogenous hypoxia marker, expression in head and neck squamous cell carcinoma and its relationship to hypoxia, necrosis and microvessel density. Cancer Res 61:5262–5267
Brahimi-Horn MC, Pouysségur J (2007) Hypoxia in cancer cell metabolism and pH regulation. Essays Biochem 43:165–178
Brennan DJ, Jirstrom K, Kronblad A et al (2006) CA IX is an independent prognostic marker in premenopausal breast cancer patients with one to three positive lymph nodes and a putative marker of radiation resistance. Clin Cancer Res 12:6421–6431
Brouwers AH, Frielink C, Oosterwijk E et al (2003) Interferons can upregulate the expression of the tumor associated antigen G250-MN/CA IX, a potential target for (radio)immunotherapy of renal cell carcinoma. Cancer Biother Radiopharm 18:539–547
Buanne P, Renzone G, Monteleone F et al (2013) Characterization of carbonic anhydrase IX interactome reveals proteins assisting its nuclear localization in hypoxic cells. J Proteome Res 12:282–292
Bui MH, Seligson D, Han KR et al (2003) Carbonic anhydrase IX is an independent predictor of survival in advanced renal clear cell carcinoma: implications for prognosis and therapy. Clin Cancer Res 9:802–811
Bui MH, Visapaa H, Seligson D et al (2004) Prognostic value of carbonic anhydrase IX and Ki67 as predictors of survival for renal clear cell carcinoma. J Urol 171:2461–2466
Buller F, Steiner M, Frey K et al (2011) Selection of carbonic anhydrase IX inhibitors from one million DNA-encoded compounds. ACS Chem Biol 6:336–344
Carta F, Maresca A, Scozzafava A et al (2012) Novel coumarins and 2-thioxo-coumarins as inhibitors of the tumor-associated carbonic anhydrases IX and XII. Bioorg Med Chem 20:2266–2273
Casey JR, Morgan PE, Vullo D et al (2004) Carbonic anhydrase inhibitors. Design of selective, membrane-impermeant inhibitors targeting the human tumor-associated isozyme IX. J Med Chem 47:2337–2347
Cecchi A, Winum JY, Innocenti A et al (2004) Carbonic anhydrase inhibitors: synthesis and inhibition of cytosolic/tumor associated carbonic anhydrase isozymes I, II, and IX with sulfonamides derived from 4-isothiocyanato-benzolamide. Bioorg Med Chem Lett 14:5775–5780
Cecchi A, Hulikova A, Pastorek J et al (2005) Carbonic anhydrase inhibitors. Design of fluorescent sulfonamides as probes of tumor-associated carbonic anhydrase IX that inhibit isozyme IX-mediated acidification of hypoxic tumors. J Med Chem 48:4834–4841
Chia SK, Wykoff CC, Watson PH et al (2001) Prognostic significance of a novel hypoxia-regulated marker, carbonic anhydrase IX, in invasive breast carcinoma. J Clin Oncol 19:3660–3668
Chiang WL, Chu SC, Yang SS et al (2002) The aberrant expression of cytosolic carbonic anhydrase and its clinical significance in human non-small cell lung cancer. Cancer Lett 188:199–205
Chiche J, Ilc K, Laferrière J et al (2009) Hypoxia-inducible carbonic anhydrase IX and XII promote tumor cell growth by counteracting acidosis through the regulation of the intracellular pH. Cancer Res 69:358–368
Chien MH, Ying TH, Hsieh YH et al (2012) Tumor-associated carbonic anhydrase XII is linked to the growth of primary oral squamous cell carcinoma and its poor prognosis. Oral Oncol 48:417–423
Cho M, Grabmaier K, Kitahori Y et al (2000) Activation of the MN/CA9 gene is associated with hypomethylation in human renal cell carcinoma cell lines. Mol Carcinog 27:184–189
Dang CV, Semenza GL (1999) Oncogenic alterations of metabolism. Trends Biochem Sci 24:68–72
Davis ID, Wiseman GA, Lee FT et al (2007) A phase I multiple dose, dose escalation study of cG250 monoclonal antibody in patients with advanced renal cell carcinoma. Cancer Immun 17(7):13
Ditte P, Dequiedt F, Svastova E et al (2011) Phosphorylation of carbonic anhydrase IX controls its ability to mediate extracellular acidification in hypoxic tumors. Cancer Res 71:7558–7567

Divgi CR, Pandit-Taskar N, Jungbluth AA et al (2007) Preoperative characterisation of clear-cell renal carcinoma using iodine-124-labelled antibody chimeric G250 (124I-cG250) and PET in patients with renal masses: a phase I trial. Lancet Oncol 8:304–310
Dogné JM, Thiry A, Pratico D et al (2007) Dual carbonic anhydrase–cyclooxygenase-2 inhibitors. Curr Top Med Chem 7:885–891
Dorai T, Sawczuk IS, Pastorek J et al (2005) The role of carbonic anhydrase IX overexpression in kidney cancer. Eur J Cancer 41:2935–2947
Doyen J, Parks SK, Marcié S et al (2012) Knock-down of hypoxia-induced carbonic anhydrases IX and XII radiosensitizes tumor cells by increasing intracellular acidosis. Front Oncol 2:199. doi:10.3389/fonc.2012.00199.
Dubois L, Douma K, Supuran CT et al (2007) Imaging the hypoxia surrogate marker CA IX requires expression and catalytic activity for binding fluorescent sulfonamide inhibitors. Radiother Oncol 83:367–373
Dubois L, Lieuwes NG, Maresca A et al (2009) Imaging of CA IX with fluorescent labelled sulfonamides distinguishes hypoxic and (re)-oxygenated cells in a xenograft tumour model. Radiother Oncol 92:423–428
Dubois L, Peeters S, Lieuwes NG et al (2011) Specific inhibition of carbonic anhydrase IX activity enhances the in vivo therapeutic effect of tumor irradiation. Radiother Oncol 99:424–431
Durdagi S, Şentürk M, Ekinci D et al (2011) Kinetic and docking studies of phenol-based inhibitors of carbonic anhydrase isoforms I, II, IX and XII evidence a new binding mode within the enzyme active site. Bioorg Med Chem 19:1381–1389
Fang JS, Gillies RD, Gatenby RA (2008) Adaptation to hypoxia and acidosis in carcinogenesis and tumor progression. Semin Cancer Biol 18:330–337
Franchi M, Vullo D, Gallori E et al (2003) Carbonic anhydrase inhibitors. Inhibition of cytosolic isozymes I and II and transmembrane, cancer-associated isozyme IX with lipophilic sulfonamides. J Enzyme Inhib Med Chem 18:333–338
Fukumura D, Jain RK (2007) Tumor microenvironment abnormalities: causes, consequences, and strategies to normalize. J Cell Biochem 101:937–949
Garaj V, Pucetti L, Fasolis G et al (2004) Carbonic anhydrase inhibitors: synthesis and inhibition of cytosolic/tumor-associated carbonic anhydrase isozymes I, II, and IX with sulfonamides incorporating 1,2,4-triazine moieties. Bioorg Med Chem Lett 14:5427–5433
Garaj V, Pucetti L, Fasolis G et al (2005) Carbonic anhydrase inhibitors: novel sulfonamides incorporating 1,3,5-triazine moieties as inhibitors of cytosolic and tumor-associated carbonic anhydrase isozymes I, II, and IX. Bioorg Med Chem Lett 15:3102–3108
Generali D, Fox SB, Berruti A et al (2006) Role of carbonic anhydrase IX expression in prediction of the efficacy and outcome of primary epirubicin/tamoxifen therapy for breast cancer. Endocr Relat Cancer 13:921–930
Giatromanolaki A, Koukourakis MI, Sivridis E et al (2001) Expression of hypoxia-inducible carbonic anhydrase-9 relates to angiogenic pathways and independently to poor outcome in non-small cell lung cancer. Cancer Res 61:7992–7998
Gieling RG, Babur M, Mamnani L (2012) Antimetastatic effect of sulfamate carbonic anhydrase IX inhibitors in breast carcinoma xenografts. J Med Chem 55:5591–5600
Gieling RG, Parker CA, De Costa LA et al (2013) Inhibition of carbonic anhydrase activity modifies the toxicity of doxorubicin and melphalan in tumour cells in vitro. J Enzyme Inhib Med Chem 28:360–369
Gnarra JR, Tory K, Weng Y et al (1994) Mutations of the VHL tumour suppressor gene in renal carcinoma. Nat Genet 7:85–90
Grabmaier K, A de Weijert MC, Verhaegh GW et al (2004) Strict regulation of CA IX(G250/MN) by HIF-1alpha in clear cell renal cell carcinoma. Oncogene 23:5624–5631
Gut MO, Parkkila S, Vernerová Z et al (2002) Gastric hyperplasia in mice with targeted disruption of the carbonic anhydrase gene Car9. Gastroenterology 123:1889–1903
Haapasalo J, Nordfors K, Hilvo M et al (2005) Expression of carbonic anhydrase IX in astrocytic tumors predicts poor prognosis. Clin. Cancer Res 12:473–477

Haapasalo J, Nordfors K, Järvelä S et al (2007) Carbonic anhydrase II in the endothelium of glial tumors: a potential target for therapy. Neuro Oncol 9:308–313

Hanahan D, Weinberg RA (2000) The hallmarks of cancer. Cell 100:57–70

Harris AL (2002) Hypoxia – a key regulatory factor in tumor growth. Nature Rev Cancer 2:38–47

Harrison SC (2008) Viral membrane fusion. Nat Struct Mol Biol 15:690–698

Helmlinger G, Sckell A, Dellian M et al (2002) Acid production in glycolysis-impaired tumors provides new insights into tumor metabolism. Clin Cancer Res 8:1284–1291

Hewett-Emmett D, Tashian RE (1996) Functional diversity, conservation, and convergence in the evolution of the alpha-, beta-, and gamma-carbonic anhydrase gene families. Mol Phylogenet Evol 5:50–77

Hilvo M, Baranauskiene L, Salzano AM (2008) Biochemical characterization of CA IX, one of the most active carbonic anhydrase isozymes. J Biol Chem 283:27799–27809

Hui EP, Chan AT, Pezzella F et al (2002) Coexpression of hypoxia-inducible factors 1alpha and 2alpha, carbonic anhydrase IX, and vascular endothelial growth factor in nasopharyngeal carcinoma and relationship to survival. Clin Cancer Res 8:2595–2504

Hulikova A, Vaughan-Jones RD, Swietach P (2011) Dual role of CO2/HCO3(-) buffer in the regulation of intracellular pH of three-dimensional tumor growths. J Biol Chem 286:13815–13826

Hussain SA, Ganesan R, Reynolds G et al (2007) Hypoxia-regulated carbonic anhydrase IX expression is associated with poor survival in patients with invasive breast cancer. Br J Cancer 96:104–109

Hutchison GJ, Valentine HR, Loncaster JA et al (2004) Hypoxia-inducible factor 1alpha expression as an intrinsic marker of hypoxia: correlation with tumor oxygen, pimonidazole measurements, and outcome in locally advanced carcinoma of the cervix. Clin Cancer Res 10:8405–8412

Hynninen P, Hamalainen JM, Pastorekova S et al (2004) Transmembrane carbonic anhydrase isozymes IX and XII in the female mouse reproductive tract. Reprod Biol Endocrinol 2:73

Hynninen P, Vaskivuo L, Saarnio J et al (2006) Expression of transmembrane carbonic anhydrases IX and XII in ovarian tumours. Histopathology 49:594–602

Iakovlev VV, Pintilie M, Morrison A et al (2007) Effect of distributional heterogeneity on the analysis of tumor hypoxia based on carbonic anhydrase IX. Lab Invest 87:1206–1217

Ihnatko R, Kubes M, Takacova M et al (2006) Extracellular acidosis elevates carbonic anhydrase IX in human glioblastoma cells via transcriptional modulation that does not depend on hypoxia. Int J Oncol 29:1925–1933

Ilie M, Mazure NM, Hofman V (2010) High levels of carbonic anhydrase IX in tumour tissue and plasma are biomarkers of poor prognostic in patients with non-small cell lung cancer. Br J Cancer 102:1627–1635

Ilie MI, Hofman V, Ortholan C (2011) Overexpression of carbonic anhydrase XII in tissues from resectable non-small cell lung cancers is a biomarker of good prognosis. Int J Cancer 128:1614–1623

Ilies MA, Vullo D, Pastorek J et al (2003) Carbonic anhydrase inhibitors. Inhibition of tumor-associated isozyme IX by halogenosulfanilamide and halogenophenylaminobenzolamide derivatives. J Med Chem 46:2187–2196

Innocenti A, Vullo D, Scozzafava A et al (2005) Carbonic anhydrase inhibitors. Interaction of isozymes I, II, IV, V, and IX with carboxylates. Bioorg Med Chem Lett 15:573–578

Ivanov SV, Kuzmin I, Wei MH et al (1998) Down-regulation of transmembrane carbonic anhydrases in renal cell carcinoma cell lines by wild-type von Hippel-Lindau transgenes. Proc Natl Acad Sci USA 95:12596–12601

Ivanov S, Liao SY, Ivanova A et al (2001) Expression of hypoxia-inducible cell-surface transmembrane carbonic anhydrases in human cancer. Am J Pathol 158:905–919

Jaiswal M, Khadikar PV, Scozzafava A et al (2004) Carbonic anhydrase inhibitors: the first QSAR study on inhibition of tumor-associated isoenzyme IX with aromatic and heterocyclic sulfonamides. Bioorg Med Chem Lett 14:3283–3290

Jakubickova L, Biesova Z, Pastorekova S et al (2005) Methylation of the CA9 promoter can modulate expression of the tumor-associated carbonic anhydrase IX in dense carcinoma cell lines. Int J Oncol 26:1121–1127

Jarvela S, Parkkila S, Bragge H et al (2008) Carbonic anhydrase IX in oligodendroglial brain tumors. BMC Cancer 8:1

Kallio H, Hilvo M, Rodriguez A et al (2010) Global transcriptional response to carbonic anhydrase IX deficiency in the mouse stomach. BMC Genomics 11:397. doi:10.1186/1471-2164-11-397

Kaluz S, Kaluzova M, Chrastina A et al (2002) Lowered oxygen tension induces expression of the hypoxia marker MN/carbonic anhydrase IX in the absence of hypoxia-inducible factor 1α stabilization : a role for phosphatidylinositol 3'-kinase. Cancer Res 62:4469–4477

Kaluz S, Kaluzova M, Stanbridge EJ (2003) Expression of the hypoxia marker carbonic anhydrase IX is critically dependent on SP1 activity. Identification of a novel type of hypoxia-responsive enhancer. Cancer Res 63:917–922

Kappler M, Taubert H, Holzhausen HJ et al (2008) Immunohistochemical detection of HIF-1alpha and CAIX in advanced head-and-neck cancer. Prognostic role and correlation with tumor markers and tumor oxygenation parameters. Strahlenther Onkol 184:393–399

Karhumaa P, Kaunisto K, Parkkila S et al (2001) Expression of the transmembrane carbonic anhydrases, CA IX and CA XII, in the human male excurrent ducts. Mol Hum Reprod 7:611–616

Karumanchi SA, Jiang L, Knebelmann B et al (2001) VHL tumor suppressor regulates Cl-/HCO3-exchange and Na+/H+ exchange activities in renal carcinoma cells. Physiol Genomics 5: 119–128

Ki DH, Jeung HC, Park CH et al (2007) Whole genome analysis for liver metastasis gene signatures in colorectal cancer. Int J Cancer 121:2005–2012

Kim JW, Dang CV (2006) Cancer's molecular sweet tooth and the Warburg effect. Cancer Res 66:8927–8930

Kim SJ, Rabbani ZN, Dewhirst MW et al (2005) Expression of HIF-1alpha, CA IX, VEGF, and MMP-9 in surgically resected non-small cell lung cancer. Lung Cancer 49:325–335

Kim JY, Shin HJ, Kim TH et al (2006) Tumor-associated carbonic anhydrases are linked to metastases in primary cervical cancer. J Cancer Res Clin Oncol 132:302–308

Kim SJ, Shin HJ, Jung KY et al (2007) Prognostic value of carbonic anhydrase IX and Ki-67 expression in squamous cell carcinoma of the tongue. Jpn J Clin Oncol 37:812–819

Kim BR, Shin HJ, Kim JY et al (2012) Dickkopf-1 (DKK-1) interrupts FAK/PI3K/mTOR pathway by interaction of carbonic anhydrase IX (CA9) in tumorigenesis. Cell Signal 24:1406–1413

Kivelä A, Parkkila S, Saarnio J et al (2000) Expression of a novel transmembrane carbonic anhydrase isozyme XII in normal human gut and colorectal tumors. Am J Pathol 156:577–584

Kivela AJ, Saarnio J, Karttunen TJ et al (2001) Differential expression of cytoplasmic carbonic anhydrases, CA I and II, and membrane-associated isozymes, CA IX and XII, in normal mucosa of large intestine and in colorectal tumors. Dig Dis Sci 46:2179–2186

Klein M, Seeger P, Schuricht B et al (2000) Polarization of Na(+)/H(+) and Cl(−)/HCO (3)(−) exchangers in migrating renal epithelial cells. J Gen Physiol 115:599–608

Kobayashi M, Matsumoto T, Ryuge S et al (2012) CAXII Is a sero-diagnostic marker for lung cancer. PLoS One 7(3):e33952. doi:10.1371/journal.pone.0033952

Kock L, Mahner S, Choschzick M et al (2011) Serum carbonic anhydrase IX and its prognostic relevance in vulvar cancer. Int J Gynecol Cancer 21:141–148

Kon-no H, Ishii G, Nagai K et al (2006) Carbonic anhydrase IX expression is associated with tumor progression and a poor prognosis of lung adenocarcinoma. Lung Cancer 54:409–418

Kopacek J, Barathova M, Dequiedt F et al (2005) MAPK pathway contributes to density- and hypoxia-induced expression of the tumor-associated carbonic anhydrase IX. Biochim Biophys Acta 1729:41–49

Koperek O, Bergner O, Pichlhöfer B et al (2011) Expression of hypoxia-associated proteins in sporadic medullary thyroid cancer is associated with desmoplastic stroma reaction and lymph node metastasis and may indicate somatic mutations in the VHL gene. J Pathol 225:63–72

Korkolopoulou P, Perdiki M, Thymara I (2007) Expression of hypoxia-related tissue factors in astrocytic gliomas. A multivariate survival study with emphasis upon carbonic anhydrase IX. Hum Pathol 38:629–638

Kowalewska M, Radziszewski J, Kulik J et al (2005) Detection of carbonic anhydrase 9-expressing tumor cells in the lymph nodes of vulvar carcinoma patients by RT-PCR. Int J Cancer 116: 957–962

Kozin SV, Shkarin P, Gerweck LE (2001) The cell transmembrane pH gradient in tumors enhances cytotoxicity of specific weak acid chemotherapeutics. Cancer Res 61:4740–4743

Kummola L, Hamalainen JM, Kivela J et al (2005) Expression of a novel carbonic anhydrase, CA XIII, in normal and neoplastic colorectal mucosa. BMC Cancer 5:41

Kuo WH, Chiang WL, Yang SF et al (2003) The differential expression of cytosolic carbonic anhydrase in human hepatocellular carcinoma. Life Sci 73:2211–2223

Le QT, Chen E, Salim A et al (2006) An evaluation of tumor oxygenation and gene expression in patients with early stage non-small cell lung cancers. Clin Cancer Res 12:1507–1514

Le QT, Kong C, Lavori PW et al (2007) Expression and prognostic significance of a panel of tissue hypoxia markers in head-and-neck squamous cell carcinomas. Int J Radiat Oncol Biol Phys 69:167–175

Leppilampi M, Karttunen TJ, Kivelä J et al (2005) Gastric pit cell hyperplasia and glandular atrophy in carbonic anhydrase IX knockout mice: studies on two strains C57/BL6 and BALB/C. Transgenic Res 14:655–663

Liao SY, Brewer C, Zavada J et al (1994) Identification of the MN antigen as a diagnostic biomarker of cervical intraepithelial squamous and glandular neoplasia and cervical carcinomas. Am J Pathol 145:598–609

Liao SY, Aurelio ON, Jan K et al (1997) Identification of the MN/CA9 protein as a reliable diagnostic biomarker of clear cell carcinoma of the kidney. Cancer Res 57:2827–2831

Loncaster JA, Harris AL, Davidson SE et al (2001) Carbonic anhydrase IX expression, a potential new intrinsic marker of hypoxia: correlations with tumour oxygen measurements and prognosis in locally advanced carcinoma of the cervix. Cancer Res 61:6394–6399

Lou Y, McDonald PC, Oloumi A et al (2011) Targeting tumor hypoxia: suppression of breast tumor growth and metastasis by novel carbonic anhydrase IX inhibitors. Cancer Res 71:3364–3376

Lounnas N, Rosilio C, Nebout M et al (2013) Pharmacological inhibition of carbonic anhydrase XII interferes with cell proliferation and induces cell apoptosis in T-cell lymphomas. Cancer Lett doi:pii: S0304–3835(13)00049-9

Luiten RM, Coney LR, Fleuren GJ et al (1996) Generation of chimeric bispecific G250/anti-CD3 monoclonal antibody, a tool to combat renal cell carcinoma. Brit J Cancer 74:735–744

Luiten RM, Warnaar SO, Schuurman J et al (1997) Chimeric immunoglobulin E reactive with tumor-associated antigen activates human Fc epsilon RI bearing cells. Hum Antibodies 8:169–180

Malentacchi F, Simi L, Nannelli C et al (2009) Alternative splicing variants of carbonic anhydrase IX in human non-small cell lung cancer. Lung Cancer 64:271–276

Malentacchi F, Vinci S, Della MA (2012) Splicing variants of carbonic anhydrase IX in bladder cancer and urine sediments. Urol Oncol 30:278–284

Mandriota SJ, Turner KJ, Davies DR et al (2002) HIF activation identifies early lesions in VHL kidneys: evidence for site-specific tumor suppressor function in the nephron. Cancer Cell 1:459–468

Maresca A, Supuran CT (2010) Coumarins incorporating hydroxy- and chloro-moieties selectively inhibit the transmembrane, tumor-associated carbonic anhydrase isoforms IX and XII over the cytosolic ones I and II. Bioorg Med Chem Lett 20:4511–4514

Maresca A, Scozzafava A, Supuran CT (2010) 7,8-disubstituted- but not 6,7-disubstituted coumarins selectively inhibit the transmembrane, tumor-associated carbonic anhydrase isoforms IX and XII over the cytosolic ones I and II in the low nanomolar/subnanomolar range. Bioorg Med Chem Lett 20:7255–7258

Martin C, Pedersen SF, Schwab A et al (2011) Intracellular pH gradients in migrating cells. Am J Physiol Cell Physiol 300:C490–C495

McIntyre A, Patiar S, Wigfield S et al (2012) Carbonic anhydrase IX promotes tumor growth and necrosis in vivo and inhibition enhances anti-VEGF therapy. Clin Cancer Res 18:3100–3111

McKiernan JM, Buttyan R, Bander NH et al (1997) Expression of the tumour-associated gene MN: a potential biomarker for human renal cell carcinoma. Cancer Res 57:2362–2365

McMurtrie HL, Cleary HJ, Alvarez BV et al (2004) The bicarbonate transport metabolon. J Enz Inhib Med Chem 19:231–236

Morgan PE, Pastoreková S, Stuart-Tilley AK et al (2007) Interactions of transmembrane carbonic anhydrase, CAIX, with bicarbonate transporters. Am J Physiol Cell Physiol 293:C738–C748
Mori M, Staniunas RJ, Barnard GF et al (1993) The significance of carbonic anhydrase expression in human colorectal cancer. Gastroenterology 105:820–826
Morimoto K, Nishimori I, Takeuchi T et al (2005) Overexpression of carbonic anhydrase-related protein XI promotes proliferation and invasion of gastrointestinal stromal tumors. Virchows Arch 447:66–73
Müller V, Riethdorf S, Rack B et al (2011) DETECT study group. Prospective evaluation of serum tissue inhibitor of metalloproteinase 1 and carbonic anhydrase IX in correlation to circulating tumor cells in patients with metastatic breast cancer. Breast Cancer Res 13:R71
Murri-Plesko MT, Hulikova A, Oosterwijk E et al (2011) Antibody inhibiting enzymatic activity of tumour-associated carbonic anhydrase isoform IX. Eur J Pharmacol 657:173–183
Neri D, Supuran CT (2011) Interfering with pH regulation in tumours as a therapeutic strategy. Nat Rev Drug Discov 10:767–777
Niemelä AM, Hynninen P, Mecklin JP et al (2007) Carbonic anhydrase IX is highly expressed in hereditary nonpolyposis colorectal cancer. Cancer Epidemiol Biomarkers Prev 16:1760–1706
Olive PL, Aquino-Parsons C, MacPhail SH et al (2001) Carbonic anhydrase 9 as an endogenous marker for hypoxic cells in cervical cancer. Cancer Res 61:8924–8929
Oosterwijk E, Ruiter DJ, Hoedemaeker PJ et al (1986) Monoclonal antibody G250 recognizes determinant present in renal cell carcinoma and absent from normal kidney. Int J Cancer 38:489–494
Opavsky R, Pastorekova S, Zelnik V et al (1996) Human MN/CA9 gene, a novel member of the carbonic anhydrase family: structure and exon to protein domain relationships. Genomics 33:480–487
Ord JJ, Streeter EH, Roberts IS et al (2005) Comparison of hypoxia transcriptome in vitro with in vivo gene expression in human bladder cancer. Br J Cancer 93:346–354
Orlowski A, De Giusti VC, Morgan PE et al (2012) Binding of carbonic anhydrase IX to extracellular loop 4 of the NBCe1 Na+/HCO3− cotransporter enhances NBCe1-mediated HCO3− influx in the rat heart. Am J Physiol Cell Physiol 303:C69–C80
Ozensoy O, Pucetti L, Fasolis G et al (2005) Carbonic anhydrase inhibitors: inhibition of the tumor-associated isoenzymes IX and XII with a library of aromatic and heteroaromatic sulfonamides. Bioorg Med Chem Lett 15:4862–4866
Pacchiano F, Carta F, McDonald PC et al (2011) Ureido-substituted benzenesulfonamides potently inhibit carbonic anhydrase IX and show antimetastatic activity in a model of breast cancer metastasis. J Med Chem 54:1896–1902
Pan PW, Parkkila AK, Autio S et al (2012) Brain phenotype of carbonic anhydrase IX-deficient mice. Transgenic Res 21:163–176
Papi A, Storci G, Guarnieri T et al (2013) Peroxisome proliferator activated receptor-α/hypoxia inducible factor-1α interplay sustains carbonic anhydrase IX and apoliprotein E expression in breast cancer stem cells. PLoS One 8(1):e54968
Parent AS, Mungenast AE, Lomniczi A et al (2007) A contactin-receptor-like protein tyrosine phosphatase beta complex mediates adhesive communication between astroglial cells and gonadotrophin-releasing hormone neurones. J Neuroendocrinol 19:847–859
Parkkila S, Innocenti A, Kallio H et al (2009) The protein tyrosine kinase inhibitors imatinib and nilotinib strongly inhibit several mammalian alpha-carbonic anhydrase isoforms. Bioorg Med Chem Lett 19:4102–4106
Parkkila S, Vullo D, Maresca A et al (2012) Serendipitous fragment-based drug discovery: ketogenic diet metabolites and statins effectively inhibit several carbonic anhydrases. Chem Commun (Camb) 48:3551–3553
Parks SK, Chiche J, Pouyssegur J (2011) pH control mechanisms of tumor survival and growth. J Cell Physiol 226:299–308
Pastorek J, Pastoreková S, Callebaut I et al (1994) Cloning and characterization of MN, a human tumor-associated protein with a domain homologous to carbonic anhydrase and a putative helix-loop-helix DNA binding segment. Oncogene 9:2877–2888

Pastorekova S, Zavadova Z, Kostal M et al (1992) A novel quasi-viral agent, MaTu, is a two-component system. Virology 187:620–626

Pastorekova S, Parkkila S, Parkkila AK et al (1997) Carbonic anhydrase IX, MN/CA IX: analysis of stomach complementary DNA sequence and expression in human and rat alimentary tracts. Gastroenterology 112:398–408

Pastorekova S, Parkkila S, Pastorek J et al (2004) Carbonic anhydrases: current state of the art, therapeutic applications and future prospects. J Enz Inhib Med Chem 19:199–229

Pastorekova S, Vullo D, Casini A et al (2005) Carbonic anhydrase inibitors: synthesis and inhibition of the tumor-associated isozymes IX and XII with polyfluorinated aromatic/heterocyclic sulfonamides. J Enzyme Inhib Med Chem 20:211–7

Pastorekova S, Ratcliffe PJ, Pastorek J (2008) Molecular mechanisms of carbonic anhydrase IX-mediated pH regulation under hypoxia. BJU Int 101(Suppl 4):8–15. doi:10.1111/j.1464-410X.2008.07642.x.

Peles E, Nativ M, Campbell PL et al (1995) The carbonic anhydrase domain of receptor tyrosine phosphatase β is a functional ligand for the axonal cell recognition molecule contactin. Cell 82:251–260

Petrul HM, Schatz CA, Kopitz CC et al (2012) Therapeutic mechanism and efficacy of the antibody-drug conjugate BAY 79-4620 targeting human carbonic anhydrase 9. Mol Cancer Ther 11:340–349

Potter C, Harris AL (2004) Hypoxia inducible carbonic anhydrase IX, marker of tumour hypoxia, survival pathway and therapy target. Cell Cycle 3:164–167

Pucetti L, Fasolis G, Cecchi A et al (2005) Carbonic anhydrase inibitors: synthesis and inhibition of cytosolic/tumor-associated carbonic anhydrase isozymes I, II and IX with sulfonamides incorporating thioureido-sulfanilyl scaffolds. Bioorg Med Chem Lett 15:2359–2364

Rademakers SE, Lok J, van der Kogel AJ et al (2011, May 12) Metabolic markers in relation to hypoxia; staining patterns and colocalization of pimonidazole, HIF-1α, CAIX, LDH-5, GLUT-1, MCT1 and MCT4. BMC Cancer 11:167. doi:10.1186/1471–2407-11-167.

Radvak P, Repic M, Svastova E et al (2013) Suppression of carbonic anhydrase IX leads to aberrant focal adhesion and decreased invasion of tumor cells. Oncol Rep 29:1147–1153

Rafajova M, Zatovicova M, Kettmann R et al (2004) Induction by hypoxia combined with low glucose or low bicarbonate and high posttranslational stability upon reoxygenation contribute to carbonic anhydrase IX expression in cancer cells. Int J Oncol 24:995–1004

Raghunand N, Gatenby RA, Gillies RJ (2003) Microenvironmental and cellular consequences of altered blood flow in tumours. Br J Radiol 76:S11–S22

Rami M, Winum JY, Innocenti A et al (2008) Carbonic anhydrase inhibitors: copper(II) complexes of polyamino-polycarboxylamido aromatic/heterocyclic sulfonamides are very potent inhibitors of the tumor-associated isoforms IX and XII. Bioorg Med Chem Lett 18:836–841

Raval RR, Lau KW, Tran MGB et al (2005) Contrasting properties of hypoxia-inducible factor 1 (HIF-1) and HIF-2 in von Hippel-Lindau-associated renal cell carcinoma. Mol Cell Biol 25:5675–5686

Reichert JM (2011) Antibody-based therapeutics to watch in 2011. MAbs 3:76–99

Saarnio J, Parkkila S, Parkkila AK et al (1998a) Immunohistochemistry of carbonic anhydrase isozyme IX (MN/CA IX) in human gut reveals polarized expression in the epithelial cells with the highest proliferative capacity. J Histochem Cytochem 46:497–504

Saarnio J, Parkkila S, Parkkila AK et al (1998b) Immunohistochemical study of colorectal tumors for expression of a novel transmembrane carbonic anhydrase, MN/CA IX, with potential value as a marker of cell proliferation. Am J Pathol 153:279–285

Saczewski F, Sławiński J, Kornicka A et al (2006) Carbonic anhydrase inhibitors. Inhibition of the cytosolic human isozymes I and II, and the transmembrane, tumor-associated isozymes IX and XII with substituted aromatic sulfonamides activatable in hypoxic tumors. Bioorg Med Chem Lett 16:4846–4851

Sandlund J, Oosterwijk E, Grankvist K et al (2007) Prognostic impact of carbonic anhydrase IX expression in human renal cell carcinoma. BJU Int 100:556–560

Sansone P, Storci G, Giovannini C et al (2007a) p66Shc/Notch-3 interplay controls self-renewal and hypoxia survival in human stem/progenitor cells of the mammary gland expanded in vitro as mammospheres. Stem Cells 25:807–815

Sansone P, Storci G, Tavolari S et al (2007b) IL-6 triggers malignant features in mammospheres from human ductal breast carcinoma and normal mammary gland. J Clin Invest 117:3988–4002

Schoppmann SF, Jesch B, Friedrich J et al (2012) Phosphorylation of signal transducer and activator of transcription 3 (STAT3) correlates with Her-2 status, carbonic anhydrase 9 expression and prognosis in esophageal cancer. Clin Exp Metastasis 29:615–624

Schwab A (2001) Function and spatial distribution of ion channels and transporters in cell migration. Am J Physiol Renal Physiol 280:F739–F747

Scozzafava A, Mastrolorenzo A, Supuran CT (2004) Modulation of carbonic anhydrase activity and its applications in therapy. Expert Opin Ther Patents 14:667–702

Shao Y, Li Y, Zhang J et al (2010) Involvement of histone deacetylation in MORC2-mediated down-regulation of carbonic anhydrase IX. Nucleic Acids Res 38:2813–2824

Shimoda LA, Fallon M, Pisarcik S et al (2006) HIF-1 regulates hypoxic induction of NHE1 expression and alkalinization of intracellular pH in pulmonary arterial myocytes. Am J Physiol Lung Cell Mol Physiol 291:L941–L949

Shin KH, Diaz-Gonzalez JA, Russell J et al (2007) Detecting changes in tumor hypoxia with carbonic anhydrase IX and pimonidazole. Cancer Biol Ther 6:70–75

Shin HJ, Rho SB, Jung DC et al (2011) Carbonic anhydrase IX (CA9) modulates tumor-associated cell migration and invasion. J Cell Sci 124:1077–1087

Siebels M, Rohrmann K, Oberneder R et al (2011) A clinical phase I/II trial with the monoclonal antibody cG250 (RENCAREX®) and interferon-alpha-2a in metastatic renal cell carcinoma patients. World J Urol 29:121–126

Smaine FZ, Winum JY, Montero JL et al (2007) Carbonic anhydrase inhibitors: selective inhibition of the extracellular, tumor-associated isoforms IX and XII over isozymes I and II with glycosyl-thioureido-sulfonamides. Bioorg Med Chem Lett 17:5096–5100

Smyth LG, O'Hurley G, O'Grady A et al (2010) Carbonic anhydrase IX expression in prostate cancer. Prostate Cancer Prostatic Dis 13:178–181

Sobhanifar S, Aquino-Parsons C, Stanbridge EJ et al (2005) Reduced expression of hypoxia-inducible factor-1alpha in perinecrotic regions of solid tumors. Cancer Res 65:7259–7266

Sørensen BS, Hao J, Overgaard J et al (2005) Influence of oxygen concentration and pH on expression of hypoxia induced genes. Radiother Oncol 76:187–193

Stillebroer AB, Mulders PF, Boerman OC et al (2010) Carbonic anhydrase IX in renal cell carcinoma: implications for prognosis, diagnosis, and therapy. Eur Urol 58:75–83

Stock C, Schwab A (2009) Protons make tumor cells move like clockwork. Pflugers Arch 458: 981–992

Stock C, Mueller M, Kraehling H et al (2007) pH nanoenvironment at the surface of single melanoma cells. Cell Physiol Biochem 20:679–686

Storci G, Sansone P, Trere D et al (2008) The basal-like breast carcinoma phenotype is regulated by SLUG gene expression. J Pathol 214:25–37

Stubbs M, McSheehy PMJ, Griffiths JR et al (2000) Causes and consequences of tumor acidity and implications for treatment. Mol Med Today 6:15–19

Supuran CT (2007) Carbonic anhydrases as drug targets—an overview. Curr Top Med Chem 7: 825–833

Supuran CT (2008) Carbonic anhydrases: novel therapeutic applications for inhibitors and activators. Nat Rev Drug Discov 7:168–181

Supuran CT, Scozzafava A (2000) Carbonic anhydrase inhibitors: aromatic sulfonamides and disulfonamides act as efficient tumor growth inhibitors. J Enzyme Inhib 15:597–610

Supuran CT, Briganti F, Tilli S et al (2001) Carbonic anhydrase inhibitors: sulfonamides as antitumor agents? Bioorg Med Chem Let 9:703–714

Svastova E, Pastorekova S (2013) Carbonic anhydrase IX: a hypoxia-controlled "catalyst" of cell migration. Cell Adh Migr 7(2):226–231

Svastova E, Zilka N, Zatovicova M et al (2003) Carbonic anhydrase IX reduces E-cadherin-mediated adhesion of MDCK cells via interaction with β-catenin. Exp Cell Res 290:332–345

Svastova E, Hulikova A, Rafajova M et al (2004) Hypoxia activates the capacity of tumor-associated carbonic anhydrase IX to acidify extracellular pH. FEBS Lett 577:439–495

Svastova E, Witarski W, Csaderova L et al (2012) Carbonic anhydrase IX interacts with bicarbonate transporters in lamellipodia and increases cell migration via its catalytic domain. J Biol Chem 287:3392–3402

Swietach P, Patiar S, Supuran CT et al (2009) The role of carbonic anhydrase 9 in regulating extracellular and intracellular ph in three-dimensional tumor cell growths. J Biol Chem 284:20299–20310

Swietach P, Hulikova A, Vaughan-Jones RD et al (2010) New insights into the physiological role of carbonic anhydrase IX in tumour pH regulation. Oncogene 29:6509–6521

Swietach P, Hulikova A, Patiar S et al (2012) Importance of intracellular pH in determining the uptake and efficacy of the weakly basic chemotherapeutic drug, doxorubicin. PLoS One 7(4):e35949. doi:10.1371/journal.pone.0035949

Swinson DE, Jones JL, Cox G et al (2004) Hypoxia-inducible factor-1 alpha in non small cell lung cancer: relation to growth factor, protease and apoptosis pathways. Int J Cancer 111:43–50

Tafreshi NK, Bui MM, Bishop K et al (2012) Noninvasive detection of breast cancer lymph node metastasis using carbonic anhydrases IX and XII targeted imaging probes. Clin Cancer Res 18:207–219. doi:10.1158/1078–0432.CCR-11-0238

Takacova M, Bartosova M, Skvarkova L et al (2013) Carbonic anhydrase IX is a clinically significant tissue and serum biomarker associated with renal cell carcinoma. Oncol Lett 5:191–197

Takahashi K, Stamenkovic I, Cutler M et al (1996) Keratan sulfate modification of CD44 modulates adhesion to hyaluronate. J Biol Chem 271:9490–9496

Teicher BA, Liu SD, Liu JT et al (1993) A carbonic anhydrase inhibitor is potential modulator of cancer therapies. Anticancer Res 13:1549–1556

Temperini C, Innocenti A, Mastrolorenzo A et al (2007) Carbonic anhydrase inhibitors. Interaction of the antiepileptic drug sulthiame with twelve mammalian isoforms: kinetic and X-ray crystallographic studies. Bioorg Med Chem Lett 17:4866–4872

Thews O, Gassner B, Kelleher DK et al (2006) Impact of extracellular acidity on the activity of P-glycoprotein and the cytotoxicity of chemotherapeutic drugs. Neoplasia 8:143–152

Thiry A, Ledecq M, Cecchi A et al (2006) Indanesulfonamides as carbonic anhydrase inhibitors. Toward structure-based design of selective inhibitors of the tumor-associated isozyme CA IX. J Med Chem 49:2743–2749

Tomes L, Emberley E, Niu Y et al (2003) Necrosis and hypoxia in invasive breast carcinoma. Breast Cancer Res Treat 81:61–69

Touisni N, Maresca A, McDonald PC et al (2011) Glycosyl coumarin carbonic anhydrase IX and XII inhibitors strongly attenuate the growth of primary breast tumors. J Med Chem 54:8271–8277

Türeci O, Sahin U, Vollmar E et al (1998) Human carbonic anhydrase XII: cDNA cloning, expression and chromosomal localization of a carbonic anhydrase gene that is overexpressed in some renal cancers. Proc Natl Acad Sci USA 95:7608–7613

Turkmen H, Dugrun M, Yilmaztekin S et al (2005) Carbonic anhydrase inibitors: novel sulfanilamide/acetazolamide derivatives obtained by the tail approach and their interaction with the cytosolic isozymes I and II and the tumor-associated isozymes IX. Bioorg Med Chem Lett 17:367–372

Turner JR, Odze RD, Crum CP et al (1997) MN antigen expression in normal, preneoplastic and neoplastic esophagus: a clinicopathological study of a new cancer-associated biomarker. Hum Pathol 28:740–744

Turner KJ, Crew JP, Wykoff CC et al (2002) The hypoxia-inducible genes VEGF and CA9 are differentially regulated in superficial vs invasive bladder cancer. Br J Cancer 86:1276–1282

Uemura H, Nakagawa Y, Yoshida K et al (1999) MN/CA IX/G250 as a potential target for immunotherapy of renal cell carcinomas. Br J Cancer 81:741–746

Ullah MS, Davies AJ, Halestrap AP (2006) The plasma membrane lactate transporter MCT4, but not MCT1, is up-regulated by hypoxia through a HIF-1alpha-dependent mechanism. J Biol Chem 281:9030–9037

van den Beucken T, Koritzinsky M, Niessen H et al (2009) Hypoxia-induced expression of carbonic anhydrase 9 is dependent on the unfolded protein response. J Biol Chem 284:24204–24212
Vander Heiden MG, Cantley LC, Thompson CB (2009) Understanding the Warburg effect: the metabolic requirements of cell proliferation. Science 324:1029–1033
Vaupel P (2010) Metabolic microenvironment of tumor cells: a key factor in malignant progression. Exp Oncol 32:125–127
Vermylen P, Roufosse C, Burny A et al (1999) Carbonic anhydrase IX antigen differentiates between preneoplastic and malignant lesions in non-small cell lung carcinomas. Eur Respir J 14:806–811
Vleugel MM, Greijer AE, Shvarts A et al (2005) Differential prognostic impact of hypoxia induced and diffuse HIF-1alpha expression in invasive breast cancer. J Clin Pathol 58:172–177
Vordermark D, Kaffer A, Riedl S et al (2005) Characterization of carbonic anhydrase IX (CA IX) as an endogenous marker of chronic hypoxia in live human tumor cells. Int J Radiat Oncol Biol Phys 61:1197–1207
Vukovic V, Tannock IF (1997) Influence of the pH on cytotoxicity of paclitaxel, mitoxanthrone and topotecan. Br J Cancer 75:1167–1172
Vullo D, Franchi M, Gallori E et al (2003) Carbonic anhydrase inhibitors. Inhibition of the tumor-associated isozyme IX with aromatic and heterocyclic sulfonamides. Bioorg Med Chem Let 13:1005–1009
Vullo D, Scozzafava A, Pastorekova S et al (2004) Carbonic anhydrase inhibitors: inhibition of the tumor-associated isozyme IX with fluorine-containing sulfonamides. The first subnanomolar CA IX inhibitor discovered. Bioorg Med Chem Lett 14:2351–2356
Weber A, Casini A, Heine A et al (2004) Unexpected nanomolar inhibition of carbonic anhydrase by COX-2-selective celecoxib: new pharmacological opportunities due to related binding site recognition. J Med Chem 47:550–557
Wiesener MS, Munchenhagen PM, Berger I et al (2001) Constitutive activation of hypoxia-inducible genes related to overexpression of hypoxia-inducible factor-1α in clear cell renal carcinomas. Cancer Res 61:5215–5222
Wingo T, Tu C, Laipis PJ et al (2001) The catalytic properties of human carbonic anhydrase IX. Biochem Biophys Res Commun 288:666–669
Winter SC, Buffa FM, Silva P et al (2007) Relation of a hypoxia metagene derived from head and neck cancer to prognosis of multiple cancers. Cancer Res 67:3441–3449
Winum JY, Vullo D, Casini A et al (2003a) Carbonic anhydrase inhibitors. Inhibition of cytosolic isozymes I and II and transmembrane, tumor-associated isozyme ix with sulfamates including EMATE also acting as steroid sulfatase inhibitors. J Med Chem 46:2197–2204
Winum JY, Vullo D, Casini A, Montero JL, Scozzafava A, Supuran CT (2003b) Carbonic anhydrase inhibitors: inhibition of transmembrane, tumor-associated isozyme IX, and cytosolic isozymes 1 and II with aliphatic sulfamates. J Med Chem 46:5471–5477
Winum JY, Pastorekova S, Jakubickova L et al (2005a) Carbonic anhydrase inibitors: synthesis and inhibition of cytosolic/tumor-associated carbonic anhydrase isozymes I, II and IX with bis-sulfamates. Bioorg Med Chem Lett 15:573–578
Winum JY, Dogne JM, Casini A et al (2005b) Carbonic anhydrase inibitors: synthesis and inhibition of cytosolic/tumor-associated carbonic anhydrase isozymes I, II and IX with sulfonamides incorporating hydrazine moieties. J Med Chem 48:2121–2125
Winum JY, Innocenti A, Nasr J et al (2005c) Carbonic anhydrase inibitors: synthesis and inhibition of cytosolic/tumor-associated carbonic anhydrase isozymes I, II and IX with N-hydroxysulfamides—a new zinc-binding function in the design of inhibitors. Bioorg Med Chem Lett 15:2353–2358
Winum JY, Cecchi A, Montero JL et al (2005d) Carbonic anhydrase inhibitors. Synthesis and inhibition of cytosolic/tumor-associated carbonic anhydrase isozymes I, II and IX with oron-containing sulfonamides, sulfamides and sulfamates: towards agents for boron neutron capture therapy of hypoxic tumors. Bioorg Med Chem Lett 15:3302–3306
Wojtkowiak JW, Verduzco D, Schramm KJ et al (2011) Drug resistance and cellular adaptation to tumor acidic pH microenvironment. Mol Pharm 8:2032–2038

Wykoff C, Beasley N, Watson P et al (2000) Hypoxia-inducible regulation of tumor-associated carbonic anhydrases. Cancer Res 60:7075–7083

Wykoff CC, Beasley N, Watson PH et al (2001) Expression of the hypoxia-inducible and tumor-associated carbonic anhydrases in ductal carcinoma in situ of the breast. Am J Pathol 158:1011–1019

Xu C, Lo A, Yammanuru A et al (2010) Unique biological properties of catalytic domain directed human anti-CAIX antibodies discovered through phage-display technology. PLoS One 5(3):e9625. doi:10.1371/journal.pone.0009625

Yoo CW, Nam BH, Kim JY et al (2010) Carbonic anhydrase XII expression is associated with histologic grade of cervical cancer and superior radiotherapy outcome. Radiat Oncol 5:101. doi:10.1186/1748-717X-5-101

Yoshiura K, Nakaoka T, Nishishita T et al (2005) Carbonic anhydrase II is a tumor vessel endothelium-associated antigen targeted by dendritic cell therapy. Clin Cancer Res 11:8201–8207

Zatovicova M, Tarabkova K, Svastova E et al (2003) Monoclonal antibodies generated in carbonic anhydrase IX-deficient mice recognize different domains of tumour-associated hypoxia-induced carbonic anhydrase IX. J Immunol Methods 282:117–134

Zatovicova M, Sedlakova O, Svastova E et al (2005) Ectodomain shedding of the hypoxia-induced carbonic anhydrase IX is a metalloprotease-dependent process regulated by TACE/ADAM17. Br J Cancer 93:1267–1276

Zatovicova M, Jelenska L, Hulikova A et al (2010) Carbonic anhydrase IX as an anticancer therapy target: preclinical evaluation of internalizing monoclonal antibody directed to catalytic domain. Curr Pharm Des 16:3255–3263

Zavada J, Zavadova Z, Pastorekova S et al (1993) Expression of MaTu-MN protein in human tumor cultures and in clinical specimens. Int J Cancer 54:268–274

Zavada J, Zavadova Z, Pastorek J et al (2000) Human tumour-associated cell adhesion protein MN/CA IX: identification of M75 epitope and of the region mediating cell adhesion. Br J Cancer 82:1808–1813

Zavada J, Zavadova Z, Zatovicova M et al (2003) Soluble form of carbonic anhydrase IX (CA IX) in the serum and urine of renal carcinoma patients. Br J Cancer 89:1067–1071

Part II
Imaging

Chapter 7
Imaging the Hypoxic Tumor Microenvironment in Preclinical Models

Mary-Keara Boss, Gregory M. Palmer and Mark W. Dewhirst

Abstract Imaging the tumor microenvironment in preclinical models has advanced the understanding of the influence of hypoxia on tumor biology and response to therapy. Tumor microenvironmental hypoxia is both spatially and temporally varying. Ideally, imaging of tumor hypoxia in the preclinical model will provide spatial and temporal information with resolution to capture these features. This chapter will discuss laboratory techniques for visualizing and studying tumor hypoxia, specifically reporter gene constructs and optical methods, such as optical spectroscopy, redox imaging, phosphorescence lifetime imaging, and photoacoustic tomography. Quantitative image analysis is focused on optical sensing, which includes quantification of the presence of fluorescence reporter genes, optical probes, hemodynamics, and other parameters in both normal and tumor tissue to characterize the microenvironment. It is through these imaging methods that a clearer picture will develop of the global consequences of oxygen depletion on neoplastic tissue in regard to the basic science of cellular signaling and protein expression. Defining, manipulating, and studying these hypoxia-induced molecular pathways experimentally, while monitoring the microenvironmental effects with innovative imaging techniques in preclinical models, are key steps to translational advancements in tumor biology and therapeutic responses.

Keywords Tumor hypoxia · Reporter gene constructs · Optical spectroscopy · Redox imaging · Phosphorescence lifetime imaging · Photoacoustic tomography

---

M. W. Dewhirst (✉) · G. M. Palmer
Department of Radiation Oncology, Duke University Medical Center,
Durham, NC, USA
e-mail: mark.dewhirst@duke.edu

M.-K. Boss
Department of Molecular Biomedical Sciences,
North Carolina State University, Raleigh, NC, USA

G. Melillo (ed.), *Hypoxia and Cancer*, Cancer Drug Discovery and Development, DOI 10.1007/978-1-4614-9167-5_7,

7.1 Introduction

The understanding of the influence of microenvironmental hypoxia on tumor biology and response to therapy has been advanced by the development of imaging techniques for preclinical models. The underlying causes for tumor hypoxia have been presented in the previous chapters and in a recent review (Dewhirst et al. 2008). Briefly, the hypoxic tumor microenvironment exists due to a combination of the following interrelated physiological effects: (Dewhirst et al. 2008) longitudinal oxygen gradient, where the intravascular partial pressure of oxygen (pO_2) drops as oxygen is unloaded from hemoglobin in red blood cells (RBCs) as they traverse distally from feeding arterioles; (Bussink et al. 2003) intravascular hypoxia, where RBCs have lost virtually all oxygen and there is no oxygen available to diffuse into the tumor; (Kallinowski et al. 1990) vascular shunts at the tumor periphery which divert blood away from the tumor; (Brizel et al. 1999) low vascular density within a tumor which limits the availability of oxygen; (Hockel et al. 1996) aberrantly oriented tumor vasculature which creates inefficiency of oxygen distribution; (Braun et al. 2001) increased oxygen consumption rate of tumor cells due to altered metabolism which depletes available oxygen; and (Graves and Giaccia 2007) intravascular hypoxia reduces RBC deformability which increases blood viscosity and reduces flow rate, exacerbating tumor hypoxia. These effects result in dynamic fluctuations in the oxygenation of the tumor microenvironment and create a state of heterogeneous hypoxia. Ideally, the imaging of tumor hypoxia in the preclinical model will provide spatial and temporal information specifically addressing the interplay of these vascular and cellular effects over a defined region of interest to investigate the changes to the microenvironment and tumor physiology. This chapter will discuss laboratory techniques for visualizing and studying tumor hypoxia.

A wide variety of imaging approaches have been developed in the past few decades for monitoring tumor oxygenation and hypoxia in vivo. We will not discuss magnetic resonance imaging (MRI) or positron emission tomography (PET) imaging methods in this chapter because they are covered elsewhere in this book. We will describe several quantitative imaging and analysis techniques for characterizing in vivo fluorescence properties and functional endpoints, including vascular morphology and oxygenation. It is through these imaging methods that a clearer picture will develop of the global consequences of oxygen depletion on neoplastic tissue in regard to the basic science of cellular signaling and protein expression. Defining, manipulating, and studying these hypoxia-induced molecular pathways experimentally, while monitoring the microenvironmental effects with innovative imaging techniques in preclinical models, are key steps to translational advancements in cancer research and oncologic therapies.

One of the most extensively used techniques to detect and quantify hypoxia in the preclinical and clinical setting is with immunohistochemistry (IHC). The immunohistochemical methods are typically based on antibodies binding to protein adducts of bioreductive nitroimidazole compounds, such as pimonidazole or EF5, that are

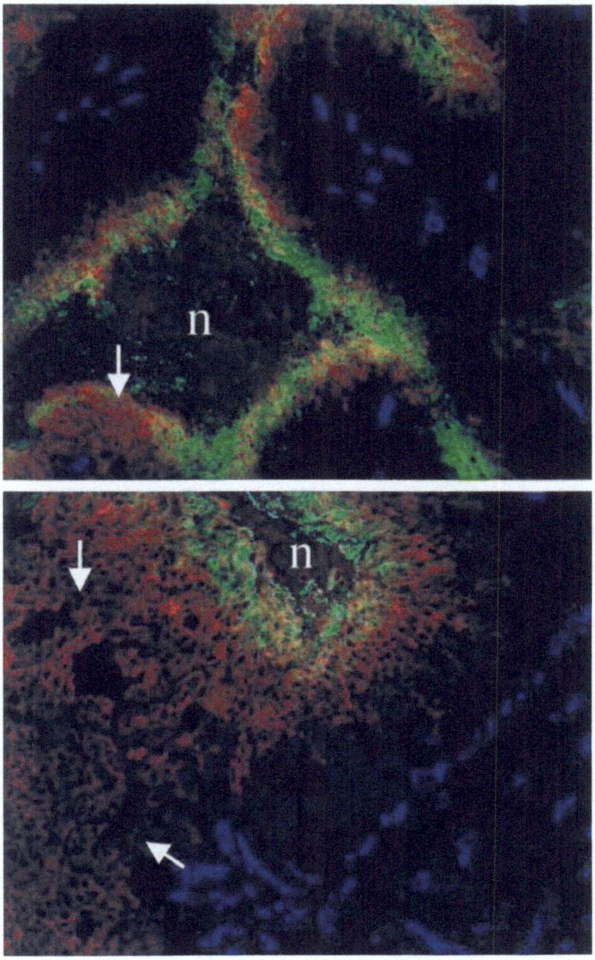

Fig. 7.1 Biopsy of two human squamous cell carcinomas of the oropharynx with triple immunofluorescent labeling of blood vessels (PAL-E, *blue*), exogenous hypoxia marker (pimonidazole, *green*), and endogenous hypoxia marker (carbonic anhydrase 9—CA9; *red*). Magnified at 100 ×. The images illustrate the colocalization, often perinecrotic (*n* = necrosis), of pimonidazole and CA9. The *arrows* indicate areas with almost no pimonidazole staining but extensive expression of CA9. CA9 expression is also often found closer to vessels than pimonidazole, illustrated in the lower photomicrograph, indicating that CA9 is expressed at higher pO_2 levels (< 20 mmHg) than the level needed for pimonidazole reduction (10 mmHg). (Reproduced from Bussink 2003 with permission from Elsevier)

systemically administered before biopsy or tumor removal is performed. Alternatively, IHC of endogenous proteins that are overexpressed under hypoxic conditions, like HIF-1α or carbonic anhydrase IX, can be used as surrogates of hypoxia (Bussink et al. 2003) (Fig. 7.1). While IHC can provide important information regarding the tumor microenvironment, the main limitations include invasiveness, limited sampling size, and the difficulty to perform repetitive measurements to monitor changes in oxygenation due to the necessity for biopsy or surgical removal of the tumor to evaluate the tissue. These results are representative of a single snapshot in time, when it is well known that tumor hypoxia is a dynamic process.

Another hypoxia detection technique that brought great progress in the 1990s in understanding the nature of solid tumor oxygenation heterogeneity in preclinical and clinical experiments was the use of oxygen probes. These probes were implanted directly into tumors to measure oxygen content and it was shown that the oxygenation

Table 7.1 Traits of the optimal preclinical hypoxia imaging system

Distinguish between hypoxia, anoxia, and necrosis
Identify cellular pO_2 values instead of vascular pO_2 values
Sensitive to pO_2 values in clinically relevant levels (0–15 mmHg)
Good spatial resolution
Short acquisition time for good temporal resolution
Ability for repeatable measurements
Noninvasive
Applicability to any tumor site (not depth limited)
Wide spatial window for measurement
Ease of use

status varies across tumor types and within the same tumor, with large heterogeneity across even short tumor distances (Kallinowski et al. 1990). When Eppendorf™ oxygen probe measurements of $pO_2 < 10$ mmHg correlated with poor local tumor control, disease-free survival, and overall survival in several clinical studies, the importance of monitoring tumor hypoxia was realized (Brizel et al. 1999; Hockel et al. 1996). Obtaining oxygen measurements with oxygen probes is invasive, requiring direct access to the solid tumor for implantation of the devices, and the probes permit only a single-site or single-line insight into tumor oxygenation. While polarography (the basis for the Eppendorf™ probe) remains a proven method for measuring tissue oxygenation, alternative probes have been developed, such as the Oxylite™ fiber optic device (Braun et al. 2001).

Noninvasive techniques that permit serial imaging of hypoxia provide valuable information because of the heterogeneous and dynamic nature of hypoxia in tumors. The requirements and interests for imaging the hypoxic tumor microenvironment vary according to whether the intended use is laboratory or clinical-based tumor imaging. For the patient, the ideal device for measuring hypoxia would allow repeated, noninvasive measurements of direct tissue pO_2 over the course of treatment with high sensitivity for tumor spatial heterogeneity. Imaging hypoxia in the laboratory requires devices that overcome an inherent challenge of studying small animal tumors. These are much smaller than most clinical tumors. As such, higher spatial resolution is needed.

Despite these application-based differences, there are general qualities that are valuable to a hypoxia imaging device. The optimal imaging system should be able to distinguish between hypoxia, anoxia, and necrosis; identify cellular pO_2 values as opposed to vascular pO_2 values; and be sensitive to pO_2 values in clinically relevant levels (0–15 mmHg). Further, aspects which will dictate the value of the modality in terms of applicability in research include good spatial resolution, short acquisition time for optimal temporal resolution, ability for repeatable measurements, noninvasiveness, applicability to any tumor site (not depth limited) with a wide spatial window for measurement, and ease of use (Table 7.1).

Multiple approaches have been developed to meet these qualities of the ideal hypoxia imaging device and the resulting forms of measurements can vary greatly, from absolute measurements of pO_2 or oxygen concentration to indirect measurements of a related parameter, such as hemoglobin saturation. In this chapter, imaging methods will be referred to by their ability to directly or indirectly quantify hypoxia. Accordingly, a direct imaging method will be one that can be calibrated to a pO_2 value, while an indirect method provides correlation to a range of pO_2 values or relative measures of oxygenation. Also, hypoxia is measured and reported under different units. The standard direct measurement is pO_2 reported in mmHg or Torr, but other units include percent oxygen (% O_2), oxygen concentration (μM), and oxygen saturation of hemoglobin. The preclinical hypoxia imaging methods differ in their approaches to oxygen measurement, as well as in which of the desired imaging device qualities they can achieve. The main hypoxia imaging devices used in the laboratory setting that will be discussed include reporter gene constructs and optical methods, such as optical spectroscopy, redox imaging, phosphorescence lifetime imaging, and photoacoustic tomography. Quantitative image analysis is focused on optical sensing, which would include quantification of the presence of fluorescence reporter genes, optical probes, hemodynamics, and other parameters in both normal and tumor tissue to characterize the microenvironment (Table 7.2).

7.2 Reporter Gene Constructs

An approach used preclinically to evaluate the effects and changes in tumor oxygenation is through in vivo molecular imaging using exogenous reporter constructs or reporter genes. The use of reporter constructs provides a large spatial view of tumor tissue while providing information on temporal fluctuations in oxygenation. Reporter genes consist of a reporter protein that can be imaged optically, such as green fluorescent protein (GFP), red fluorescent protein (RFP) from the gene dsRed, and the family of luciferase enzymes (Graves and Giaccia 2007). In regard to hypoxia detection, the reporter genes are created to contain gene targets, known as hypoxia-responsive elements (HRE), which normally exist in the promoter region of genes that are regulated by the heterodimeric hypoxia-inducible transcription factor, HIF-1. Under hypoxic conditions, hypoxia-inducible factor-1α (HIF-1α) dimerizes with constitutively expressed HIF-1β, translocates to the nucleus, and this dimerized HIF-1 transcription factor binds to HRE, driving the expression of various genes (Semenza 2010). As such, reporter constructs aid in observing in vivo HIF-1 expression by way of activation of the HRE. HIF-1 expression leads to increased transcription of the reporter gene and therefore increased fluorescence intensity, which can be quantified in the small animal tumor model. The first use of this HIF-1 reporter construct method to study tumor hypoxia was described by Vordermark and Brown (Vordermark and Brown 2003).

Another approach to measure HIF-1 levels is through the creation of reporter proteins with the proline-containing oxygen-dependent degradation domain (ODD) of HIF-1α. Under normoxic conditions, HIF-1α is targeted for degradation by a family of prolyl hydroxylases, which hydroxylate the prolyl residues on the ODD of HIF-1α. This activates HIF-1α to bind to the von Hippel–Lindau (VHL) E3 ubiquitin ligase complex, allowing HIF-1α to be degraded (Maxwell et al. 1999). Reporter

Table 7.2 Comparison of the preclinical in vivo hypoxia imaging techniques according to optimal traits

	Reporter gene constructs	Optical spectroscopy	Phosphorescence lifetime imaging	Photoacoustic tomography
Chemical probe	Exogenous reporter constructs	None	Exogenous phosphors	None
Direct/indirect	Indirect	Indirect	Direct	Indirect
Basis of oxygen measurement	Fluorescence intensity of accumulated proteins	Optical absorbance of hemoglobin	Decay of lifetime of phosphor upon interaction with O_2	Optical irradiation forms photoacoustic waves
Invasive/noninvasive	Invasive	Noninvasive	Minimally invasive	Noninvasive
Sensing depth	cm	mm–cm	mm–cm	cm
Vascular/tissue pO_2	Tissue	Vascular	Both	Vascular
Temporal resolution	min–h	s	s	s
Range	Not applicable	0–100 % Hb saturation	Probe dependent	0–100 % Hb saturation
Sensitivity	Not applicable	High sensitivity to hemoglobin	< 1 mmHg	High sensitivity to hemoglobin
Spatial resolution	Varies with imaging device	mm–cm	< mm–cm	< mm
Ability to monitor changes in hypoxia	Varies with reporter half-life	Real time	min	Real time
Cost (1–5; high–low)	3	3	3	5
Availability (1–5; high–low)	2	2	2	4

proteins with the proline-containing ODD exploit HIF-1 oxygen-dependent degradation, leading to posttranslational regulation of the reporter by oxygen (Liu et al. 2005). Under hypoxic conditions, the reporter protein accumulates and can be detected and quantified, while normoxia will induce ubiquitination of the protein and loss of signal. The reporter gene could be luciferase or a fluorescent reporter protein, such as GFP or RFP.

As useful as these reporter constructs can be in preclinical tumor hypoxia imaging techniques, a main disadvantage to these approaches is that the reporter proteins can be highly sensitive to cellular oxygen concentrations. For example, luciferase reacts with the substrate luciferin through an oxygen-dependent reaction generating a photon that can be detected externally and subsequently localized and quantified to reveal the amount of expressed luciferase present in the experimental subject (Contag et al. 1998). In the absence of oxygen, luciferase has been shown to have marked signal loss (Cecic et al. 2007). Similarly, GFP requires an oxygen cofactor for sufficient fluorescence activity to be detected and quantified.

In addition to an awareness of how the absence of oxygen may alter the fluorescence of the reporter constructs, careful consideration must be taken to understand how HRE or ODDs might be influenced by changes in HIF-1 that are not directly linked to hypoxia. The association between HIF-1α and hypoxic conditions is clear; however, expression of HIF-1 is not always correlated with low pO_2. Genetic events can influence HIF expression, such as the loss of function of the VHL tumor suppressor protein (Semenza 2010; Kaelin 2007). Further, cells that are deficient in VHL function have impaired degradation of HIF-1α and this can result in accumulation and increased expression of an HIF-1 reporter construct, regardless of the oxygenation status of the tumor microenvironment. Because neoplastic cells can develop deficiencies in VHL function, as has been documented in clear-cell renal carcinomas and cerebellar hemangiomas, HIF-1 reporter constructs used to document hypoxia in these examples would be inappropriate (Maxwell et al. 1999; Zhong et al. 1999). Additionally, Moeller et al. demonstrated that increased HIF expression occurs in the tumor following exposure to ionizing radiation due to reoxygenation, the presence of reactive oxygen species, and stress granule depolymerization (Moeller et al. 2004).

Regardless, exogenous reporter genes provide opportunities for the advancement of scientific insight into the molecular consequences of hypoxia on a cellular level. The reporter constructs can serve as a critical adjunct to other imaging modalities.

7.3 Optical Spectroscopy

Optical techniques represent an emerging field for the characterization of hypoxia, and optical spectroscopy enables the quantification of intrinsic sources of optical absorption, scattering, and fluorescence in tissue. There are two primary sources of intrinsic contrast for determining tissue hypoxia in vivo: hemoglobin saturation and fluorescence redox ratio.

Hemoglobin is the dominant tissue absorber throughout the visible spectrum. Optical spectroscopy is sensitive to hemoglobin saturation due to the differing absorption spectra of oxygenated hemoglobin (HbO_2) and deoxygenated hemoglobin (Hb) (Zijlstra et al. 1991). Typically, a mixture of HbO_2 and Hb is present in the blood. The heme groups of hemoglobin bind oxygen, which changes their absorption spectrum. It is possible to determine what fraction of the heme-binding sites of hemoglobin is bound to oxygen by characterizing the tissue absorbance as a function of wavelength. This allows the measurement of hemoglobin oxygen saturation (Fig. 7.2).

Once hemoglobin saturation has been determined, it is possible to relate this value directly to the tissue $pO_2$by using the hemoglobin dissociation curve (Severinghaus

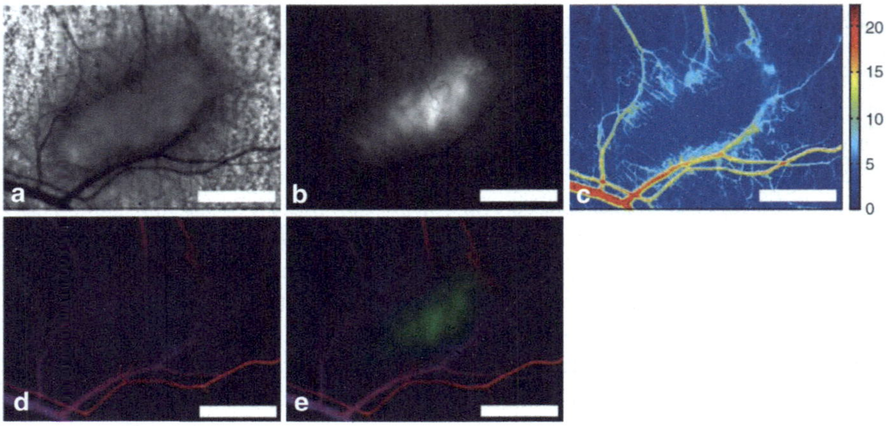

Fig. 7.2 Functional preclinical hypoxia imaging in dorsal skin-fold window chamber. Magnified (2.5 ×) raw and processed images of a 4T1 tumor expressing green fluorescence protein (GFP) concurrently with hypoxia inducible factor-1 (HIF-1α). **a** Bright-field transmission image of the central tumor and surrounding normal tissue. **b** Grayscale image of raw fluorescence intensity. **c** Mapping of relative variations in total hemoglobin. Total hemoglobin is proportional to hematocrit. **d** Simultaneous visualization of inverse total hemoglobin absorption and hemoglobin saturation. Increased pixel brightness represents an increased hemoglobin component at that pixel location. Hemoglobin saturation is mapped on a blue–red color gradient, with the deoxygenated hemoglobin component represented in the blue color channel, and the oxygenated hemoglobin component represented in the red color channel. **e** Simultaneous visualization of total hemoglobin absorption (brightness of red–blue signal), hemoglobin saturation (red–blue color scale), and GFP expression (green channel). Scale bars, 1 mm. (Reproduced from Palmer 2011 with permission from Nature Publishing Group)

1979; Leow 2007). This direct correlation can be skewed, however, because the affinity of hemoglobin for oxygen is affected by a variety of microenvironmental factors, such as pH, carbon dioxide, temperature, and the allosteric effector 2,3-diphosphoglycerate (Ganong 1987; Grippi 1995), all of which facilitate the effective delivery of oxygen under variable tissue conditions. In understanding the influence of these factors, an absolute calibration of tissue pO_2 is difficult and data obtained via optic spectroscopy are presented and analyzed instead only in terms of hemoglobin saturation. Also, it is not possible to measure oxygenation outside of the vasculature with optical spectroscopy; therefore, this technique is limited to providing information regarding vascular $pO_2$and not tissue pO_2.

As an additional optical spectroscopy technique relevant to hypoxia, fluorescence redox imaging utilizes the intrinsic contrast of electron carriers, nicotinamide adenine dinucleotide (phosphate) (NAD(P)) and flavin adenine dinucleotide (FAD), within tissue (Chance et al. 1979). NAD(P) is nonfluorescent, while its reduced form, NAD(P)H, fluoresces when excited with ultraviolet (UV) light. The reverse is true of FAD, which is fluorescent in the oxidized (FAD), but not the reduced ($FADH_2$) form (Chance et al. 1979). It is in measuring the fluorescence of both these parameters that a representation of the redox status of a biologic sample can be obtained. Commonly, a fluorescence "redox ratio" is calculated, defined as FAD/(FAD + NAD(P)H) or alternatively as NAD(P)H/FAD. The ratio of the autofluorescence contributed by these compounds can be used as a ratiometric indicator of tissue metabolic activity, an indirect measure of oxidative metabolism, and therefore hypoxia. With this approach, it is possible to investigate the physiologic response to hypoxia; however, it does not directly reflect tissue pO_2. Another limitation is the penetration depth. Since these fluorophores are excited in the UV range, the penetration depth in the tissue is to the order of millimeters.

Optical spectroscopy techniques have been developed which make use of these sources of intrinsic contrast (hemoglobin saturation and fluorescence redox) to further the understanding of tissue hypoxia in vivo. The four main applications of optical spectroscopy principles are diffuse reflectance, hyperspectral imaging, and diffuse optic topography and tomography.

7.3.1 Diffuse Reflectance Spectroscopy

Diffuse reflectance spectroscopy, also known as light-scattering spectroscopy, enables the quantification of bulk tissue absorption and scattering. Measuring diffuse reflectance as a function of wavelength enables the quantification of tissue absorption and scattering using quantitative models of light–tissue interaction (Brown et al. 2009). These measurements are commonly made using a fiber-optic probe placed in contact with the tissue surface. This allows a spot-based modality for quantification of hemoglobin saturation and fluorescence redox ratio.

The most basic approach to measuring hemoglobin saturation in vivo is to measure the diffuse reflectance as a function of wavelength. This is achieved by illumination of the tissue and the backscattered light is measured as a function of wavelength, usually with light coupled to a light source and detector via a fiber optic probe. The resulting diffuse reflectance spectrum can be modeled and related to the underlying absorption and scattering properties of tissue; this, then, can be related to hypoxia. For a single measurement site, which comprises a point measurement of tissue physiology, a range of modeling solutions is available, including empirical or simple analytical expressions (Pfefer et al. 2003; Yang et al. 2002; Amelink et al. 2004; Mourant et al. 1997; Chance et al. 1988), analytical (commonly diffusion based) approximations of light transport (Farrell et al. 1992; Zonios et al. 1999; Ghosh et al. 2001; Finley and Foster 2004; Cerussi et al. 2007; Shah et al. 2005), or Monte Carlo modeling of light transport (Palmer and Ramanujam 2006; Thueler et al. 2003). It is possible to obtain two-dimensional (2D) representations of tissue optical properties by acquiring an array of measurement sites or by sequentially moving the fiber optic probe to achieve spatial information. Scanning or wide-field approaches using noncontact methods or placement of an array of sensors enables the extension of this approach to imaging (Brown et al. 2009). As an example of the utility of this technique, Vishwanath et al. recently demonstrated that early changes in oxygenation in response to radiation treatment are predictive of local tumor control in a xenograft model (Vishwanath et al.

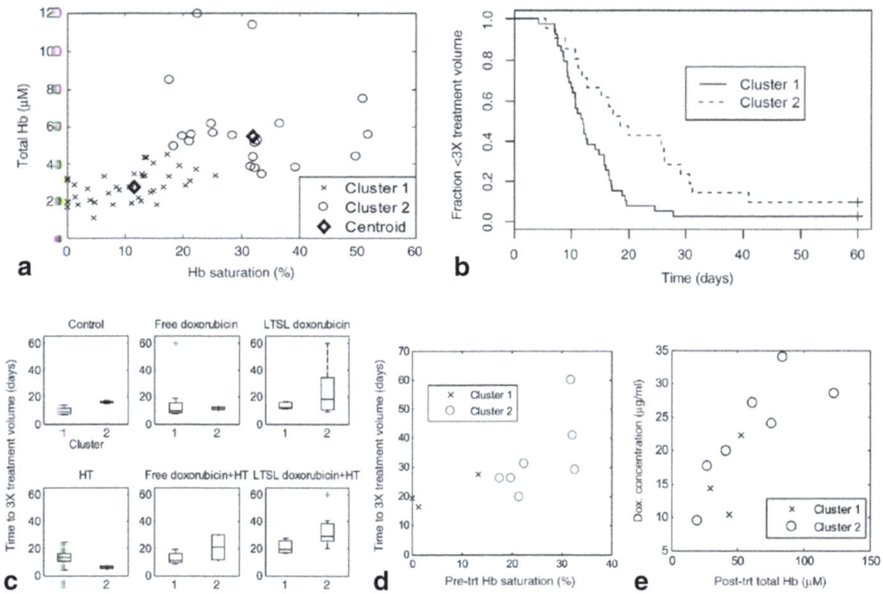

Fig. 7.3 Optical spectroscopy to identify tumor phenotypes based on physiologic data via k-means cluster analysis of the hemoglobin saturation and total hemoglobin, measured at baseline for all animals. **a** Two clusters were identified; *cluster 1* having relatively low oxygenation and low hemoglobin content, and *cluster 2* having the opposite. There was a significantly worse prognosis for cluster 1 (hypoxic) relative to the control ($p = 0.003$). **b** Survival curve for the two physiologic clusters indicating that the more hypoxic cluster has a worse prognosis. **c** Box plots of the time to reach three times the pretreatment volume for the two clusters, indicating that, for most groups, the median time is longer in cluster 2, consistent with a better prognosis. **d–e** Highly significant correlations between pretreatment Hb saturation and time to reach three times the pretreatment volume and the posttreatment total Hb and doxorubicin concentration ($r = 0.8$, $p = 0.01$, and $r = 0.89$, $p = 0.001$, respectively, by Spearman rank correlation). (Reproduced from Palmer 2010 with permission from Elsevier)

2009) In another study, noninvasive monitoring of intra-tumor drug concentration using fluorescence diffuse reflectance spectroscopy predicted the tumor response to doxorubicin in vivo (Palmer et al. 2010a) (Fig. 7.3). A main limitation of this technique is that the ability to obtain depth-resolved information is restricted.

7.3.2 *Hyperspectral Imaging*

Hyperspectral imaging allows quantification of transmitted or reflected light as a function of wavelength, commonly using multiple emission filters or a tunable filter. This imaging method enables the acquisition of fluorescence emission or reflectance/transmission spectra on a pixel-by-pixel basis. Two-dimensional imaging of tumor hemoglobin saturation has been possible with hyperspectral imaging. This has been achieved in vivo with window chamber models, and with this experimental approach, the high-resolution imaging of hemoglobin saturation is feasible (Pittman and Duling 1975; Shonat et al. 1997a; Sorg et al. 2005). The window chamber tumor model involves the surgical attachment of a titanium frame which encloses an implanted tumor; typically, the window chamber is placed in a dorsal skin fold (Algire 1943), but fixation to orthotopic sites has also been possible (Yuan et al. 1994; Shan et al. 2003). The window chamber allows direct visualization of the tumor microvasculature with a commercially available microscope. Having the microscope equipped with a tunable optic filter enables measurement of the transmitted or reflected light as a function of wavelength. Researchers using the window chamber tumor model have used a liquid-crystal tunable filter for such work, which permits a continuously tunable, narrow band-pass transmission wavelength (Sorg et al. 2005). The primary advantage of this is that it enables quantitative separation of multiple sources of contrast using chemometric algorithms (Andersson and Bro 2000). Measurements of the transmitted or reflected light can then be related to the wavelength-dependent absorption coefficient to yield hemoglobin saturation (Sorg et al. 2005; Shonat et al. 1997b; Styp-Rekowska et al. 2007). The advantages of the hyperspectral imaging method to study hypoxia are that (Dewhirst et al. 2008) it is capable of high resolution and (Bussink et al. 2003) it can be used to image oxygen saturation of individual blood vessels. This level of resolution enables the modeling of oxygen transport in the microvasculature. Further, hyperspectral imaging in conjunction with imaging the fluorescence redox ratio, providing dual-modality characterization of oxygen supply and metabolism, has been performed recently in the window chamber tumor model (Skala et al. 2010) (Fig. 7.4).

Additionally, hyperspectral imaging has been used to study dynamic tumor microenvironmental oxygenation via dual-emissive fluorescent/phosphorescent boron nanoparticles (BNPs), which serve as ratiometric indicators of tissue oxygen tension (Palmer et al. 2010b; Pfister et al. 2008). Iodide-substituted difluoroboron dibenzoylmethane-polylactide (BF_2dbm(I)PLA) is a light-emitting biomaterial that offers many advantages for optical hypoxia imaging (Zhang et al. 2009). It is multiemissive, exhibiting both short-lived fluorescence (F) and long-lived fluorescence (P) after one- or two-photon excitation. The long-lived fluorescence properties of BF_2dbm(I)PLA are maintained at room temperature and 37 °C, as well as in aqueous environments (Pfister et al. 2008). These traits make BNPs uniquely useful for oxygen sensing via dynamic quenching mechanisms (Fig. 7.5).

A standard fluorescent microscope can be used for hyperspectral imaging of oxygen-sensitive nanoparticles; specifically, a 4′,6-diamidino-2-phenylindole (DAPI) excitation filter cube is used for excitation, and a liquid-crystal tunable filter is used to isolate the emission wavelengths (Palmer et al. 2010b). Through ratiometric sensing, a simple ratio of F/P intensities at the respective emission maxima provides information about relative O_2 levels, whereas calibration allows for absolute pO_2measurement (Palmer et al. 2010b). Ratiometric methods also eliminate the effects of concentration and fluctuations in light intensity or detector sensitivity. Further, since the BNP serves as both the standard (F) and the oxygen sensor (P) at once, if a sensor molecule degrades, the intrinsically coupled F and P are equally affected.

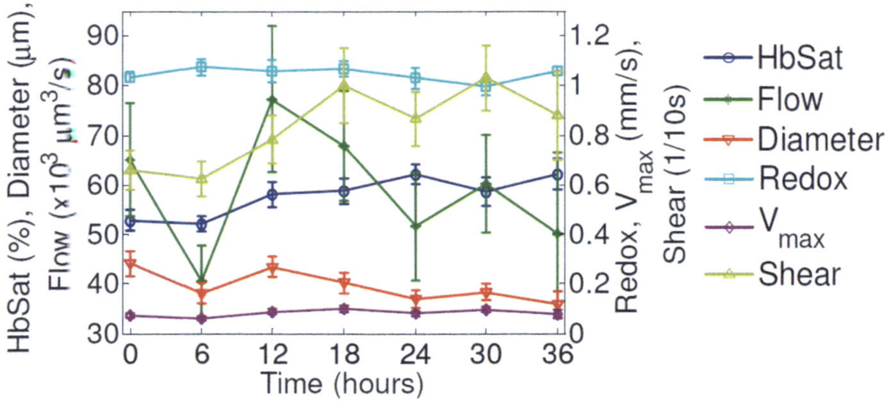

Fig. 7.4 Longitudinal imaging of tumor metabolic demand, hemoglobin oxygen saturation, and blood flow as a function of time to investigate mechanisms and distribution of cycling hypoxia in tumors. Values at each time point were averaged across all regions of interest and animals for the hemoglobin saturation (*HbSat*), vessel flow (*Flow*), vessel diameter (*Diameter*), vessel maximum velocity (V_{max}), and vessel shear rate (*Shear*) ($n = 21$), as well as for the redox ratio (*Redox*) ($n = 14$). The linear mixed model determined that all biomarkers significantly change ($p < 0.05$) with time, and that the vessel diameter is associated with the redox ratio of the adjacent tissue ($p < 0.05$) after adjusting for the effect of time. Error bars are standard error. (Reproduced from Skala 2010 with permission from SPIE)

The long P lifetimes ($\tau = \sim 5$ ms; vs. F: $\tau = 0.5$ ns) correlate with high oxygen sensitivity which underscores the utility of these materials for hypoxia imaging.

Coupling BNP emission and hemoglobin-based hyperspectral imaging techniques is advantageous to monitor dynamic changes in tissue and vascular oxygenation, respectively, at high spatial and temporal resolutions. These techniques are implemented using a simple modification to a standard fluorescent microscope, enabling a wide range of potential applications.

7.3.3 Diffuse Optical Topography and Tomography

Diffuse optical topography and tomography techniques have been developed to obtain deeper tissue Hb saturation information than the previously mentioned superficial optical techniques. Diffuse optical topography and tomography involve acquiring measurements through the use of an array of light sources/detectors spanning a relatively deep (tens of millimeters) region of tissue (Koizumi et al. 2003). Diffuse optical tomography is similar to diffuse optical topography, but it uses more widely spaced sources and detectors to enable even greater penetration depth. Because of the need for deep-penetrating photons, near-infrared (NIR) light is used. Tissue absorption is exceptionally low in this spectral range (Jobsis-vanderVliet 1999), so it is commonly referred to as the near-infrared window. Light is absorbed primarily

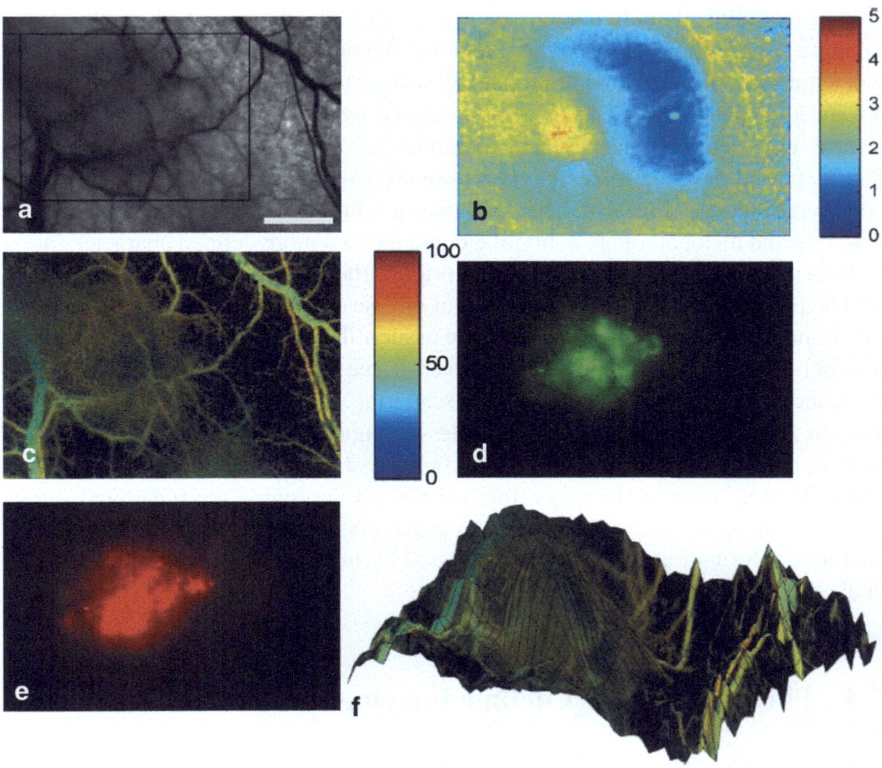

Fig. 7.5 Optical data demonstrating quantitative imaging of tumor hypoxia and vascular function. **a** Bright-field transmission image of the tumor and its vasculature. **b** Oxygen tension as percent oxygen (% O_2) extracted from the boron nanoparticle (BNP) spectrum. There is a region in the tumor with relatively high oxygen tension, while relatively low oxygen tension appears in adjacent regions. This poorly oxygenated zone appears to be poorly vascularized and downstream of the tumor blood supply. **c** Hemoglobin saturation on a color scale from *blue* (deoxygenated) to *red* (oxygenated). The color encoding at each pixel is multiplied by the total hemoglobin content extracted from the same pixel, such that large vessels appear brighter, while nonvascularized pixels appear *black* due to low hemoglobin content. The tumor appears well vascularized with heterogeneous oxygenation. **d** Green fluorescence protein–hypoxia inducible factor-1 (GFP–HIF-1) fluorescence intensity shows distinct pockets of intense GFP fluorescence within the central part of the tumor. **e** Constitutively active red fluorescence protein (RFP) showing the extent of the tumor. **f** Combined plot shows hemoglobin oxygen saturation plotted as in (**c**) with a three-dimensional (3D) visualization of the relationship between hemoglobin oxygen saturation, hemoglobin content, and tissue oxygen tension. The hemoglobin oxygen saturation color map is overlaid on a 3D projection of the tissue oxygen tension plotted on the surface map, with the oxygen tension appearing on the vertical axis. Thus, oxygen gradients can be visualized as slopes on the plot, with a peak appearing on one side of the tumor close to a large artery on the lower left side. These images were acquired toward the end of the imaging sequence ($t = 55$ min). Scale bar seen in (**a**), 1 mm. (Reproduced from Palmer 2010 with permission from SPIE)

by hemoglobin, but small amounts are also absorbed by myoglobin found in muscle and cytochrome *c* oxidase (Boushel et al. 2001).

A recent clinical study demonstrates the utility of diffuse optical spectroscopic imaging. Concentrations of oxyhemoglobin, deoxyhemoglobin, total Hb, and oxygen saturation in tumor and contralateral normal tissue from patients with locally advanced primary breast cancer were measured using diffuse optical spectroscopy prior to treatment with neoadjuvant chemotherapy (Ueda et al. 2012). It was found that increased baseline tumor oxygen saturation was correlated with a pathologic complete response (Fig. 7.6). The combination of noninvasive diffuse optical spectroscopy and histopathology subtyping could aid in a more defined characterization of individual patients with breast cancer prior to therapy.

The primary difficulty of the methods of diffuse optical topography and tomography is that, while the absorption is low, the tissue still scatters light strongly, resulting in relatively poor resolution. It enables deep-tissue imaging (~ 10-cm depth); however, the resolution is limited by tissue scattering, typically being on the order of 1 cm at a depth of about 10 cm. Regardless, a significant advantage of these optic techniques is that they are able to provide functional information regarding the tumor tissue. Current research has been focused toward improving the functional significance of these methods, as opposed to attempting to achieve the high-resolution anatomic information that is better addressed by other imaging modalities (Gibson et al. 2005).

7.4 Phosphorescence Lifetime Imaging

Another preclinical hypoxia imaging technique is phosphorescence lifetime imaging. This method studies phosphors that are encapsulated in a water-soluble dendrimer and injected into the vasculature of the tissue region of interest. The dendrimer shields and reduces the sensitivity of the phosphor to the microenvironment. These phosphors absorb in the NIR region of 620–1,000 nm (Wilson et al. 1996). This spectral window proves to be advantageous since there is little absorbance from natural body pigments in this region, resulting in high specificity for this method. A light guide focuses the excited light from the phosphor to the surface of the tissue where it is then detected by a phosphorometer. The phosphor most commonly used in vivo for this method is Pcl-porphryin; other phosphors available include OxyphorG2 and Green2W (Dunphy et al. 2002).

Advantages of phosphorescence lifetime imaging include the ability to provide direct measurements on oxygen content throughout the entire tissue volume in absolute units with calibration. Importantly, this method has high signal-to-noise ratio in low pO_2 environments. The NIR light used for this imaging modality results in tissue penetration depths of a few millimeters to a centimeter; therefore, the resolution of phosphorescence lifetime imaging will depend on the detection technique used. A temporal resolution of seconds or less allows almost real-time data acquisition and the ability for repeated measurements (Vinogradov et al. 2003).

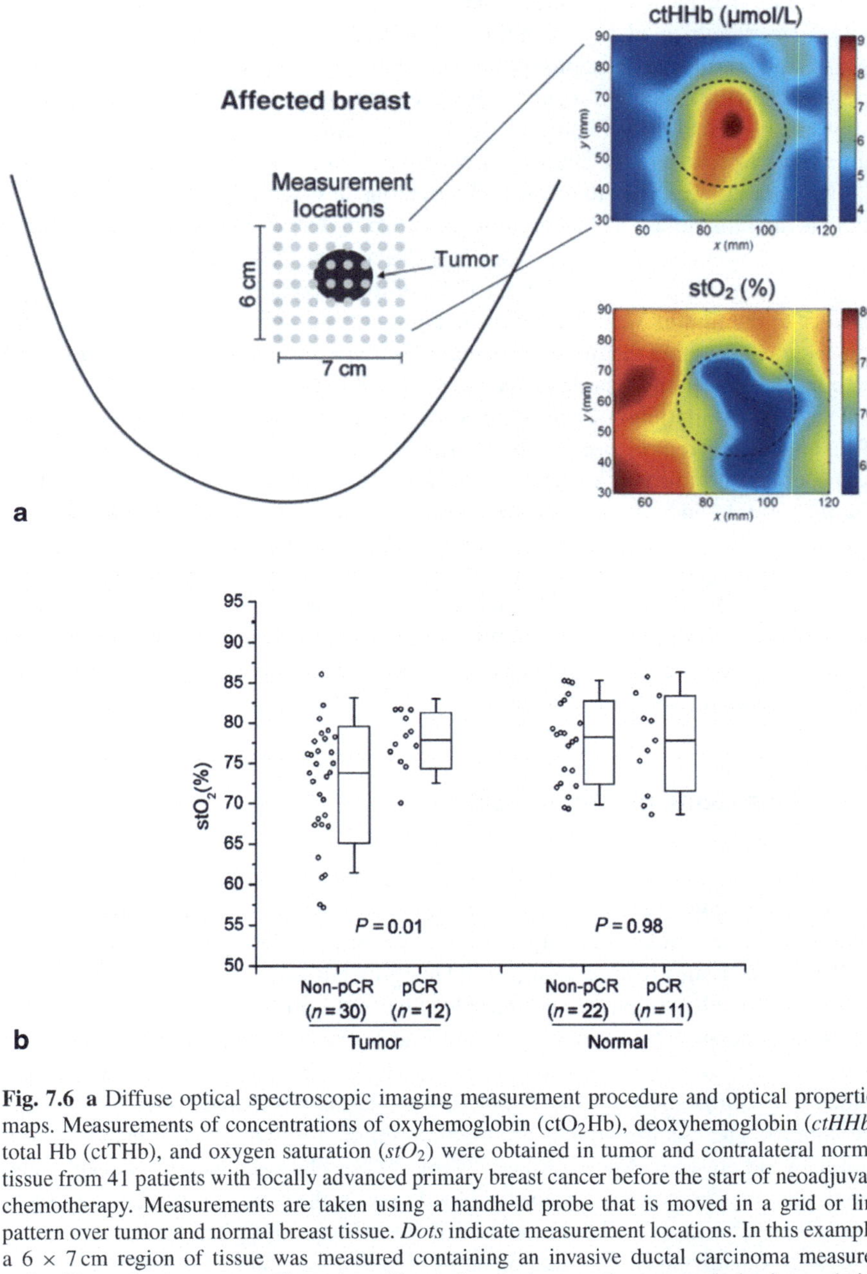

Fig. 7.6 **a** Diffuse optical spectroscopic imaging measurement procedure and optical properties maps. Measurements of concentrations of oxyhemoglobin (ctO_2Hb), deoxyhemoglobin (*ctHHb*), total Hb (ctTHb), and oxygen saturation (stO_2) were obtained in tumor and contralateral normal tissue from 41 patients with locally advanced primary breast cancer before the start of neoadjuvant chemotherapy. Measurements are taken using a handheld probe that is moved in a grid or line pattern over tumor and normal breast tissue. *Dots* indicate measurement locations. In this example, a 6×7 cm region of tissue was measured containing an invasive ductal carcinoma measured to be 27 mm in the greatest dimension. Maps of optical properties are made by interpolating data values between measurement points. In this example, both ctHHb and stO_2 are shown. In both maps, which are from identical tissue locations, the *dotted circle* indicates the approximate tumor location determined by ultrasound and palpation. This subject did not achieve a pathologic complete response. Note the relatively low oxygen saturation in the tumor region compared with the surrounding normal tissue. **b** Box-and-whisker plots showing the difference in tumor stO_2 levels between pathologic complete response (*pCR*) and nonpathologic complete response (*Non-pCR*)

Phosphors can be detected using a variety of optical imaging modalities enabling a wide range of multi-modality capabilities. In regard to preclinical in vivo applications used to study tissue hypoxia, phosphorescence imaging has become increasingly important due to the capacity for successful three-dimensional (3D) spatial registration using confocal imaging (Plant and Burns 1993; Koo et al. 2004) and diffuse tomography (Apreleva et al. 2006). For example, phosphorescence lifetime imaging via Oxyphor G2 was used to measure fluctuations in vascular pO_2 in fibrosarcomas, gliomas, and mammary adenocarcinomas grown in dorsal skin-fold window chambers every 2.5 min for 60–90 min (Cardenas-Navia et al. 2008). Instabilities in tumor oxygenation over time were a characteristic of the three tumor types, where oxygen delivery to the tumors was constantly changing, resulting in continuous reoxygenation events throughout the tumor. Vascular pO_2 maps revealed significant spatial and temporal heterogeneity (Cardenas-Navia et al. 2008) (Fig. 7.7).

The main limitations of phosphorescence lifetime imaging are administration of an intravascular injection of the phosphor and the requirement of specialized light sources and imaging systems to characterize the lifetime of these compounds. The preparation and handling of the phosphors require specialized techniques and expertise, although some are now commercially available. This method primarily measures and reports vascular pO_2, although there are phosphor probes that can penetrate into tissue which would characterize the tissue pO_2 status. Integration or comparison of phosphorescence lifetime imaging measurements with other imaging techniques is required for a comprehensive interpretation of the status of the tumor microenvironment when the phosphors are confined to the vasculature.

7.5 Photoacoustic Tomography

An imaging technique which has bypassed the fundamental depth limitations of other preclinical imaging methods is photoacoustic tomography. This imaging system combines strong optical contrast and high ultrasonic resolution in a single modality to allow deep-tissue cross-sectional or 3D imaging based on the photoacoustic effect (Wang 2008). Photoacoustic tomography involves optical irradiation by a short-pulsed laser beam to produce acoustic impulse responses. These propagate in tissue as ultrasonic waves, known as photoacoustic waves. The photoacoustic waves are detected by ultrasonic transducers placed outside the tissue and converted to electric signals. The electric signals are then amplified, digitized, and transferred to a computer, where an image is formed (Wang 2008). Photoacoustic tomography can simultaneously image cross sections of blood vessels, the concentration and oxygenation of hemoglobin, as well as blood flow in vivo (Wang 2008). These parameters can be used to quantify oxygen metabolism in vivo at a high spatial resolution without the requirement of an exogenous contrast agent.

Fig. 6 tumors (*left*; median, 77.8 % vs. 72.3 %; $p = 0.01$, Wilcoxon) and the lack of difference in stO_2 levels between contralateral normal tissues (*right*; median 77.7 % vs. 78.1 %; $p = 0.98$, Wilcoxon). (Reproduced from Ueda 2012 with permission from the American Association for Cancer Research)

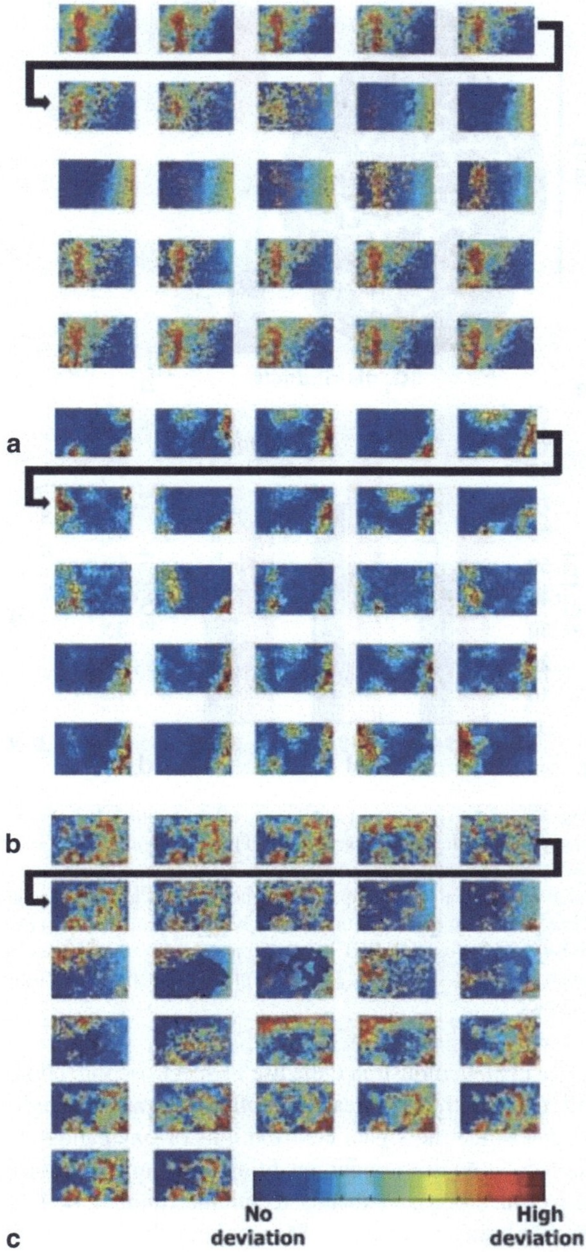

Fig. 7.7 Phosphorescence lifetime imaging to measure fluctuations in vascular pO_2 in rat fibrosarcomas (**a**), 9 L gliomas (**b**), and R3230 mammary adenocarcinomas (**c**) grown in dorsal skin-fold window chambers ($n = 6$ for each tumor type) and imaged every 2.5 min for a duration of 60–90 min. Representative watershed segmentation results for each 2.5-min time point displayed with time increased from left to right, top to bottom, in 2.5-min increments. Watershed segmentation creates boundaries at sharp gradients in pO_2; segmented regions can be thought of as pO_2 isobars. Segments are color coded by their deviations from the median $pO_2$of the image. *Red*, high deviations from the median; *blue*, no deviation from the median. (Reproduced from Cardenas-Navia 2008 with permission from the American Association for Cancer Research)

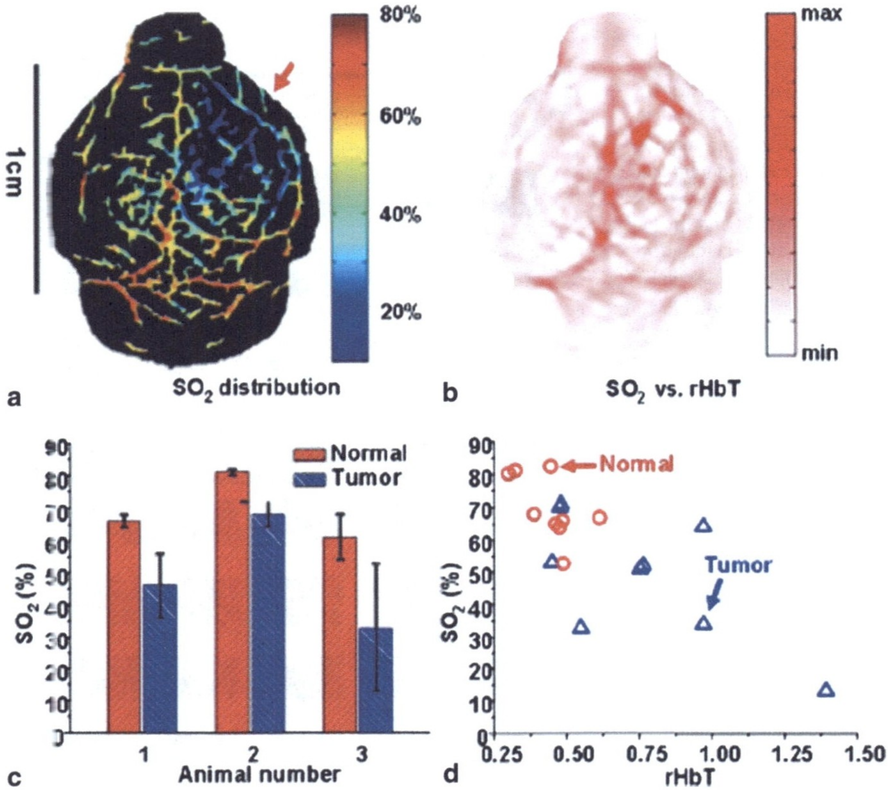

Fig. 7.8 In vivo imaging of a nude mouse brain with a U87 glioblastoma xenograft by spectroscopic phosphoacoustic tomography. **a** Blood oxygen saturation (SO_2) image of a nude mouse brain. The *red arrow* indicates the hypoxic region. **b** Total hemoglobin concentration (HbT) image of the same nude mouse brain. **c** Comparison of normal and tumor vasculatures in $SO_2$from triplicate mice. Three normal and three tumor vessels were chosen from each SO_2image for the calculation. **d** The SO_2versus the relative HbT in normal and tumor vasculatures from the same triplicate mice as in (**b**). (Reproduced from Li 2008 with permission from the Institute of Electrical and Electronics Engineers)

Li et al. demonstrated the use of spectroscopic photoacoustic tomography, which offers both strong optical absorption contrast and high ultrasonic spatial resolution, to determine the signal contributions of oxyhemoglobin, deoxyhemoglobin, and a molecular contrast agent, enabling simultaneous molecular and functional imaging (Li et al. 2008). With this technique, human U87 glioblastomas grown in nude mouse brains were imaged to quantify the hemoglobin oxygen saturation and the total hemoglobin concentration of the vasculature, which revealed hypoxia in tumor neovasculature (Li et al. 2008) (Fig. 7.8).

Some disadvantages of photoacoustic tomography include the complexity of the instrumentation and modeling algorithms, the need for a high-power source for sample illumination, and long imaging times; however, advancements in hardware continue to be made as the use of, as well as improving upon the technique of photoacoustic tomography has become a very active area of research.

7.6 Conclusion

Optical techniques for functional imaging in mice have a number of key advantages over other common imaging modalities such as MRI, PET, or computed tomography, including high resolution, low cost, and an extensive library of available contrast agents and reporter genes. A major challenge to such work is the limited penetration depth imposed by tissue turbidity. With the advent of fluorescent reporter genes and advancements in fluorescence probes and imaging techniques, there is a vast array of potential functional and molecular information that can be obtained in vivo using these techniques (Salazar Vazquez et al. 2010), including angiogenesis and vascular function (Cao et al. 2005; Skala et al. 2009; Vakoc et al. 2009), vascular permeability (Dreher et al. 2006), molecular response to therapy (Moeller et al. 2004), and tissue oxygen tension and hemoglobin oxygenation (Sorg et al. 2005; Zhang et al. 2009).

Acknowledgments This work was supported in part by NIH NCI CA40355 (Dewhirst) and T32 RR024394 (Boss).

References

Algire G (1943) An adaptation of the transparent chamber technique to the mouse. J Natl Cancer Inst Aug-Jun 4:1–11

Amelink A, Sterenborg H, Bard M, Burgers S (2004) In vivo measurement of the local optical properties of tissue by use of differential path-length spectroscopy. Opt Lett May 15 29(10):1087–1089

Andersson CA, Bro R (2000) The N-way toolbox for MATLAB. Chemometrics Intellig Lab Syst Aug 14 52(1)

Apreleva SV, Wilson DF, Vinogradov SA (2006) Tomographic imaging of oxygen by phosphorescence lifetime. Appl Opt Nov 20 45(33):8547–8559

Boushel R, Langberg H, Olesen J, Gonzales-Alonzo J, Bulow J, Kjaer M (2001) Monitoring tissue oxygen availability with near infrared spectroscopy (NIRS) in health and disease. Scand J Med Sci Sports Aug 11(4)

Braun R, Lanzen J, Snyder S, Dewhirst M (2001) Comparison of tumor and normal tissue oxygen tension measurements using oxyLite or microelectrodes in rodents. Am J Physiol Jun 280(6):H2533–H2544

Brizel D, Dodge R, Clough R, Dewhirst M (1999) Oxygenation of head and neck cancer: changes during radiotherapy and impact on treatment outcome. Radiother Oncol Nov 53(2):113–117

Brown JQ, Vishwanath K, Palmer GM, Ramanujam N (2009) Advances in quantitative UV-visible spectroscopy for clinical and pre-clinical application in cancer. Curr Opin Biotechnol Feb 20(1)

Bussink J, Kaanders J, Kogel A van der (2003) Tumor hypoxia at the micro-regional level: Clinical relevance and predictive value of exogenous and endogenous hypoxic cell markers. Radiother Oncol Apr 67(1):3–15

Cao YT, Li CY, Moeller BJ, Yu DH, Zhao YL, Dreher MR et al (2005) Observation of incipient tumor angiogenesis that is independent of hypoxia and hypoxia inducible factor-1 activation. Cancer Res Jul 1 65(13)

Cardenas-Navia LI, Mace D, Richardson RA, Wilson DF, Shan S, Dewhirst MW (2008) The pervasive presence of fluctuating oxygenation in tumors. Cancer Res Jul 15 68(14):5812–5819

Cecic I, Chan DA, Sutphin PD, Ray P, Gambhir SS, Giaccia AJ et al (2007) Oxygen sensitivity of reporter genes: implications for preclinical imaging of tumor hypoxia. Molecular Imaging Jul-Aug 6(4):219–228

Cerussi A, Hsiang D, Shah N, Mehta R, Durkin A, Butler J et al (2007) Predicting response to breast cancer neoadjuvant chemotherapy using diffuse optical spectroscopy. Proc Natl Acad Sci U S A Mar 6 104(10):4014–4019

Chance B, Schoener B, Oshino R, Itshak F, Nakase Y (1979) Oxidation-reduction ratio studies of mitochondria in freeze-trapped samples—nadh and flavoprotein fluorescence signals. J Biol Chem 254(11):4764–4771

Chance B, Leigh JS, Miyake H, Smith DS, Nioka S, Greenfeld R et al (1988) Comparison of time-resolved and time-unresolved measurements of deoxyhemoglobin in brain. Proc Natl Acad Sci U S A Jul 85(14)

Contag P, Olomu I, Stevenson D, Contag C (1998) Bioluminescent indicators in living mammals. Nat Med Feb 4(2):245–247

Dewhirst MW, Cao Y, Moeller B (2008) Cycling hypoxia and free radicals regulate angiogenesis and radiotherapy response. Nature Reviews Cancer Jun 8(6):425–437

Dreher MR, Liu WG, Michelich CR, Dewhirst MW, Yuan F, Chilkoti A (2006) Tumor vascular permeability, accumulation, and penetration of macromolecular drug carriers. J Natl Cancer Inst Mar 1 98(5)

Dunphy I, Vinogradov S, Wilson D (2002) Oxyphor R2 and G2: phosphors for measuring oxygen by oxygen-dependent quenching of phosphorescence. Anal Biochem Nov 15 310(2):191–198

Farrell T, Patterson M, Wilson B (1992) A diffusion-theory model of spatially resolved, steady-state diffuse reflectance for the noninvasive determination of tissue optical-properties invivo. Med Phys Jul-Aug 19(4):879–888

Finlay J, Foster T (2004) Hemoglobin oxygen saturations in phantoms and in vivo from measurements of steady-state diffuse reflectance at a single, short source-detector separation. Med Phys Jul 31(7):1949–1959

Ganong WF (ed) (1987) Review of medical physiology. 13th ed. Appleton & Lange, Norwalk, CT

Ghosh N, Mohanty S, Majumder S, Gupta P (2001) Measurement of optical transport properties of normal and malignant human breast tissue. Appl Opt Jan 1 40(1):176–184

Gibson A, Hebden J, Arridge S (2005) Recent advances in diffuse optical imaging. Phys Med Biol Feb 21 50(4):R1–R43

Graves EE, Giaccia AJ (2007) Imaging tumoral hypoxia: oxygen concentrations and beyond. Oncology-New York Mar 21(3):368–376

Grippi MA (ed) (1995) Pulmonary pathophysiology, 1st edn. Lippincott, Philadelphia, PA

Hockel M, Schlenger K, Aral B, Mitze M, Schaffer U, Vaupel P (1996) Association between tumor hypoxia and malignant progression in advanced cancer of the uterine cervix. Cancer Res Oct 1 56(19):4509–4515

Jobsis-vanderVliet F (1999) Discovery of the near-infrared window into the body and the early development of near-infrared spectroscopy. J Biomed Opt Oct 4(4):392–396

Kaelin WG Jr (2007) The von hippel-lindau tumor suppressor protein: an update. Oxygen Biol Hypoxia 435:371–383

Kallinowski F, Zander R, Hoeckel M, Vaupel P (1990) Tumor-tissue oxygenation as evaluated by computerized-Po2-histography. International Journal of Radiation Oncology Biology Physics Oct 19(4):953–961

Koizumi H, Yamamoto T, Maki A, Yamashita Y, Sato H, Kawaguchi H et al (2003) Optical topography: practical problems and new applications. Appl Opt Jun 1 42(16):3054–3062

Koo Y, Cao Y, Kopelman R, Koo S, Brasuel M, Philbert M (2004) Real-time measurements of dissolved oxygen inside live cells by organically modified silicate fluorescent nanosensors. Anal Chem May 1 76(9):2498–2505

Leow MK (2007) Configuration of the hemoglobin oxygen dissociation curve demystified: a basic mathematical proof for medical and biological sciences undergraduates. Adv Physiol Educ Jun 31(2):198–201

Li M, J Oh, Xie X, Ku G, Wang W, Li C et al (2008) Simultaneous molecular and hypoxia imaging of brain tumors in vivo using spectroscopic photoacoustic tomography. Proc IEEE Mar 96(3)

Liu JY, Qu RJ, Ogura M, Shibata T, Harada H, Hiraoka M (2005) Real-time imaging of hypoxia-inducible factor-1 activity in tumor xenografts. J Radiat Res Mar 46(1)

Maxwell PH, Wiesener MS, Chang GW, Clifford SC, Vaux EC, Cockman ME et al (1999) The tumour suppressor protein VHL targets hypoxia-inducible factors for oxygen-dependent proteolysis. Nature May 20:399(6733)

Moeller B, Cao Y, Li C, Dewhirst M (2004) Radiation activates HIF-1 to regulate vascular radiosensitivity in tumors: role of reoxygenation, free radicals, and stress granules. Cancer Cell May 5(5):429–441

Mourant J, Bigio I, Jack D, Johnson T, Miller H (1997) Measuring absorption coefficients in small volumes of highly scattering media: source-detector separations for which path lengths do not depend on scattering properties. Appl Opt Aug 1 36(22):5655–5661

Palmer G, Ramanujam N (2006) Monte carlo-based inverse model for calculating tissue optical properties. part I: theory and validation on synthetic phantoms. Appl Opt Feb 10 45(5):1062–1071

Palmer GM, Boruta RJ, Viglianti BL, Lan L, Spasojevic I, Dewhirst MW (2010a) Non-invasive monitoring of intra-tumor drug concentration and therapeutic response using optical spectroscopy. J Controlled Release Mar 19 142(3):457–464

Palmer GM, Fontanella AN, Zhang G, Hanna G, Fraser CL, Dewhirst MW (2010b) Optical imaging of tumor hypoxia dynamics. J Biomed Opt Nov-Dec 15(6):066021

Pfefer T, Matchette L, Bennett C, Gall J, Wilke J, Durkin A et al (2003) Reflectance-based determination of optical properties in highly attenuating tissue. J Biomed Opt Apr 8(2):206–215

Pfister A, Zhang G, Zareno J, Horwitz AF, Fraser CL (2008) Boron polylactide nanoparticles exhibiting fluorescence and phosphorescence in aqueous medium. Acs Nano Jun 2(6):1252–1258

Pittman R, Duling B (1975) Measurement of percent oxyhemoglobin in microvasculature. J Appl Physiol 38(2):321–327

Plant R, Burns D (1993) Quantitative, depth-resolved imaging of oxygen concentration by phosphorescence lifetime measurement. Appl Spectrosc Oct 47(10):1594–1599

Salazar Vazquez BY, Hightower CM, Sapuppo F, Tartakovsky DM, Intaglietta M (2010) Functional optical imaging at the microscopic level. J Biomed Opt Jan-Feb 15(1):011102

Semenza GL (2010) Defining the role of hypoxia-inducible factor 1 in cancer biology and therapeutics. Oncogene Feb 4 29(5):625–634

Severinghaus J (1979) Simple, accurate equations for human-blood O2 dissociation computations. J Appl Physiol 46(3):599–602

Shah N, Gibbs J, Wolverton D, Cerussi A, Hylton N, Tromberg B (2005) Combined diffuse optical spectroscopy and contrast-enhanced magnetic resonance imaging for monitoring breast cancer neoadjuvant chemotherapy: a case study. J Biomed Opt Sep-Oct 10(5):051503

Shan S, Sorg B, Dewhirst M (2003) A novel rodent mammary window of orthotopic breast cancer for intravital microscopy. Microvasc Res Mar 65(2):109–117

Shonat R, Wachman E, Niu W, Koretsky A, Farkas D (1997a) Near-simultaneous hemoglobin saturation and oxygen tension maps in mouse brain using an AOTF microscope. Biophys J Sep 73(3):1223–1231

Shonat RD, Wachman ES, Niu WH, Koretsky AP, Farkas DL (1997b) Near-simultaneous hemoglobin saturation and oxygen tension maps in mouse brain using an AOTF microscope. Biophys J Sep 73(3)

Skala MC, Fontanella A, Hendargo H, Dewhirst MW, Izatt JA (2009) Combined hyperspectral and spectral domain optical coherence tomography microscope for noninvasive hemodynamic imaging. Opt Lett Feb 1 34(3)
Skala MC, Fontanella A, Lan L, Izatt JA, Dewhirst MW (2010) Longitudinal optical imaging of tumor metabolism and hemodynamics. J Biomed Opt Jan-Feb 15(1):011112
Sorg B, Moeller B, Donovan O, Cao Y, Dewhirst M (2005) Hyperspectral imaging of hemoglobin saturation in tumor microvasculature and tumor hypoxia development. J Biomed Opt Jul-Aug 10(4):044004
Styp-Rekowska B, Disassa NM, Reglin B, Ulm L, Kuppe H, Secomb TW et al (2007) An imaging spectroscopy approach for measurement of oxygen saturation and hematocrit during intravital microscopy. Microcirculation Apr-May 14(3):207–221
Thueler P, Charvet I, Bevilacqua F, St Ghislain M, Ory G, Marquet P et al (2003) In vivo endoscopic tissue diagnostics based on spectroscopic absorption, scattering, and phase function properties. J Biomed Opt Jul 8(3):495–503
Ueda S, Roblyer D, Cerussi A, Durkin A, Leproux A, Santoro Y et al (2012) Baseline tumor oxygen saturation correlates with a pathologic complete response in breast cancer patients undergoing neoadjuvant chemotherapy. Cancer Res Sep 1 72(17):4318–4328
Vakoc BJ, Lanning RM, Tyrrell JA, Padera TP, Bartlett LA, Stylianopoulos T et al (2009) Three-dimensional microscopy of the tumor microenvironment in vivo using optical frequency domain imaging. Nat Med Oct 15(10)
Vinogradov S, Grosul P, Rozhkov V, Dunphy I, Shuman L, Dugan B et al (2003) Oxygen distributions in tissue measured by phosphorescence quenching. Oxygen transport to tissue volume Xxiii: oxygen Measurements in the 21st Century: Basic Techniques and Clinical Relevance 510:181–185
Vishwanath K, Klein D, Chang K, Schroeder T, Dewhirst MW, Ramanujam N (2009) Quantitative optical spectroscopy can identify long-term local tumor control in irradiated murine head and neck xenografts. J Biomed Opt Sep-Oct 14(5):054051
Vordermark D, Brown J (2003) Evaluation of hypoxia-inducible factor-1 alpha (HIF-1 alpha) as an intrinsic marker of tumor hypoxia in U87 MG human glioblastoma: in vitro and xenograft studies. International Journal of Radiation Oncology Biology Physics Jul 15 56(4):1184–1193
Wang LV (2008) Prospects of photoacoustic tomography. Med Phys Dec 35(12):5758–5767
Wilson D, Vinogradov S, Lo L, Huang L (1996) Oxygen dependent quenching of phosphorescence—a status report. Oxygen Transport to Tissue Xvii. Adv Exp Med Biol 388:101–107
Yang Y, Celmer E, Koutcher J, Alfano R (2002) DNA and protein changes caused by disease in human breast tissues probed by the kubelka-munk spectral function. Photochem Photobiol Jun 75(6):627–632
Yuan F, Salehi H, Boucher Y, Vasthare U, Tuma R, Jain R (1994) Vascular-permeability and microcirculation of gliomas and mammary carcinomas transplanted in rat and mouse cranial windows. Cancer Res Sep 1 54(17):4564–4568
Zhang G, Palmer GM, Dewhirst M, Fraser CL (2009) A dual-emissive-materials design concept enables tumour hypoxia imaging. Nature Materials Sep 8(9)
Zhong H, De Marzo A, Laughner E, Lim M, Hilton D, Zagzag D et al (1999) Overexpression of hypoxia-inducible factor 1 alpha in common human cancers and their metastases. Cancer Res Nov 15 59(22):5830–5835
Zijlstra W, Buursma A, Meeuwsenvanderroest W (1991) Absorption-spectra of human fetal and adult oxyhemoglobin, de-oxyhemoglobin, carboxyhemoglobin, and methemoglobin. Clin Chem Sep 37(9):1633–1638
Zonios G, Perelman L, Backman V, Manoharan R, Fitzmaurice M, Van DJ (1999) Diffuse reflectance spectroscopy of human adenomatous colon polyps in vivo. Appl Opt Nov 1 38(31):6628–6637

Chapter 8
Clinical Imaging of Hypoxia

Michael R. Horsman, Morten Busk, Thomas Nielsen, Marianne Nordsmark and Jens Overgaard

Abstract Areas of reduced oxygenation (hypoxia) have been identified in most solid tumours. Viable tumour cells can exist under such conditions and are known to be a source of resistance to conventional therapy, especially radiation and certain types of chemotherapy. Such hypoxic cells also play an important role in influencing malignant progression in terms of both enhancing growth of the primary tumour and increasing metastatic spread. Numerous attempts have been made to try to identify the presence of such cells in tumours so that one can predict outcome to therapy and where necessary select additional treatments to eliminate hypoxic cells. None of these techniques have become established in routine clinical use because they were too invasive, not easy to use on a routine basis or did not necessarily indicate the relevant hypoxic cells. What is needed now are non-invasive approaches that can accurately and reliably image hypoxia in tumours, especially using techniques that are routinely available in the clinic, such as positron emission tomography, magnetic resonance and computer tomography. The aim of this review is to critically assess the potential use of imaging to monitor hypoxia and how such techniques should be used to influence cancer therapy.

Keywords Tumour hypoxia · Non-invasive imaging · Positron emission tomography · Magnetic resonance · Computer tomography · Therapeutic outcome

8.1 Introduction

Hypoxia is a characteristic feature of most animal (Moulder and Rockwell 1984) and human (Vaupel et al.1989) solid tumours. Broadly defined, hypoxia occurs when the oxygen content of tissues falls below normal levels. It arises in tumours because the

---

M. R. Horsman (✉) · M. Busk · T. Nielsen · J. Overgaard
Department of Experimental Clinical Oncology, Aarhus University Hospital,
Nørrebrogade 44, Bldg. 5, 8000 Aarhus C, Denmark
e-mail: mike@oncology.dk

M. Nordsmark
Department of Oncology, Aarhus University Hospital,
Nørrebrogade 44, Bldg. 5, 8000 Aarhus C, Denmark

G. Melillo (ed.), *Hypoxia and Cancer,* Cancer Drug Discovery and Development, DOI 10.1007/978-1-4614-9167-5_8,

oxygen demands of the growing tumour mass exceed the oxygen supply (Dewhirst et al 2008). The growth and development of solid tumours require the formation of a functional vascular supply (Brem et al. 1976), which evolves from the normal host vascular network by the process of angiogenesis (Bergers and Benjamin 2003). However, the tumour neo-vasculature that develops is a chaotic and primitive system that exhibits a number of structural and functional abnormalities when compared to the normal host tissues from which it is arises (Vaupel et al. 1989). As a result, oxygen gradients are established, which eventually lead to the development of diffusion-limited chronic hypoxia (Horsman et al. 2012), as first proposed by Thomlinson and Gray (1955) based on their histological observations in carcinoma of the bronchus. Nutrient and pH gradients also exist; thus, chronically hypoxic cells have a tendency to be associated with nutrient deprivation, energy depletion and acidity (Vaupel 2004). In the 1980s, it was suggested (Brown 1979) and then confirmed (Chaplin et al. 1987) that a second type of hypoxia could exist in tumours and one that was somewhat transient in nature. This was shown to be the result of fluctuations in tumour blood flow; hence, it is generally referred to as perfusion-limited acute hypoxia, although the term "acute" is often interchanged with "cycling" or "fluctuating". The original observations were based on complete shut-downs in blood flow (Chaplin and Trotter 1991), but additional studies have now shown that blood flow can fluctuate (Kimura et al. 1996), often leading to partial flow reductions; tumour cells exposed to a total flow reduction would be starved of oxygen and nutrients, but with partial shut-downs one could have plasma flow but no red blood cells and therefore the cells would be only oxygen deprived. Fluctuating perfusion/hypoxia has now been observed in human tumours (Powell et al. 1997; Begg et al. 2001; Janssen et al. 2002).

The significance of hypoxia in cancer comes from the findings that cells that exist under such oxygen-deprived conditions are known to be viable and a source of resistance to conventional therapy, especially radiation and certain types of chemotherapy (Horsman et al. 2009). Hypoxic cells are also believed to play an important role in influencing malignant progression both in terms of increasing the growth of primary tumours and their metastatic spread (Bussink et al. 2003). More recent studies actually suggest that hypoxia is a "niche"where the cancer stem cells responsible for tumour growth and development can actually hide (Baumann et al. 2008). As a result of its importance, considerable effort has been made in the last five decades to find techniques that can identify those patients with tumours that contain significant levels of hypoxia. Such data will allow us to determine those patients who are likely to have a poorer prognosis due to the presence of hypoxia and thus require some additional therapy designed to eliminate the negative influence of hypoxia. In this chapter, we critically discuss the various techniques that have been applied to identify hypoxia in patients (as listed in Table 8.1) with a focus on the use of non-invasive imaging.

Table 8.1 Methods used for monitoring tumor hypoxia in patients

Vascular based methods	*Exogenous markers*
Biopsy/immunohistochemistry	Biopsy/immunohistochemistry
Intercapillary distance	Nitroimidazole markers
Vascular density	Imaging approaches
Cell to nearest vessel distance	PET (nitroimidazoles)
HbO_2 saturation	CuATSM
Imaging approaches	MRI (nitroimidazoles)
HbO_2 saturation (NIRS/BOLD)	SPECT (nitroimidazoles/technetium)
PET (^{15}Oxygen labelled water)	*Surrogate markers*
CT perfusion	Metabolism
DCE-MRI	Biopsy/bioluminescence (Lactate/ATP/glucose)
Direct Oxygen measurements	Imaging (FDG/MRS)
Electrodes (Glass/Eppendorf)	DNA damage
EPR	Comet assay
Endogenous markers	H2AX phosphorylation
Individual genes/proteins	Interstitial fluid pressure
Biopsy studies	Probes
Serum/Plasma	DCE-MRI
Gene signatures	

8.2 Techniques used to Monitor Hypoxia in Patients

8.2.1 Vasculature-Based Parameters

Some of the earliest attempts to determine hypoxia in human tumours focused on the vascular supply, as it was only via the tumour vasculature that oxygen could be delivered. The endpoints included measurements of intercapillary distance (Kolstad 1968; Awwad et al. 1986), vascular density (Delides et al. 1988; Révész et al. 1989; Streffer et al. 1989) and the distance from tumour cells to the nearest blood vessel (Lauk et al. 1989). Positive correlations were found between the degree of vascularisation and outcome to radiation therapy in carcinoma of the cervix (Kolstad 1968; Awwad et al. 1986), nasopharyngeal tumours (Delides et al. 1988) and rectal carcinomas (Streffer et al. 1989), consistent with better vascularised tumours being less hypoxic. However, the study in oral squamous cell carcinomas (Lauk et al. 1989) actually showed that the tumours that were better vascularised had a poorer treatment outcome. This inconsistency is perhaps not surprising as measurements of vasculature alone give no information regarding the functionality of the vessels or the oxygen-carrying capacity of the blood. Attempts to address the latter issue have suggested using oxyhaemoglobin (HbO_2) saturation as a more relevant endpoint (Mueller-Kleiser et al. 1981). Cryophotometric measurements of this parameter on biopsies from solid tumours of the oral cavity were certainly consistent with the degree of tumour vascularisation (Mueller-Kleiser et al. 1981). Information about the oxygen-carrying capacity of the blood can also be obtained non-invasively using either near infrared spectroscopy (NIRS) or blood oxygen level-dependent (BOLD) contrast magnetic resonance imaging (MRI). NIRS uses visible light in the near infrared region and the NIRS signal is primarily obtained from absorption of light by haemoglobin (Vikram et al. 2007), while BOLD measures the levels of paramagnetic deoxyhaemoglobin within red blood cells (Padhani et al. 2007). The spatial resolution of NIRS is limited and, therefore, although it has been used in normal tissues and a variety of pathological diseases (Boushel et al. 2001), its clinical use in cancer has been limited to tissues most accessible to light penetration, such as breast cancer (Ripoll et al. 2008). BOLD MRI has been used more widely but primarily for monitoring changes in vascular oxygenation as seen when following the inhalation of high oxygen-content gas (Hallac et al. 2012). Additional studies in prostate cancer were able to correlate the BOLD MRI evaluation with histological assessment of the binding of the classical hypoxic cell marker pimonidazole (Hoskin et al. 2007) and more direct oxygen measurements with electrodes (Chopra et al. 2009). However, it has been suggested that using the BOLD MRI signal one may be able to obtain information about the original tumour oxygenation status (Rodrigues et al. 2004; Chopra et al. 2009); but, as yet, no attempts have been undertaken to relate such measurements to outcome. As BOLD MRI can be used to monitor changes in oxygenation in the blood supply, this technique may actually be useful in monitoring fluctuating hypoxia. Indeed, the preliminary results from one study in glioma patients suggest that such an approach is feasible (Krohn et al. 2012).

Other non-invasive vasculature-based techniques used to try to obtain information about tumour oxygenation status have concentrated on tumour perfusion. It is likely that blood perfusion and hypoxia are related as oxygen delivery occurs via the vascular supply and therefore should influence chronic hypoxia, and changes in perfusion are clearly responsible for fluctuating hypoxia. Perfusion can be estimated with positron emission tomography (PET) by using [^{15}O]-labelled water. However, clinical studies using this technique in head and neck cancer have produced conflicting results; in one study, [^{15}O]-labelled water perfusion estimates and oxygenation measurements with [^{18}F]-labelled EF5, a known marker for hypoxic cells, were reasonably well inversely correlated (Komar et al. 2008); yet in another study poor local tumour control and survival after radiation therapy were associated with high blood perfusion (Lehtio et al. 2004) rather than the low perfusion that one would expect to be indicative of hypoxia. PET perfusion estimates obtained with [^{15}O]-labelled water have also been compared to those estimated with computed tomography (CT) and similar results obtained (Ng et al. 2009). As CT perfusion is more widely available in the clinic, it may be a more useful method for measuring perfusion-related hypoxia. Limited studies have tried to relate CT perfusion to outcome in head and neck cancer (Hermans et al. 1999; Bisdas et al. 2009; Truong et al. 2011) and the results suggested that poorly perfused tumours, and thus more hypoxic, had a reduced response to radiotherapy (Figs. 8.1a and 8.2). However, the only direct attempt to compare CT perfusion with the more classical hypoxic marker pimonidazole in head and neck cancer actually found no significant correlation (Newbold et al. 2009); hence, the potential of CT perfusion to truly monitor hypoxia is debatable. Another popular technique for monitoring tumour perfusion-related parameters is dynamic contrast-enhanced magnetic resonance imaging (DCE-MRI). This involves injecting intravenously a contrast agent and then monitoring its extravasation over several minutes from a region of interest (Nielsen et al. 2012). Clinical studies reported that measurements with DCE-MRI correlated with oxygen electrode measurements in cervix (Cooper et al. 2000; Lyng et al. 2001); pimonidazole binding in head and neck

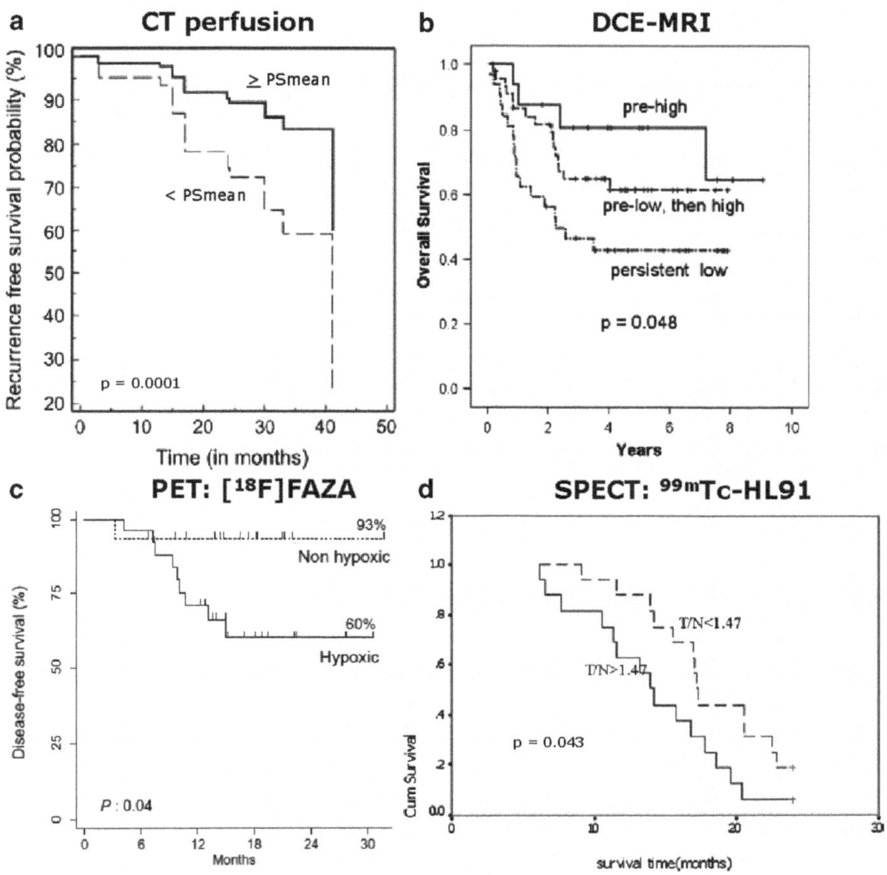

Fig. 8.1 Results from four different clinical trials showing the relationship between hypoxia imaging and outcome to therapy. **a** Recurrence-free survival in head and neck cancer patients based on the median value of the mean permeability surface area (PSmean = 18 mL/min/100 g) estimated by CT; **b** disease-specific survival in cervical cancer patients based on the level of perfusion measured with DCE-MRI either before or before and during radiation treatment; **c** disease-free survival in head and neck cancer patients based on the pre-radiation therapy estimate of hypoxia as determined by a tumour-to-muscle ratio of ≥ 1.4 from $[^{18}F]$ FAZA PET measurements; and **d** overall survival in non-small-cell lung cancer patients in which the tumour-to-normal tissue (T/N) ratio measured with ^{99m}Tc-HL91 SPECT before radiation therapy was above/below 1.47. The figures have been adapted from Bisdas et al. 2009 (**a**), Mayr et al. 2010 (**b**), Mortensen et al. 2012 (**c**), and Li et al. 2006 (**d**)

(Newbold et al. 2009; Donaldson et al. 2011); and $[^{18}F]$-labelled misonidazole uptake in glioblastoma multiforme (Swanson et al. 2009) and head and neck nodal metastases (Jansen et al. 2010). Several studies have also attempted to correlate the DCE-MRI measurements with radiotherapy outcome in patients with cervical cancer (Loncaster et al. 2002; Mayr et al. 2010; Andersen et al. 2012) and found that patients with more hypoxic tumours had a poorer response to radiotherapy (Figs. 8.1b and 8.2).

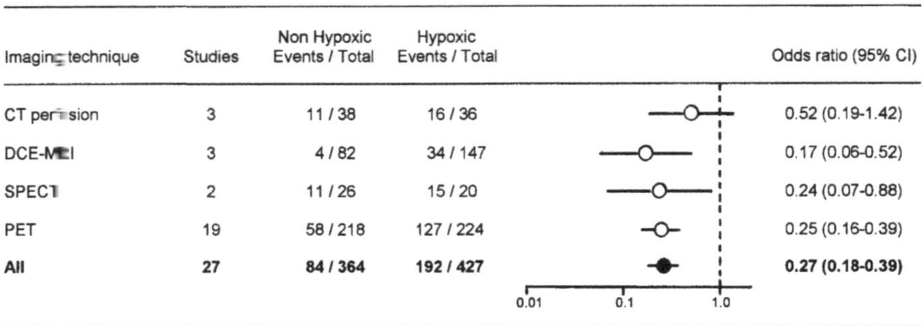

Fig. 8.2 Forest plot of the relationship between imaging of hypoxia and outcome to radiation therapy. Hypoxia imaging was performed using CT perfusion, DCE-MRI, SPECT or PET. These studies (27) include a variety of tumour sites and the 21 studies involving SPECT/PET measurements were performed using a range of imaging agents. The odds ratio (with 95 % confidence limits) shows significant differences between the hypoxic and non-hypoxic groups based on the number of events out of the total number of patients (values below 1.0 indicate that patients with hypoxic tumours did worse). Modified from Horsman et al. (2012)

8.2.2 Direct Oxygen Measurements

The most direct method for identifying hypoxia in human tumours is to determine oxygen partial pressure (pO_2) using polarographic electrodes. Early attempts to achieve this used "home-made" glass electrodes, which were cumbersome and fragile, and only a few pO_2 values 3–4 mm below the surface of the tumour were possible. Nevertheless, clinical data were obtained in cervix (Kolstad 1968) and head and neck cancers (Gatenby et al. 1988) that clearly demonstrated a relationship between such oxygenation measurements and outcome to radiation therapy, in that those patients with tumours that were better oxygenated had a significantly superior local response to irradiation.

This whole area was revolutionised with the development of the Eppendorf histograph, which had two distinct improvements. The first was having the oxygen sensor inside a metal needle and the second was the attachment of this needle to a stepping motor that allowed for multiple measurements along the needle track through the tumour. Numerous clinical studies were thus undertaken in a variety of human tumour types. The results clearly showed that hypoxia was to be found in virtually all human tumours investigated, although its degree could be variable (Vaupel et al. 1989). Probably the most significant finding from these studies was the confirmation that hypoxia influenced outcome to therapy. This has been reported for head and neck (Nordsmark et al. 1996; Brizel et al. 1997; Stadler et al. 1999; Rudat et al. 2001; Nordsmark et al. 2005), cervix (Hoeckel et al. 1993, 1996; Knocke et al. 1999; Fyles et al. 1998, 2006; Lyng et al. 2000) and soft tissue sarcomas (Brizel et al. 1996; Nordsmark et al. 2001). The most striking results were those made in cancer of the cervix and sarcomas that showed hypoxia influencing outcome in patients in whom surgery was the primary or only treatment (Hoeckel et al. 1996; Nordsmark et al. 2001), suggesting that hypoxia could also influence malignant progression, especially metastatic spread. In fact, one other study in cancer of the cervix was able to show that the primary tumours of patients with metastases were indeed less oxygenated than those of patients without metastases (Sundfør et al. 1998). Today, the Eppendorf electrode is no longer commercially available and there are a number of reasons for this. Without using concurrent imaging during the oxygen measurements, it was impossible to state whether the values obtained were from viable tissue, and even where this was done one could not state whether the cells in the hypoxic regions were clonogenic; the tumours themselves had generally to be easily accessible; the technique was invasive; and despite the positive correlations between the Eppendorf measurements and treatment outcome, the machine was never predictive of response on a patient-to-patient basis. Despite these limitations, the results obtained from the Eppendorf studies have to be considered positive in that they supplied us with a tremendous level of information about tumour hypoxia and its importance.

Fibre-optic probes have also been developed which, unlike the Eppendorf polarographic electrode, do not consume oxygen with each measurement; thus, continuous observations of oxygenation in the same tumour region are possible (Griffiths and Robinson 1999). Pre-clinical studies comparing the commercially available Oxylite sensor with the Eppendorf electrode (Collingridge et al. 1997; Braun et al. 2001; Seddon et al. 2001) or paired survival curve estimates of radiobiological hypoxia (Urano et al. 2002) reported both similarities and differences depending on the tissue type, tumour size or hypoxic modification used. Despite substantial pre-clinical evaluation, these probes have not been used clinically in cancer.

Other less invasive attempts to directly measure tumour oxygenation have involved phosphorescence tomography or MR-based approaches. The former requires the infusion of water-soluble phosphor probes into the vasculature (Vikram et al. 2007) and has been used to map oxygen concentration in pre-clinical tumour models (Wilson and Cerniglia 1992; Helmlinger et al. 1997) but, again, it has not been used in patients. The MR approaches include monitoring oxygen-sensitive reporter molecules (^{19}F-oximetry). Several such molecules have been developed including perflurochemical emulsions and hexafluorobenzene (Pacheco-Torres et al. 2011). The latter approach allowed for actual quantification of the MR signals and conversion into oxygen concentrations at the pixel level (Zhao et al. 2005). However, systemic toxicity required the imaging agent to be injected directly into tumours, limiting its potential clinical application. An alternative MR method is electron paramagnetic resonance (EPR), which detects paramagnetic materials that have been injected into tissues (Krishna et al. 2012). It can provide quantitative and repeated three-dimensional (3-D) estimates of oxygenation and has been extensively used in pre-clinical studies and even in patients for a range of different clinical problems (Swartz et al. 2004). Although many of the pre-clinical studies have focused on tumour hypoxia, the clinical application of EPR in cancer has, however, been somewhat limited (Krishna et al. 2012).

8.2.3 Exogenous Markers

One of the more widely studied methods used for detecting tumour hypoxia involves the administration of exogenous compounds that, under hypoxic conditions, undergo a chemical change from a non-reactive to a highly reactive product that will then bind to macromolecules within the cell. By applying techniques that can verify the bound product, one should then be able to identify the presence of hypoxia. The most popular agents used in this context have been 2-nitroimidazole-based markers. These nitroimidazole compounds were originally developed as hypoxic cell radiosensitisers, with the 2-nitroimidazoles being the most effective radiosensitizers in pre-clinical models (Adams and Cooke 1969). Such compounds are characterized by having an NO_2 grouping attached to the imidazole ring structure and this NO_2 group can undergo a six-electron intracellular reduction to produce NH_2. While the NO_2 and NH_2 moieties are generally inactive, one of the intermediates formed is highly reactive and can bind to any macromolecule (Horsman et al. 2012). The hypoxia specificity comes from the fact that in the presence of oxygen, typically above 10 mmHg, the first electron reduction species reacts with oxygen and returns to the NO_2 moiety with the subsequent production of oxygen radicals that ultimately form hydrogen peroxide. This bound product formed under hypoxia can be identified either using an antibody to the product or by radioactively labelling the original compound. The most commonly used nitroimidazole is pimonidazole, the binding of which correlated with radiobiological hypoxia in pre-clinical studies (Raleigh et al. 1999). Clinical studies showed that the degree of pimonidazole binding was related to radiation-induced local tumour control in head and neck cancer (Kaanders et al. 2002), but not cancer of the cervix (Nordsmark et al. 2006). Positive clinical findings between hypoxia levels and outcome in head and neck cancer were found with another nitroimidazole marker, EF5 (Evans et al. 1995).

Labelling the nitroimidazole compound with ^{18}F allows for the bound product, and thus hypoxia, to be identified using PET. The first tracer developed for hypoxia PET imaging was a [^{18}F]-fluorinated version of the radiosensitizer misonidazole (FMISO) and it was found to be capable of identifying hypoxia in a range of human tumours (Rasey et al. 1996). The tracer was followed by a group of compounds based on another radiosensitizer, etanidazole (i.e., EF3/5). These markers have a relatively high lipophilicity, which ensured easy penetration of cell membranes and diffusion into tumour tissue, but simultaneously limited the clearance of unbound tracer leading to relatively low tumour-to-reference tissue ratios. Other fluorinated compounds have been developed that were more soluble in water than FMISO and thus easier to clear from non-hypoxic tissue. These included [^{18}F] fluoroetanidazole (FETA), [^{18}F] fluoroerythronitroimidazole (FETNIM), [^{18}F] fluoroazomycinarabinofuranoside (FAZA) and HX4, of which the latter two are currently in clinical evaluation (Schuetz et al. 2010; van Loon et al. 2010). To date, there has not been any systematic examination of all the 2-nitroimidazoles tracers in the same tumour model or patient group, so that it is difficult to say whether one tracer is superior to another in identifying tumour hypoxia. The ideal tracer would be one in which clearance of unbound tracer is complete at the time of imaging; thus, only bound material indicative of hypoxia is measured. This can take hours or days to achieve, but such measurements have to take into account decay of the radioactive marker and normal clinical schedules. Thus typically, static scans are made 2–4 h after tracer injection resulting in low inter-tissue and intra-tumour contrast. An alternative approach involves labelling the nitroimidazoles with long-lived radionuclides (e.g., [^{124}I]-iodoazomycin arabinoside (^{124}I-AZA) and [^{124}I]-iodoazomycin galactoside (^{124}I-AZG)), which allows delayed scans after 24 and 48 h. Unfortunately, the results have been disappointing with no improvement in image contrast and poor counting statistics (Rischin et al. 2006; Reischl et al. 2007).

Pharmacokinetic analysis of the shape of tumour time–activity curves (TACs), as deduced from dynamic scans, may allow further separation of tumours/tumour sub-volumes that are inseparable based on static scans. Accordingly, it has been demonstrated that analysis of TAC shapes increased prognostic accuracy compared to traditional analysis of static PET images in a small study on head and neck cancer patients (Thorwarth et al. 2005). However, dynamic scans are cumbersome, expensive, cause inconvenience to patients, are analytically complex and it is not clear whether different kinetic models provide similar estimates of hypoxia. Regardless of whether static or dynamic assessment is applied, one of the major issues with PET markers is that the cells must be hypoxic for significant time periods in order to be detected, which means that such markers are more likely to identify chronic rather than acute hypoxia. Another problem facing the application of PET hypoxia markers is resolution, where the voxel sizes identified in the PET scan are much larger than most of the hypoxic structures (Horsman et al. 2012). Thus, the actual PET image does not accurately reflect the true hypoxia heterogeneity at the micro-regional level.

Clinically, only FMISO, FETNIM and FAZA have undergone evaluation in terms of correlating hypoxia with outcome following radiation therapy. The majority of studies involved FMISO in head and neck cancer (Rajendran et al. 2006; Rischin et al. 2006; Thorwarth et al. 2006; Eschmann et al. 2007; Dirix et al. 2009; Lee et al. 2009; Kikuchi et al. 2011; Zips et al. 2012). Two other studies used FETNIM (Lehtio et al. 2004) and FAZA (Mortensen et al. 2012) in head and neck cancer. FETNIM was also used in cancer of the lung (Li et al. 2010) and of the oesophagus (Yue et al. 2012); FAZA in sarcoma (Khamly et al. 2008) and cancer of the cervix (Schuetz et al. 2010); and FMISO in tumours of the central nervous system (CNS; Spence et al. 2008). The results obtained with FAZA in head and neck patients (Mortensen et al. 2012), showing a clear separation between the hypoxic and non-hypoxic tumours in terms of disease-free survival, are illustrated in Fig. 8.1c. When all the clinical data are combined, a highly significant influence of hypoxia on outcome was observed (Fig. 8.2).

The finding that [^{18}F]-labelled nitroimidazoles could be identified with PET led to the possibility that, by using alternative radioactive labels, other non-invasive imaging techniques could be used to designate hypoxia. One approach employed [^{19}F] which could be detected with MR. Two such [^{19}F]-labelled nitroimidazoles were developed ([^{19}F]-EF5 and [^{19}F]-SR 4554), although only the latter underwent some clinical evaluation (Seddon et al. 2003), but to date there has not been any real follow-up. A number of [^{123}I/^{125}I]-iodoazomycin derivatives have also been developed which can be detected using single-photon emission computed tomography (SPECT). Only [^{123}I]-iodoazomycin arabinoside (I-AZA) has undergone clinical evaluation (Urtasun et al. 1996) and it was found that head and neck cancer patients with positive I-AZA scans had a poorer outcome to radiotherapy than those with negative scans. Other non-nitroimidazole SPECT markers have been developed that include both ^{99m}Tc-labelled compounds (e.g., BMS 181321 and BRU59-21) and complex ligands (e.g., HL91). BRU59-21 and HL91 have undergone clinical evaluation and the results included in Fig. 8.2. In the case of BRU59-21, a phase I study in head and neck cancer showed a significant correlation with pimonidazole binding (Hoebers et al. 2002). HL91 was investigated in patients with non-small cell lung cancer prior to radiation therapy and those patients with the highest uptake had a significantly poorer response (Li et al. 2006), as shown in Fig. 8.1d.

Cu-ATSM (Cu (II)-diacetyl-*bis*(N4-methylthiosemicarbazone)) belongs to a chemically different group of putative PET hypoxia markers labelled with a positron emitting isotope of copper. Unlike the 2-nitroimidazoles, the exact retention mechanism of Cu-ATSM is still incompletely understood. It has been shown to have high membrane permeability and fast tumour uptake, thus allowing for rapid imaging after injection (Dearling et al. 1998). However, while the Cu^{2+} moiety can easily pass across cell membranes; under low oxygen conditions, the Cu^{2+} is converted to Cu^{1+} which is then trapped within cells (Vävere and Lewis 2007); measuring this trapped material should be indicative of hypoxia. Pre-clinical studies have shown good correlation between Cu-ATSM and pO_2 measurements (Lewis and Welch 2001; O'Donoghue et al. 2005) and nitroimidazole-based hypoxia markers (O'Donoghue et al. 2005; Dence et al. 2008), although the O'Donoghue (2005) studies showed these effects to be time and tumour dependent. Clinical studies in lung (Dehdashti et al. 2003), cervix (Dehdashti et al. 2008), rectal (Dietz et al. 2008) and head and neck cancer (Minagawa et al. 2011) support the potential of Cu-ATSM to be used as a marker of outcome to radiotherapy. However, recent studies have shown Cu-ATSM to be affected by mechanisms other than hypoxia per se (Yuan et al. 2006) and that it is insensitive to treatment that affects oxygenation (Yuan et al. 2006), suggesting that additional studies are needed before it can be marketed as a specific marker of tumour hypoxia.

8.2.4 Endogenous Markers

Hypoxia is well known to increase the cellular expression of a wide variety of genes and proteins (Harris 2002). This has led to suggestions that measurements of these elevated expression patterns could be indicative of hypoxia and thus prognostic for outcome. Measurements have been primarily made on biopsy material and the principal markers have included hypoxia inducible factor 1 (HIF-1), carbonic anhydrase IX (CAIX), the glucose transporters GLUT-1 and GLUT-3, and osteopontin (OPN). Attempts to relate expression levels with more established assays for hypoxia reported mixed results. Thus, HIF-1 expression was shown to weakly correlate with pimonidazole binding in cancer of the cervix (Hutchinson et al. 2004; Jankovic et al. 2006); yet, compared to Eppendorf pO_2 measurements in other cervix trials, it was significantly associated (Haugland et al. 2002), weakly associated (Hutchinson et al. 2004) or showed no association (Mayer et al. 2004). Studies in cancer of the cervix with CAIX reported expression to be correlated with pimonidazole binding (Olive et al. 2001a; Jankovic et al. 2006) and pO_2 estimates (Loncaster et al. 2001; Iakovlev et al. 2007) and with pO_2 in non-small cell lung cancer (Le et al. 2006), but no correlation was seen with pO_2 in two other cervix trials (Hedley et al. 2003; Mayer et al. 2005a). The expression of GLUT-1 was observed to correlate with pimonidazole binding (Airley et al. 2003), weakly correlated to pO_2 (Airley et al. 2001), and not correlated to pO_2 (Mayer et al. 2005b), again all investigated in cancer of the cervix. OPN was demonstrated to correlate with pO_2 estimates regardless of whether the OPN measurements were performed on head and neck histological sections (Le et al. 2007) or plasma samples from both head and neck (Le et al. 2003b) and non-small cell lung cancer (Le et al. 2006). These inconsistencies are perhaps not entirely unexpected as the expression of many endogenous hypoxia markers is up-regulated by factors other than hypoxia, and even where hypoxia is involved the up-regulation is often seen at intermediate levels of hypoxia where oxygen concentrations are above those causing radiobiological hypoxia (Sørensen et al. 2007). Such effects may ultimately limit the applicability of this approach for determining hypoxia.

Despite this, several studies have been able to correlate increased expression with outcome (Bussink et al. 2003), although such effects may be unrelated to hypoxia. One study that attempted to correlate expression of an endogenous marker with outcome involved a retrospective analysis of OPN levels in plasma samples from a Danish Head and Neck Cancer trial (DAHANCA 5). This study involved patients included in a randomised double-blind placebo-controlled trial between radiation alone and radiation combined with the hypoxic cell radiosensitiser nimorazole and found that high levels of OPN were associated with poor local control and survival after radiation (Overgaard et al. 2005). On the other hand, a recent international trial (Trans-Tasmanian Radiation Oncology Group; TROG 02.02), also in head and neck cancer, reported that plasma levels of OPN were not associated with outcome and could not predict for hypoxia-targeted therapy using the bioreductive drug tirapazamine (Lim et al. 2011). However, a more recent analysis of the DAHANCA 5 patient population applied a hypoxic gene signature and demonstrated that it could predict for those patients who had hypoxic tumours and thus benefited from the application of nimorazole (Toustrup et al. 2011); thus, the potential of measurements of elevated expression patterns to be indicative of hypoxia and prognostic for outcome is still unclear.

Recent studies have now shown that imaging hypoxia-regulated genes is possible. One study reported that a ^{89}Zr-labelled monoclonal antibody fragment against CAIX allowed non-invasive PET-based quantification and mapping of CAIX expression in a human xenograft tumour model (Hoeben et al. 2010). Sulphonamides, which are potent inhibitors of CAIX, have been fluorescently labelled and optically imaged in pre-clinical studies (Dubois et al. 2009) and efforts to label sulphonamides with PET radionuclides are ongoing (Asakawa et al. 2011).

8.2.5 Surrogate Markers

A number of techniques that measure hypoxia-related aspects have been proposed as possible surrogate hypoxic markers. Hypoxic cells, especially those distant from functional blood vessels, often have a tendency to be associated with nutrient deprivation and acidity (Vaupel et al. 1989). The lack of oxygen and nutrients shifts the balance of cellular energy production towards anaerobic glycolysis, with lactate being a by-product, which is probably one of the major contributors to the acidic conditions. It has, therefore, been proposed that measurement of the metabolic micromilieu could serve as a surrogate for hypoxia. Indeed, clinical studies in which tumour biopsies from both cervix (Walenta et al. 2000) and head and neck (Brizel et al. 2001) cancer patients were analysed for lactate levels using a bioluminescence technique reported that elevated concentrations were associated with both metastatic disease (Walenta et al. 2000; Brizel et al. 2001) and disease recurrence (Walenta et al. 2000). An additional study showed not only that lactate could be measured non-invasively using ^{1}H MR spectroscopy but that high lactate levels correlated with hypoxia estimated with the Eppendorf histograph (Star-Lack et al. 2000). A more recent pre-clinical study using a variety of head and neck human tumour xenografts indicated a stronger correlation between glucose concentration and the pimonidazole hypoxic fraction than when the latter was compared with lactate (Yaromina et al. 2009). As fluorodeoxyglucose (FDG) is an indicator of glucose metabolism and the most commonly used PET marker in oncology, it perhaps gives the possibility to apply another non-invasive imaging approach for measuring tumour hypoxia . In fact, FDG uptake is often stimulated in hypoxic areas both in experimental tumours (Busk et al. 2008) and in patients (Rajendran et al. 2004). However, studies that have tried to link overall FDG uptake with hypoxia have found no correlation with FMISO in head and neck (Gagel et al. 2007) or non-small cell lung cancer (Vera et al. 2011); with CuATSM in cancer of the cervix (Dehdashti et al. 2008); or with oxygen electrode measurements in head neck cancer (Gagel et al. 2007). These findings may be related to the fact that increased glucose uptake and accumulation of lactate can occur in tumours under normoxic conditions (Gulledge and Dewhirst 1996), which would suggest that the use of lactate and glucose metabolism as specific tracers for hypoxia appears limited.

Another potential surrogate marker is DNA damage. It is known that hypoxic cells are 2–3 times more resistant to radiation than normoxic cells (Horsman et al. 2009), and by measuring the degree of DNA damage after administering a specific radiation dose one should be able to separate the radiobiologically hypoxic cells from those that are non-hypoxic. Several techniques could be used here including alkaline elution, the comet assay and histone (H2AX) phosphorylation (Olive 2009). Clinical studies using a variety of tumours reported good correlations between comet assay estimates of the percentage of hypoxic cells and hypoxia determined using the Eppendorf electrode (Aquino-Parsons et al. 1999; Olive et al. 2001b) or pimonidazole binding (Olive et al. 2001a). However, one study in head and neck patients found no correlation between comet results and Eppendorf pO_2 measurements (Le et al. 2003a). Another investigation using patients with advanced carcinoma of the cervix showed that pimonidazole-positive and CAIX-positive cells were more likely to exhibit endogenous γH2AX foci (Olive et al. 2010). Additional clinical investigations showed that comet assay estimates of DNA damage could detect the effect of hypoxic modifiers (McLaren et al. 1997; Aquino-Parsons et al. 1999; Dorie et al. 1999; Olive et al. 2001b; Partridge et al. 2001). However, despite a majority of positive findings, the use of methods to estimate DNA damage has not become established in the clinic.

Preliminary investigations in cervical carcinoma in which interstitial fluid pressure (IFP) was estimated using a wick-in-needle apparatus and oxygenation status determined with the Eppendorf histograph indicated a potential relationship between the measurements by these techniques, thus suggesting the use of IFP as another surrogate marker for hypoxia (Milosevic et al. 1998). Additional pre-clinical and clinical examinations indicate that IFP can be determined using non-invasive DCE-MRI (Haider et al. 2007; Nielsen et al. 2010; Gulliksrud et al. 2011; Hompland et al. 2012). However, other clinical studies found that while IFP could predict survival in cervical cancer patients, this was actually independent of oxygenation measurements (Milosevic et al. 2001; Fyles et al. 2006); hence, the potential of IFP in this context remains uncertain.

8.3 Conclusions and Perspectives

The presence of hypoxic cells within tumours is clearly a clinically relevant issue that has a negative influence on outcome. As a result, substantial effort has been made to find clinically relevant methods that can identify this hypoxia. Unfortunately, none has become established in routine clinical practice due to a variety of reasons including the invasive nature of many as well as being generally not easy to use routinely. What are now needed are non-invasive approaches that give quantitative spatial and temporal estimates of hypoxia on an individual patient basis. In that context, there are several possibilities being investigated based on techniques that are routinely used clinically in most hospitals, specifically PET, SPECT, CT and MRI. As shown in Figs. 8.1 and 8.2, results obtained using these approaches are certainly consistent with them being able to detect hypoxia. However, the question is whether they actually give information about the relevant hypoxic cell population. One major focus of current cancer research is the role of cancer stem cells in tumourigenesis and therapy. Such cells amount to around 1–25 % of the total viable tumour cell population (Hill et al. 2009), but they are the cells that must be eliminated to obtain response (Baumann et al. 2008). It is now suggested that hypoxia is a "niche" for such stem cells (Baumann et al. 2008); thus, hypoxia imaging must identify these relevant areas; identifying hypoxia has no relevance if the cells in that area have no clonogenic potential. Although hypoxic cells have been identified in virtually all tumour types in which they have been looked for, the degree and type of hypoxia are also extremely heterogeneous. Regarding therapy, we know that chronic and acute hypoxic cells will influence radiation damage and thus decrease local tumour control, but it has been suggested that acutely hypoxic cells may actually be more resistant to radiation than chronically hypoxic ones (Denekamp and Dasu 1999) because the former tend not to be nutrient deprived and thus more likely to invoke enzymatic processing of the radiation-induced lesions and thus capable of repairing the damage. For other conventional treatments, such as chemotherapy, it is not known whether chronically or acutely hypoxic cells respond equally. With respect to overall survival, the situation is more complex. There is clear evidence, at least in squamous cell carcinomas, that a relationship exists between local control and overall survival (Overgaard 2007). However, survival will also be influenced by malignant progression, both in terms of growth of the primary tumour and metastatic spread, which is again affected by hypoxia. However, in this situation acute hypoxia may play a more significant role (Lunt et al. 2009) and other micro-environmental factors such as intermediate hypoxia, low pH and glucose deprivation can also be important (Lunt et al. 2009). Currently available imaging techniques are not able to identify all the critical factors. This is shown from pre-clinical measurements of radiobiological hypoxia, pimonidazole binding and DCE-MRI estimates (Fig. 8.3) made in animals receiving either no hypoxic modification or hypoxia decreased by injecting nicotinamide or breathing carbogen gas. In this particular tumour model, nicotinamide prevents the transient changes in blood perfusion, thus only decreasing acute hypoxia, while carbogen only reduces the levels of chronic hypoxia (Horsman 1995). While both nicotinamide and carbogen decreased radiobiological hypoxia, pimonidazole could only detect the effect of carbogen; thus, imaging agents of this type, including most of the PET markers, simply monitor chronic hypoxia, whereas the effect of nicotinamide was only observed with DCE-MRI consistent with the effects being perfusion based. These results not only illustrate the deficiencies of the more popular imaging approaches; more importantly they argue for the use of combined imaging methods to get a better indicator for relevant hypoxia.

Once information about the degree and distribution of hypoxia within tumours is obtained, there is the question of how to use it. The obvious choice is to use some form of modification. These could include increasing oxygen delivery (e.g., carbogen gas breathing, haemoglobin modification or nicotinamide), preferentially killing the hypoxic cells (e.g., bioreductive drugs or hyperthermia) or where the conventional therapy is radiation one can apply hypoxic cell radiosensitisers (e.g., nitroimidazole compounds or again hyperthermia). All of these have been shown to work in pre-clinical models (Horsman et al. 2011) and many of them have been successfully applied clinically (Horsman and Overgaard 2007; Overgaard 2007). These treatments often affect specific cell populations; thus, the most suitable hypoxia-targeting agents must be used in connection with the appropriate imaging assessment; it is pointless to use an agent that targets chronic hypoxia if the imaging technique identifies only acute hypoxia. Where radiation therapy is applied, there is also the possibility to increase the radiation dose delivered to the tumour (Ling et al. 2000; Søvik et al. 2009; Bentzen and Gregoire 2011). This can involve an increase of the gross tumour volume (GTV) in which a substantial hypoxic volume has been identified, although this raises the potential of increasing the dose to surrounding normal tissues, which would proportionally enhance the risk of adverse reactions. This is likely to be a less of a problem if one uses the imaging data to define a biological target volume (BTV) and thus simply increase the dose to just the hypoxic area. Whether one uses a GTV or a BTV approach, one must be sure that the areas receiving the increased radiation actually contain hypoxic stem cells (Horsman et al. 2012).

a Radiation response

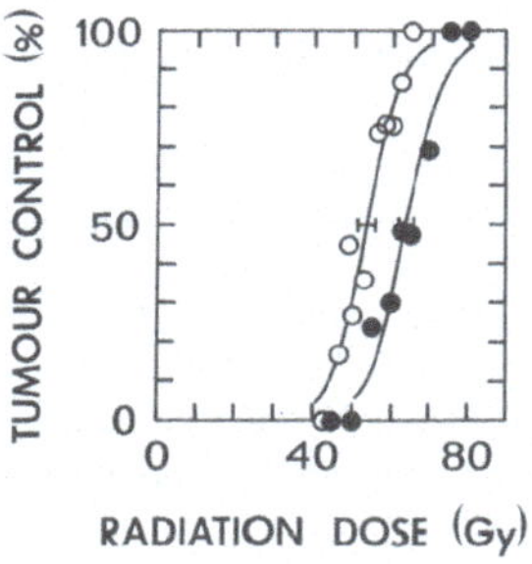

b Pimonidazole binding

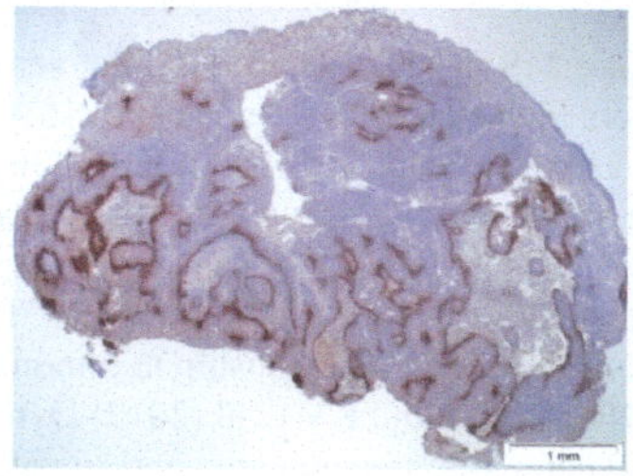

c DCE-MRI

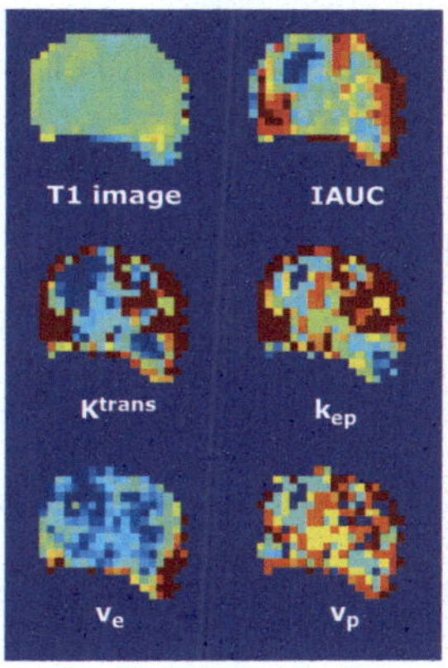

d Results obtained

	Control	Nicotinamide	Carbogen
HF-rad	23% ± 8	7% ± 4*	6% ± 3*
HF-pimo	8.1% ± 1.7	8.1% ± 2.0	1.4% ± 0.9*
IAUC	0.96 ± 0.06	1.02 ± 1.70	1.17 ± 0.91
K^{trans}	0.83 ± 0.07	0.87 ± 0.12	1.09 ± 0.13
k_{ep}	1.18 ± 0.10	0.77 ± 0.08*	1.31 ± 0.08
v_e	0.66 ± 0.03	0.95 ± 0.11*	0.70 ± 0.05
v_p	0.69 ± 0.06	0.66 ± 0.08	0.78 ± 0.05

Fig. 8.3 Effect of nicotinamide or carbogen breathing on radiation response, pimonidazole binding and DCE-MRI images in a murine C3H mammary carcinoma. **a** Local tumour control 90 days after irradiating with graded radiation doses administered under normal air breathing conditions (○) or when tumours were clamped (●); points are from an average of 10 mice/group with the lines fitted following logit analysis; **b** representative image of a pimonidazole-stained histological section from a control tumour measured 90 min after injection (pimonidazole is shown by the brown colour); **c** DCE-MRI images from a control tumour showing a T1 image, IAUC (the initial area under the uptake curve), K^{trans} (the transfer constant), k_{ep} (rate constant), v_e (extravascular extracellular space) and v_p (plasma volume fraction) obtained with a 3-tesla magnet following injection of gadolinium-diethylene triamine pentaacetic acid (gadolinium-DTPA); and **d** summary of the hypoxic fractions (HFs) calculated from the radiation data (HF-rad; mathematically determined from the difference between the normal and clamped dose–response curves) or degree of pimonidazole binding (HF-pimo), and the DCE-MRI parameters, as measured in control mice, 30 min after intraperitoneally injecting nicotinamide (1,000 mg/kg) or breathing carbogen (95 % O_2 + 5 % CO_2) for 5 min before and during measurement. The same animals were used for the pimonidazole and DCE-MRI estimates. Values show means (± 1 standard error, S.E.) with those significantly different from controls indicated*

Acknowledgements The authors would like to thank the following organizations for financial support: the Danish Agency for Science Technology and Innovation; the Danish Cancer Society; the EC FP7 project METOXIA (project no. 222741); and CIRRO—the Lundbeck Foundation Center for Interventional Research in Radiation Oncology and the Danish Council for Strategic Research.

References

Adams GE, Cooke MS (1969) Electron–affinic sensitization. I. A structural basis for chemical radiosensitizers in bacteria. Int J Radiat Biol 15:457–471

Airley RE, Loncaster J, Davidson S et al (2001) Glucose transporter glut-1 expression correlates with tumor hypoxia and predicts metastasis-free survival in advanced carcinoma of the cervix. Clin Cancer Res 7:928–934

Airley RE, Loncaster J, Raleigh J et al (2003) GLUT-1 and CAIX as intrinsic markers of hypoxia in carcinoma of the cervix: relationship to pimonidazole binding. Int J Cancer 104:85–91

Andersen EKF, Hole KH, Lund KV et al (2012) Dynamic contrast-enhanced MRI of cervical cancers: temporal percentile screening of contrast enhancement identifies parameters for prediction of chemoradioresistance. Int J Radiat Oncol Biol Phys 82:485–492

Aquino-Parsons C, Luo C, Vise CM et al (1999) Comparison between the comet assay and the oxygen microelectrode for measurement of tumor hypoxia. Radiother Oncol 51:179–185

Asakawa C, Ogawa M, Kumata K et al (2011) Radiosynthesis of three [11C]ureido-substituted benzenesulfonamides as PET probes for carbonic anhydrase IX in tumors. Bioorg Med Chem Lett 21:7017–7020

Awwad HK, Nagger M, Mocktar N et al (1986) Intercapillary distance measurement as an indicator of hypoxia in carcinoma of the cervix uteri. Int J Radit Oncol Biol Phys 12:1329–1333

Baumann M, Krause M, Hill R (2008) Exploring the role of cancer stem cells in radioresistance. Nat Revs cancer 8:545–554

Begg AC, Janssen H, Sprong D et al (2001) Hypoxia and perfusion measurements in human tumours—initial experience with pimonidazole and IUdR. Acta Oncol 40:924–928

Bentzen SM, Gregoire V (2011) Molecular imaging-based dose painting: a novel paradigm for radiation therapy prescription. Sem Radiat Oncol 21:101–110

Bergers G, Benjamin LE (2003) Tumorigenesis and the angiogenic switch. Nat Revs Cancer 3:401–410

Bisdas S, Ngyuyen SA, Anand SK et al (2009) Outcome prediction after surgery and chemoradiation of squamous cell carcinoma in te oral cavity, oropharynx, and hypopharynx: use of baseline perfusion CT microcirculatory parameters vs. tumor volume. Int J Radiat Oncol Biol Phys 73:1313–1318

Boushel R, Langberg H, Olesen J et al (2001) Monitoring tissue oxygen availability with near infrared spectroscopy (NIRS) in health and disease. Scand J Med Sci Sports 11:213–222

Braun RD, Lanzen JL, Snyder SA et al (2001) Comparison of tumor and normal tissue oxygen tension measurements using Oxylite or microelectrodes in rodents. Am J Physiol Heart Circ Physiol 280:H2533–H2544

Brem S, Brem H, Folkman J et al (1976) Prolonged tumor dormancy by prevention of neovascularization in the vitreous. Cancer Res 36:2807–2812

Brizel DM, Schroeder T, Scher RL et al (2001) Elevated tumor lactate concentrations predict for an increased risk of metastases in head-and-neck cancer. Int J Radiat Oncol Biol Phys 51:349–353

Brizel DM, Scully SP, Harrelson JM et al (1996) Tumor oxygenation predicts for the likelihood of distant metastases in human soft tissue sarcoma. Cancer Res 56:941–943

Brizel DM, Sibley GS, Prosnitz LR et al (1997) Tumor hypoxia adversely affects the prognosis of carcinoma of the head and neck. Int J Radiat Oncol Biol Phys 38:285–289

Brown JM (1979) Evidence for acutely hypoxic cells in mouse tumours, and a possible mechanism of reoxygenation. Br J Radiol 52:650–656

Busk M, Horsman MR, Kristjansen PEG et al (2008) Aerobic glycolysis in cancers: implications for the usability of oxygen-responsive genes and fluorodeoxyglucose-PET as markers of tissue hypoxia. Int J Cancer 122:2726–2734

Bussink J, Kaanders JHAM, Kogel AJ van der (2003) Tumor hypoxia at the micro-regional level: clinical relevance and predictive value of exogenous and endogenous hypoxic cell markers. Radiother Oncol 67:3–15

Chaplin DJ, Olive PL, Durand RE (1987) Intermittent blood flow in a murine tumor: radiobiological effects. Cancer Res 47:597–601

Chaplin DJ, Trotter MJ (1991) Chemical modifiers of tumor blood flow. In: Vaupel P, Jain RK (eds) Tumor blood supply and metabolic microenvironment. Gustav Fischer Verlag, Stuttgart

Chopra S, Foltz WD, Milosevic MF et al (2009) Comparing oxygen-sensitive MRI (BOLD R2*) with oxygen electrode measurements: a pilot study in men with prostate cancer. Int J Radiat Biol 85:805–813

Collingridge DR, Young WK, Vojnovic B et al (1997) Measurement of tumor oxygenation: a comparison between polarographic needle electrodes and a time-resolved luminescence-based optical sensor. Radiat Res 147:329–334

Cooper RA, Carrington BM, Loncaster JA et al (2000) Tumour oxygenation levels correlate with dynamic contrast-enhanced magnetic resonance imaging parameters in carcinoma of the cervix. Radiother Oncol 57:53–59

Dearling JL, Lewis JS, Mullen GE et al (1998) Design of hypoxia-targeting radiopharmaceuticals: selective uptake of copper-64 complexes in hypoxic cells in vitro. Eur J Nucl Med 25:788–792

Dehdashti F, Mintun MA, Lewis JS et al (2003) In vivo assessment of tumor hypoxia in lung cancer with ^{60}Cu-ATSM. Eur J Nucl Med Mol Imaging 30:844–850

Dehdashti F, Grigsby PW, Lewis JS et al (2008) Assessing tumor hypoxia in cervical cancer by PET with 60Cu-labeled diacetyl-bis(N4-methylthiosemicarbazone). J Nucl Med 49:201–205

Delides GS, Venizelos J, Révész L (1988) Vascularization and curability of stage III and IV nasopharyngeal tumours. J Cancer Res Clin Oncol 114:321–323

Dence CS, Ponde DE, Welch MJ et al (2008) Autoradiographic and small-animal PET comparisons between ^{18}F-FMISO, ^{18}F-FDG, ^{18}F-FLT and the hypoxic selective ^{64}Cu-ATSM in a rodent model of cancer. Nucl Med Biol 35:713–720

Denekamp J, Dasu A (1999) Inducible repair and the two forms of tumour hypoxia—time for a paradigm shift. Acta Oncol 38:903–918

Dewhirst MW, Cao Y, Moeller B (2008) Cycling hypoxia and free radicals regulate angiogenesis and radiotherapy response. Nat Revs Cancer 8:425–437

Dietz DW, Dehdashti F, Grigsby PW et al (2008) Tumor hypoxia detected by positron emission tomography with 60Cu-ATSM as a predictor of response and survival in patients undergoing Neoadjuvant chemoradiotherapy for rectal carcinoma: a pilot study. Dis Colon Rectum 51:1641–1648

Dirix P, Vandecaveye V, De Keyzer F et al (2009) Dose painting in radiotherapy for head and neck squamous cell carcinoma: value of repeated functional imaging with ^{18}F-FDG PET, ^{18}F-fluoromisonidazole PET, diffusion-weighted MRI, and dynamic contrast-enhanced MRI. J Nucl Med 50:1020–1027 (2009)

Donaldson SB, Betts G, Bonington SC et al (2011) Perfusion estimated with rapid dynamic contrast-enhanced magnetic resonance imaging correlates inversely with vascular endothelial growth factor expression and pimonidazole staining in head-and-neck cancer: a pilot study. Int J Radiat Oncol Biol Phys 81:1176–1183

Dorie MJ, Kovacs MS, Gabalski EC et al (1999) DNA damage measured by the comet assay in head and neck cancer patients treated with tirapazamine. Neoplasia 1:461–467

Dubois L, Lieuwes NG, Maresca A et al (2009) Imaging of CA IX with fluorescent labelled sulphonamides distinguishes hypoxic and (re)-oxygenated cells in a xenograft tumour model. Radiother Oncol 92:423–428

Eschmann SM, Paulsen F, Bedeshem C et al (2007) Hypoxia-imaging with ^{18}F-Misonidazole and PET: changes of kinetics during radiotherapy of head-and-neck cancer. Radiother Oncol 83:406–410

Evans SM, Joiner B, Jenkins WT et al (1995) Identification of hypoxia in cells and tissues of epigastric 9 L rat glioma using EF5 [2-(2-nitro-1H-imidazol-1-yl)-N-(2,2,3,3,3-pentafluoropropyl) acetamide]. Br J Cancer 72:875–882

Fyles A, Milosevic M, Pintilie M et al (2006) Long-term performance of interstitial fluid pressure and hypoxia as prognostic factors in cervix cancer. Radiother Oncol 80:132–137

Fyles AW, Milosevic M, Wong R et al (1998) Oxygenation predicts radiation response and survival in patients with cervix cancer. Radiother Oncol 48:149–156

Gagel B, Piroth M, Pinkawa M et al (2007) pO polarography, contrast enhanced color duplex sonography (CDS), [^{18}F] fluoromisonidazole and [^{18}F] fluorodeoxyglucose positron emission tomography: validated methods for the evaluation of therapy-relevant tumor oxygenation or only bricks in the puzzle of tumor hypoxia. BMC Cancer 7:113 (2007)

Gatenby RA, Kessler HB, Rosenblum JS et al (1988) Oxygen distribution in squamous cell carcinoma metastases and its relationship to outcome of radiation therapy. Int J Radiat Oncol Biol Phys 14:831–838

Griffiths JR, Robinson SP (1999) The oxylite: a fibre-optic oxygen sensor. Br J Radiol 72:627–630

Gulledge CJ, Dewhirst MW (1996) Tumor oxygenation: a matter of supply and demand. Anticancer Res 16:741–749

Gulliksrud K, Hompland T, Galappathi K et al (2011) Assessment of tumor hypoxia and interstitial fluid pressure by gadomelitol-based dynamic contrast-enhanced magnetic resonance imaging. Radiother Oncol 101:217–222

Haider MA, Sitartchouk I, Roberts TP et al (2007) Correlations between dynamic contrast-enhanced magnetic resonance imaging-derived measures of tumor microvasculature and interstitial fluid pressure in patients with cervical cancer. J Magn Reson Imaging 25:153–159

Hallac RR, Ding Y, Yuan Q et al (2012) Oxygenation in cervical cancer and normal uterine cervix assessed using blood oxygenation level-dependent (BOLD) MRI at 3T. NMR Biomed 25:1321–1330

Harris AL (2002) Hypoxia—a key regulatory factor in tumour growth. Nat Revs Cancer 2:38–47

Haugland HK, Vukovic V, Pintilie M et al (2002) Expression of hypoxia-inducible factor-1alpha in cervical carcinomas: correlation with tumor oxygenation. Int J Radiat Oncol Biol Phys 53:854–861

Hedley D, Pintilie M, Woo J et al (2003) Carbonic anhydrase IX expression, hypoxia, and prognosis in patients with uterine cervical carcinomas. Clin Cancer Res 9:5666–5674

Helmlinger G, Yuan F, Dellian M et al (1997) Interstitial pH and pO_2 gradients in solid tumors in vivo: high-resolution measurements reveal a lack of correlation. Nat Med 3:177–182

Hermans R, Lambin P, Goten A van der et al (1999) Tumoural perfusion as measured by dynamic computed tomography in head and neck carcinoma. Radiother Oncol 53:105–111

Hill RP, Marie-Egyptienne DT, Hedley DW (2009) Cancer stem cells, hypoxia and metastasis. Sem Radiat Oncol 19:106–111

Hoeben BA, Kaanders JH, Franssen GM et al (2010) PET of hypoxia with 89Zr-labeled cG250-F(ab')2 in head and neck tumors. J Nucl Med 51:1076–1083

Hoebers FJP, Janssen HLK, Valdés Olmos RA et al (2002) Phase 1 study to identify tumour hypoxia in patients with head and neck cancer using technetium-99m BRU 59-21. Eur J Nucl Med 29:1206–1211

Hoeckel M, Knoop C, Schlenger K et al (1993) Intratumoral pO2 predicts survival in advanced cancer of the uterine cervix. Radiother Oncol 26:45–50

Hoeckel M, Schlenger K, Aral B et al (1996) Association between tumor hypoxia and malignant progression in advanced cancer of the uterine cervix. Cancer Res 56:4509–4515

Hompland T, Ellingsen C, ØvrebøKM et al (2012) Interstitial fluid pressure and associated lymph node metastasis revealed in tumors by dynamic contrast-enhanced MRI. Cancer Res 72:4899–4908

Horsman MR (1995) Nicotinamide and other benzamide analogs as agents for overcoming hypoxic cell radiation resistance in tumours. Acta Oncol 34:571–587

Horsman MR, Overgaard J (2007) Hyperthermia: a potent enhancer of radiotherapy. Clin Oncol 19:418–426

Horsman MR, Wouters BG, Joiner MC et al (2009) The oxygen effect and fractionated radiotherapy. In: van der Kogel AJ, Joiner MC (eds) Basic clinical radiobiology, 4th edn. Hodder Arnold, London

Horsman MR, Overgaard J, Siemann DW (2011) Impact on radiotherapy. In: Siemann DW (ed) Tumor microenvironment. John Wiley & Sons Ltd., Chichester

Horsman MR, Mortensen LS, Petersen JB et al (2012) Imaging hypoxia to improve radiotherapy outcome. Nat Rev Clin Oncol 9:674–687

Hoskin PJ, Carnell DM, Taylor NJ et al (2007) Hypoxia in prostate cancer: correlation of BOLD-MRI with pimonidazole immunohistochemistry—initial observations. Int J Radiat Oncol Biol Phys 68:1065–1071

Hutchinson GJ, Valentine HR, Loncaster JA et al (2004) Hypoxia-inducible factor 1alpha expression as an intrinsic marker of hypoxia: correlation with tumor oxygen, pimonidazole measurements, and outcome in locally advanced carcinoma of the cervix. Clin Cancer Res 10:8405–8412

Iakovlev VV, Pintilie M, Morrison A et al (2007) Effect of distributional heterogeneity on the analysis of tumor hypoxia based on carbonic anhydrase IX. Lab Invest 87:1206–1217

Jankovic B, Aquino-Parsons C, Raleigh JA et al (2006) Comparison between pimonidazole binding, oxygen electrode measurements, and expression of endogenous hypoxia markers in cancer of the uterine cervix. Cytometry B Clin Cytom 70:45–55

Jansen JFA, Schöder H, Lee NY et al (2010) Noninvasive assessment of tumor microenvironment using dynamic contrast-enhanced magnetic resonance imaging and ^{18}F-fluoromisonidazole positron emission tomography imaging in neck nodal metastases. Int J Radiat Oncol Biol Phys 77:1403–1410

Janssen HL, Haustermans KM, Sprong D et al (2002) HIF-1A, pimonidazole, and iododeoxyuridine to estimate hypoxia and perfusion in human head-and-neck tumors. Int J Radiat Oncol Biol Phys 54:1537–1549

Kaanders JH, Wijffels KI, Marres HA et al (2002) Pimonidazole binding and tumor vascularity predict for treatment outcome in head and neck cancer. Cancer Res 62:7066–7074

Khamly K, Choong P, Ngan S et al (2008) Hypoxia in soft-tissue sarcomas on [^{18}F]-fluoroazomycin arabinoside positron emission tomography (FAZA-PET) powerfully predicts response to radiotherapy and early relapse (abstract 35029). Presented at the 14th Connective Tissue Oncology Society Annual Meeting, November 13–15 (London)

Kikuchi M, Yamane T, Shinoharas S et al (2011) 18F-fluoromisonidazole positron emission tomography before treatment is a predictor of radiotherapy outcome and survival prognosis in patients with head and neck squamous cell carcinoma. Ann Nucl Med 25:625–633

Kimura H, Braun RD, Ong ET et al (1996) Fluctuations in red cell flux in tumor microvessels can lead to transient hypoxia and reoxygenation in tumor parenchyma. Cancer Res 56:5522–5528

Knocke TH, Weitmann HD, Feldmann HJ et al (1999) Intratumoral pO2-measurements as predictive assay in the treatment of carcinoma of the uterine cervix. Radiother Oncol 53:99–104

Kolstad P (1968) Intercapillary distance, oxygen tension and local recurrence in cervix cancer. Scand J Clin Lab Invest Suppl 106:145–157

Komar G, Seppänen M, Eskola O et al (2008) ^{18}F-EF5: a new PET tracer for imaging hypoxia in head and neck cancer. J Nucl Med 49:1944–1951

Krishna MC, Matsumoto S, Yasui H et al (2012) Electron paramagnetic resonance imaging of tumor pO_2. Radiat Res 177:376–386

Krohn KA, Muzi M, Link JM et al (2012) Role of hypoxia imaging with [F-18]FMISO-PET and BOLD-MRI in clinical trials. Eur J Cancer 48:141 (abstract)

Lauk S, Skates S, Goodman M et al (1989) A morphometric study of the vascularity of oral squamous cell carcinomas and its relation to outcome of radiation therapy. Eur J Cancer Clin Oncol 25:1431–1440

Le QT, Kovacs MS, Dorie MJ et al (2003a) Comparison of the comet assay and the oxygen microelectrode for measurming tumor oxygenation in head-and-neck cancer patients. Int J Radiat Oncol Biol Phys 56:375–383

Le QT, Sutphin PD, Raychaudhuri S et al (2003b) Identification of osteopontin as a prognostic plasma marker for head and neck squamous cell carcinomas. Clin Cancer Res 9:59–67

Le QT, Chen E, Salim A et al (2006) An evaluation of tumor oxygenation and gene expression in patients with early stage non-small cell lung cancers. Clin Cancer Res 12:1507–1514

Le QT, Kong C, Lavori PW et al (2007) Expression and prognostic significance of a panel of tissue hypoxia markers in head-and-neck squamous cell carcinomas. Int J Radiat Oncol Biol Phys 69:167–175

Lee N, Nehmeh S, Schöder H et al (2009) Prospective trial incorporating pre-/mid-treatment [^{18}F]-misonidazole positron emission tomography for head-and-neck cancer patients undergoing concurrent chemoradiotherapy. Int J Radiat Oncol Biol Phys 75:101–108

Lehtio K, Eskola O, Vijanen T et al (2004) Imaging perfusion and hypoxia with PET to predict radiotherapy response in head-and-neck cancer. Int J Radiat Oncol Biol Phys 59:971–982

Lewis JS, Welch MJ (2001) PET imaging of hypoxia. Quant J Nucl Med 45:183–188

Li L, Yu J, Xing L et al (2006) Serial hypoxia imaging with ^{99m}Tc-HL91 SPECT to predict radiotherapy response in non small cell lung cancer. Amer J Clin Oncol 29:628–633

Li L, Hu M, Zhu H et al (2010) Comparison of ^{18}F-Fluoroerythronitroimidazole and ^{18}F-fluorodeoxyglucose positron emission tomography and prognostic value in locally advanced non-small-cell lung cancer. Clin Lung Cancer 11:335–340

Lim AM, Rischin D, Fisher R et al (2011) Prognostic significance of plasma osteopontin in patients with locoregionally advanced head and neck squamous cell carcinoma treated on TROG 02.02 phase III trial. Clin Cancer Res 18:301–307

Ling CC, Humm J, Larson S et al (2000) Towards multidimensional radiotherapy (MD-CRT): biological imaging and biological conformality. Int J Radiat Oncol Biol Phys 47:551–560

Loncaster JA, Harris AL, Davidson SE et al (2001) Carbonic anhydrase (CA IX) expression, a potential new intrinsic marker of hypoxia: correlations with tumor oxygen measurements and prognosis in locally advanced carcinoma of the cervix. Cancer Res 61:6394–6399

Loncaster JA, Carrington BM, Sykes JR et al (2002) Prediction of radiotherapy outcome using dynamic contrast enhanced MRI of carcinoma of the cervix. Int J Radiat Oncol Biol Phys 54:759–767

Lunt SJ, Chaudary N, Hill RP (2009) The tumor microenvironment and metastatic disease. Clin Exp Mets 26:19–34

Lyng H, Sundfør K, Tropé C et al (2000) Disease control of uterine cervical cancer: relationships to tumor oxygen tension, vascular density, cell density, and frequency of mitosis and apoptosis measured before treatment and during radiotherapy. Clin Cancer Res 6:1104–1112

Lyng H, Vorren AO, Sundfør K et al (2001) Assessment of tumor oxygenation in human cervical carcinoma by use of dynamic Gd-DTPA-enhanced MR imaging. J Magn Reson Imaging 14:750–756

Mayer A, Wree A, Höckel M et al (2004) Lack of correlation between expression of HIF-1alpha protein and oxygenation status in identical tissue areas of squamous cell carcinomas of the uterine cervix. Cancer Res 64:5876–5881

Mayer A, Höckel M, Vaupel P (2005a) Carbonic anhydrase IX expression and tumor oxygenation status do not correlate at the microregional level in locally advanced cancers of the uterine cervix. Clin Cancer Res 11:7220–7225

Mayer A, Höckel M, Wree A et al (2005b) Microregional expression of glucose transporter-1 and oxygenation status: lack of correlation in locally advanced cervical cancers. Clin Cancer Res 11:2768–2773

Mayr NA, Wang JZ, Zhang D et al (2010) Longitudinal changes in tumor perfusion pattern during the radiation therapy course and its clinical impact in cervical cancer. Int J Radiat Oncol Biol Phys 77:502–508

McLaren DB, Pickles T, Thomson T et al (1997) Impact of nicotinamide on human tumour hypoxic fraction measured using the comet assay. Radiother Oncol 45:175–182

Milosevic MF, Fyles AW, Wong W et al (1998) Interstitial fluid pressure in cervical carcinoma: within tumor heterogeneity, and relation to oxygen tension. Cancer 82:2418–2426

Milosevic MF, Fyles AW, Hedley D et al (2001) Interstitial fluid pressure predicts survival in patients with cervix cancer independent of clinical prognostic factors and tumor oxygen measurements. Cancer Res 61:6400–6405

Minagawa Y, Shizukuishi K, Koike I et al (2011) Assessment of tumor hypoxia by ^{62}Cu-ATSM PET/CT as a predictor of response in head and neck cancer: a pilot study. Ann Nucl Med 25:339–345

Mortensen LS, Johansen J, Kallehauge J et al (2012) FAZA PET/CT hypoxia imaging in patients with squamous cell carcinoma of the head and neck treated with radiotherapy: results from the DAHANCA 24 trial. Radiother Oncol 105:14–20

Moulder JE, Rockwell S (1984) Hypoxic fractions of solid tumour. Int J Radiat Oncol Biol Phys 10:695–712

Mueller-Kleiser W, Vaupel P, Manz R et al (1981) Intracapillary oxyhemoglobin saturation of malignant tumors in humans. Int J Radiat Oncol Biol Phys 7:1397–1404

Newbold K, Castellano I, Charles-Edwards E et al (2009) An exploratory study into the role of dynamic contrast-enhanced magnetic resonance imaging or perfusion computed tomography for detection of intratumoral hypoxia in head-and-neck cancer. Int J Radiat Oncol Biol Phys 74:29–37

Ng CS, Kodama Y, Mullani NA et al (2009) Tumor blood flow measured by perfusion computed tomography and ^{15}O-labeled water positron emission tomography: a comparison study. J Comput Assist Tomogr 33:460–465

Nielsen T, Murata R, Maxwell RJ et al (2010) Non-invasive imaging of combretastatin activity in different tumour models: association with more invasive estimates. Acta Oncol 49:906–913

Nielsen T, Wittenborn T, Horsman MR (2012) Dynamic contrast-enhanced magnetic resonance imaging (DCE-MRI) in preclinical studies of antivascular treatments. Pharmaceutics 4:563–589

Nordsmark M, Overgaard M, Overgaard J (1996) Pretreatment oxygenation predicts radiation response in advanced squamous cell carcinoma of the head and neck. Radiother Oncol 41:31–39

Nordsmark M, Alsner J, Keller J et al (2001) Hypoxia in human soft tissue sarcomas: adverse impact on survival and no association with p53 mutations. Br J Cancer 84:1070–1075

Nordsmark M, Bentzen SM, Rudat V et al (2005) Prognostic value of tumor oxygenation in 397 head and neck tumors after primary radiation therapy. An international multi-center study. Radiother Oncol 77:18–24

Nordsmark M, Loncaster J, Aquino-Parsons C et al (2006) The prognostic value of pimonidazole and tumour pO2 in human cervix carcinomas after radiation therapy: a prospective international multi-center study. Radiother Oncol 80:123–131

O'Donoghue JA, Zanzonico P, Pugachev A et al (2005) Assessment of regional tumor hypoxia using ^{18}F-fluoromisonidazole and ^{64}Cu(II)-diacetyl-bis(N4-methylthiosemicarbazone) positron emission tomography: comparative study featuring microPET imaging, Po_2 probe measurement, autoradiography, and fluorescent microscopy in the R3327-AT and FaDu rat tumor models. Int J Radiat Oncol Biol Phys 61:1493–1502

Olive PL (2009) Impact of the comet assay in radiobiology. Mutation Res 681:13–23

Olive PL, Aquino-Parsons C, MacPhail SH et al (2001a) Carbonic anhydrase 9 as an endogenous marker for hypoxic cells in cervical cancer. Cancer Res 61:8924–8929

Olive PL, Banáth JP, Aquino-Parsons C (2001b) Measuring hypoxia in solid tumors. Is there a gold standard? Acat Oncol 40:917–923

Olive PL, Banuelos CA, Durand RE et al (2010) Endogenous and radiation-induced expression of γH2AX in biopsies from patients treated for carcinoma of the uterine cervix. Radiother Oncol 94:82–89

Overgaard J (2007) Hypoxic radiosensitization: adored and ignored. J Clin Oncol 25:4066–4074

Overgaard J, Eriksen JG, Nordsmark M et al (2005) Plasma osteopontin, hypoxia, and response to the hypoxia sensitiser nimorazole in radiotherapy of head and neck cancer: results from the DAHANCA 5 randomized double-blind placebo-controlled trial. Lancet Oncol 6:757–764

Pacheco-Torres J, López-Larrubia P, Ballesteros P et al (2011) Imaging tumor hypoxia by magnetic resonance methods. NMR Biomed 24:1–16

Padhani AR, Krohn KA, Lewis JS et al (2007) Imaging oxygenation of human tumours. Eur Radiol 17:861–872

Partridge SE, Aquino-Parsons C, Luo C et al (2001) A pilot study comparing intratumoral oxygenation using the comet assay following 2.5 % and 5 % carbogen and 100 % oxygen. Int J Radiat Oncol Biol Phys 49:575–580

Powell ME, Hill SA, Saunders MI et al (1997) Human tumor blood flow is enhanced by nicotinamide and carbogen breathing. Cancer Res 57:5261–5264

Rajendran JG, Mankoff DA, O'Sullivan F et al (2004) Hypoxia and glucose metabolism in malignant tumors: evaluation by [^{18}F]fluoromisonidazole and [^{18}F]fluorodeoxyglucose positron emission tomography imaging. Clin Cancer Res 10:2245–2252

Rajendran JG, Schwartz DL, O'Sullivan J et al (2006) Tumor hypoxia imaging with [F-18] fluoromisonidazole positron emission tomography in head and neck cancer. Clin Cancer Res 12:5435–5441

Raleigh JA, Chou SC, Arteel GE et al (1999) Comparisons among pimonidazole binding, oxygen electrode measurements and radiation response in C3H mouse tumors. Radiat Res 151:580–589

Rasey JS, Koh WJ, Evans ML et al (1996) Quantifying regional hypoxia in human tumors with positron emission tomography of [^{18}F]fluoromisonidazole: a pretherapy study of 37 patients. Int J Radiat Oncol Biol Phys 36:417–428

Reischl G, Dorow DS, Cullinane C et al (2007) Imaging of tumor hypoxia with [^{124}I]IAZA in comparison with [^{18}F]FMISO and [^{18}F]FAZA–first small animal PET results. J Pharm Pharm Sci 10:203–211

Révész L, Siracka E, Siracky J et al (1989) Variation of vascular density within and between tumors of the uterine cervix and its predictive value for radiotherapy. Int J Radiat Oncol Biol Phys 16:1161–1163

Ripoll J, Ntziachristos V, Cannet C et al (2008) Investigating pharmacology in vivo using magnetic resonance and optical imaging. Drugs R D 9:277–306

Rischin D, Hicks RJ, Fisher R et al (2006) Prognostic significance of [^{18}F]-misonidazole positron emission tomography-detected tumor hypoxia in patients with advanced head and neck cancer randomly assigned to chemoradiation with or without tirapazamine: a substudy of Trans-Tasman Radiation Oncology Group Study 98.02. J Clin Oncol 24:2098–2104

Rodrigues LM, Howe FA, Griffiths JR et al (2004) Tumor R_2* is a prognostic indicator of acute radiotherapy response in rodent tumors. J Magn Reson Imaging 19:482–488

Rudat V, Stadler P, Becker A et al (2001) Predictive value of the tumor oxygenation by means of pO2 histography in patients with advanced head and neck cancer. Strahlenther Onkol 177:462–468

Schuetz M, Schmid MP, Pötter R et al (2010) Evaluating repetitive 18F-fluoroazomycin-arabinoside (18FAZA) PET in the setting of MRI guided adaptive radiotherapy in cervical cancer. Acta Oncol 49:941–947

Seddon BM, Honess DJ, Vojnovic B et al (2001) Measurement of tumor oxygenation: in vivo comparison of a luminescence fiber-optic sensor and a polarographic electrode in the p22 tumor. Radiat Res 155:837–846

Seddon BM, Payne GS, Simmons L et al (2003) A phase I study of SR-4554 via intravenous administration for noninvasive investigation of tumor hypoxia by magnetic resonance spectroscopy in patients with malignancy. Clin Cancer Res 9:5101–5112

Sørensen BS, Alsner J, Overgaard J et al (2007) Hypoxia induced expression of endogenous markers in vitro is highly influenced by pH. Radiother Oncol 83:362–366

Søvik Å, Malinen E, Olsen DR (2009) Strategies for biologic image-guided dose escalation: a review. Int J Radiat Oncol Biol Phys 73:650–658

Spence AM, Muzi M, Swanson KR et al (2008) Regional hypoxia in glioblastoma multiforme quantified with [18F]fluoromisonidazole positron emission tomography before radiotherapy: correlation with time to progression and survival. Clin Cancer Res 14:2623–2630

Stadler P, Becker A, Feldmann HJ et al (1999) Influence of the hypoxic subvolume on the survival of patients with haed and neck cancer. Int J Radiat Oncol Biol Phys 44:749–754

Star-Lack JM, Adalsteinsson E, Adam MF et al (2000) In vivo ^{1}H MR spectroscopy of human head and neck lymph node metastasis and comparison with oxygen tension measurements. AJNR Am J Neuroradiol 21:183–193

Streffer C, Beuningen D van, Gross E et al (1989) Determination of DNA, micronuclei and vascular density in human rectum carcinomas. In: Chapman JD, Peters LJ, Withers HR (eds) Prediction of tumour treatment response. Pergamon, London

Sundfør K, Lyng H, Rofstad EK (1998) Tumour hypoxia and vascular density as predictors of metastasis in squamous cell carcinoma of the uterine cervix. Br J Cancer 78:822–827

Swanson KR, Chakraborty G, Wang CH et al (2009) Complementary but distinct roles for MRI and ^{18}F-fluoromisonidazole PET in the assessment of human glioblastomas. J Nucl Med 50:36–44

Swartz HM, Khan N, Buckey J et al (2004) Clinical applications of EPR: overview and perspectives. NMR Biomed 17:335–351

Thomlinson RH, Gray LH (1955) The histological structure of some human lung cancers and the possible implications for radiotherapy. Br J Cancer 9:539–549

Thorwarth D, Eschmann SM, Scheiderbauer J et al (2005) Kinetic analysis of dynamic ^{18}F-fluoromisonidazole PET correlates with radiation treatment outcome in head-and-neck cancer. BMC Cancer 5:152

Thorwarth D, Eschmann SM, Holzner F et al (2006) Combined uptake of [^{18}F]FDG and [^{18}F]FMISO correlates with radiation therapy outcome in head-and-neck cancer patients. Radiother Oncol 80:151–156

Toustrup K, Sørensen BS, Nordsmark M et al (2011) Development of a hypoxia gene expression classifier with predictive impact for hypoxic modification of radiotherapy in head and neck cancer. Cancer Res 71:5923–5931

Truong MT, Saito N, Ozonoff A et al (2011) Prediction of locoregional control in head and neck squamous cell carcinoma with serial CT perfusion during radiotherapy. Am J Neuroradiol 32:1195–1201

Urano M, Chen Y, Humm J et al (2002) Measurements of tumor tissue oxygen tension using a time-resolved luminescence-based optical oxylite probe: comparison with a paired survival assay. Radiat Res 158:167–173

Urtasun RC, Parliament MB, McEwan AJ et al (1996) Measurement of hypoxia in human tumours by non-invasive spect imaging of iodoazomycin arabinoside. Br J Cancer 74(Suppl.):S209–S212

van Loon J, Janssen MHM, Öllers M et al (2010) PET imaging of hypoxia using [^{18}F] HX4: a phase I trial. Eur J Nucl Med Mol Imaging 37:1663–1668

Vaupel P (2004) Tumor microenvironmental physiology and its implications for radiation oncology. Sem Radiat Oncol 14:198–206

Vaupel P, Kallinowski F, Okunieff P (1989) Blood flow, oxygen and nutrient supply, and metabolic micro-environment of human tumors: a review. Cancer Res 49:6449–6465

Vävere AL, Lewis JS (2007) Cu-ATSM: a radiopharmaceutical for the PET imaging of hypoxia. Dalton Trans 21:4893–4902

Vera P, Bohn P, Edet-Sanson A et al (2011) Simultaneous positron emission tomography (PET) assessment of metabolism with ^{18}F-fluoro-deoxy-D-glucose (FDG), proliferation with ^{18}F-fluoro-thymidine (FLT), and hypoxia with ^{18}F-fluoro-misonidazole (F-miso) before and during radiotherapy in patients with non-small-cell lung cancer (NSCLC): a pilot study. Radiother Oncol 98:109–116

Vikram DS, Zweier JL, Kuppusamy P (2007) Methods for noninvasive imaging of tissue hypoxia. Antioxid Redox Signal 9:1745–1756

Walenta S, Wetterling M, Lehrke M et al (2000) High lactate levels predict likelihood of metastases, tumor recurrence, and restricted patient survival in human cervical cancers. Cancer Res 60:916–921

Wilson DF, Cerniglia GJ (1992) Localization of tumors and evaluation of their state of oxygenation by phosphorescence imaging. Cancer Res 52:3988–3993

Yaromina A, Quennet V, Zips D et al (2009) Co-localisation of hypoxia and perfusion markers with parameters of glucose metabolism in human squamous cell carcinomas (hSCC) xenografts. Int J Radiat Biol 85:972–980

Yuan H, Schroeder T, Bowsher JE et al (2006) Intertumoral differences in hypoxia selectivity of the PET imaging agent ^{64}Cu(II)-diacetyl-bis(N^4-methylthiosemicarbazone). J Nucl Med 47:989–998

Yue J, Yang Y, Cabrera AR et al (2012) Measuring tumor hypoxia with ^{18}F-FETNIM PET in esophageal squamous cell carcinoma: a pilot clinical study. Dis Esophagus 25:54–61

Zhao D, Jiang L, Hahn EW et al (2005) Tumor physiologic response to combretastatin A4 phosphate assessed by MRI. Int J Radiat Oncol Biol Phys 62:872–880

Zips D, Zöphel K, Abolmaali N et al (2012) Exploratory prospective trial of hypoxia imaging during radiochemotherapy in patients with locally advanced head-and-neck cancer. Radiother Oncol 105:21–28

Part III
Therapy

Chapter 9
Hypoxia, Metastasis, and Antiangiogenic Therapies

Dietmar W. Siemann, Yao Dai and Michael R. Horsman

Abstract It is well known that aberrant microenvironmental conditions in solid tumors negatively impact their response to conventional anticancer therapies. Indeed the presence of hypoxia in the primary tumor has been identified as a prognostic indicator of treatment outcome in a number of clinical settings. Hypoxic conditions can arise at the limits of oxygen diffusion from blood vessels (diffusion or chronic hypoxia) or as a consequence of vascular collapse (perfusion or acute hypoxia). Both types of hypoxia upregulate key signaling pathways associated with metastasis and related tumor cell behavior and function; and although the impact of acute and chronic oxygen deprivation on these pathways may differ markedly, the effect of such exposures is to significantly enhance the metastatic phenotype. In lieu of the critical role of metastasis in treatment failures, there has been significant interest in developing molecular targeting strategies that impair signaling pathways associated with tumor cell dissemination. Many of these targeted tumor cell associated functions are enhanced by hypoxia, including invasion and angiogenesis. Agents targeting angiogenesis are of interest because they could interfere with the progression of established microtumor deposits to macroscopic size by interfering with blood vessel development but they also raise the caveat that the use of such agents might induce hypoxia in the primary tumor which could conceivably enhance the dissemination of tumor cells. This chapter discusses the impact of oxygen deficiencies on the maintenance of stem-like characteristics of tumor initiating cells, the potential interplay between genomic markers and hypoxia/HIFs, and the effect of hypoxia on signaling pathway critical to the metastatic process, including VEGF, c-Src, and the HGF/c-Met axis. It further explores how the interference with signaling molecules may have significant therapeutic potential in the development of novel therapeutic intervention approaches for the treatment of cancer metastasis.

D. W. Siemann (✉) · Y. Dai
Department of Radiation Oncology, University of Florida Shands Cancer Center,
2000 SW Archer Road, Gainesville, FL 32610, USA
e-mail: siemadw@ufl.edu

M. R. Horsman
Department of Experimental Clinical Oncology, Aarhus University Hospital,
Nørrebrogade 44, Bldg. 5, 8000 Aarhus C, Denmark
e-mail: mike@oncology.dk

G. Melillo (ed.), *Hypoxia and Cancer*, Cancer Drug Discovery
and Development, DOI 10.1007/978-1-4614-9167-5_9,

Keywords Tumor-initiating cells (TIC) · Angiogenesis inhibitors · HIF-1 · VEGF · SOX2 · c-Src · c-Met · Acute and chronic hypoxia

9.1 Cancer Metastasis

Despite improvements in early detection, more refined diagnostic modalities, and a better understanding of the natural history of the disease, metastatic cancers remain largely incurable and cause nearly 90 % of all cancer-related deaths. The presence of metastases results in a substantial reduction of the quality of life and the drastic worsening of the prognosis. For example, in prostate cancer, a disease that is highly curable if locally confined, nearly 90 % of cancer patients with advanced diseases develop skeletal metastases, a result which lowers the 5-year survival rate to ~31 % (Jemal et al. 2010). Similarly, the 5-year survival for breast cancer patients drops from 75.8 to 8.3 % for patients with bone metastases (Yong et al. 2011). Given the serious consequences of cancer cell dissemination, novel treatment strategies aimed at disrupting metastasis will be required to improve the survival of cancer patients.

9.2 Hypoxia and Metastasis

The metastatic cascade is a complex process that involves multiple stages, including escape from the primary tumor, survival in the circulation, extravasation at the secondary site, initiation of angiogenesis, and growth to a macroscopic tumor (Gupta and Massague 2006; Steeg 2006; Kirsch et al. 2004). It is now generally recognized that metastasis is not only due to malignant properties of the tumor cells themselves but also a consequence of altered tumor microenvironments. Hypoxia is a common feature of the microenvironment of solid tumors (Brown and Wilson 2004). Its occurrence has long been known to be a detriment to conventional cancer treatment and cytotoxic drug delivery (Wilson and Hay 2011), and it is now also recognized as a key contributor to metastasis (Bristow and Hill 2008). In the case of prostate cancer, several studies have correlated hypoxia with biochemical failure in patients undergoing radiotherapy (Milosevic et al. 2007; Movsas et al. 2002; Vergis et al. 2008; Milosevic et al. 2012). The association between hypoxia and local recurrence and poor prognosis (Movsas et al. 2002; Vergis et al. 2008; Milosevic et al. 2012; Parker et al. 2004) not only supports the role of hypoxia in radioresistance but also suggests its ability to enhance metastatic potential. Of particular interest, hypoxia (Fyles et al. 2002; Pitson et al. 2001) or its associated markers (Vergis et al. 2008; Brennan et al. 2006; Bos et al. 2003) have been identified as independent predictors for disease progression and negative clinical outcomes irrespective of previous treatments, further supporting the notion that deleterious microenvironments impact tumor cell dissemination.

9.2.1 Hypoxia inducible factor 1 (HIF-1)

In preclinical cancer models, hypoxia-associated therapeutic resistance (Hsieh et al. 2010; Sullivan et al. 2008; Frederiksen et al. 2003) and metastatic phenotypes (Bristow and Hill 2008; Dai et al. 2011; Ruan et al. 2009) have been intimately linked to hypoxia-inducible factor 1 (HIF-1) (Semenza 2003), a central controller of gene expression and cellular behavior in response to low oxygen exposures. In patients, HIF-1 has been associated with aggressiveness of disease and increased mortality in several cancer types (Semenza 2003). HIF-1α has also been suggested as a possible marker for premalignant lesions (Zhong et al. 1999) since hypoxia is considered an early event in tumorigenesis (Pipinikas et al. 2008). Hypoxia triggers expression of >800 human genes, most of which are driven by HIF-1 (Semenza 2012). Under normoxic conditions, HIF-1α is rapidly degraded by hydroxylation by prolyl hydroxylase domain protein (PHD) followed by binding to the von Hippel–Lindau protein (VHL). However, when oxygen is deprived, HIF-1α is stabilized and forms a heterodimer with HIF-1β to functionally promote transcription of an array of genes involved in angiogenesis, metabolism, survival, and invasion, such as carbonic anhydrase-9 (CA-9), vascular endothelial growth factor (VEGF), and lysyl oxidase (LOX) (Fig. 9.1a; Dai et al. 2011). In some cancer types, HIF-1 can also be activated in an oxygen-independent manner as gain of function for oncogenes (such as epithelial growth factor receptor, Src, and Akt) or loss of function for tumor suppressor genes (PTEN, VHL) (Giaccia et al. 2003). In prostate cancer for example, HIF-1α is frequently overexpressed due to PTEN loss which leads to hyperactivation of the Akt/mTOR pathway which promotes HIF-1 signaling (Hudson et al. 2002).

9.2.2 Hypoxia Modulates Metastatic Phenotypes

In support of clinical observations that hypoxic tumor cells have a higher tendency to metastasize (Fyles et al. 2006), preclinical findings have demonstrated that hypoxic exposure elevates tumor metastases formation in experimental and spontaneous metastasis models (Zhang and Hill 2007). Indeed, HIF-1-induced gene products are widely involved in metastasis-associated functions, such as invasion and angiogenesis (Gort et al. 2008; Sullivan and Graham 2007; Chan and Giaccia 2007). Cancer cells exposed to hypoxia also undergo epithelial-to-mesenchymal transition (EMT) which features E-cadherin loss and elevated expression of EMT-associated genes such as SNAIL and SLUG (Mak et al. 2010; Huang et al. 2009).

Hypoxia also increases cancer cell invasiveness by activating several key proteolytic enzymes responsible for extracellular matrix degradation. These include matrix metalloproteinase-2 (MMP-2), urokinase plasminogen activator receptor (uPAR), and LOX, although some of these genes are not purely HIF-1-dependent (Sullivan and Graham 2007). LOX is a typical HIF-1-dependent gene that is able to mediate hypoxia-induced invasion of tumor cells and formation of metastatic lesions in mice (Sion and Fig 2006; Erler et al. 2006). HIF-1α also augments pancreatic cancer cell

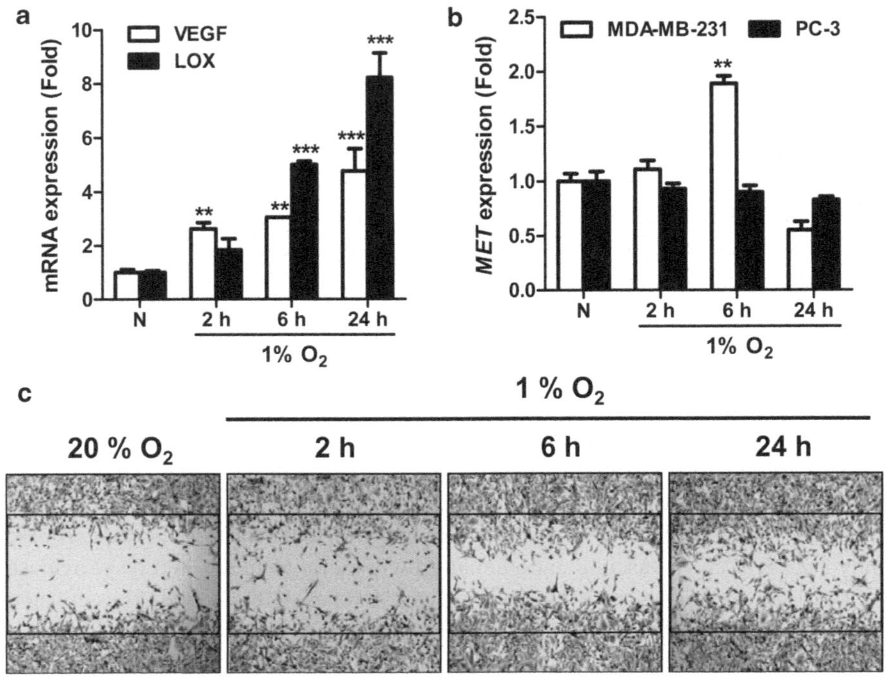

Fig. 9.1 Impact of hypoxic exposures on gene expression and cell function. **a** Oxygen deprivation induces HIF-1-dependent gene expression in MDA-MB-231 breast cancer cells. **b** *MET* gene expression was upregulated by short-term (6 h) hypoxia in MDA-MB-231 cells but not by any exposure time tested in prostate cancer PC-3 cells. Cells were exposed to 1 % O_2 for 2, 6, and 24 h and mRNA levels of VEGF (**a**), LOX (**a**), and *MET* (**b**) were determined by qPCR followed by normalizing to ACTIN. *Columns*, mean; *bars*, SD ($n = 3$). $^{**}p < 0.01$; $^{***}p < 0.001$ vs. normoxia (*N*). **c** MDA-MB-231 cell monolayers were scratched and incubated under hypoxia (1 % O_2) for 2, 6, or 24 h, followed by reculturing under normoxic conditions (20 % O_2) for 22, 18, or 0 h, respectively. Typical photographs are shown (original magnification, X 50)

invasion under hypoxia, by upregulating several MMPs including MMP-2, MMP-7, and membrane type 1 (MT1) -MMP (Ide et al. 2006). Cuvier et al. suggested that transient hypoxia could increase the invasiveness partially mediated by other extracellular proteases—cathepsins of tumor cells resulting in tumor progression (Cuvier et al. 1997).

Angiogenesis is another metastasis-related phenotype that is directly associated with intratumoral hypoxia. VEGF is a major HIF-1-targeted pro-angiogenic gene, which is known to stimulate endothelial cell proliferation, survival, and migration, as well as the disruption of endothelial cell walls leading to increased vessel permeability (Dvorak et al. 1999). In addition to VEGF, Rho family GTPases, Rac1 and Cdc42, may also contribute to hypoxia-mediated angiogenesis by promoting locomotive ability of endothelial cells (Xue et al. 2006). Taken together, hypoxia not only promotes the key steps of the metastatic cascade but also potentiates the aggressive state of tumor cells.

9.2.3 Metastasis-associated Signaling Pathways that are Upregulated by Hypoxia

Hypoxia has been shown to activate many pro-invasive pathways, including the c-Met pathway. c-Met is an oncogene-encoded receptor tyrosine kinase whose cognate ligand is hepatocyte growth factor (HGF). The oncogenic HGF/c-Met axis is widely implicated in various human malignancies and its dysregulation has been correlated with disease progression and poor prognosis in patients (Trusolino et al. 2010; Gentile et al. 2008). *MET* gene-activating mutations associate with extensive metastases (Lorenzato et al. 2002), and the existence of HGF/c-Met autocrine loops in malignant cells can lead to the generation of invasive tumors (Otsuka et al. 1998). Therefore, a strong association exists between dysregulated c-Met signaling and disease progression that renders *MET* a prognostic indicator for cancer progression (Maulik et al. 2002). Hypoxia has been shown to promote motility and invasion in tumor cells by inducing c-Met activity (Pennacchietti et al. 2003; Hara et al. 2006; Knudsen and Vande Woude 2008) and further enhancing HGF-stimulated functional behaviors (Pennacchietti et al. 2003). Since the HGF/c-Met signaling axis upregulates both *MET* gene expression and HIF-1 activity (Sullivan and Graham 2007), amplification of HGF-induced c-Met signaling by hypoxia indicates a positive feedback loop between HIF-1 and c-Met that could drive oxygen-deprived cancer cells toward a more metastatic state. However, it should be noted that direct upregulation of the *MET* gene by oxygen deletion is not a general phenomenon among different cancer cell types (Fig. 9.1b; Toda et al. 2005; Eckerich et al. 2007; Wang et al. 2007).

In addition to c-Met signaling, c-Src is another deregulated oncogenic protein that has been widely accepted as a fundamental cause of metastasis. This non-receptor tyrosine kinase is frequently hyperactivated in primary tumors (Verbeek et al. 1996; Tatarov et al. 2009; Bolen et al. 1987) and has been shown to contribute to increased aggressiveness and adverse outcomes in patients (Summy and Gallick 2003; Guarino 2010). Functionally, c-Src signaling participates in many of the steps of tumor cell dissemination including migration, invasion, survival, and angiogenesis (Fizazi 2007; Finn 2008; Chambers et al. 2002; Kim et al. 2009). Src family kinases transmit signals promoting cell survival and proliferation, and in addition, exert a profound effect on the reorganization of the cytoskeleton and the adhesion systems that underpin cell motility and invasion (Guarino 2010). In the invasion process, c-Src is involved in matrix remodeling and basement membrane degradation (Frame 2002; Allgayer et al. 1999), amplification of focal adhesion kinase (FAK) activity (Guarino 2010; Schlaepfer and Mitra 2004), and regulation of MMPs (Guarino 2010). Src also promotes angiogenesis by modulating VEGF expression in tumor cells (Kim et al. 2009; Park et al. 2007) and enhancing endothelial cell functions (Park et al. 2007). In bone metastases, aberrant c-Src activity is a major contributor to bone resorption and morbidity (Aleshin and Finn 2010; Myoui et al. 2003). Oxygen deprivation has been shown to promote c-Src activation in cancer cells (Lluis et al. 2007; Mukhopadhyay et al. 1995) and c-Src protein levels have been observed to be higher in chronically hypoxic regions of tumors (Pham et al. 2009). Furthermore, Src-dependent hypoxia-induced VEGF expression may be regulated by HIF-1α (Gray et al. 2005). Although c-Src has been shown to be phosphorylated by oxygen deletion, the exact mechanism of hypoxia-mediated Src activation remains to be elucidated.

9.3 Acute Versus Chronic Hypoxia

Two types of hypoxia exist in solid tumors (Brown and Wilson 2004; Bristow and Hill 2008). Tumor cells at the limits of oxygen diffusion from blood vessels may experience chronic or "diffusion-limited" hypoxia. Such hypoxic conditions usually last for relatively long periods of hours or days (Brown and Wilson 2004). In contrast, tumor cells exposed to transient hypoxia as a consequence of intermittent blood flow fluctuations are considered to experience acute or "perfusion-limited" hypoxia (Brown and Wilson 2004) typically characterized by relatively short hypoxic–oxic cycles lasting from minutes to hours (Kimura et al. 1996). It is likely that there is no discrete temporal boundary between these two states in the tumor, rather it is a continuum (Chaudary and Hill 2007).

The relative contributions of acute versus chronic hypoxia on cancer progression and therapy resistance remain controversial. Some studies suggest chronic hypoxia to be the key determinant (Thews et al. 2004; Alqawi et al. 2007; Yamasaki et al. 2012), while others support acute or cycling hypoxia as equally or more important in mediating angiogenic and metastatic phenotypes (Rofstad et al. 2007; Cairns et al. 2001; Cairns and Hill 2004; Rofstad et al. 2010; Dewhirst et al. 2008). For example, breast cancer cells that have experienced both short (6 h) and prolonged (24 h) periods of oxygen depletion yield enhanced locomotive ability (Fig. 9.1c). Also, one study has proposed that long-term hypoxia may achieve a more aggressive phenotype than short-term hypoxia (Alqawi et al. 2007), whereas others reported that chronic hypoxic exposure resulted in reduced invasion (Ackerstaff et al. 2007) and increased apoptosis (McKenzie et al. 2008). Furthermore, mice exposed to cyclic short-term intervals of low oxygen presented with an increased frequency of lung metastases (Cairns et al. 2001; Hill et al. 2001). In terms of therapeutic resistance, both long-term and acutely hypoxic cells demonstrate increased resistance to ionizing radiation and chemotherapy (Dewhirst et al. 2008; Vordermark et al. 2004; Rofstad et al. 2000; Chan et al. 2007; Chou et al. 2012; Horsman and Overgaard 1992).

9.4 Influence of Hypoxia on Tumor-initiating Cells

Recent evidence indicates that hypoxia not only facilitates metastasis-associated functions but can also contribute to stem cell maintenance (Mohyeldin et al. 2010). Cancer stem cells or tumor-initiating cells (TICs) are a subpopulation of tumor cells that selectively possess tumor initiation and self-renewal capacity and the ability to give rise to bulk populations of non-tumorigenic cancer cell progeny through differentiation (Jordan et al. 2006; Gupta et al. 2009). Such cells were initially described in leukemia (Lapidot et al. 1994) but subsequently have been identified in a variety of solid tumors (Patrawala et al. 2007). Further, it has been suggested that TICs exist in hypoxic regions of tumors (Baumann et al. 2008) and low oxygen tensions drive and maintain the stemness phenotype (Li et al. 2009; Liang et al. 2012), majorly mediated by HIF-2α, another hypoxia-regulated transcriptional factor that shares a similar sequence and regulatory machinery with HIF-1α (Li et al. 2009; Keith and Simon 2007). HIF-2α drives several key genes associated with stem cell self-renewal and multipotency (Li et al. 2009; Covello et al. 2006), including Oct3/4 (Liang et al. 2012), SOX2 (Mathieu et al. 2011), and Nanog (Ma et al. 2011). These core factors are recognized to sustain TIC pluripotency and self-renewal (Takahashi et al. 2007) that play essential roles in tumor initiation and progression. Interestingly, non-TIC cells that have experienced oxygen depletion may gain stem-like characteristics (Li et al. 2009; Heddleston et al. 2009), suggesting a determining role for hypoxia in modulating non-TIC and TIC populations in tumors. The alteration of expression of stem cell markers by hypoxia implies an adaptive response of tumor cells to aberrant tumor microenvironments and reveals a novel mechanism by which cells acquire a more aggressive behavior and/or stem cell-like characteristics after exposure to low oxygen tensions. Indeed, hypoxia can selectively enhance the metastatic ability of TIC (Salnikov et al. 2012; Li et al. 2012), suggesting a positive environmental regulation of stem cell-like subpopulations.

9.5 Targeting the Invasive Metastatic Phenotype

Intervention strategies designed to target key signaling pathways associated with metastasis have been intensively studied in preclinical models, and lead agents are in clinical development in a number of solid tumor settings. Among the pathways associated with invasive phenotypes, c-Met and c-Src signaling have been most extensively targeted by small molecule inhibitors. In the case of the HGF/c-Met signaling axis, therapeutic strategies include antagonistic compounds, monoclonal antibodies, and small-molecule kinase inhibitors (Cecchi et al. 2010). Although neutralizing antibodies targeting either HGF or c-Met have shown promising effects, Met kinase inhibitors are proving to be effective in cancers driven by both the receptor and the ligand (Qian et al. 2009), hence perhaps exhibiting a broader range of application. While clinical studies with such agents have focused on primary tumor growth inhibition, preclinical investigations have indicated a possible role for them as antimetastatic agents. BMS-777607 (ASLAN002) is one of the potent Met kinase inhibitors under clinical evaluation that not only delays xenografted tumor growth (Schroeder et al. 2009) but also inhibits various metastasis-related functions in HGF-dependent (Fig. 9.2a; Dai and Siemann 2010) and -independent manners in vitro (Dai and Siemann 2010, 2012; Dai et al. 2012) and impedes the formation of pulmonary metastases in vivo (Dai and Siemann 2012).

Increasing evidence also suggests c-Src as a suitable target for therapeutic interventions of tumor cell dissemination. As single agents, Src inhibitors are undergoing investigation in many tumor types. Currently, three small-molecule adenosine triphosphate (ATP) competitors, dasatinib, saracatinib, and bosutinib, are being actively evaluated in a variety of solid tumor clinical trials (Kim et al. 2009). Preclinical studies indicate that Src kinase inhibitors decrease not only tumor growth but also the development of metastases (Rucci et al. 2006; Nam et al. 2002). Dasatinib inhibits

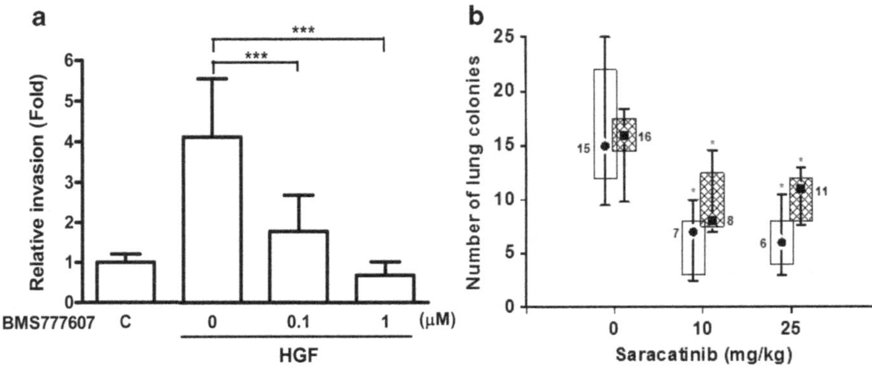

Fig. 9.2 Therapeutic effects of small-molecule inhibitors targeting metastasis-associated pathways. **a** The c-Met inhibitor BMS-777607 inhibits HGF-mediated cell invasion. PC-3 prostate cancer cells were seeded into Matrigel-coated transwell inserts in the presence or absence of BMS-777607. HGF (25 ng/ml) was loaded on the bottom chamber as a chemo-attractant. Invaded cells were scored 24 h after cell seeding. Data are shown as mean ± SD ($n = 3$); ***$p < 0.001$; $C =$ control. **b** The c-Src inhibitor saracatinib suppresses experimental lung metastases. KHT murine sarcoma cells were injected intravenously. Beginning after cell injection, the mice were treated daily with saracatinib or vehicle control for 21 days. Data represent two independent experiments. The number shown indicates median. *$p < 0.05$ vs. vehicle control (Wilcoxon rank-sum test). (Data are adapted from Siemann et al. (125) (**a**) and from Dai et al. (109) (**b**))

cell survival (Tsao et al. 2007; Johnson et al. 2005; Rice et al. 2012), migration (Tsao et al. 2007; Rice et al. 2012; Nautiyal et al. 2009; Morton et al. 2010), and invasion (Tsao et al. 2007; Johnson et al. 2005; Rice et al. 2012; Nautiyal et al. 2009; Morton et al. 2010; Nam et al. 2005) in vitro and inhibits lymph node (Park et al. 2008), lung (Fraser et al. 2010), and bone (Koreckij et al. 2009) metastases in vivo. Saracatinib (Yang et al. 2010; Dong et al. 2010; Siemann et al. 2012; Hannon et al. 2012) and bosutinib (Rabbani et al. 2010; Vultur et al. 2008; Jallal et al. 2007; Golas et al. 2005) also suppress the metastasis-associated tumor cell functions in vitro and metastases formation in vivo. For example, saracatinib significantly suppresses secondary lung nodules in a murine fibrosarcoma model (Fig. 9.2b). Moreover, Src inhibitors have been shown to significantly reduce tumor cell-induced blood vessel formation (Rice et al. 2012; Nautiyal et al. 2009), thus raising the possibility that such agents are capable of impacting two key steps in the metastatic process, namely invasion and angiogenesis.

9.6 Antiangiogenic Therapies and Metastasis

Angiogenesis contributes to tumor cell dissemination both in the primary tumor and at the site of secondary tumor growth. In the primary tumor, the expanding blood vessel network, which is typically a disorganized labyrinth of immature, tortuous vessels which lack conventional blood vessel hierarchy (Konerding and Fait 2001),

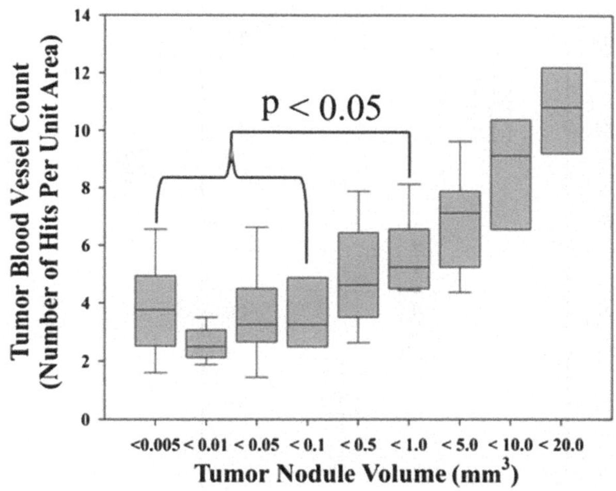

Fig. 9.3 Relationship between lung colony growth and vascular development. Progressive growth of tumor cells seeding to the lungs coincides with the development of an expanding blood vessel network. At various times after cell injection, lungs were removed and serial sections were taken throughout the lobes for morphometric analysis. Tumor volumes were calculated using the adapted "Prolate Spheroid" model from measurements of the same nodule on two H&E-stained sections that were a known distance apart. Sections adjacent to those used for size determination were stained with CD31 to identify blood vessels associated with each nodule. Vascularity, determined by the Chalkley counting method, is shown in SCCVII tumor lung nodules as a function of size ($n = 156$). Data represent the median (*line*), 25–75 % (*box*) and 10–90 % (*error bars*) values. (Data are adapted from Salmon et al. 2012)

offers a means of escape for neoplastic cells. Furthermore, tumor blood vessels are of inconsistent diameter and uneven shape with abnormal bulges and blind ends, arteriolar–venous shunts, and plasma channels lacking red blood cells (Konerding and Fait 2001). Their irregular shape and diameter coupled with their immature nature lead to impaired blood flow and a diminished ability to deliver nutrients, and consequently, to areas within the tumor that are oxygen deprived (hypoxic) and nutrient deficient (Konerding and Fait 2001; Vaupel 2004). These aberrant microenvironments enhance the metastatic phenotype through a variety of key signaling pathways described above. At the secondary site, the initiation of angiogenesis is essential to the initiation and progressive growth of a developing metastasis. This can be readily demonstrated by following the development of metastatic lung lesions and their associated vasculature in mice injected with tumor cells via the lateral tail vein. The results (Salmon et al. 2012) showed that while lung metastases < 0.5 mm^3 can rely on preexisting lung vasculature, nodule growth beyond this size requires the induction of new tumor vessels (Fig. 9.3).

The observation that tumors cannot grow beyond a size of a few millimeters without the support of neovascularization (Folkman 1971), coupled with the recognition of the impact of the primary tumor's vascular network on the metastatic phenotype, has led to the clinical development of a plethora of angiogenesis-inhibiting agents (Carmeliet and Jain 2000; Hicklin and Ellis 2005). VEGF has emerged as primary

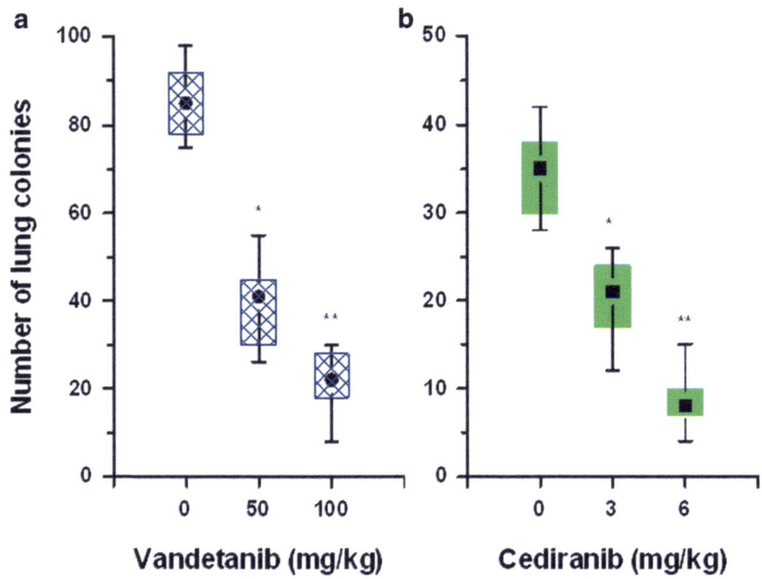

Fig. 9.4 Impact of anti-VEGF therapy on lung colony formation. Selective VEGFR2-associated tyrosine kinase inhibition impairs the ability of blood-borne tumor cells to form lung colonies. Immediately following IV injection of KHT sarcoma cells, mice were treated with once-daily doses of vandetanib (*left*) or cediranib (*right*). Tumor foci on the fixed lung tissues were scored 3 weeks later. Data points, median ($n=9$). $^*p<0.05$; $^{**}p<0.01$. (Wilcoxon rank-sum test)

target due to its critical role in the angiogenic process (Hicklin and Ellis 2005; Ferrara and Kerbel 2005). A wide variety of strategies have been devised to interfere with the VEGF signaling cascade typically focused on targeting the VEGF protein or the VEGF family receptors (Hicklin and Ellis 2005). Indeed, the combination of the VEGF targeting monoclonal antibody bevacizumab with chemotherapy provided the first clinical validation of the value of antiangiogenic cancer therapy (Hurwitz et al. 2005). A great number of preclinical studies subsequently have shown that antiangiogenic therapies applied alone or in conjunction with conventional anticancer treatments, including chemotherapy or radiotherapy, result in improved tumor responses (Siemann and Horsman 2009; Eskens 2004; Horsman and Siemann 2006; Mazeron et al. 2011). Targeting the angiogenesis-driven sprouting of new vessels alone or in combination with cytotoxic cancer therapies also has been shown to yield significant inhibition of tumor metastases (Gerber and Ferrara 2005; Kosharskyy et al. 2006). For example, daily treatment with the VEGFR2 tyrosine kinase inhibitors vandetanib or cediranib significantly impaired the ability of tumor cells to establish lung metastases (Fig. 9.4). However, reductions in metastases have not always been observed. Most disconcertingly, there have also been reports that VEGF-targeted drugs may promote tumor invasiveness and metastasis (Ebos et al. 2009; Paez-Ribes et al. 2009). The mechanism responsible for this conflicting observation has typically been ascribed to angiogenesis inhibitor (AI)-induced tumor hypoxia.

9.6.1 Antiangiogenic Therapies and Tumor Oxygenation

The effects of AI treatments on the tumor microenvironment are complex and controversial. One train of thought is that targeting angiogenesis could reduce or abolish vascular abnormalities leading to vessel stabilization/normalization and a more efficient vasculature similar to that seen in normal tissues (Jain 2001). As a consequence, AI treatment would lead to improved oxygen delivery and a reduction in tumor hypoxia. And indeed, several studies have reported that treatment with a range of AIs can give rise to a decrease in tumor hypoxia (Table 9.1). Typically, the improvement in oxygenation associated with vessel normalization only lasted for a few days despite the drug treatment being continued. Such a short "normalization window" has been seen in studies with a variety of AIs including DC101 (Winkler et al. 2004), thalidomide (Ansiaux et al. 2005; Segers et al. 2006), SU5416 (Ansiaux et al. 2006), a nucleolin antagonist (Fogal et al. 2009), and bevacizumab (Dings et al. 2007; McGee et al. 2010; Myers et al. 2010; Vangestel et al. 2012), regardless of the technique used to monitor the changes in oxygenation. The transient nature of the normalization suggests that unless the timing of measurements is optimal, the improvement in oxygenation by AIs could easily be missed. This importance of timing is further stressed by two studies in which both a decrease and an increase in hypoxia were noted depending on the time of measurement after DC101 (Hansen-Algenstaedt et al. 2000) or bevacizumab (Dings et al. 2007) therapy. Investigators also have suggested that anti-VEGF agents could improve tumor oxygenation not by vascular remodeling but by decreasing tumor cell respiration (Ansiaux et al. 2006, 2009). However, other findings contradict the importance of oxygen consumption (Hansen-Algenstaedt et al. 2000).

Several studies have reported no change in tumor oxygenation following AI exposure (Table 9.1), even though such treatments led to growth inhibitions (Lee et al. 2000; Gong et al. 2003; Williams et al. 2004) as well as reductions in tumor vascularity (Lee et al. 2000; Gong et al. 2003) and tumor perfusion (Gong et al. 2003).

Finally, antiangiogenic therapy also has the potential to increase tumor hypoxia. Table 9.1 lists several studies with a variety of AIs were reported to decrease tumor oxygenation as measured using polarographic oxygen electrodes (Gong et al. 2003; Ding et al. 2003), hypoxic cell markers (Paez-Ribes et al. 2009; Vangestel et al. 2012; Fenton et al. 2003, 2004, 2005; Franco et al. 2006; Pore et al. 2006; Tailor et al, 2010; Maione et al. 2012; Ou et al. 2009; Riesterer et al. 2006), or classical radiation response assays (Murata et al. 1997; Leith et al. 1992). In addition, some reports indicate that AIs could alter tumor blood perfusion (Williams et al. 2004; Griffin et al. 2002; Keunen et al. 2011) which, while not a direct measure of oxygenation, reflects such changes.

Taken together, the data illustrated in Table 9.1 indicate that the results of the AI treatment on tumor oxygenation are highly variable. Interestingly, many agents can both increase and decrease tumor hypoxia. These include TNP470, Suramin, DC101, endostatin, bevacizumab, and sunitinib. It could be argued that these differences are the result of different drug doses and treatment schedules, or the time of hypoxia assessment. However, given that the same agent, treatment, and assessment schedule could lead to improved oxygenation (Winkler et al. 2004) and increased hypoxia (Fenton et al. 2004) it strongly suggests that the effects are tumor-dependent phenomena.

Overall, we can conclude that the effect of AI treatment on tumor oxygenation status in animal models can be highly variable with 18 studies showing a decrease in hypoxia, 5 having no effect, and 19 reporting an increase in hypoxia (Table 9.1). Clinical data on this subject are far more limited, but the same controversy exists with both a decrease (Hugonnet et al. 2011) and an increase (Hattingen et al. 2011) in tumor hypoxia having been reported. Clearly, no generalized statements concerning the effects of antiangiogenic therapy on tumor hypoxia can be made.

9.7 Conclusions

In the metastatic setting, conventional anticancer therapies offer only modest benefits and are primarily palliative. Clearly, a better understanding of the mechanisms underlying the metastatic process will aid the development of improved therapeutic interventions. Hypoxia, a common feature of most solid tumors, has long been associated with enhanced metastasis and poor patient outcome. Hypoxia arises in tumors because the aggressive growth of neoplastic cells and associated over-expression of pro-angiogenic factors leads to aberrant vascular networks that are incapable of adequately delivering nutrients and removing waste products. Tumor cells that experience oxygen deprivation display a more metastatic phenotype. Evidence also indicates that such oxygen deficiencies aid the maintenance of stem-like characteristics of TICs, cells considered to be critically important for tumor proliferation, invasion, and metastasis. Our studies have recently demonstrated a correlation between the TIC marker, SOX2, and HIF1, suggesting the potential interplay between this genomic marker and hypoxia/HIFs. Hypoxia/HIFs also impact other signaling pathways critical to the metastatic process, including VEGF, c-Src, and the HGF/c-Met axis. The deregulated activation of these signaling pathways orchestrates a wide variety of cellular functions associated with tumor cell dissemination. Thus, the combined interference with key signaling molecules should maximize the therapeutic potential against cancer metastasis. We believe that treatment options that interfere with the process of metastasis will have a significant impact on cancer therapy outcomes and related mortality. Furthermore, gaining a better understanding of how the varying oxygen deprivations that can occur in solid tumors impact metastasis-associated signaling and behavior should, in the future, lead to the development of novel treatment options.

Table 9.1 Effect of angiogenesis-inhibiting agents on the oxygenation status of tumors

Inhibitor	Tumor type	Assay[a]	Hypoxia[b]	Reference
TNP-470	9 L rat gliosarcoma	Eppendorf (pO_2)	Decreased	Teicher et al. 1995
Suramin	E106 human glioblastoma	Hypoxic marker (PIMO)	Decreased	Bernsen et al. 1999
Anti-VEGF antibody	U87 glioblastoma	Eppendorf (pO_2)	Decreased	Lee et al. 2000
Endostatin	MCa4 mammary carcinoma	Hypoxic marker (EF5)	Decreased	Fenton et al. 2003
Thrombospondin	D-12 human melanoma	Radiation (SF/TG)	Decreased	Rofstad et al. 2003
DC101	U87 glioma	Hypoxic marker (PIMO)	Decreased	Winkler et al. 2004
Thalidomide	FSaII fibrosarcoma	EPR oximetry	Decreased	Ansiaux et al. 2005
Thalidomide	TLT mouse liver tumor	EPR oximetry	Decreased	Segers et al. 2006
SU5416	TLT mouse liver tumor/FSaII	EPR oximetry	Decreased	Ansiaux et al. 2006
Vandetanib (ZD6474)	TLT mouse liver tumor	EPR oximetry	Decreased	Ansiaux et al. 2009
Nucleolin antagonist	MDA-MB-435 mammary	Hypoxic marker (PIMO)	Decreased	Fogal et al. 2009
Bevacizumab	U87 glioblastoma	Hypoxic marker (PIMO)	Decreased	McGee et al. 2010
Bevacizumab	Alveolar rhabdomyosarcoma	Oxylite	Decreased	Myers et al. 2010
Bevacizumab	OVCAR-3 ovarian carcinoma	Hypoxic marker (PIMO)	Decreased	Martinez-Poveda et al. 2011
Sunitinib	C6 glioma	FMISO-PET	Decreased	Valable et al. 2011
Sunitinib	SCCVII	EPR oximetry	Decreased	Matsumoto et al. 2011
Bevacizumab	MA148 ovarian/B16F10 melanoma/SCK mammary	Eppendorf (pO_2)	Decreased and Increased	Dings et al. 2007
DC101	Shionogi mammary carcinoma	Phosphorous quenching	Increased and Decreased	Hansen-Algenstaedt et al. 2000
Suramin	DLD-2 human colon	Radiation (SF)	Increased	Leith et al. 1992
TNP-470	C3H mammary carcinoma	Radiation (TC)	Increased	Murata et al. 1997
DC101	WAC2 neuroblastoma	Eppendorf (pO_2)	Increased	Gong et al. 2003
DC101	MCa4/MCa35 mammary ca.	Hypoxic marker (EF5)	Increased	Fenton et al. 2004
Anginex	SCK mammary carcinoma	Eppendorf (pO_2)	Increased	Dings et al. 2005
DC101	Spontaneous mammary ca.	Hypoxic marker (EF5)	Increased	Fenton et al. 2004

Table 9.1 (continued)

Inhibitor	Tumor type	Assay[a]	Hypoxia[b]	Reference
Endostatin	Spontaneous mammary ca.	Hypoxic marker (EF5)	Increased	Fenton et al. 2004
DC101	MDA-MB-435	Hypoxic marker (PIMO)	Increased	Franco et al. 2006
Nelfinavir	TLT mouse liver tumor	Hypoxic marker (EF5)	Increased	Pore et al. 2006
PTK787/ZK222584	Murine mammary carcinoma	FMISO-PET	Increased	Riesterer et al. 2006
AG-013736	DU145 human prostate	Hypoxic marker (EF5)	Increased	Fenton and Paoni 2007
Bevacizumab	NCI-H441 human lung adenoca.	Hypoxic marker (PIMO)	Increased	Ou et al. 2009
DC101	Pancreatic neuroendocrine	Hypoxic marker (PIMO)	Increased	Paez-Ribes et al. 2009
Sunitinib	Pancreatic neuroendocrine	Hypoxic marker (PIMO)	Increased	Paez-Ribes et al. 2009
Pazopanib	A549 human NSCLC	Hypoxic marker (EF5)	Increased	Tailor et al. 2010
Bevacizumab	Colo 205	Hypoxic marker (PIMO)	Increased	Vangestel et al. 2012
DC101	Cervical cancer	Hypoxic marker (PIMO)	Increased	Maione et al. 2012
Sunitinib	Cervical cancer	Hypoxic marker (PIMO)	Increased	Maione et al. 2012
SU6668	FSaII + SCK + CFPAC tumors	Perfusion (Rb)	"Increased"	Griffin et al. 2002
Vandetanib/ZD6474	CaLu 6 NSCLC	Perfusion (H.33342)	"Increased"	Williams et al. 2004
Bevacizumab	Orthotopic glioblastoma spheroids	MRS	"Increased"	Keunen et al. 2011
Anti-VEGF antibody	LS1747 colon adenocarcinoma	Eppendorf (pO_2)	No change	Lee et al. 2000
DC101	54A SCLC	Eppendorf (pO_2)/radiation (TG)	No change	Kozin et al. 2001
Arginine deiminase	WAC2 neuroblastoma	Eppendorf (pO_2)	No change	Gong et al. 2003
Endostatin	MCa35 mammary carcinoma	Hypoxic marker (EF5)	No change	Fenton et al. 2003
SU5416	E106 human glioblastoma	Hypoxic marker (PIMO)	No change	Schuuring et al. 2005

[a] Assays were radiation response measured using the paired survival curve (SF), clamped tumor control (TC), or clamped tumor growth (TG); Eppendorf polarographic oxygen electrode; Oxylite oxygen electrode; hypoxic markers, pentafluorinated derivative of etanidazole (EF5), or pimonidazole (PIMO); phosphorous quenching imaging; electronic paramagnetic resonance (EPR) oximetry; 18F-labeled misonidazole-positron emission spectroscopy (FMISO-PET); magnetic resonance spectroscopy (MRS); and blood perfusion measured using either RbCl uptake (Rb) or Hoechst 33342 (H.33342) staining.

[b] Tumor hypoxia was decreased, increased, or unchanged (no change); the use of the term "Increased" indicates that tumor hypoxia was not directly measured directly, but what was to be expected based on the changes in perfusion and MRS obtained.

References

Ackerstaff E, Artemov D, Gillies RJ, Bhujwalla ZM (2007) Hypoxia and the presence of human vascular endothelial cells affect prostate cancer cell invasion and metabolism. Neoplasia 9(12):1138–1151

Aleshin A, Finn RS (2010) SRC: a century of science brought to the clinic. Neoplasia 12(8):599–607

Allgayer H, Wang H, Gallick GE et al (1999) Transcriptional induction of the urokinase receptor gene by a constitutively active Src. Requirement of an upstream motif (-152/-135) bound with Sp1. J Biol Chem 274(26):18428–18437

Alqawi O, Wang HP, Espiritu M, Singh G (2007) Chronic hypoxia promotes an aggressive phenotype in rat prostate cancer cells. Free Radic Res 41(7):788–797

Ansiaux R, Baudelet C, Jordan BF et al (2005) Thalidomide radiosensitizes tumors through early changes in the tumor microenvironment. Clin Cancer Res 11(2 Pt 1):743–750

Ansiaux R, Baudelet C, Jordan BF et al (2006) Mechanism of reoxygenation after antiangiogenic therapy using SU5416 and its importance for guiding combined antitumor therapy. Cancer Res 66(19):9698–9704

Ansiaux R, Dewever J, Gregoire V, Feron O, Jordan BF, Gallez B (2009) Decrease in tumor cell oxygen consumption after treatment with vandetanib (ZACTIMA; ZD6474) and its effect on response to radiotherapy. Radiat Res 172(5):584–591

Baumann M, Krause M, Hill R (2008) Exploring the role of cancer stem cells in radioresistance. Nat Rev Cancer 8(7):545–554

Bernsen HJ, Rijken PF, Peters JP et al (1999) Suramin treatment of human glioma xenografts; effects on tumor vasculature and oxygenation status. J Neurooncol 44(2):129–136

Bolen JB, Veillette A, Schwartz AM, DeSeau V, Rosen N (1987) Activation of pp60c-src protein kinase activity in human colon carcinoma. Proc Natl Acad Sci USA 84(8):2251–2255

Bos R, van der Groep P, Greijer AE et al (2003) Levels of hypoxia-inducible factor-1alpha independently predict prognosis in patients with lymph node negative breast carcinoma. Cancer 97(6):1573–1581

Brennan DJ, Jirstrom K, Kronblad A et al (2006) CA IX is an independent prognostic marker in premenopausal breast cancer patients with one to three positive lymph nodes and a putative marker of radiation resistance. Clin Cancer Res 12(21):6421–6431

Bristow RG, Hill RP (2008) Hypoxia and metabolism. Hypoxia, DNA repair and genetic instability. Nat Rev Cancer 8(3):180–192

Brown JM, Wilson WR (2004) Exploiting tumour hypoxia in cancer treatment. Nat Rev Cancer 4(6):437–447

Cairns RA, Hill RP (2004) Acute hypoxia enhances spontaneous lymph node metastasis in an orthotopic murine model of human cervical carcinoma. Cancer Res 64(6):2054–2061

Cairns RA, Kalliomaki T, Hill RP (2001) Acute (cyclic) hypoxia enhances spontaneous metastasis of KHT murine tumors. Cancer Res 61(24):8903–8908

Carmeliet P, Jain RK (2000) Angiogenesis in cancer and other diseases. Nature 407(6801):249–257

Cecchi F, Rabe DC, Bottaro DP (2010) Targeting the HGF/Met signalling pathway in cancer. Eur J Cancer 46(7):1260–1270

Chambers AF, Groom AC, MacDonald IC (2002) Dissemination and growth of cancer cells in metastatic sites. Nat Rev Cancer 2(8):563–572

Chan DA, Giaccia AJ (2007) Hypoxia, gene expression, and metastasis. Cancer Metastasis Rev 26(2):333–339

Chan N, Milosevic M, Bristow RG (2007) Tumor hypoxia, DNA repair and prostate cancer progression: new targets and new therapies. Future Oncol 3(3):329–341

Chaudary N, Hill RP (2007) Hypoxia and metastasis. Clin Cancer Res 13(7):1947–1949

Chou CW, Wang CC, Wu CP et al (2012) Tumor cycling hypoxia induces chemoresistance in glioblastoma multiforme by upregulating the expression and function of ABCB1. Neuro Oncol 14(10):1227–1238

Covello KL, Kehler J, Yu H et al (2006) HIF-2alpha regulates Oct-4: effects of hypoxia on stem cell function, embryonic development, and tumor growth. Genes Dev 20(5):557–570
Cuvier C, Jang A, Hill RP (1997) Exposure to hypoxia, glucose starvation and acidosis: effect on invasive capacity of murine tumor cells and correlation with cathepsin (L + B) secretion. Clin Exp Metastasis 15(1):19–25
Dai Y, Siemann DW (2010) BMS-777607, a small-molecule met kinase inhibitor, suppresses hepatocyte growth factor-stimulated prostate cancer metastatic phenotype in vitro. Mol Cancer Ther 9(6):1554–1561
Dai Y, Siemann DW (2012) Constitutively active c-Met kinase in PC-3 cells is autocrine-independent and can be blocked by the Met kinase inhibitor BMS-777607. BMC Cancer 12:198
Dai Y, Bae K, Siemann DW (2011) Impact of hypoxia on the metastatic potential of human prostate cancer cells. Int J Radiat Oncol Biol Phys 81(2):521–528
Dai Y, Bae K, Pampo C, Siemann DW (2012) Impact of the small molecule Met inhibitor BMS-777607 on the metastatic process in a rodent tumor model with constitutive c-Met activation. Clin Exp Metastasis 29(3):253–261
Dewhirst MW, Cao Y, Moeller B (2008) Cycling hypoxia and free radicals regulate angiogenesis and radiotherapy response. Nat Rev Cancer 8(6):425–437
Dings RP, Williams BW, Song CW, Griffioen AW, Mayo KH, Griffin RJ (2005) Anginex synergizes with radiation therapy to inhibit tumor growth by radiosensitizing endothelial cells. Int J Cancer 115(2):312–319
Dings RP, Loren M, Heun H et al (2007) Scheduling of radiation with angiogenesis inhibitors anginex and Avastin improves therapeutic outcome via vessel normalization. Clin Cancer Res 13(11):3395–3402
Dong M, Rice L, Lepler S, Pampo C, Siemann DW (2010) Impact of the Src inhibitor saracatinib on the metastatic phenotype of a fibrosarcoma (KHT) tumor model. Anticancer Res 30:4405–4414
Dvorak HF, Nagy JA, Feng D, Brown LF, Dvorak AM (1999) Vascular permeability factor/vascular endothelial growth factor and the significance of microvascular hyperpermeability in angiogenesis. Curr Top Microbiol Immunol 237:97–132
Ebos JM, Lee CR, Cruz-Munoz W, Bjarnason GA, Christensen JG, Kerbel RS (2009) Accelerated metastasis after short-term treatment with a potent inhibitor of tumor angiogenesis. Cancer Cell 15(3):232–239
Eckerich C, Zapf S, Fillbrandt R, Loges S, Westphal M, Lamszus K (2007) Hypoxia can induce c-Met expression in glioma cells and enhance SF/HGF-induced cell migration. Int J Cancer 121(2):276–283
Erler JT, Bennewith KL, Nicolau M et al (2006) Lysyl oxidase is essential for hypoxia-induced metastasis. Nature 440(7088):1222–1226
Eskens FA (2004) Angiogenesis inhibitors in clinical development; where are we now and where are we going? Br J Cancer 90(1):1–7
Fenton BM, Paoni SF (2007) The addition of AG-013736 to fractionated radiation improves tumor response without functionally normalizing the tumor vasculature. Cancer Res 67(20):9921–9928
Fenton BM, Paoni SF, Grimwood BG, Ding I (2003) Disparate effects of endostatin on tumor vascular perfusion and hypoxia in two murine mammary carcinomas. Int J Radiat Oncol Biol Phys 57(4):1038–1046
Fenton BM, Paoni SF, Ding I (2004) Pathophysiological effects of vascular endothelial growth factor receptor-2-blocking antibody plus fractionated radiotherapy on murine mammary tumors. Cancer Res 64(16):5712–5719
Fenton BM, Paoni SF, Grimwood B, Ding I (2005) Varied response of spontaneous tumors to antiangiogenic agents. Adv Exp Med Biol 566:59–65
Ferrara N, Kerbel RS (2005) Angiogenesis as a therapeutic target. Nature 438(7070):967–974
Finn RS (2008) Targeting Src in breast cancer. Ann Oncol 19(8):1379–1386
Fizazi K (2007) The role of Src in prostate cancer. Ann Oncol 18(11):1765–1773
Fogal V, Sugahara KN, Ruoslahti E, Christian S (2009) Cell surface nucleolin antagonist causes endothelial cell apoptosis and normalization of tumor vasculature. Angiogenesis 12(1):91–100

Folkman J (1971) Tumor angiogenesis: therapeutic implications. N Engl J Med 285(21):1182–1186
Frame MC (2002) Src in cancer: deregulation and consequences for cell behaviour. Biochim Biophys Acta 1602(2):114–130
Franco M, Man S, Chen L et al (2006) Targeted anti-vascular endothelial growth factor receptor-2 therapy leads to short-term and long-term impairment of vascular function and increase in tumor hypoxia. Cancer Res 66(7):3639–3648
Fraser CK, Lousberg EL, Guerin LR et al (2010) Dasatinib alters the metastatic phenotype of B16-OVA melanoma in vivo. Cancer Biol Ther 10(7):715–727
Frederiksen LJ, Siemens DR, Heaton JP, Maxwell LR, Adams MA, Graham CH (2003) Hypoxia induced resistance to doxorubicin in prostate cancer cells is inhibited by low concentrations of glyceryl trinitrate. J Urol 170(3):1003–1007
Fyles A, Milosevic M, Hedley D et al (2002) Tumor hypoxia has independent predictor impact only in patients with node-negative cervix cancer. J Clin Oncol 20(3):680–687
Fyles A, Milosevic M, Pintilie M et al (2006) Long-term performance of interstial fluid pressure and hypoxia as prognostic factors in cervix cancer. Radiother Oncol 80(2):132–137
Gentile A, Trusolino L, Comoglio PM (2008) The Met tyrosine kinase receptor in development and cancer. Cancer Metastasis Rev 27(1):85–94
Gerber HP, Ferrara N (2005) Pharmacology and pharmacodynamics of bevacizumab as monotherapy or in combination with cytotoxic therapy in preclinical studies. Cancer Res 65(3):671–680
Giaccia A, Siim BG, Johnson RS (2003) HIF-1 as a target for drug development. Nat Rev Drug Discov 2(10):803–811
Golas JM, Lucas J, Etienne C et al (2005) SKI-606, a Src/Abl inhibitor with in vivo activity in colon tumor xenograft models. Cancer Res 65(12):5358–5364
Gong H, Pottgen C, Stuben G, Havers W, Stuschke M, Schweigerer L (2003) Arginine deiminase and other antiangiogenic agents inhibit unfavorable neuroblastoma growth: potentiation by irradiation. Int J Cancer 106(5):723–728
Gort EH, Groot AJ, van der Wall E, van Diest PJ, Vooijs MA (2008) Hypoxic regulation of metastasis via hypoxia-inducible factors. Curr Mol Med 8(1):60–67
Gray MJ, Zhang J, Ellis LM et al (2005) HIF-1alpha, STAT3, CBP/p300 and Ref-1/APE are components of a transcriptional complex that regulates Src-dependent hypoxia-induced expression of VEGF in pancreatic and prostate carcinomas. Oncogene 24(19):3110–3120
Griffin RJ, Williams BW, Wild R, Cherrington JM, Park H, Song CW (2002) Simultaneous inhibition of the receptor kinase activity of vascular endothelial, fibroblast, and platelet-derived growth factors suppresses tumor growth and enhances tumor radiation response. Cancer Res 62(6):1702–1706
Guarino M (2010) Src signaling in cancer invasion. J Cell Physiol 223(1):14–26
Gupta GP, Massague J (2006) Cancer metastasis: building a framework. Cell 127(4):679–695
Gupta PB, Chaffer CL, Weinberg RA (2009) Cancer stem cells: mirage or reality? Nat Med 15(9):1010–1012
Hannon RA, Finkelman RD, Clack G et al (2012) Effects of Src kinase inhibition by saracatinib (AZD0530) on bone turnover in advanced malignancy in a Phase I study. Bone 50(4):885–892
Hansen-Algenstaedt N, Stoll BR, Padera TP et al (2000) Tumor oxygenation in hormone-dependent tumors during vascular endothelial growth factor receptor-2 blockade, hormone ablation, and chemotherapy. Cancer Res 60(16):4556–4560
Hara S, Nakashiro K, Klosek SK, Ishikawa T, Shintani S, Hamakawa H (2006) Hypoxia enhances c-Met/HGF receptor expression and signaling by activating HIF-1alpha in human salivary gland cancer cells. Oral Oncol 42(6):593–598
Hattingen E, Jurcoane A, Bahr O et al (2011) Bevacizumab impairs oxidative energy metabolism and shows antitumoral effects in recurrent glioblastomas: a 31P/1H MRSI and quantitative magnetic resonance imaging study. Neuro Oncol 13(12):1349–1363
Heddleston JM, Li Z, McLendon RE, Hjelmeland AB, Rich JN (2009) The hypoxic microenvironment maintains glioblastoma stem cells and promotes reprogramming towards a cancer stem cell phenotype. Cell Cycle 8(20):3274–3284

Hicklin DJ, Ellis LM (2005) Role of the vascular endothelial growth factor pathway in tumor growth and angiogenesis. J Clin Oncol 23(5):1011–1027
Hill RP, De Jaeger K, Jang A, Cairns R (2001) pH, hypoxia and metastasis. Novartis Found Symp 240:154–165; discussion 65–68
Horsman MR, Overgaard J (1992) Overcoming tumour radiation resistance resulting from acute hypoxia. Eur J Cancer 28A(12):2084–2085
Horsman MR, Siemann DW (2006) Pathophysiologic effects of vascular-targeting agents and the implications for combination with conventional therapies. Cancer Res 66(24):11520–11539
Hsieh CH, Lee CH, Liang JA, Yu CY, Shyu WC (2010) Cycling hypoxia increases U87 glioma cell radioresistance via ROS induced higher and long-term HIF-1 signal transduction activity. Oncol Rep 24(6):1629–1636
Huang CH, Yang WH, Chang SY et al (2009) Regulation of membrane-type 4 matrix metalloproteinase by SLUG contributes to hypoxia-mediated metastasis. Neoplasia 11(12):1371–1382
Hudson CC, Liu M, Chiang GG et al (2002) Regulation of hypoxia-inducible factor 1alpha expression and function by the mammalian target of rapamycin. Mol Cell Biol 22(20):7004–7014
Hugonnet F, Fournier L, Medioni J et al (2011) Metastatic renal cell carcinoma: relationship between initial metastasis hypoxia, change after 1 month's sunitinib, and therapeutic response: an 18F-fluoromisonidazole PET/CT study. J Nucl Med 52(7):1048–1055
Hurwitz HI, Fehrenbacher L, Hainsworth JD et al (2005) Bevacizumab in combination with fluorouracil and leucovorin: an active regimen for first-line metastatic colorectal cancer. J Clin Oncol 23(15):3502–3508
Ide T, Kitajima Y, Miyoshi A et al (2006) Tumor-stromal cell interaction under hypoxia increases the invasiveness of pancreatic cancer cells through the hepatocyte growth factor/c-Met pathway. Int J Cancer 119(12):2750–2759
Jain RK (2001) Normalizing tumor vasculature with anti-angiogenic therapy: a new paradigm for combination therapy. Nat Med 7(9):987–989
Jallal H, Valentino ML, Chen G, Boschelli F, Ali S, Rabbani SA (2007) A Src/Abl kinase inhibitor SKI-606, blocks breast cancer invasion, growth, and metastasis in vitro and in vivo. Cancer Res 67(4):1580–1588
Jemal A, Siegel R, Xu J, Ward E (2010) Cancer statistics. CA Cancer J Clin 60(5):277–300
Johnson FM, Saigal B, Talpaz M, Donato NJ (2005) Dasatinib (BMS-354825) tyrosine kinase inhibitor suppresses invasion and induces cell cycle arrest and apoptosis of head and neck squamous cell carcinoma and non-small cell lung cancer cells. Clin Cancer Res 11(19 Pt 1):6924–6932
Jordan CT, Guzman ML, Noble M (2006) Cancer stem cells. N Engl J Med 355(12):1253–1261
Keith B, Simon MC (2007) Hypoxia-inducible factors, stem cells, and cancer. Cell 129(3):465–472
Keunen O, Johansson M, Oudin A et al (2011) Anti-VEGF treatment reduces blood supply and increases tumor cell invasion in glioblastoma. Proc Natl Acad Sci USA 108(9):3749–3754
Kim LC, Song L, Haura EB (2009) Src kinases as therapeutic targets for cancer. Nat Rev Clin Oncol 6(10):587–595
Kimura H, Braun RD, Ong ET et al (1996) Fluctuations in red cell flux in tumor microvessels can lead to transient hypoxia and reoxygenation in tumor parenchyma. Cancer Res 56(23):5522–5528
Kirsch M, Schackert G, Black PM (2004) Metastasis and angiogenesis. Cancer Treat Res 117:285–304
Knudsen BS, Vande Woude G (2008) Showering c-MET-dependent cancers with drugs. Curr Opin Genet Dev 18(1):87–96
Konerding MA, Fait E (2001) Gaumann A. 3D microvascular architecture of pre-cancerous lesions and invasive carcinomas of the colon. Br J Cancer 84(10):1354–1362
Koreckij T, Nguyen H, Brown LG, Yu EY, Vessella RL, Corey E (2009) Dasatinib inhibits the growth of prostate cancer in bone and provides additional protection from osteolysis. Br J Cancer 101(2):263–268
Kosharskyy B, Solban N, Chang SK, Rizvi I, Chang Y, Hasan T (2006) A mechanism-based combination therapy reduces local tumor growth and metastasis in an orthotopic model of prostate cancer. Cancer Res 66(22):10953–10958

Kozin SV, Boucher Y, Hicklin DJ, Bohlen P, Jain RK, Suit HD (2001) Vascular endothelial growth factor receptor-2-blocking antibody potentiates radiation-induced long-term control of human tumor xenografts. Cancer Res 61(1):39–44

Lapidot T, Sirard C, Vormoor J et al (1994) A cell initiating human acute myeloid leukaemia after transplantation into SCID mice. Nature 367(6464):645–648

Lee CG, Heijn M, di Tomaso E et al (2000) Anti-Vascular endothelial growth factor treatment augments tumor radiation response under normoxic or hypoxic conditions. Cancer Res 60(19):5565–5570

Leith JT, Papa G, Quaranto L, Michelson S (1992) Modification of the volumetric growth responses and steady-state hypoxic fractions of xenografted DLD-2 human colon carcinomas by administration of basic fibroblast growth factor or suramin. Br J Cancer 66(2):345–348

Li Z, Bao S, Wu Q et al (2009) Hypoxia-inducible factors regulate tumorigenic capacity of glioma stem cells. Cancer Cell 15(6):501–513

Li J, Zucker S, Pulkoski-Gross A et al (2012) Conversion of stationary to invasive tumor initiating cells (TICs): role of hypoxia in membrane type 1-matrix metalloproteinase (MT1-MMP) trafficking. PLoS One 7(6):e38403

Liang D, Ma Y, Liu J et al (2012) The hypoxic microenvironment upgrades stem-like properties of ovarian cancer cells. BMC Cancer 12:201

Lluis JM, Buricchi F, Chiarugi P, Morales A, Fernandez-Checa JC (2007) Dual role of mitochondrial reactive oxygen species in hypoxia signaling: activation of nuclear factor-{kappa}B via c-SRC and oxidant-dependent cell death. Cancer Res 67(15):7368–7377

Lorenzato A, Olivero M, Patane S et al (2002) Novel somatic mutations of the MET oncogene in human carcinoma metastases activating cell motility and invasion. Cancer Res 62(23):7025–7030

Ma Y, Liang D, Liu J et al (2011) Prostate Cancer Cell Lines under Hypoxia Exhibit Greater Stem-Like Properties. PLoS One 6(12):e29170

Maione F, Capano S, Regano D et al (2012) Semaphorin 3A overcomes cancer hypoxia and metastatic dissemination induced by antiangiogenic treatment in mice. J Clin Invest 122(5):1832–1848

Mak P, Leav I, Pursell B et al (2010) ERbeta impedes prostate cancer EMT by destabilizing HIF-1alpha and inhibiting VEGF-mediated snail nuclear localization: implications for Gleason grading. Cancer Cell 17(4):319–332

Martinez-Poveda B, Gomez V, Alcaide-German M et al (2011) Non-invasive monitoring of hypoxia-inducible factor activation by optical imaging during antiangiogenic treatment in a xenograft model of ovarian carcinoma. Int J Oncol 39(3):543–552

Mathieu J, Zhang Z, Zhou W et al (2011) HIF induces human embryonic stem cell markers in cancer cells. Cancer Res 71(13):4640–4652

Matsumoto S, Batra S, Saito K et al (2011) Antiangiogenic agent sunitinib transiently increases tumor oxygenation and suppresses cycling hypoxia. Cancer Res 71(20):6350–6359

Maulik G, Shrikhande A, Kijima T, Ma PC, Morrison PT, Salgia R (2002) Role of the hepatocyte growth factor receptor, c-Met, in oncogenesis and potential for therapeutic inhibition. Cytokine Growth Factor Rev 13(1):41–59

Mazeron R, Anderson B, Supiot S, Paris F, Deutsch E (2011) Current state of knowledge regarding the use of antiangiogenic agents with radiation therapy. Cancer Treat Rev 37(6):476–486

McGee MC, Hamner JB, Williams RF et al (2010) Improved intratumoral oxygenation through vascular normalization increases glioma sensitivity to ionizing radiation. Int J Radiat Oncol Biol Phys 76(5):1537–1545

McKenzie S, Sakamoto S, Kyprianou N (2008) Maspin modulates prostate cancer cell apoptotic and angiogenic response to hypoxia via targeting AKT. Oncogene 27(57):7171–7179

Milosevic M, Chung P, Parker C et al (2007) Androgen withdrawal in patients reduces prostate cancer hypoxia: implications for disease progression and radiation response. Cancer Res 67(13):6022–6025

Milosevic M, Warde P, Menard C et al (2012) Tumor hypoxia predicts biochemical failure following radiotherapy for clinically localized prostate cancer. Clin Cancer Res 18(7):2108–2114

Mohyeldin A, Garzon-Muvdi T, Quinones-Hinojosa A (2010) Oxygen in stem cell biology: a critical component of the stem cell niche. Cell Stem Cell 7(2):150–161

Morton JP, Karim SA, Graham K et al (2010) Dasatinib inhibits the development of metastases in a mouse model of pancreatic ductal adenocarcinoma. Gastroenterology 139(1):292–303

Movsas B, Chapman JD, Hanlon AL et al (2002) Hypoxic prostate/muscle pO2 ratio predicts for biochemical failure in patients with prostate cancer: preliminary findings. Urology 60(4):634–639

Mukhopadhyay D, Tsiokas L, Zhou XM, Foster D, Brugge JS, Sukhatme VP (1995) Hypoxic induction of human vascular endothelial growth factor expression through c-Src activation. Nature 375(6532):577–581

Murata R, Nishimura Y, Hiraoka M (1997) An antiangiogenic agent (TNP-470) inhibited reoxygenation during fractionated radiotherapy of murine mammary carcinoma. Int J Radiat Oncol Biol Phys 37(5):1107–1113

Myers AL, Williams RF, Ng CY, Hartwich JE, Davidoff AM (2010) Bevacizumab-induced tumor vessel remodeling in rhabdomyosarcoma xenografts increases the effectiveness of adjuvant ionizing radiation. J Pediatr Surg 45(6):1080–1085

Myoui A, Nishimura R, Williams PJ et al (2003) C-SRC tyrosine kinase activity is associated with tumor colonization in bone and lung in an animal model of human breast cancer metastasis. Cancer Res 63(16):5028–5033

Nam JS, Ino Y, Sakamoto M, Hirohashi S (2002) Src family kinase inhibitor PP2 restores the E-cadherin/catenin cell adhesion system in human cancer cells and reduces cancer metastasis. Clin Cancer Res 8(7):2430–2436

Nam S, Kim D, Cheng JQ et al (2005) Action of the Src family kinase inhibitor, dasatinib (BMS-354825), on human prostate cancer cells. Cancer Res 65(20):9185–9189

Nautiyal J, Majumder P, Patel BB, Lee FY, Majumdar AP (2009) Src inhibitor dasatinib inhibits growth of breast cancer cells by modulating EGFR signaling. Cancer Lett 283(2):143–151

Otsuka T, Takayama H, Sharp R et al (1998) c-Met autocrine activation induces development of malignant melanoma and acquisition of the metastatic phenotype. Cancer Res 58(22):5157–5167

Ou G, Itasaka S, Zeng L et al (2009) Usefulness of HIF-1 imaging for determining optimal timing of combining bevacizumab and radiotherapy. Int J Radiat Oncol Biol Phys 75(2):463–467

Paez-Ribes M, Allen E, Hudock J et al (2009) Antiangiogenic therapy elicits malignant progression of tumors to increased local invasion and distant metastasis. Cancer Cell 15(3):220–231

Park SI, Shah AN, Zhang J, Gallick GE (2007) Regulation of angiogenesis and vascular permeability by Src family kinases: opportunities for therapeutic treatment of solid tumors. Expert Opin Ther Targets 11(9):1207–1217

Park SI, Zhang J, Phillips KA et al (2008) Targeting SRC family kinases inhibits growth and lymph node metastases of prostate cancer in an orthotopic nude mouse model. Cancer Res 68(9):3323–3333

Parker C, Milosevic M, Toi A et al (2004) Polarographic electrode study of tumor oxygenation in clinically localized prostate cancer. Int J Radiat Oncol Biol Phys 58(3):750–757

Patrawala L, Calhoun-Davis T, Schneider-Broussard R, Tang DG (2007) Hierarchical organization of prostate cancer cells in xenograft tumors: the CD44 + alpha2beta1 + cell population is enriched in tumor-initiating cells. Cancer Res 67(14):6796–6805

Pennacchietti S, Michieli P, Galluzzo M, Mazzone M, Giordano S, Comoglio PM (2003) Hypoxia promotes invasive growth by transcriptional activation of the met protooncogene. Cancer Cell 3(4):347–361

Pham NA, Magalhaes JM, Do T et al (2009) Activation of Src and Src-associated signaling pathways in relation to hypoxia in human cancer xenograft models. Int J Cancer 124(2):280–286

Pipinikas CP, Carter ND, Corbishley CM, Fenske CD (2008) HIF-1alpha mRNA gene expression levels in improved diagnosis of early stages of prostate cancer. Biomarkers 13(7):680–691

Pitson G, Fyles A, Milosevic M, Wylie J, Pintilie M, Hill R (2001) Tumor size and oxygenation are independent predictors of nodal diseases in patients with cervix cancer. Int J Radiat Oncol Biol Phys 51(3):699–703

Pore N, Gupta AK, Cerniglia GJ et al (2006) Nelfinavir down-regulates hypoxia-inducible factor 1alpha and VEGF expression and increases tumor oxygenation: implications for radiotherapy. Cancer Res 66(18):9252–9259

Qian F, Engst S, Yamaguchi K et al (2009) Inhibition of tumor cell growth, invasion, and metastasis by EXEL-2880 (XL880, GSK1363089), a novel inhibitor of HGF and VEGF receptor tyrosine kinases. Cancer Res 69(20):8009–8016

Rabbani SA, Valentino ML, Arakelian A, Ali S, Boschelli F (2010) SKI-606 (Bosutinib) blocks prostate cancer invasion, growth, and metastasis in vitro and in vivo through regulation of genes involved in cancer growth and skeletal metastasis. Mol Cancer Ther 9(5):1147–1157

Rice L, Lepler S, Pampo C, Siemann DW (2012) Impact of the SRC inhibitor dasatinib on the metastatic phenotype of human prostate cancer cells. Clin Exp Metastasis 29:133–142

Riesterer O, Honer M, Jochum W, Oehler C, Ametamey S, Pruschy M (2006) Ionizing radiation antagonizes tumor hypoxia induced by antiangiogenic treatment. Clin Cancer Res 12(11 Pt 1):3518–3524

Rofstad EK, Sundfor K, Lyng H, Trope CG (2000) Hypoxia-induced treatment failure in advanced squamous cell carcinoma of the uterine cervix is primarily due to hypoxia-induced radiation resistance rather than hypoxia-induced metastasis. Br J Cancer 83(3):354–359

Rofstad EK, Henriksen K, Galappathi K, Mathiesen B (2003) Antiangiogenic treatment with thrombospondin-1 enhances primary tumor radiation response and prevents growth of dormant pulmonary micrometastases after curative radiation therapy in human melanoma xenografts. Cancer Res 63(14):4055–4061

Rofstad EK, Galappathi K, Mathiesen B, Ruud EB (2007) Fluctuating and diffusion-limited hypoxia in hypoxia-induced metastasis. Clin Cancer Res 13(7):1971–1978

Rofstad EK, Gaustad JV, Egeland TA, Mathiesen B, Galappathi K (2010) Tumors exposed to acute cyclic hypoxic stress show enhanced angiogenesis, perfusion and metastatic dissemination. Int J Cancer 127(7):1535–1546

Ruan K, Song G, Ouyang G (2009) Role of hypoxia in the hallmarks of human cancer. J Cell Biochem 107(6):1053–1062

Rucci N, Recchia I, Angelucci A et al (2006) Inhibition of protein kinase c-Src reduces the incidence of breast cancer metastases and increases survival in mice: implications for therapy. J Pharmacol Exp Ther 318(1):161–172

Salmon HW, Guha A, Rojiani AM, Siemann DW (2012) Vascular development in mouse lung metastases. Am J Cancer Res 2(5):581–588

Salnikov AV, Liu L, Platen M et al (2012) Hypoxia Induces EMT in Low and Highly Aggressive Pancreatic Tumor Cells but Only Cells with Cancer Stem Cell Characteristics Acquire Pronounced Migratory Potential. PLoS One 7(9):e46391

Schlaepfer DD, Mitra SK (2004) Multiple connections link FAK to cell motility and invasion. Curr Opin Genet Dev 14(1):92–101

Schroeder GM, An Y, Cai ZW et al (2009) Discovery of N-(4-(2-Amino-3-chloropyridin-4-yloxy)-3-fluorophenyl)-4-ethoxy-1-(4-fluor ophenyl)-2-oxo-1,2-dihydropyridine-3-carboxamide (BMS-777607), a Selective and Orally Efficacious Inhibitor of the Met Kinase Superfamily. J Med Chem 52(5):1251–1254

Schuuring J, Bussink J, Bernsen HJ, Peeters W, Der KAJ van (2005) Irradiation combined with SU5416: microvascular changes and growth delay in a human xenograft glioblastoma tumor line. Int J Radiat Oncol Biol Phys 61(2):529–534

Segers J, Di Fazio V, Ansiaux R et al (2006) Potentiation of cyclophosphamide chemotherapy using the anti-angiogenic drug thalidomide: importance of optimal scheduling to exploit the 'normalization' window of the tumor vasculature. Cancer Lett 244(1):129–135

Semenza GL (2003) Targeting HIF-1 for cancer therapy. Nat Rev Cancer 3(10):721–732

Semenza GL (2012) Hypoxia-inducible factors: mediators of cancer progression and targets for cancer therapy. Trends Pharmacol Sci 33(4):207–214

Siemann DW, Horsman MR (2009) Vascular targeted therapies in oncology. Cell Tissue Res 335(1):241–248

Siemann DW, Dong M, Pampo C, Shi W (2012) Src-signaling interference impairs the dissemination of blood-borne tumor cells. Cell Tissue Res 349(2):541–550

Sion AM, Figg WD (2006) Lysyl oxidase (LOX) and hypoxia-induced metastases. Cancer Biol Ther 5(8):909–911

Steeg PS (2006) Tumor metastasis: mechanistic insights and clinical challenges. Nat Med 12(8):895–904

Sullivan R, Graham CH (2007) Hypoxia-driven selection of the metastatic phenotype. Cancer Metastasis Rev 26(2):319–331

Sullivan R, Pare GC, Frederiksen LJ, Semenza GL, Graham CH (2008) Hypoxia-induced resistance to anticancer drugs is associated with decreased senescence and requires hypoxia-inducible factor-1 activity. Mol Cancer Ther 7(7):1961–1973

Summy JM, Gallick GE (2003) Src family kinases in tumor progression and metastasis. Cancer Metastasis Rev 22(4):337–358

Tailor TD, Hanna G, Yarmolenko PS et al (2010) Effect of pazopanib on tumor microenvironment and liposome delivery. Mol Cancer Ther 9(6):1798–1808

Takahashi K, Tanabe K, Ohnuki M et al (2007) Induction of pluripotent stem cells from adult human fibroblasts by defined factors. Cell 131(5):861–872

Tatarov O, Mitchell TJ, Seywright M, Leung HY, Brunton VG, Edwards J (2009) SRC family kinase activity is up-regulated in hormone-refractory prostate cancer. Clin Cancer Res 15(10):3540–3549

Teicher BA, Holden SA, Ara G et al (1995) Influence of an anti-angiogenic treatment on 9 L gliosarcoma: oxygenation and response to cytotoxic therapy. Int J Cancer 61(5):732–737

Thews O, Wolloscheck T, Dillenburg W et al (2004) Microenvironmental adaptation of experimental tumours to chronic vs acute hypoxia. Br J Cancer 91(6):1181–1189

Toda S, Yamada S, Aoki S, Inokuchi A, Sugihara H (2005) Air-liquid interface promotes invasive growth of laryngeal squamous cell carcinoma with or without hypoxia. Biochem Biophys Res Commun 326(4):866–872

Trusolino L, Bertotti A, Comoglio PM (2010) MET signalling: principles and functions in development, organ regeneration and cancer. Nat Rev Mol Cell Biol 11(12):834–848

Tsao AS, He D, Saigal B et al (2007) Inhibition of c-Src expression and activation in malignant pleural mesothelioma tissues leads to apoptosis, cell cycle arrest, and decreased migration and invasion. Mol Cancer Ther 6(7):1962–1972

Valable S, Petit E, Roussel S et al (2011) Complementary information from magnetic resonance imaging and (18)F-fluoromisonidazole positron emission tomography in the assessment of the response to an antiangiogenic treatment in a rat brain tumor model. Nucl Med Biol 38(6):781–793

Vangestel C, Van de Wiele C, Mees G et al (2012) Single-Photon Emission Computed Tomographic Imaging of the Early Time Course of Therapy-Induced Cell Death Using Technetium 99m Tricarbonyl His-Annexin A5 in a Colorectal Cancer Xenograft Model. Mol Imaging 11(2):135–147

Vaupel P (2004) Tumor microenvironmental physiology and its implications for radiation oncology. Semin Radiat Oncol 14(3):198–206

Verbeek BS, Vroom TM, Adriaansen-Slot SS et al (1996) c-Src protein expression is increased in human breast cancer. An immunohistochemical and biochemical analysis. J Pathol 180(4):383–388

Vergis R, Corbishley CM, Norman AR et al (2008) Intrinsic markers of tumour hypoxia and angiogenesis in localised prostate cancer and outcome of radical treatment: a retrospective analysis of two randomised radiotherapy trials and one surgical cohort study. Lancet Oncol 9(4):342–351

Vordermark D, Katzer A, Baier K, Kraft P, Flentje M (2004) Cell type-specific association of hypoxia-inducible factor-1 alpha (HIF-1 alpha) protein accumulation and radiobiologic tumor hypoxia. Int J Radiat Oncol Biol Phys 58(4):1242–1250

Vultur A, Buettner R, Kowolik C et al (2008) SKI-606 (bosutinib), a novel Src kinase inhibitor, suppresses migration and invasion of human breast cancer cells. Mol Cancer Ther 7(5):1185–1194

Wang T, Niki T, Goto A et al (2007) Hypoxia increases the motility of lung adenocarcinoma cell line A549 via activation of the epidermal growth factor receptor pathway. Cancer Sci 98(4):506–511

Williams KJ, Telfer BA, Brave S et al (2004) ZD6474, a potent inhibitor of vascular endothelial growth factor signaling, combined with radiotherapy: schedule-dependent enhancement of antitumor activity. Clin Cancer Res 10(24):8587–8593

Wilson WR, Hay MP (2011) Targeting hypoxia in cancer therapy. Nat Rev Cancer 11(6):393–410

Winkler F, Kozin SV, Tong RT et al (2004) Kinetics of vascular normalization by VEGFR2 blockade governs brain tumor response to radiation: role of oxygenation, angiopoietin-1, and matrix metalloproteinases. Cancer Cell 6(6):553–563

Xue Y, Bi F, Zhang X et al (2006) Role of Rac1 and Cdc42 in hypoxia induced p53 and von Hippel-Lindau suppression and HIF1alpha activation. Int J Cancer 118(12):2965–2972

Yamasaki M, Nomura T, Sato F, Mimata H (2012) Chronic hypoxia induces androgen-independent and invasive behavior in LNCaP human prostate cancer cells. Urol Oncol

Yang JC, Bai L, Yap S, Gao AC, Kung HJ, Evans CP (2010) Effect of the specific Src family kinase inhibitor saracatinib on osteolytic lesions using the PC-3 bone model. Mol Cancer Ther 9(6):1629–1637

Yong M, Jensen AO, Jacobsen JB, Norgaard M, Fryzek JP, Sorensen HT (2011) Survival in breast cancer patients with bone metastases and skeletal-related events: a population-based cohort study in Denmark (1999–2007). Breast Cancer Res Treat 129(2):495–503

Zhang L, Hill RP (2007) Hypoxia enhances metastatic efficiency in HT1080 fibrosarcoma cells by increasing cell survival in lungs, not cell adhesion and invasion. Cancer Res 67(16):7789–7797

Zhong H, De Marzo AM, Laughner E et al (1999) Overexpression of hypoxia-inducible factor 1alpha in common human cancers and their metastases. Cancer Res 59(22):5830–5835

Chapter 10
The Unfolded Protein Response and Therapeutic Opportunities

Carly M. Sayers, Souvik Dey, Stacey L. Lehman and Constantinos Koumenis

Abstract Tumor cells employ multiple elaborate, evolutionarily conserved mechanisms that enable them to respond to stress conditions in the tumor microenvironment, including hypoxia. Although the most studied cellular signaling pathway induced by hypoxia is mediated by the transcriptional activity of hypoxia-inducible factors (HIFs), several HIF-independent mechanisms have been implicated in hypoxic adaptation, especially in the regulation of macromolecular synthesis. One such mechanism, known as the unfolded protein response (UPR), encompasses a trio of cellular signaling cascades. The UPR is activated by the accumulation of misfolded or unfolded proteins in the endoplasmic reticulum (ER). Numerous *in vitro* and *in vivo* studies using genetic and pharmacological modifications of UPR signaling components have demonstrated an important role for the UPR in determining tumor cell survival following transient and chronic hypoxia. This review summarizes the important aspects of UPR signaling and the role of the UPR in determining tumor cell survival or death under hypoxic stress. We also discuss novel pharmacological approaches for targeting critical UPR components as potential anti-tumor strategies.

Keywords Unfolded protein response · Hypoxia · PERK · ATF4 · IRE1 · ATF6 · Translation initiation factor · Autophagy · GCN2 · Hypoxia tolerance

C. Koumenis (✉) · C. M. Sayers · S. Dey · S. L. Lehman
Department of Radiation Oncology, Smilow Center for Translational Research,
Perelman School of Medicine, University of Pennsylvania, Room 8-124,
3400 Civic Center Blvd., Bldg 421, Philadelphia PA 19104-5156, USA
e-mail: koumenis@xrt.upenn.edu

C. M. Sayers
Pharmacology Graduate Group, Perelman School of Medicine, University of Pennsylvania,
Philadelphia, PA 19104, USA
e-mail: carlymsayers@gmail.com

S. L. Lehman
Cell and Molecular Biology Graduate Group, Perelman School of Medicine,
University of Pennsylvania, Philadelphia, PA 19104, USA
e-mail: lehmanst@mail.med.upenn.edu

S. Dey
e-mail: deys@mail.med.upenn.edu

G. Melillo (ed.), *Hypoxia and Cancer,* Cancer Drug Discovery
and Development, DOI 10.1007/978-1-4614-9167-5_10,

10.1 Introduction

A common characteristic of most solid tumors is the presence of hypoxic regions, where the oxygen concentration can be as low as 0.01 %, while oxygen levels in normal tissue range from 3.1 to 8.7 % (Kizaka-Kondoh et al. 2003). Chronic hypoxia occurs when the tumor outgrows the existing vasculature, increasing the distance oxygen must diffuse to reach portions of the tumor. Tumors can also experience acute, or intermittent, hypoxia, which is caused by abnormal blood flow dynamics in the newly forming vessels (Kizaka-Kondoh et al. 2003). Tumor cells can survive adverse hypoxic conditions for extended periods of time depending on factors such as expression of anti-apoptotic genes, mutations in proapoptotic genes, and activation of prosurvival programs. Clinically, hypoxic tumors are more resistant to both chemotherapy, due to limited drug diffusion and cell-cycle dysregulation, and radiation therapy, due to chemical repair of DNA damage produced by free radicals (Brown and Wilson 2004). Moreover, the metastatic potential of cancer cells has been shown to be increased by hypoxia (Brown 1990; Cairns and Hill 2004; Chang et al. 2011). It is not surprising then that hypoxia has been associated with poor local tumor control and reduced overall survival (Höckel and Vaupel 2001; Vaupel et al. 2001; Koumenis 2006).

Cells adapt to hypoxic stress using hypoxia-inducible factor (HIF)-dependent and HIF-independent pathways (Koumenis 2006; Bertout et al. 2008; Semenza 2009). The HIF-dependent pathway regulates the expression of proteins important for anaerobic glycolysis, angiogenesis, and cell survival and is discussed at greater length in Chap. 2. HIF-independent effects are designed to curtail oxygen consumption by energy-expensive processes, such as DNA replication and protein synthesis. In particular, the unfolded protein response (UPR) is upregulated by hypoxia to decrease global messenger RNA (mRNA) translation (Koumenis 2006).

10.2 Overview of the UPR

The UPR is a coordinated cellular program induced by cells to adapt to transient and chronic endoplasmic reticulum (ER) stress caused by factors such as unfolded proteins, hypoxia, nutrient deprivation, and oncogenic transformation (Kaufman 2002; Schönthal 2012). Induction of the UPR relies on three ER transmembrane proteins: PKR-like endoplasmic reticulum kinase (PERK), inositol-requiring enzyme 1 (IRE1), and activating transcription factor 6 (ATF6) (Fig. 10.1). During unstressed conditions, these signaling transducers are maintained in an inactive state by an ER lumenal chaperone protein called GRP78 (glucose-regulated protein, 78 kDa; also known as BiP). When misfolded proteins begin to accrue, GRP78 dissociates from PERK, IRE1, and ATF6 to aid in protein folding and degradation, which allows the activation of these three proteins (Bertolotti et al. 2000; Shen et al. 2002). Activation of PERK involves homodimerization and transautophosphorylation. One major target of PERK is eukaryotic translation initiation factor 2 α-subunit (eIF2α) (Harding

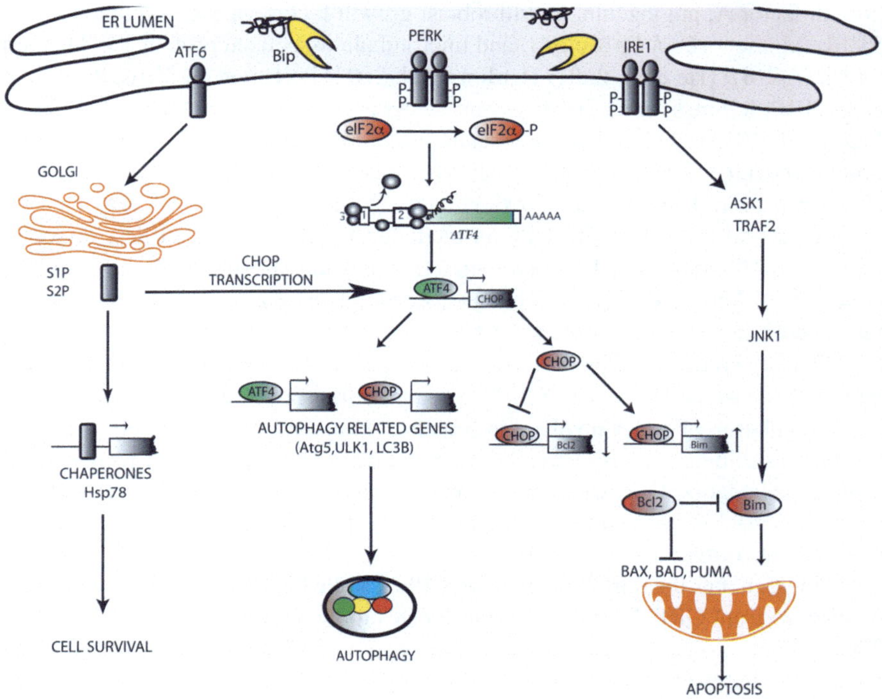

Fig. 10.1 Consequences of the activation of the unfolded protein response (UPR) following hypoxia. Hypoxia-induced *ROS* disrupt proper protein folding, resulting in an accumulation of unfolded proteins in the endoplasmic reticulum (ER). *ER* stress is detected by three transmembrane UPR sensors—PERK, IRE1, and ATF6—by disassociation of the chaperone GRP78/BiP. Activation of the UPR leads to the upregulation of protein chaperones and other proteins important for neutralizing the cellular stress. Hypoxia-induced UPR signaling also mediates a cytoprotective autophagy response. However, chronic activation of PERK and IRE1 following hypoxia can lead to cellular apoptosis by various mechanisms

et al. 1999; Harding et al. 2000b). Normally, eIF2α binds guanosine triphosphate (GTP) as part of the initiation of cap-dependent translation. To begin the elongation phase of translation, eIF2α-GTP is hydrolyzed to eIF2α-guanosine diphosphate (GDP), which then disassociates from the ribosome. When PERK phosphorylates eIF2α at Ser51, the exchange of GDP for GTP is inhibited. This leads to a global reduction in translation, preventing further accumulation of unfolded proteins in the ER. Paradoxically, the translation of a few mRNAs is substantially elevated under these conditions. The most extensively studied of these mRNAs is that encoding activating transcription factor 4 (*ATF4*) (Shi et al. 1998; Harding et al. 2000a; Vattem and Wek 2004). ATF4 is involved in the transcription of genes encoding proteins that assist the cell in neutralizing the stress, such as chaperones (*GRP78*) and proteins important for amino acid metabolism (asparagine synthetase (*ASNS*), glycine transporter 1, and several transfer RNA (tRNA) synthetases), redox homeostasis (cystathionine γ-lyase and heme oxygenase 1), angiogenesis (vascular endothelial growth factor A, angiogenin, and fibroblast growth factor 2), and autophagy (unc-51-like kinase 1 (*ULK1*), beclin 1, and microtubule-associated protein 1 light chain 3 beta (*LC3B*)) (He et al. 2001; Harding et al. 2003; Pereira et al. 2010; Rouschop et al. 2010; Rzymski et al. 2010; Avivar-Valderas et al. 2011; Dickhout et al. 2012; Pike et al. 2013). ATF4 also upregulates the transcription of C/EBP homologous protein (*CHOP*), a proapoptotic protein that will be discussed in more detail later (Friedman 1996; Harding et al. 2000a).

The second branch of the UPR is mediated by IRE1. Like PERK, dissociation of GRP78 allows IRE1 homodimerization and autophosphorylation (Bertolotti et al. 2000). Activated IRE1 also possesses endonuclease activity, which processes unspliced X-box binding protein 1 (*XBP1*) mRNA by removing an inhibitory intron (Tirasophon et al. 1998; Yoshida et al. 2001). The spliced mRNA can then be translated to the functional XBP1-s protein, a transcription factor important for the upregulation of genes involved in protein folding and disulfide bond formation (protein disulfide isomerase-P5 and protein kinase inhibitor p58) and ER-associated protein degradation (ER degradation enhancer mannosidase alpha-like 1 (*EDEM*) and homocysteine-induced ER protein) (Yoshida et al. 2001; Calfon et al. 2002; Lee et al. 2003a; Yamamoto et al. 2004).

ATF6 regulates the third branch of the UPR. Unlike PERK and IRE1, when ATF6 is released from GRP78 binding, it translocates to the Golgi apparatus (Chen et al. 2002). Once in the Golgi, two resident proteases, S1P and S2P, cleave ATF6 to release the cytosolic transcription factor domain (Ye et al. 2000). This activated ATF6 domain then translocates to the nucleus, where it stimulates the transcription of XBP1 and genes encoding protein chaperones (*GRP78, GRP94*, and calreticulin) (Haze et al. 1999; Yoshida et al. 2001; Okada et al. 2002).

Cells also cope with ER stress by activating ER-associated degradation (ERAD). ERAD is a quality control pathway which allows cells to eliminate, as well as degrade, any terminally misfolded proteins from the ER and avoid proteotoxicity. The process of ERAD is regulated via two regulatory proteins in the lumen of the ER, namely EDEM and Yos9p. These proteins can recognize the irreversibly misfolded protein and target it for degradation (Smith et al. 2011; Walter and Ron 2011) the 26S proteasome in the cytosol with the help of critical ER membrane proteins Hrd1p and Hrd3p (hydroxymethylglutaryl-coenzyme A (HMG-CoA) reductase degradation proteins), which possess really interesting new gene (RING) domains as present in various ubiquitin ligases (Bays et al. 2001).

UPR activation leads to an increase in oxidative stress due to an increase in protein folding—a process further accentuated by ER oxidoreductase 1α (ERO1α) (Walter and Ron 2011). ERO1α oxidizes protein disulfide isomerase (PDI), which introduces new disulfide bonds in nascent or misfolded proteins to reduce ER load. Along with oxidizing PDI, ERO1α reduces oxygen to form H_2O_2, a primary causative agent for oxidative stress (Frand and Kaiser 1999; Ron and Walter 2007). However, PERK can neutralize the increasing oxidative nature of the ER lumen to maintain ER homeostasis by directly phosphorylating and activating Nrf2 (nuclear factor, erythroid 2-like 2), a transcription factor which induces antioxidant genes (Cullinan et al. 2003).

10.3 Activation of the UPR and the Integrated Stress Response by Hypoxia and Other Stresses

Although the mechanism is not well understood, hypoxia is thought to activate the UPR by disrupting the redox balance of cells. The ER contains a variety of chaperones that assist in the folding of nascent polypeptides. In budding yeast, the enzyme Ero1p initiates polypeptide folding by oxidizing PDI, which then directly interacts with client proteins to promote proper disulfide bond formation (Tu and Weissman 2004). Importantly, this cascade of folding events requires molecular oxygen as the final electron acceptor (Tu and Weissman 2004). Thus, chaperone activity is compromised under hypoxia, leading to an accumulation of unfolded proteins. Although a similar pathway has not been formally identified in mammals, it is hypothesized that chaperone activity in mammalian cells is also oxygen dependent.

Hypoxia has also been shown to activate the UPR through the generation of reactive oxygen species (ROS). Liu and colleagues demonstrated that exposure of cells to the reactive oxygen species H_2O_2 results in phosphorylation of PERK and eIF2α (Liu et al. 2008). Treatment with catalase diminishes hypoxia-induced phosphorylation of eIF2α, indicating that H_2O_2 generated under hypoxia contributes toward activation of the UPR (Liu et al. 2008). Future studies are required to determine the mechanism through which ROS disrupt protein folding.

In addition to hypoxia, the tumor microenvironment is often characterized by shortages of other nutrients such as glucose and amino acids. One pathway tumor cells utilize to cope with these stressors is the integrated stress response (ISR). The ISR consists of four kinases, including PERK. The three remaining kinases are general control nonderepressible 2 (GCN2), RNA-dependent protein kinase (PKR), and heme-regulated inhibitor (HRI) (Wek et al. 2006). The GCN2 arm of the ISR is particularly important for tumor cells to respond to nutrient stress (Ye et al. 2010). Under conditions of amino acid deprivation, uncharged tRNAs accumulate in cells. GCN2 binds to uncharged tRNAs, causing it to undergo a conformational change and become catalytically active (Wek et al. 1989; Padyana et al. 2005). Activated GCN2 phosphorylates eIF2α to attenuate translation to conserve amino acids (Harding et al. 2000a). Additionally, upregulation of ATF4 by GCN2-mediated eIF2α phosphorylation initiates a transcriptional program to induce genes such as amino acid biosynthetic enzymes and amino acid transporters to promote recovery from nutrient stress (Harding et al. 2003). GCN2 is also activated under conditions of glucose deprivation because cells consume amino acids as an energy source when glucose is unavailable (Ye et al. 2010). Interestingly, PERK also contributes to ISR activation under glucose deprivation. Glycosylation is one important protein modification that occurs in the ER. If glucose is unavailable, glycosylation is reduced and unmodified proteins accumulate in the ER and activate the UPR (Lee 1992). Thus, the glycosylation inhibitor tunicamycin is frequently used as an experimental inducer of the UPR in the laboratory.

It should be noted that although hypoxia and nutrient deprivation are major sources of noncell autonomous activation of the UPR, cell intrinsic stresses can also induce this pathway. Our lab has recently identified the c-Myc oncogene as a cell autonomous activator of the UPR in cancer cells (Hart et al. 2012). As one of its many functions as a transcription factor, c-Myc upregulates genes involved in ribosome biogenesis (Meyer and Penn 2008). Oncogenic activation of c-Myc places increased demands on the protein synthetic machinery of a cell, resulting in dependence on the UPR to manage the folding of a large number of client proteins in the ER. Activation of the UPR is required for c-Myc-dependent tumorigenesis, as inhibition of this pathway through genetic or pharmacological means greatly reduces tumor cell viability *in vitro* and compromises the ability of cells to form tumors *in vivo* (Hart et al. 2012).

10.4 Evidence of *In Vivo* Activation of the UPR and ISR in Human Tumors

A wealth of clinical data has established that all three branches of the UPR are activated in human tumors. Evidence for PERK pathway activation under hypoxia comes from cervical carcinoma samples. Here, high levels of ATF4 and CHOP colocalize with staining for pimonidazole, a hypoxia-sensitive dye (Bi et al. 2005). It has also been demonstrated in breast carcinoma that ATF4 expression is increased in areas of necrosis known to be anoxic (Ameri et al. 2004). Additionally, ATF4 overexpression has been found in brain, breast, cervical, and skin cancers as compared to normal tissue controls (Bi et al. 2005). Activation of the IRE1 arm of the UPR has been found in lymphoma, liver, and breast cancer patient samples. In B-cell lymphoma and liver cancer, increased levels of spliced XBP1 are found in malignant tissue as compared to adjacent normal tissue, while levels of total XBP1 are elevated in breast cancer (Fujimoto et al. 2003; Shuda et al. 2003; Hart et al. 2012). Hepatocellular carcinoma samples also express higher levels of *ATF6* mRNA than normal tissue (Shuda et al. 2003). Additionally, the increased expression of the ER chaperone GRP78 is found in a variety of tumor types, including stomach, breast, and lung cancer, relative to normal tissue controls (Fernandez et al. 2000; Uramoto et al. 2005; Wang et al. 2005; Zheng et al. 2008).

10.4.1 GRP78, Hypoxia, and Tumor Progression

Several models have now been developed to better understand the consequences of UPR inhibition in tumor cells. One of the first components of the UPR studied *in vivo* was GRP78. Previous work in cell culture demonstrated that knocking down GRP78 sensitized tumor cells to hypoxia (Koong et al. 1994). Jamora and colleagues extended these studies to an animal model by demonstrating that B/C10ME fibrosarcoma cells expressing an anti-sense construct against GRP78 failed to form tumors as efficiently as their wild-type counterparts. When injected subcutaneously into mice, these cells either completely failed to form tumors or formed tumors that regressed after a period of time (Jamora et al. 1996).

10.4.2 The PERK Arm of the UPR in Malignancy

Perhaps one of the most extensively studied arms of the UPR *in vivo* is the PERK pathway. In cell culture, $PERK^{-/-}$ mouse embryonic fibroblasts (MEFs) fail to phosphorylate eIF2α in response to hypoxia and exhibit reduced clonogenic survival after hypoxic stress (Bi et al. 2005). When Ras-transformed $PERK^{-/-}$ MEFs are subcutaneously injected into mice, they grow more slowly and exhibit lower levels of p-eIF2α than $PERK^{+/+}$cells (Bi et al. 2005). Additionally, $PERK^{-/-}$ tumors have smaller areas of hypoxia that overlap with areas of apoptosis, indicating that the PERK arm of the UPR supports tumor cell survival under hypoxia *in vivo* (Bi et al. 2005). Similar results were found with HT29 human colorectal carcinoma cell lines stably expressing a dominant-negative PERK construct (Bi et al. 2005).

PERK has also been studied in genetic mouse models of cancer. Mammary carcinoma-prone MMTV-Neu transgenic mice crossed to $PERK^{fl/fl}$/MMTV-Cre mice demonstrated that deletion of PERK in the mammary gland slows tumor growth, resulting in improved overall survival (Bobrovnikova-Marjon et al. 2010). Mice with deletion of PERK in the mammary gland also have fewer lung metastases than their wild-type counterparts. Interestingly, MMTV-Neu/$PERK^{\Delta/\Delta}$ tumor cells have significantly higher levels of ROS and DNA damage, indicating that PERK can also promote tumor growth by regulating the redox balance of cells. However, this is a double-edged sword as MMTV-Neu/$PERK^{\Delta/\Delta}$ mice develop spontaneous mammary carcinomas as they age, due to an increased level of genomic instability (Bobrovnikova-Marjon et al. 2010).

Gupta and colleagues also characterized the role of PERK in a genetically engineered mouse model of insulinoma (Gupta et al. 2009). In this model, PERK knockout mice develop fewer and smaller insulinomas due to slower rates of proliferation rather than increased rates of apoptosis. Interestingly, PERK knockout mice also have more poorly vascularized tumors than wild-type mice. This observation corroborates research performed *in vitro* demonstrating that PERK regulates the angiogenic switch and controls the expression of VEGF and other proangiogenic factors (Ghosh et al. 2010; Wang et al. 2012). Thus, PERK may not only control tumor cells' response to hypoxia through cell autonomous activation of the UPR but also through noncell autonomous mechanisms by controlling oxygen delivery to the tumor by promoting angiogenesis.

10.4.3 UPR and Tumor Immunity

The PERK arm of the UPR has also been shown to influence tumor growth by promoting an anti-tumor immune response. Particular kinds of stressors, such as ultraviolet C (UV-C) irradiation and anthracycline treatment, induce exposure of the ER chaperone calreticulin on the plasma membrane of cells (Obeid et al. 2007). Exposed calreticulin serves as an engulfment signal for dendritic cells, thus eliciting an immune response. Knocking down calreticulin blocks immunogenic cell death (Obeid et al. 2007). Furthermore, cells that lack ERp57, a disulfide isomerase required for calreticulin exposure, become resistant to chemotherapy because they are unable to elicit an immune response in the host (Panaretakis et al. 2008).

Stressors that induce calreticulin exposure have been shown to induce both PERK and eIF2α phosphorylation. Cells with PERK knockdown or a nonphosphorylatable knock-in mutation of eIF2α, in which serine 51 is converted to alanine, fail to expose calreticulin in response to stress (Panaretakis et al. 2009). However, knockdown of ATF6 or knockdown of IRE1 does not affect calreticulin exposure, indicating that this stress response is specific to the PERK arm of the UPR. Vaccination with CT26 murine colon cancer cells treated with chemotherapeutics that induce calreticulin exposure was able to prevent tumor growth in mice after a later challenge of live, untreated CT26 cells. This effect was attenuated with vaccination of chemotherapy-treated CT26 cells with PERK knockdown, demonstrating that PERK can also exert noncell autonomous effects on tumor growth through inducing an immune response to cancer cells (Panaretakis et al. 2009).

10.5 Role of PERK Pathway Effectors in Hypoxic and Nutrient Stress Adaptation and Tumor Growth

The consequences of PERK pathway inhibition have also been studied at the level of eIF2α phosphorylation and ATF4 induction. Similar to PERK$^{-/-}$ MEFs, MEFs containing the eIF2α S51A knock-in mutation experience reduced clonogenic survival after exposure to extreme hypoxia (Bi et al. 2005). These results were recapitulated *in vivo*: Ras-transformed eIF2α S51A MEFs form much smaller tumors than their wild-type counterparts (Bi et al. 2005). Additionally, a large majority of apoptotic cells in eIF2α S51A tumors are found within hypoxic areas, while in wild-type tumors, apoptotic cells are mostly located within areas of necrosis. As expected, ATF4$^{-/-}$ MEFs experience increased levels of apoptosis under both moderate and severe hypoxia (Bi et al. 2005). *In vivo*, HT1080 human fibrosarcoma cells stably transfected with a short hairpin RNA (shRNA) against ATF4 form much smaller tumors than cells transfected with a nontargeting shRNA when injected subcutaneously into nude mice (Ye et al. 2010).

However, not all effects of ATF4 on tumor growth can be attributed to PERK. As previously mentioned, GCN2, another kinase member of the ISR, phosphorylates eIF2α and induces ATF4 in response to amino acid deprivation. When grown in culture, HT1080 shATF4 cells exhibit increased apoptosis that is rescued by addition of the nonessential amino acid asparagine (Ye et al. 2010). This effect can be attributed to the ATF4 target gene *ASNS*, which transfers an amino group from glutamine to aspartate, forming asparagine. Indeed, overexpression of ASNS in HT1080 shATF4 cells partially rescues the ability of these cells to form tumors in mice (Ye et al. 2010). These results highlight the variety of stressors that exist in the tumor microenvironment and the necessity to study the interplay between various survival pathways in tumor cells.

10.6 The IRE1 and ATF6 Arms of the UPR in Hypoxic Adaptation and Tumor Progression

The IRE1 arm of the UPR has also been studied extensively *in vivo*. Work from the Koong lab was the first to establish that this pathway is required for tumor growth (Romero-Ramirez et al. 2004). MEFs lacking XBP1 have greatly reduced clonogenic survival after exposure to severe hypoxia. These cells also fail to form tumors when injected subcutaneously into mice. Tumors formed from XBP1 wild-type MEFs showed excellent colocalization of XBP1 with the hypoxia-sensitive dye EF5, indicating that XBP1 is expressed in hypoxic areas of tumors. Similar results were obtained with HT1080 cells with knockdown of XBP1 (Chen et al. 2005). Since XBP1 is required for plasma cell differentiation, it has been frequently studied in multiple myeloma (Iwakoshi et al. 2003). In fact, sustained expression of spliced XBP1 in the B-cell compartment of mice recapitulates the human disease (Carrasco et al. 2007). Lee and colleagues demonstrated that inhibition of XBP1 through RNA interference or expression of a dominant negative mutant rendered multiple myeloma cells sensitive to ER stress (Lee et al. 2003b).

Finally, ATF6 also shows promise as an anti-tumor target, but it has not been as extensively characterized as the other members of the UPR. Work in HEp3 human epidermoid carcinoma cells shows that ATF6 is constitutively activated in dormant, but not tumorigenic cells (Schewe and Aguirre-Ghiso 2008). This effect is attributed to ATF6-dependent induction of Rheb (Ras homolog enriched in brain), which increases mechanistic target of rapamycin (mTOR) activity, thus promoting tumor survival during periods of dormancy. Mice injected with dormant HEp3 cells with shATF6 experienced increased tumor-free survival as compared to mice injected with dormant HEp3 cells with an empty vector control. Further studies are required to characterize the role that ATF6 plays in survival under hypoxia and the consequences of inhibition of ATF6 signaling.

10.7 The ISR: Players and Consequences of its Activation

PERK-mediated phosphorylation of eIF2α following hypoxia leads to preferential translation of ATF4, which ultimately leads to remediation of the stress condition or to apoptosis (Harding et al. 2003; Blais et al. 2004; Ron and Walter 2007). The expression of the protein from the *ATF4* mRNA following cellular stress is regulated by a translation control mechanism involving the 5′ untranslated region (UTR). The 5′-UTR of *ATF4* contains two upstream open reading frames (uORFs) that function differentially to regulate stress-specific enhancement of ATF4 expression (Fig. 10.1). The proximal uORF1, which is three amino acid residues in length, acts as a 'positive element' by promoting ribosome reinitiation at downstream start codons, while the distal uORF2, which overlaps the start codon, acts as a negative element for mRNA translation. During an abundance of oxygen supply, high eIF2-GTP levels cause ribosome reinitiation at uORF2, resulting in inhibition of *ATF4* translation. However, hypoxia-mediated eIF2α phosphorylation reduces eIF2α-GTP levels, causing delayed ribosome reinitiation and bypass of the negative uORF2. As a result, the 40S and the 60S ribosomal subunits can reinitiate at the ATF4 start codon, leading to increased expression (Blais et al. 2004; Vattem and Wek 2004). Expression of ATF4 is also shown to be modulated by hypoxia through increased mRNA stability (Ameri and Harris 2008). It has been previously shown that several mRNAs that encode critical members of the ISR, including ATF4, are highly labile and undergo rapid degradation by a mechanism involving nonsense-mediated mRNA decay (NMD) (Gardner 2010). Hypoxia-induced eIF2α phosphorylation leads to inhibition of NMD—a mechanism which involves localization of important factors of the NMD machinery to cytoplasmic dense aggregates known as stress granules, which act as storehouses for several mRNAs as well as components of the translation preinitiation complex. This sequestration of the NMD machinery allows stabilization of *ATF4* mRNA following hypoxic stress (Gardner 2008).

Hypoxic induction of ATF4 has been shown to be solely dependent on PERK-induced eIF2α phosphorylation, as $PERK^{-/-}$ and the eIF2α S51A knock-in mutant MEFs fail to induce ATF4 expression (Bi et al. 2005; Koumenis et al. 2007). Induction of ATF4 following hypoxia and anoxia has been shown to be independent of HIF1 expression and involves a mechanism of protein stabilization by the oxygen sensor prolyl hydroxylase 3 (PHD3) (Ameri et al. 2004; Koditz et al. 2007). Once expressed, ATF4 homodimerizes or heterodimerizes with other basic leucine zipper domain (bZIP) transcription factors such as CCAAT-enhancer-binding protein β (C/EBPβ) to upregulate genes involved in remediation of oxidative stress, amino acid metabolism, differentiation, and hypoxic tolerance (Harding et al. 2003).

One of the genes that is transcriptionally upregulated by ATF4, is *CHOP* which has major implications in determining cellular fate (Fawcett et al. 1999; Harding et al. 2000a). CHOP transcriptionally activates several key genes that lead to cell death following chronic hypoxic ER stress. Analysis of human, as well as mouse, tumor tissues has shown colocalization of ATF4 and CHOP expression with hypoxic regions (Bi et al. 2005). Acute hypoxia has been shown to be a major inducer of CHOP (Bi et al. 2005). This induction of CHOP was ATF4-dependent, as silencing of ATF4 significantly reduces its induction. Expression of CHOP following induction of the UPR has also been shown to be under the transcriptional control of ATF6 (Ma et al. 2002).

ATF3 is another member of the ATF/CREB family of transcription factors that has been shown to be overexpressed in various human tumors and mouse cancer models (Janz et al. 2006; Pelzer et al. 2006). Although the expression of ATF3 is thought to be primarily controlled by the binding of ATF4 to its promoter to upregulate its transcription, several studies have shown that non-ISR pathways including nuclear factor-kappa B (NF-κB), p53, and c-Jun N-terminal kinase (JNK) play important roles in the regulation of its expression (Thompson et al. 2009). Along with the above mentioned transcriptional control, ATF3 was also shown to be regulated by mRNA stabilization via a mechanism involving recruitment of RNA binding proteins to the 3′ untranslated region of the mRNA (Pan et al. 2005). However, the regulation of ATF3 following hypoxia has been debated. Several studies have shown that ATF3 is expressed only following anoxia but not during hypoxic conditions. Such expression of ATF3 following anoxia is independent of HIF1α and p53 and involves stabilization of the mRNA (Ameri et al. 2007). In contrast, circulating tumor cells (CTCs), which are associated with highly aggressive metastatic tumors, were highly hypoxic but had increased expression of ATF3 (Ameri et al. 2010).

Another transcriptional target of ATF4 that has been shown to be upregulated following nutrient stress (low amino acids and glucose) is *ASNS*. Intriguingly, increased *ASNS* expression has also been observed in hypoxic CTCs (Ameri et al. 2010). L-asparaginase is currently used as treatment in patients with childhood acute lymphoblastic leukemia and acute myeloblastic leukemia (Richards and Kilberg 2006). ATF4 transcriptionally regulates *ASNS* by binding to the C/EBP-ATF response elements (CAREs) in its promoter, which are comprised of two nutrient-sensing response elements (NSRE-I and NSRE-II) (Zhong et al. 2003). In response to ER stress, ATF4 binds to the NSREs, resulting in increased rate of *ASNS* transcription. However, with sustained ER stress, ATF3 and C/EBPβ replace ATF4 from the CARE elements, resulting in decreased promoter activity. This type of feedback inhibition following chronic ER stress is regarded as self-limiting regulation of ATF4 (Su and Kilberg 2008).

10.8 Hypoxia-Induced UPR and the Decision Between Cell Survival and Apoptosis

Stimulation of prolonged ER stress causes apoptosis in cells via hyperactivation of CHOP and IRE1, in addition to other various mechanisms (Fig. 10.1). CHOP interacts with liver inhibitory protein (LIP), an inhibitory isoform of C/EBPβ, following chronic ER stress. Interaction with LIP allows CHOP to translocate to the nucleus and repress transcription of *BCL2* (Chiribau et al. 2010). BCL2 interacts with several proteins from the BH3 family, such as BAD, NOXA, and PUMA (BCL2-associated agonist of cell death, phorbol-12-myristate-13-acetate-induced protein 1, BCL2 binding component 3, respectively), and sequesters these factors from BAX (BCL2-associated X protein)/BAD-mediated permeabilization of the mitochondrial membrane, leading to increased mitochondrial apoptosis (Tabas and Ron 2011). CHOP can also induce apoptosis in a more direct mechanism by heterodimerizing with C/EBPα and binding to the *BIM* (BCL2-like 11 (apoptosis facilitator)) promoter to increase its expression (Puthalakath et al. 2007). Thus, decreased BCL2 and increased BIM expression following chronic ER stress contribute to increased cellular apoptosis. In contrast, hypoxia-mediated chronic activation of ATF4 is beneficial for cellular survival, and ablation of ATF4 results in increased tumor cell sensitivity *in vitro* and *in vivo* (Bi et al. 2005). It is possible that the proapoptotic properties of CHOP may be modest compared to the much stronger anti-apoptotic activity of ATF4, which has multiple transcriptional targets including autophagy genes (see section on autophagy below).

Chronic ER stress is often associated with increased oxidative stress due to an increase in aberrant misfolding of proteins in the lumen of the ER. An important factor that contributes to the hyperoxidation of the ER is ERO1α, which is transcriptionally upregulated by CHOP (Marciniak et al. 2004). The mechanism through which CHOP-activated ERO1α increases cell death is via activation of inositol triphosphate receptor 1(IP3R1), an ER calcium release channel (Li et al. 2009). Hypoxia causes rapid efflux of ER Ca^{2+} into the cytoplasm by a similar activation of IP3R1 as observed in PC12 rat adrenal pheochromocytoma cells and cerebellar Purkinje cells (Patterson et al. 2004). Increased cytoplasmic Ca^{2+} levels activate Ca^{2+}/calmodulin-dependent protein kinase II (CaMKII), a calcium-activated kinase, which in turn upregulates NADPH oxidase subunit (NOX2). This ultimately leads to generation of ROS, leading to cell death (Tabas and Ron 2011). Chronic ER stress also contributes to upregulation (via ATF4 and CHOP) of death receptor 5 (DR5) and tribbles 3 (TRB3)—two factors with proapoptotic functions (Yamaguchi and Wang 2004; Ohoka et al. 2005). Finally, overexpression of the CHOP target GADD34 (growth arrest and DNA damage-inducible protein 34), a phosphatase cofactor responsible for dephosphorylating eIF2α, can result in premature resumption of protein synthesis, which, during extended periods of ER stress, has been suggested to cause cellular apoptosis due to overloading of the ER (Marciniak et al. 2004).

Even though CHOP is often regarded as the primary factor responsible for inducing cellular apoptosis, other arms of the UPR, especially IRE1, have also been linked to cell death. The apoptotic properties of IRE1 have been mostly attributed to the fact that IRE1 interacts with tumor necrosis factor receptor-associated factor 2 (TRAF2), an adapter protein that can stimulate cell death (Urano et al. 2000). There has also been evidence from coimmunoprecipitation experiments showing that activated IRE1 can directly interact with BAK and BAX to stimulate mitochondria-induced apoptosis (Hetz et al. 2006). Finally, chronic ER stress has been shown to induce ER-associated mRNA degradation via the endoribonuclease activity of IRE1 (Hollien et al. 2009). Hyperactivation of the endonuclease function of IRE1 has been shown to induce apoptosis by a mechanism which has not been clearly explored (Han et al. 2009).

High levels of ATF6 have been detected in regions of ischemic tissue which have an activated UPR (Doroudgar et al. 2009). Unlike the other two arms of the UPR, the role of ATF6 following chronic ER stress is not well understood. This is partly because ATF6 has been shown to regulate the induction of several cytoprotective chaperones, such as BiP, as well as apoptotic factors such as CHOP (Yoshida et al. 2000).

10.9 Hypoxia-Induced UPR and Autophagy

Eukaryotic cells have evolved several adaptive pathways to cope with various environmental and intracellular stress conditions. One of these key pathways is autophagy—a self-eating catabolic process that involves the formation of double membrane vesicles known as autophagosomes, which engulf cellular organelles and mediate their lysosomal breakdown. Thus, autophagy acts as a major regulatory mechanism to sequester harmful components of the cells and to replenish essential components, maintaining cellular homeostasis. Autophagy has been shown to be upregulated in highly aggressive and metastatic tumors (Kenific et al. 2010; Ravikumar et al. 2010; Rubinsztein et al. 2012). It has been suggested that autophagy represents an important mechanism in tumor progression by enabling cancer cells to survive when they encounter prolonged metabolic and hypoxic stress (Rubinsztein et al. 2012). In fact, several clinical trials have been initiated to test if inhibitors of autophagy sensitize cancer cells to chemotherapeutic agents (Yang et al. 2011; Rubinsztein et al. 2012).

Even though autophagy was discovered in lower eukaryotes as a mechanism to cope with nutritional stress conditions primarily, the UPR has now also been linked to the autophagic process. Of the three branches of the UPR, the PERK pathway has been shown to be the primary inducer of autophagy (Fig. 10.1) (He and Klionsky 2009; Rouschop and Wouters 2009; Hart et al. 2012). Cells which lack PERK or express mutant eIF2α (S51A) fail to induce a proper autophagic response. This inability to induce autophagy following chronic ER stress leads to increased cellular apoptosis (Ogata et al. 2006). Several studies suggest direct regulation of autophagy by UPR components that transcriptionally regulate several key autophagy-related genes (Atg). Hypoxia-mediated activation of the UPR leads to upregulation of LC3 and Atg5—a process under direct transcriptional regulation of ATF4 and CHOP (Rouschop et al. 2010). Recently, ATF4 was also shown to transcriptionally upregulate ULK1, a serine/threonine kinase which acts upstream of phosphatidylinositol-4,5-bisphosphate 3-kinase (PI3K) and induces formation of autophagosomes (Pike et al. 2012). Therefore, if global protein synthesis is downregulated following chronic hypoxic stress, the ISR allows the cells to preferentially upregulate genes required for autophagy, presumably to replenish cellular nutrient pools.

Enhanced tumorigenesis is often characterized by hyperactivation of oncogenes that upregulate cell cycle progression, proliferation, and protein synthesis. One such oncogene that is hyperactivated in cancers such as Burkitt's lymphoma is c-Myc. Even though chronic hypoxia has been shown to block c-Myc-mediated gene transcription by a mechanism that involves HIF1α-mediated disruption of the active c-Myc/Max complex, transient hypoxia, followed by reoxygenation, leads to its activation (Huang 2008). As c-Myc enhances protein synthesis by increasing ribosome biogenesis, overexpression of c-Myc results in activation of the UPR and an accompanying PERK-mediated cytoprotective autophagy response (Hart et al. 2012). UPR-mediated autophagy by overexpression of c-Myc was also observed *in vivo* as isolated B-cell lymphocytes from transgenic E_{μ}-Myc mice, which are characterized by hyperactive c-Myc, showed increased UPR activation and a concomitant induction of autophagy (Hart et al. 2012). Even though PERK enhances the levels of autophagy in cells, the IRE1-dependent arm of the UPR has been shown to prevent autophagy induction. Inhibition of IRE1 and its downstream target XBP1 were shown to induce basal autophagy *in vitro* and *in vivo* (Hetz et al. 2009; Kroemer et al. 2010). However, the mechanism by which IRE1 inhibits autophagy has yet to be elucidated.

Cellular autophagy is negatively regulated by activation of mTOR. As a consequence, cells have developed several negative feedback mechanisms to prevent mTOR activation in response to cellular stress conditions such as hypoxia (Rubinsztein et al. 2012). As mentioned before, chronic hypoxia leads to activation of JNK, which is mediated by IRE1/TRAF2. Activated JNK phosphorylates and inactivates insulin receptor substrate 1 (IRS1), an inducer of mTOR activity (Wouters and Koritzinsky 2008). Alternatively, mTOR can also be negatively regulated by hypoxia-mediated increased efflux of calcium ions from the ER to the cytoplasm. Increased calcium concentration in the cytosol activates CAMKK2 (calcium/calmodulin-dependent protein kinase kinase 2, beta), which in turn activates AMPK (protein kinase, AMP-activated), leading to inhibition of mTOR activity (Wouters and Koritzinsky 2008).

Hypoxia-induced autophagy may also play a role in breast tumor development. Hypoxia has been detected in early breast cancer lesions. Activation of the UPR, and specifically PERK, was shown to be important for effective acini formation from early breast cell carcinomas (Avivar-Valderas et al. 2011). PERK was shown to initiate a cytoprotective autophagic response in these cells that provides resistance to anoikis-mediated cell death—a phenomenon which causes cell death following detachment from the extracellular matrix. The absence of an effective UPR in these cells prevents acinar formation and sensitizes cells to anoikis-mediated cell death (Avivar-Valderas et al. 2011).

10.10 Therapeutic Targeting of the UPR

Therapeutic targeting of the UPR is an attractive approach for cancer treatment, since this pathway is frequently activated in animal and human tumors but less so in normal tissues (with exceptions being cells with high secretory capacity such as pancreatic β cells and B lymphocytes) (Todd et al. 2008; Fonseca et al. 2010; Li et al. 2011). Strategies for doing so fall into two main categories: inhibition of the UPR components or overactivation of the UPR pathway. It is well established that tumor cells rely on the UPR to cope with the nutrient-deprived, hypoxic microenvironment, so that inhibiting one or more branches of this pathway would impair the ability of the tumor cells to counteract these stresses, leading to cell death. On the other hand, overactivating the UPR by pharmacological means would also be a feasible method for specifically targeting tumor cells because, as explained earlier, prolonged or overwhelming stresses shift the balance in favor of the proapoptotic functions of this signaling cascade (Fels and Koumenis 2006).

10.10.1 Small-Molecule Inhibitors of UPR Components

One approach for inhibiting the UPR is to develop small-molecule drugs that would target the initial activators of the pathway, preventing the tumor cell from coping with hypoxic stress. Indeed, a group from GlaxoSmithKline recently reported having identified an orally available PERK inhibitor, GSK2606414A, that exhibited potent antitumor activity against human pancreatic xenograft tumors (Axten et al. 2012). Another group has characterized a compound that inhibited the endonuclease activity of IRE1 without affecting its kinase activity and reduced the tumor burden in mice subcutaneously implanted with human multiple myeloma cells (Papandreou et al. 2011). Other inhibitors of IRE1 endonuclease activity, such as trierixin, could also prove to be clinically useful (Tashiro et al. 2007).

A different method for blocking the prosurvival function of the UPR would be to target GRP78 activity. Epigallocatechin gallate (EGCG), which targets the adenosine triphosphate (ATP)-binding domain of GRP78, and versipelostatin, which downregulates transcription of GRP78, both induce cytotoxicity in tumor cells *in vitro* (Park et al. 2004; Ermakova et al. 2006; Matsuo et al. 2009). The catalytic A subunit of bacterial AB5 subtilase cytotoxin selectively cleaves and inactivates GRP78 leading to tumor cell death *in vivo* and *in vitro*; however, subtilase causes unresolved ER stress and is lethal in mice, indicating it would not be a useful cancer therapy (Paton et al. 2006). Versipelostatin has the most promise as a therapeutic strategy though, as it appears to be specific for GRP78, whereas EGCG is not (Healy et al. 2009).

10.10.2 Overactivation of the UPR

The other approach for preferential killing of hypoxic tumor cells would be to apply additional ER stress to further activate the UPR and tip the balance toward the proapoptotic signaling function of the UPR. One extensively studied class of compounds shown to activate the UPR is proteasome inhibitors (Fribley et al. 2004; Schönthal 2012). Inhibiting the proteasome prevents protein degradation and results in an accumulation of unfolded and misfolded proteins, thereby activating the UPR. Our lab has previously demonstrated that the reversible proteasome inhibitor bortezomib preferentially kills hypoxic over normoxic tumor cells (Fels et al. 2008). In addition, treating multiple myeloma cells, which have a constitutively high level of UPR activation due to their role as secretory cells, with bortezomib induced apoptosis (Obeng et al. 2006). In fact, several clinical trials are ongoing to test the efficacy of various proteasome inhibitors for treatment of multiple myeloma (Li et al. 2011). One example is carfilzomib (CFZ), a peptide epoxyketone that irreversibly inhibits the proteasome, which was recently approved by the Food and Drug Administration (FDA) for patients with relapsed and refractory multiple myeloma. In addition, CFZ is currently being examined in clinical trials for the treatment of acute myeloid leukemia, acute lymphocytic leukemia, and chronic lymphocytic lymphoma (Crawford et al. 2011; Goldberg 2012). Other proteasome inhibitors, such as ONX0912,

MLN9708, and NPI-0052, are in various phases of clinical trials for treatment of solid tumors, multiple myeloma, and lymphoma, respectively (Crawford et al. 2011; Goldberg 2012), and could also demonstrate synergistic effects when combined with UPR inhibitors.

As discussed earlier, activation of the UPR induces autophagy to facilitate degradation of the misfolded proteins and inhibition of autophagy aggravates ER stress, so another tactic for exploiting the UPR in hypoxic tumor cells would be to inhibit autophagy. Chloroquine, which blocks autophagy by inhibiting lysosome acidification, has been shown to sensitize tumor cells to hypoxia and to decrease the hypoxic fraction of xenograft tumors while increasing tumor response to radiation (Bertolotti et al. 2000; Rouschop et al. 2010). Other agents that inhibit autophagy, such as bafilomycin A or 3-methyladenine, could also be useful agents for inducing overactivation of the UPR in hypoxic cells (Aronson and Davies 2012).

Several other strategies could also be utilized to hyperactivate the UPR in conjunction with hypoxia. For instance, the cyclooxygenase 2 inhibitor celecoxib was shown to cause ER stress and CHOP induction by inhibiting ER calcium pumps responsible for maintaining the calcium concentration within the ER (Tsutsumi et al. 2004; Pyrko et al. 2007). Both celecoxib and its noncoxib analog dimethyl-celecoxib act synergistically with bortezomib to induce tumor cell killing (Kardosh et al. 2008). Another approach that has shown some promise is inhibition of ER-associated protein degradation. Eeyarestatin I, a compound identified in a screen for inhibitors of proteasome-independent misfolded protein degradation, potentiates the cytotoxic effects of bortezomib *in vitro* and reduces tumor burden *in vivo* (Fiebiger et al. 2004; Wang et al. 2009; Valle et al. 2011). Additionally, our group also recently published a study describing the identification and characterization of a small-molecule compound, E235. While the mechanism by which it exerts its effects remains unclear, E235 upregulates ATF4 expression only in tumor cells, suggesting that it could have specific cytotoxic effects on hypoxic tumor cells (Sayers et al. 2013).

In summary, there have been a multitude of studies showing promising antitumor effects via inhibition or hyperactivation of the UPR. In addition, combining two agents that overactivate the UPR has also yielded encouraging results. Therefore, ER stress activators or inhibitors could have selective cytotoxic effects in hypoxic cells, which rely on UPR signaling for survival.

Acknowledgments The authors are supported by NIH R01 CA094214 and NIH R01 CA139362. C.M.S. and S.L.L. are supported by the National Institutes of Health National Cancer Institute Fellowship Training Grant [5-T32-CA-009677-20].

References

Ameri K, Harris AL (2008) Activating transcription factor 4. Int J Biochem Cell Biol 40:14–21

Ameri K, Lewis CE, Raida M, Sowter H, Hai T, Harris AL (2004) Anoxic induction of ATF-4 through HIF-1–independent pathways of protein stabilization in human cancer cells. Blood 103:1876–1882

Ameri K, Hammond EM, Culmsee C, Raida M, Katschinski DM, Wenger RH, Wagner E, Davis RJ, Hai T, Denko N, Harris AL (2007) Induction of activating transcription factor 3 by anoxia is independent of p53 and the hypoxic HIF signalling pathway. Oncogene 26:284–289

Ameri K, Luong R, Zhang H, Powell AA, Montgomery KD, Espinosa I, Bouley DM, Harris AL, Jeffrey SS (2010) Circulating tumour cells demonstrate an altered response to hypoxia and an aggressive phenotype. Br J Cancer 102:561–569

Aronson LI, Davies FE (2012) DangER: protein ovERload. Targeting protein degradation to treat myeloma. Haematologica 97:1119–1130

Avivar-Valderas A, Salas E, Bobrovnikova-Marjon E, Diehl JA, Nagi C, Debnath J, Aguirre-Ghiso JA (2011) PERK Integrates Autophagy and Oxidative Stress Responses To Promote Survival during Extracellular Matrix Detachment. Mol Cell Biol 31:3616–3629

Axten JM, Medina JR, Feng Y, Shu A, Romeril SP, Grant SW, Li WHH, Heerding DA, Minthorn E, Mencken T, Atkins C, Liu Q, Rabindran S, Kumar R, Hong X, Goetz A, Stanley T, Taylor JD, Sigethy SD, Tomberlin GH, Hassell AM, Kahler KM, Shewchuk LM, Gampe RT (2012) Discovery of 7-Methyl-5-(1-{[3-(trifluoromethyl)phenyl]acetyl}-2,3-dihydro-1H-indol-5-yl)-7H-pyrrolo[2,3-d]pyrimidin-4-amine (GSK2606414), a Potent and Selective First-in-Class Inhibitor of Protein Kinase R (PKR)-like Endoplasmic Reticulum Kinase (PERK). J Med Chem 55:7193–7207

Bays NW, Gardner RG, Seelig LP, Joazeiro CA, Hampton RY (2001) Hrd1p/Der3p is a membrane-anchored ubiquitin ligase required for ER-associated degradation. Nat Cell Biol 3:24–29

Bertolotti A, Zhang Y, Hendershot LM, Harding HP, Ron D (2000) Dynamic interaction of BiP and ER stress transducers in the unfolded-protein response. Nat Cell Biol 2:326–332

Bertout JA, Patel SA, Simon MC (2008) The impact of O2 availability on human cancer. Nat Rev Cancer 8:967–975

Bi M, Naczki C, Koritzinsky M, Fels D, Blais J, Hu N, Harding H, Novoa I, Varia M, Raleigh J, Scheuner D, Kaufman RJ, Bell J, Ron D, Wouters BG, Koumenis C (2005) ER stress-regulated translation increases tolerance to extreme hypoxia and promotes tumor growth. EMBO J 24:3470–3481

Blais JD, Filipenko V, Bi MX, Harding HP, Ron D, Koumenis C, Wouters BG, Bell JC (2004) Activating transcription factor 4 is translationally regulated by hypoxic stress. Mol Cell Biol 24:7469–7482

Bobrovnikova-Marjon E, Grigoriadou C, Pytel D, Zhang F, Ye J, Koumenis C, Cavener D, Diehl JA (2010) PERK promotes cancer cell proliferation and tumor growth by limiting oxidative DNA damage. Oncogene 29:3881–3895

Brown JM (1990) Tumor Hypoxia, Drug Resistance, and Metastases. J Natl Cancer Inst 82:338–339

Brown JM, Wilson WR (2004) Exploiting tumour hypoxia in cancer treatment. Nat Rev Cancer 4:437–447

Cairns RA, Hill RP (2004) Acute Hypoxia Enhances Spontaneous Lymph Node Metastasis in an Orthotopic Murine Model of Human Cervical Carcinoma. Cancer Res 64:2054–2061

Calfon M, Zeng H, Urano F, Till JH, Hubbard SR, Harding HP, Clark SG, Ron D (2002) IRE1 couples endoplasmic reticulum load to secretory capacity by processing the XBP-1 mRNA. Nature 415:92–96

Carrasco DR, Sukhdeo K, Protopopova M, Sinha R, Enos M, Carrasco DE, Zheng M, Mani M, Henderson J, Pinkus GS, Munshi N, Horner J, Ivanova EV, Protopopov A, Anderson KC, Tonon G, DePinho RA (2007) The differentiation and stress response factor XBP-1 drives multiple myeloma pathogenesis. Cancer Cell 11:349–360

Chang Q, Jurisica I, Do T, Hedley DW (2011) Hypoxia Predicts Aggressive Growth and Spontaneous Metastasis Formation from Orthotopically Grown Primary Xenografts of Human Pancreatic Cancer. Cancer Res 71:3110–3120

Chen X, Shen J, Prywes R (2002) The Luminal Domain of ATF6 Senses Endoplasmic Reticulum (ER) Stress and Causes Translocation of ATF6 from the ER to the Golgi. J Biol Chem 277:13045–13052

Chen Y, Feldman DE, Deng C, Brown JA, De Giacomo AF, Gaw AF, Shi G, Le QT, Brown JM, Koong AC (2005) Identification of Mitogen-Activated Protein Kinase Signaling Pathways That Confer Resistance to Endoplasmic Reticulum Stress in Saccharomyces cerevisiae. Mol Cancer Res 3:669–677

Chiribau CB, Gaccioli F, Huang CC, Yuan CL, Hatzoglou M (2010) Molecular symbiosis of CHOP and C/EBP beta isoform LIP contributes to endoplasmic reticulum stress-induced apoptosis. Mol Cell Biol 30:3722–3731

Crawford LJ, Walker B, Irvine AE (2011) Proteasome inhibitors in cancer therapy. J Cell Commun Signal 5:101–110

Cullinan SB, Zhang D, Hannink M, Arvisais E, Kaufman RJ, Diehl JA (2003) Nrf2 is a direct PERK substrate and effector of PERK-dependent cell survival. Mol Cell Biol 23:7198–7209

Dickhout JG, Carlisle RE, Jerome DE, Mohammed-Ali Z, Jiang H, Yang G, Mani S, Garg SK, Banerjee R, Kaufman RJ, Maclean KN, Wang R, Austin RC (2012) Integrated Stress Response Modulates Cellular Redox State via Induction of Cystathionine γ-Lyase: Cross-Talk Between Integrated Stress Response and Thiol Metabolism. J Biol Chem 287:7603–7614

Doroudgar S, Thuerauf DJ, Marcinko MC, Belmont PJ, Glembotski CC (2009) Ischemia Activates the ATF6 Branch of the Endoplasmic Reticulum Stress Response. J Biol Chem 284:29735–29745

Ermakova SP, Kang BS, Choi BY, Choi HS, Schuster TF, Ma W-Y, Bode AM, Dong Z (2006) (–)-Epigallocatechin Gallate Overcomes Resistance to Etoposide-Induced Cell Death by Targeting the Molecular Chaperone Glucose-Regulated Protein 78. Cancer Res 66:9260–9269

Fawcett TW, Martindale JL, Guyton KZ, Hai T, Holbrook NJ (1999) Complexes containing activating transcription factor (ATF)/cAMP-responsive-element-binding protein (CREB) interact with the CCAAT/enhancer-binding protein (C/EBP)-ATF composite site to regulate Gadd153 expression during the stress response. Biochem J 339:135–141

Fels D, Koumenis C (2006) The PERK/eIF2α/ATF4 module of the UPR in hypoxia resistance and tumor growth. Cancer Biol Ther 5:723–728

Fels DR, Ye J, Segan AT, Kridel SJ, Spiotto M, Olson M, Koong AC, Koumenis C (2008) Preferential Cytotoxicity of Bortezomib toward Hypoxic Tumor Cells via Overactivation of Endoplasmic Reticulum Stress Pathways. Cancer Res 68:9323–9330

Fernandez PM, Tabbara SO, Jacobs LK, Manning FC, Tsangaris TN, Schwartz AM, Kennedy KA, Patierno SR (2000) Overexpression of the glucose-regulated stress gene GRP78 in malignant but not benign human breast lesions. Breast Cancer Res Treat 59:15–26

Fiebiger E, Hirsch C, Vyas JM, Gordon E, Ploegh HL, Tortorella D (2004) Dissection of the Dislocation Pathway for Type I Membrane Proteins with a New Small Molecule Inhibitor, Eeyarestatin. Mol Biol Cell 15:1635–1646

Fonseca SG, Urano F, Burcin M, Gromada J (2010) Stress hypERactivation in the β-cell. Islets 2:1–9

Frand AR, Kaiser CA (1999) Ero1p oxidizes protein disulfide isomerase in a pathway for disulfide bond formation in the endoplasmic reticulum. Mol Cell 4:469–477

Fribley A, Zeng Q, Wang C-Y (2004) Proteasome Inhibitor PS-341 Induces Apoptosis through Induction of Endoplasmic Reticulum Stress-Reactive Oxygen Species in Head and Neck Squamous Cell Carcinoma Cells. Mol Cell Biol 24:9695–9704

Friedman AD (1996) GADD153/CHOP, a DNA Damage-inducible Protein, Reduced CAAT/Enhancer Binding Protein Activities and Increased Apoptosis in 32D cl3 Myeloid Cells. Cancer Res 56:3250–3256

Fujimoto T, Onda M, Nagai H, Nagahata T, Ogawa K, Emi M (2003) Upregulation and overexpression of human X-box binding protein 1 (hXBP-1) gene in primary breast cancers. Breast Cancer 10:301–306

Gardner LB (2008) Hypoxic inhibition of nonsense-mediated RNA decay regulates gene expression and the integrated stress response. Mol Cell Biol 28:3729–3741

Gardner LB (2010) Nonsense-mediated RNA decay regulation by cellular stress: implications for tumorigenesis. Mol Cancer Res 8:295–308

Ghosh R, Lipson KL, Sargent KE, Mercurio AM, Hunt JS, Ron D, Urano F (2010) Transcriptional regulation of VEGF-A by the unfolded protein response pathway. PLoS One 5:e9575

Goldberg AL (2012) Development of proteasome inhibitors as research tools and cancer drugs. J Cell Biol 199:583–588

Gupta S, McGrath B, Cavener DR (2009) PERK regulates the proliferation and development of insulin-secreting beta-cell tumors in the endocrine pancreas of mice. PLoS One 4:e8008

Han D, Lerner AG, Vande Walle L, Upton JP, Xu W, Hagen A, Backes BJ, Oakes SA, Papa FR (2009) IRE1alpha kinase activation modes control alternate endoribonuclease outputs to determine divergent cell fates. Cell 138:562–575

Harding HP, Zhang Y, Ron D (1999) Protein translation and folding are coupled by an endoplasmic-reticulum-resident kinase. Nature 397:271–274

Harding HP, Novoa I, Zhang Y, Zeng H, Wek R, Schapira M, Ron D (2000a) Regulated Translation Initiation Controls Stress-Induced Gene Expression in Mammalian Cells. Mol Cell 6:1099–1108

Harding HP, Zhang Y, Bertolotti A, Zeng H, Ron D (2000b) Perk Is Essential for Translational Regulation and Cell Survival during the Unfolded Protein Response. Mol Cell 5:897–904

Harding HP, Zhang Y, Zeng H, Novoa I, Lu PD, Calfon M, Sadri N, Yun C, Popko B, Paules R, Stojdl DF, Bell JC, Hettmann T, Leiden JM, Ron D (2003) An integrated stress response regulates amino acid metabolism and resistance to oxidative stress. Mol Cell 11:619–633

Hart LS, Cunningham JT, Datta T, Dey S, Tameire F, Lehman SL, Qiu B, Zhang H, Cerniglia G, Bi M, Li Y, Gao Y, Liu H, Li C, Maity A, Thomas-Tikhonenko A, Perl AE, Koong A, Fuchs SY, Diehl JA, Mills IG, Ruggero D, Koumenis C (2012) ER stress-mediated autophagy promotes Myc-dependent transformation and tumor growth. J Clin Invest 122:4621–4634

Haze K, Yoshida H, Yanagi H, Yura T, Mori K (1999) Mammalian Transcription Factor ATF6 Is Synthesized as a Transmembrane Protein and Activated by Proteolysis in Response to Endoplasmic Reticulum Stress. Mol Biol Cell 10:3787–3799

He C, Klionsky DJ (2009) Regulation mechanisms and signaling pathways of autophagy. Annu Rev Genet 43:67–93

He CH, Gong P, Hu B, Stewart D, Choi ME, Choi AMK, Alam J (2001) Identification of Activating Transcription Factor 4 (ATF4) as an Nrf2-interacting Protein: Implication for Heme Oxygenase-1 Gene Regulation. J Biol Chem 276:20858–20865

Healy SJM, Gorman AM, Mousavi-Shafaei P, Gupta S, Samali A (2009) Targeting the endoplasmic reticulum-stress response as an anticancer strategy. Eur J Pharmacol 625:234–246

Hetz C, Bernasconi P, Fisher J, Lee A-H, Bassik MC, Antonsson B, Brandt GS, Iwakoshi NN, Schinzel A, Glimcher LH, Korsmeyer SJ (2006) Proapoptotic BAX and BAK Modulate the Unfolded Protein Response by a Direct Interaction with IRE1α. Science 312:572–576

Hetz C, Thielen P, Matus S, Nassif M, Court F, Kiffin R, Martinez G, Cuervo AM, Brown RH, Glimcher LH (2009) XBP-1 deficiency in the nervous system protects against amyotrophic lateral sclerosis by increasing autophagy. Genes Dev 23:2294–2306

Höckel M, Vaupel P (2001) Biological consequences of tumor hypoxia. Semin Oncol 28. Supplement 8:36–41

Hollien J, Lin JH, Li H, Stevens N, Walter P, Weissman JS (2009) Regulated Ire1-dependent decay of messenger RNAs in mammalian cells. J Cell Biol 186:323–331

Huang LE (2008) Carrot and stick: HIF-alpha engages c-Myc in hypoxic adaptation. Cell Death Differ 15:672–677

Iwakoshi NN, Lee A-H, Glimcher LH (2003) The X-box binding protein-1 transcription factor is required for plasma cell differentiation and the unfolded protein response. Immunol Rev 194:29–38

Jamora C, Dennert G, Lee AS (1996) Inhibition of tumor progression by suppression of stress protein GRP78/BiP induction in fibrosarcoma B/C10ME. Proc Natl Acad Sci USA 93:7690–7694

Janz M, Hummel M, Truss M, Wollert-Wulf B, Mathas S, Johrens K, Hagemeier C, Bommert K, Stein H, Dorken B, Bargou RC (2006) Classical Hodgkin lymphoma is characterized by high constitutive expression of activating transcription factor 3 (ATF3), which promotes viability of Hodgkin/Reed-Sternberg cells. Blood 107:2536–2539

Kardosh A, Golden EB, Pyrko P, Uddin J, Hofman FM, Chen TC, Louie SG, Petasis NA, Schönthal AH (2008) Aggravated Endoplasmic Reticulum Stress as a Basis for Enhanced Glioblastoma Cell Killing by Bortezomib in Combination with Celecoxib or Its Non-Coxib Analogue, 2,5-Dimethyl-Celecoxib. Cancer Res 68:843–851

Kaufman RJ (2002) Orchestrating the unfolded protein response in health and disease. J Clin Invest 110:1389–1398

Kenific CM, Thorburn A, Debnath J (2010) Autophagy and metastasis: another double-edged sword. Curr Opin Cell Biol 22:241–245

Kizaka-Kondoh S, Inoue M, Harada H, Hiraoka M (2003) Tumor hypoxia: A target for selective cancer therapy. Cancer Sci 94:1021–1028

Koditz J, Nesper J, Wottawa M, Stiehl DP, Camenisch G, Franke C, Myllyharju J, Wenger RH, Katschinski DM (2007) Oxygen-dependent ATF-4 stability is mediated by the PHD3 oxygen sensor. Blood 110:3610–3617

Koong AC, Auger EA, Chen EY, Giaccia AJ (1994) The regulation of GRP78 and messenger RNA levels by hypoxia is modulated by protein kinase C activators and inhibitors. Radiat Res 138:S60–S63

Koumenis C (2006) ER stress, hypoxia tolerance and tumor progression. Curr Mol Med 6:55–69

Koumenis C, Bi M, Ye J, Feldman D, Koong AC (2007) Hypoxia and the unfolded protein response. Methods Enzymol 435:275–293

Kroemer G, Marino G, Levine B (2010) Autophagy and the integrated stress response. Mol Cell 40:280–293

Lee AH, Iwakoshi NN, Glimcher LH (2003a) XBP-1 Regulates a Subset of Endoplasmic Reticulum Resident Chaperone Genes in the Unfolded Protein Response. Mol Cell Biol 23:7448–7459

Lee AH, Iwakoshi NN, Anderson KC, Glimcher LH (2003b) Proteasome inhibitors disrupt the unfolded protein response in myeloma cells. Proc Natl Acad Sci USA 100:9946–9951

Lee AS (1992) Mammalian stress response: induction of the glucose-regulated protein family. Curr Opin Cell Biol 4:267–273

Li G, Mongillo M, Chin KT, Harding H, Ron D, Marks AR, Tabas I (2009) Role of ERO1-alpha-mediated stimulation of inositol 1,4,5-triphosphate receptor activity in endoplasmic reticulum stress-induced apoptosis. J Cell Biol 186:783–792

Li X, Zhang K, Li Z (2011) Unfolded protein response in cancer: the Physician's perspective. J Hematol Oncol 4:8

Liu L, Wise DR, Diehl JA, Simon MC (2008) Hypoxic Reactive Oxygen Species Regulate the Integrated Stress Response and Cell Survival. J Biol Chem 283:31153–31162

Ma Y, Brewer JW, Diehl JA, Hendershot LM (2002) Two distinct stress signaling pathways converge upon the CHOP promoter during the mammalian unfolded protein response. J Mol Biol 318:1351–1365

Marciniak SJ, Yun CY, Oyadomari S, Novoa I, Zhang Y, Jungreis R, Nagata K, Harding HP, Ron D (2004) CHOP induces death by promoting protein synthesis and oxidation in the stressed endoplasmic reticulum. Genes Dev 18:3066–3077

Matsuo J, Tsukumo Y, Sakurai J, Tsukahara S, Park H-R, Shin-ya K, Watanabe T, Tsuruo T, Tomida A (2009) Preventing the unfolded protein response via aberrant activation of 4E-binding protein 1 by versipelostatin. Cancer Sci 100:327–333

Meyer N, Penn LZ (2008) Reflecting on 25 years with MYC. Nat Rev Cancer 8:976–990

Obeid M, Tesniere A, Ghiringhelli F, Fimia GM, Apetoh L, Perfettini J-L, Castedo M, Mignot G, Panaretakis T, Casares N, Metivier D, Larochette N, Endert P van, Ciccosanti F, Piacentini M, Zitvogel L, Kroemer G (2007) Calreticulin exposure dictates the immunogenicity of cancer cell death. Nat Med 13:54–61

Obeng EA, Carlson LM, Gutman DM, Harrington WJ, Lee KP, Boise LH (2006) Proteasome inhibitors induce a terminal unfolded protein response in multiple myeloma cells. Blood 107:4907–4916

Ogata M, Hino S, Saito A, Morikawa K, Kondo S, Kanemoto S, Murakami T, Taniguchi M, Tanii I, Yoshinaga K, Shiosaka S, Hammarback JA, Urano F, Imaizumi K (2006) Autophagy is activated for cell survival after endoplasmic reticulum stress. Mol Cell Biol 26:9220–9231

Ohoka N, Yoshii S, Hattori T, Onozaki K, Hayashi H (2005) TRB3, a novel ER stress-inducible gene, is induced via ATF4-CHOP pathway and is involved in cell death. EMBO J 24:1243–1255

Okada T, Yoshida H, Akazawa R, Negishi M, Mori K (2002) Distinct roles of activating transcription factor 6 (ATF6) and double-stranded RNA-activated protein kinase-like endoplasmic reticulum kinase (PERK) in transcription during the mammalian unfolded protein response. Biochem J 366:585–594

Padyana AK, Qiu H, Roll-Mecak A, Hinnebusch AG, Burley SK (2005) Structural basis for autoinhibition and mutational activation of eukaryotic initiation factor 2alpha protein kinase GCN2. J Biol Chem 280:29289–29299

Pan YX, Chen H, Kilberg MS (2005) Interaction of RNA-binding proteins HuR and AUF1 with the human ATF3 mRNA 3'-untranslated region regulates its amino acid limitation-induced stabilization. J Biol Chem 280:34609–34616

Panaretakis T, Joza N, Modjtahedi N, Tesniere A, Vitale I, Durchschlag M, Fimia GM, Kepp O, Piacentini M, Froehlich KU, Endert P van, Zitvogel L, Madeo F, Kroemer G (2008) The cotranslocation of ERp57 and calreticulin determines the immunogenicity of cell death. Cell Death Differ 15:1499–1509

Panaretakis T, Kepp O, Brockmeier U, Tesniere A, Bjorklund AC, Chapman DC, Durchschlag M, Joza N, Pierron G, Endert P van, Yuan J, Zitvogel L, Madeo F, Williams DB, Kroemer G (2009) Mechanisms of pre-apoptotic calreticulin exposure in immunogenic cell death. EMBO J 28:578–590

Papandreou I, Denko NC, Olson M, Van Melckebeke H, Lust S, Tam A, Solow-Cordero DE, Bouley DM, Offner F, Niwa M, Koong AC (2011) Identification of an Ire1alpha endonuclease specific inhibitor with cytotoxic activity against human multiple myeloma. Blood 117:1311–1314

Park H-R, Tomida A, Sato S, Tsukumo Y, Yun J, Yamori T, Hayakawa Y, Tsuruo T, Shin-ya K (2004) Effect on Tumor Cells of Blocking Survival Response to Glucose Deprivation. J Natl Cancer Inst 96:1300–1310

Paton AW, Beddoe T, Thorpe CM, Whisstock JC, Wilce MCJ, Rossjohn J, Talbot UM, Paton JC (2006) AB5 subtilase cytotoxin inactivates the endoplasmic reticulum chaperone BiP. Nature 443:548–552

Patterson RL, Boehning D, Snyder SH (2004) Inositol 1,4,5-trisphosphate receptors as signal integrators. Annu Rev Biochem 73:437–465

Pelzer AE, Bektic J, Haag P, Berger AP, Pycha A, Schafer G, Rogatsch H, Horninger W, Bartsch G, Klocker H (2006) The expression of transcription factor activating transcription factor 3 in the human prostate and its regulation by androgen in prostate cancer. J Urol 175:1517–1522

Pereira ER, Liao N, Neale GA, Hendershot LM (2010) Transcriptional and Post-Transcriptional Regulation of Proangiogenic Factors by the Unfolded Protein Response. PLoS One 5:e12521

Pike LR, Phadwal K, Simon AK, Harris AL (2012) ATF4 orchestrates a program of BH3-only protein expression in severe hypoxia. Mol Biol Rep 39:10811–10822

Pike LRG, Singleton DC, Buffa F, Abramczyk O, Phadwal K, Li JL, Simon AK, Murray JT, Harris AL (2013) Transcriptional up-regulation of ULK1 by ATF4 contributes to cancer cell survival. Biochem J 449:389–400

Puthalakath H, O'Reilly LA, Gunn P, Lee L, Kelly PN, Huntington ND, Hughes PD, Michalak EM, McKimm-Breschkin J, Motoyama N, Gotoh T, Akira S, Bouillet P, Strasser A (2007) ER stress triggers apoptosis by activating BH3-only protein Bim. Cell 129:1337–1349

Pyrko P, Kardosh A, Liu Y-T, Soriano N, Xiong W, Chow RH, Uddin J, Petasis NA, Mircheff AK, Farley RA, Louie SG, Chen TC, Schönthal AH (2007) Calcium-activated endoplasmic reticulum stress as a major component of tumor cell death induced by 2,5-dimethyl-celecoxib, a non-coxib analogue of celecoxib. Mol Cancer Ther 6:1262–1275

Ravikumar B, Sarkar S, Davies JE, Futter M, Garcia-Arencibia M, Green-Thompson ZW, Jimenez-Sanchez M, Korolchuk VI, Lichtenberg M, Luo S, Massey DC, Menzies FM, Moreau K, Narayanan U, Renna M, Siddiqi FH, Underwood BR, Winslow AR, Rubinsztein DC (2010) Regulation of mammalian autophagy in physiology and pathophysiology. Physiol Rev 90:1383–1435

Richards NG, Kilberg MS (2006) Asparagine synthetase chemotherapy. Annu Rev Biochem 75:629–654

Romero-Ramirez L, Cao H, Nelson D, Hammond E, Lee A-H, Yoshida H, Mori K, Glimcher LH, Denko NC, Giaccia AJ, Le Q-T, Koong AC (2004) XBP1 Is Essential for Survival under Hypoxic Conditions and Is Required for Tumor Growth. Cancer Res 64:5943–5947

Ron D, Walter P (2007) Signal integration in the endoplasmic reticulum unfolded protein response. Nat Rev Mol Cell Biol 8:519–529

Rouschop KM, Wouters BG (2009) Regulation of autophagy through multiple independent hypoxic signaling pathways. Curr Mol Med 9:417–424
Rouschop KMA, Beucken T van den, Dubois L, Niessen H, Bussink J, Savelkouls K, Keulers T, Mujcic H, Landuyt W, Voncken JW, Lambin P, Kogel AJ van der, Koritzinsky M, Wouters BG (2010) The unfolded protein response protects human tumor cells during hypoxia through regulation of the autophagy genes MAP1LC3B and ATG5. J Clin Invest 120:127–141
Rubinsztein DC, Codogno P, Levine B (2012) Autophagy modulation as a potential therapeutic target for diverse diseases. Nat Rev Drug Discov 11:709–730
Rzymski T, Milani M, Pike L, Buffa F, Mellor HR, Winchester L, Pires I, Hammond E, Ragoussis I, Harris AL (2010) Regulation of autophagy by ATF4 in response to severe hypoxia. Oncogene 29:4424–4435
Sayers CM, Papandreou I, Guttmann DM, Maas NL, Diehl JA, Witze ES, Koong AC, Koumenis C (2013) Identification and Characterization of a Potent Activator of p53-Independent Cellular Senescence via a Small-Molecule Screen for Modifiers of the Integrated Stress Response. Mol Pharmacol 83:594–604
Schewe DM, Aguirre-Ghiso JA (2008) ATF6α-Rheb-mTOR signaling promotes survival of dormant tumor cells in vivo. Proc Natl Acad Sci USA 105:10519–10524
Schönthal AH (2012) Pharmacological targeting of endoplasmic reticulum stress signaling in cancer. Biochem Pharmacol
Semenza GL (2009) Regulation of Oxygen Homeostasis by Hypoxia-Inducible Factor 1. Physiology 24:97–106
Shen J, Chen X, Hendershot L, Prywes R (2002) ER Stress Regulation of ATF6 Localization by Dissociation of BiP/GRP78 Binding and Unmasking of Golgi Localization Signals. Dev Cell 3:99–111
Shi Y, Vattem KM, Sood R, An J, Liang J, Stramm L, Wek RC (1998) Identification and Characterization of Pancreatic Eukaryotic Initiation Factor 2 α-Subunit Kinase, PEK, Involved in Translational Control. Mol Cell Biol 18:7499–7509
Shuda M, Kondoh N, Imazeki N, Tanaka K, Okada T, Mori K, Hada A, Arai M, Wakatsuki T, Matsubara O, Yamamoto N, Yamamoto M (2003) Activation of the ATF6, XBP1 and grp78 genes in human hepatocellular carcinoma: a possible involvement of the ER stress pathway in hepatocarcinogenesis. J Hepatol 38:605–614
Smith MH, Ploegh HL, Weissman JS (2011) Road to Ruin: Targeting Proteins for Degradation in the Endoplasmic Reticulum. Science 334:1086–1090
Su N, Kilberg MS (2008) C/EBP homology protein (CHOP) interacts with activating transcription factor 4 (ATF4) and negatively regulates the stress-dependent induction of the asparagine synthetase gene. J Biol Chem 283:35106–35117
Tabas I, Ron D (2011) Integrating the mechanisms of apoptosis induced by endoplasmic reticulum stress. Nat Cell Biol 13:184–190
Tashiro E, Hironiwa N, Kitagawa M, Futamura Y, Suzuki S, Nishio M, Imoto M (2007) Trierixin, a Novel Inhibitor of ER Stress-induced XBP1 Activation from Streptomyces sp. J Antibiot 60:547–553
Thompson MR, Xu D, Williams BR (2009) ATF3 transcription factor and its emerging roles in immunity and cancer. J Mol Med (Berl) 87:1053–1060
Tirasophon W, Welihinda AA, Kaufman RJ (1998) A stress response pathway from the endoplasmic reticulum to the nucleus requires a novel bifunctional protein kinase/endoribonuclease (Ire1p) in mammalian cells. Genes Dev 12:1812–1824
Todd DJ, Lee A-H, Glimcher LH (2008) The endoplasmic reticulum stress response in immunity and autoimmunity. Nat Rev Immunol 8:663–674
Tsutsumi S, Gotoh T, Tomisato W, Mima S, Hoshino T, Hwang HJ, Takenaka H, Tsuchiya T, Mori M, Mizushima T (2004) Endoplasmic reticulum stress response is involved in nonsteroidal anti-inflammatory drug-induced apoptosis. Cell Death Differ 11:1009–1016
Tu BP, Weissman JS (2004) Oxidative protein folding in eukaryotes: mechanisms and consequences. J Cell Biol 164:341–346

Uramoto H, Sugio K, Oyama T, Nakata S, Ono K, Yoshimastu T, Morita M, Yasumoto K (2005) Expression of endoplasmic reticulum molecular chaperone Grp78 in human lung cancer and its clinical significance. Lung Cancer 49:55–62

Urano F, Wang X, Bertolotti A, Zhang Y, Chung P, Harding HP, Ron D (2000) Coupling of Stress in the ER to Activation of JNK Protein Kinases by Transmembrane Protein Kinase IRE1. Science 287:664–666

Valle CW, Min T, Bodas M, Mazur S, Begum S, Tang D, Vij N (2011) Critical Role of VCP/p97 in the Pathogenesis and Progression of Non-Small Cell Lung Carcinoma. PLoS One 6:e29073

Vattem KM, Wek RC (2004) Reinitiation involving upstream ORFs regulates ATF4 mRNA translation in mammalian cells. Proc Natl Acad Sci USA 101:11269–11274

Vaupel P, Kelleher DK, Höckel M (2001) Oxygenation status of malignant tumors: Pathogenesis of hypoxia and significance for tumor therapy. Semin Oncol 28. Supplement 8:29–35

Walter P, Ron D (2011) The unfolded protein response: from stress pathway to homeostatic regulation. Science 334:1081–1086

Wang Q, He Z, Zhang J, Wang Y, Wang T, Tong S, Wang L, Wang S, Chen Y (2005) Overexpression of endoplasmic reticulum molecular chaperone GRP94 and GRP78 in human lung cancer tissues and its significance. Cancer Detect Prev 29:544–551

Wang Q, Mora-Jensen H, Weniger MA, Perez-Galan P, Wolford C, Hai T, Ron D, Chen W, Trenkle W, Wiestner A, Ye Y (2009) ERAD inhibitors integrate ER stress with an epigenetic mechanism to activate BH3-only protein NOXA in cancer cells. Proc Natl Acad Sci USA 106:2200–2205

Wang Y, Alam GN, Ning Y, Visioli F, Dong Z, Nör JE, Polverini PJ (2012) The Unfolded Protein Response Induces the Angiogenic Switch in Human Tumor Cells through the PERK/ATF4 Pathway. Cancer Res 72:5396–5406

Wek RC, Jackson BM, Hinnebusch AG (1989) Juxtaposition of domains homologous to protein kinases and histidyl-tRNA synthetases in GCN2 protein suggests a mechanism for coupling GCN4 expression to amino acid availability. Proc Natl Acad Sci USA 86:4579–4583

Wek RC, Jiang HY, Anthony TG (2006) Coping with stress: eIF2 kinases and translational control. Biochem Soc Trans 34:7–11

Wouters BG, Koritzinsky M (2008) Hypoxia signalling through mTOR and the unfolded protein response in cancer. Nat Rev Cancer 8:851–864

Yamaguchi H, Wang HG (2004) CHOP is involved in endoplasmic reticulum stress-induced apoptosis by enhancing DR5 expression in human carcinoma cells. J Biol Chem 279:45495–45502

Yamamoto K, Yoshida H, Kokame K, Kaufman RJ, Mori K (2004) Differential Contributions of ATF6 and XBP1 to the Activation of Endoplasmic Reticulum Stress-Responsive cis-Acting Elements ERSE, UPRE and ERSE-II. J Biochem 136:343–350

Yang ZJ, Chee CE, Huang S, Sinicrope FA (2011) The role of autophagy in cancer: therapeutic implications. Mol Cancer Ther 10:1533–1541

Ye J, Rawson RB, Komuro R, Chen X, Davé UP, Prywes R, Brown MS, Goldstein JL (2000) ER Stress Induces Cleavage of Membrane-Bound ATF6 by the Same Proteases that Process SREBPs. Mol Cell 6:1355–1364

Ye J, Kumanova M, Hart LS, Sloane K, Zhang H, De Panis DN, Bobrovnikova-Marjon E, Diehl JA, Ron D, Koumenis C (2010) The GCN2-ATF4 pathway is critical for tumour cell survival and proliferation in response to nutrient deprivation. EMBO J 29:2082–2096

Yoshida H, Okada T, Haze K, Yanagi H, Yura T, Negishi M, Mori K (2000) ATF6 activated by proteolysis binds in the presence of NF-Y (CBF) directly to the cis-acting element responsible for the mammalian unfolded protein response. Mol Cell Biol 20:6755–6767

Yoshida H, Matsui T, Yamamoto A, Okada T, Mori K (2001) XBP1 mRNA Is Induced by ATF6 and Spliced by IRE1 in Response to ER Stress to Produce a Highly Active Transcription Factor. Cell 107:881–891

Zheng HC, Takahashi H, Li XH, Hara T, Masuda S, Guan YF, Takano Y (2008) Overexpression of GRP78 and GRP94 are markers for aggressive behavior and poor prognosis in gastric carcinomas. Hum Pathol 39:1042–1049

Zhong C, Chen C, Kilberg MS (2003) Characterization of the nutrient-sensing response unit in the human asparagine synthetase promoter. Biochem J 372:603–609

Chapter 11
Small Molecules Targeting the VHL/Hypoxic Phenotype

Raymond J. Louie, Mercè Padró, Amato J. Giaccia and Denise A. Chan

Abstract The hypoxic phenotype, characterized by a variety of adaptations that cells make to low oxygen conditions, is a hallmark of solid tumors that serves as both a major challenge and a substantial opportunity in the treatment of cancer. These adaptations influence a wide range of cellular processes, including cell cycle, metabolism, oxygen delivery, proliferation, energy production, differentiation, replication, and sensitivity to growth and death signals. From a clinical perspective, the hypoxic phenotype results in more aggressive tumors that exhibit greater resistance to chemotherapy and radiation therapy and enhanced propensity to metastasize. At the cellular and molecular level, many of the adaptations to hypoxia are mediated by a single transcription factor, the hypoxia-inducible factor (HIF). HIF is predominantly regulated by the von Hippel–Lindau (VHL) tumor suppressor gene. Mutation of VHL results in constitutive activation of HIF and a pseudo-hypoxic phenotype. Here, we review efforts to target either VHL or the hypoxic phenotype to identify pathways and agents as potential cancer therapeutics. Particular focus is given to distinguish between HIF-dependent and HIF-independent therapies.

Keywords Hypoxia · Synthetic lethality · Von Hippel–Lindau · Small molecules · Hypoxia · inducible factors · Renal cell carcinoma · Kidney cancer

---

Raymond J. Louie and Mercè Padró contributed equally to this work.

D. A. Chan (✉) · R. J. Louie · M. Padró
Department of Radiation Oncology, University of California,
San Francisco, CA 94117, USA
e-mail: chand@radonc.ucsf.edu

A. J. Giaccia
Department of Radiation Oncology, Stanford University School of Medicine,
Stanford, CA 94305, USA

G. Melillo (ed.), *Hypoxia and Cancer,* Cancer Drug Discovery and Development, DOI 10.1007/978-1-4614-9167-5_11,

11.1 Introduction

Hypoxia, or conditions of low oxygen tensions, is a feature of solid tumors, occurring because of malformed and inadequate vasculature. The rapid growth of tumor cells outpaces the ability of the existing blood supply to provide oxygen and nutrients and to eliminate waste products. Tumors respond to hypoxia in a variety of ways, including, most prominently, adapting to low oxygen conditions and sending signals to promote oxygen delivery. Hypoxic cells, for example, in addition to cycling more slowly and decreasing energy production, induce angiogenesis and red blood cell production (Semenza 2002). Frequently, the newly formed blood vessels of rapidly expanding tumors are haphazardly constructed, irregular, and inefficient, unable to meet the demands of tumors and resulting in chronically poor perfusion and hypoxia. Because adaptation to persistent hypoxia is a distinctive feature that differentiates tumor cells from the majority of normal cells, it holds substantial appeal as a potential target for cancer therapy.

At the molecular level, the response to hypoxia converges on a single transcription factor, which serves as a master regulator of the cellular response, the hypoxia-inducible factor (HIF). HIF is a heterodimeric transcription factor, composed of an oxygen-labile α-subunit and a constitutive β-subunit (also known as ARNT), and its stabilization in response to low-oxygen conditions is a key step in enabling cells to adapt to hypoxia. HIF controls a number of processes in the hypoxic response, influencing metabolism, cell proliferation, tissue remodeling, glycolysis, erythropoiesis, and oxygen homeostasis (Chan and Giaccia 2007). As a result of this broad impact, tight regulation of HIF is necessary to prevent inappropriate hypoxic responses.

Molecular and cellular studies of kidney cancers were critical in identifying the key components and regulators of the hypoxic response, in large part, due to a strong genetic association between the development of kidney cancer and the dysregulation of the hypoxic response (Ohh et al. 2000). The large majority of kidney cancers are classified as renal clear cell carcinoma and the vast majority of these have mutations in the von Hippel–Lindau (VHL) tumor suppressor gene. VHL is a classical tumor suppressor gene that follows the Knudsen two-hit hypothesis, which states that multiple mutations to DNA are required to cause cancer. VHL disease is characterized by a distinct subset of tumor types, notably in the central nervous system (CNS) (Kaelin 2007). These tumors include hemangioblastomas, pheochromocytomas, and, beyond the nervous system, renal clear cell carcionoma, all of which are highly vascular. These tumors are angiogenic in large part because of the molecular consequences of VHL mutation. VHL is an E3 ubiquitin ligase, which targets the alpha subunit of HIF for oxygen-dependent degradation by the proteasome. Disruption of the VHL function can stabilize HIF even in the absence of hypoxia, leading to proangiogenic transcription. VHL mutation, implicated in approximately 80 % of hereditary and spontaneous kidney cancers, therefore, serves as a link between dysregulation of the hypoxic response and the development of cancer.

Kidney cancer presents a particular set of clinical challenges for diagnosis and treatment. Because of a lack of early symptoms, the diagnosis of kidney cancer is generally due to incidental findings. As a result, kidney cancer is frequently not diagnosed until after it has spread beyond the kidney (Kattan et al. 2001). Compounding this challenge, this cancer is particularly intractable, resistant to standard chemotherapies. However, recent targeted therapies against the tumor vasculature offer some benefit and have become the standard of care. Current therapies include the small molecules temsirolimus (Bellmunt et al. 2008), everolimus (Motzer et al. 2008), sunitinib (Motzer et al. 2006), sorafenib (Escudier et al. 2007), and biologic bevacizumab (Rini 2007), all of which target angiogenic pathways. Both temsirolimus and everolimus are mammalian target of rapamycin (mTOR) inhibitors, interfering with protein translation and vascular endothelial growth factor (VEGF) signaling. Sunitinib and sorafenib are tyrosine protein kinase inhibitors, targeting both vascular endothelial growth factor receptor (VEGFR) and platelet-derived growth factor receptor (PDGFR), each of which contributes to angiogenesis. Sorafenib also interferes with Raf kinases, whereas sunitinib also inhibits KIT and flt3. While these agents target a particular consequence of VHL mutation, namely unregulated angiogenesis, ongoing efforts aim to more broadly target cells with altered VHL function or hypoxic responses through both HIF-dependent and HIF-independent consequences of VHL disruption.

11.2 Modulating the Master Transcription Factor, HIF1, and Its Target Genes

Since VHL and the hypoxic phenotype are hallmarks of solid tumors, much effort has been made to identify agents that specifically target these pathways. Such potential agents would have the potential to target tumor cells specifically, while sparing normal tissue from their cytotoxic effects. A variety of efforts have been used, including chemical synthetic lethality screening, RNA interference libraries, and genomic analysis, to classify new therapies and pathways important for targeting the VHL/hypoxic phenotype. Recently, several novel compounds have been identified to inhibit the HIF/VHL pathway. Although these small molecules were initially characterized as VHL-dependent inhibitors, their primary function is to broadly target cells with high levels of HIF, such as during hypoxia, implicating them as potential tools against all solid tumors.

11.3 Cha3 Reduces Tumor Growth in VHL-Deficient Tumors

Among the early drugs found to selectively target cells lacking VHL was a naturally occurring small molecule, chromomycin A3 (ChA3)(Sutphin et al. 2007). Genomic analysis coupled with drug sensitivity of the NCI-60 panel of cancer cells revealed that ChA3, an antibiotic from *Streptomyces griseus*, selectively kills VHL-deficient cells, while sparing cells with wild-type VHL. When VHL-deficient cells (RCC4 and

786-O, two commonly used kidney cancer cell lines lacking functional VHL) were exposed to this compound, they exhibited reduced metabolic activity and reduced clonogenic survival, indicative of impaired proliferation and increased cytotoxicity. The antitumor effects of ChA3 were tested in vivo, showing a significant reduction in tumor growth in VHL-deficient RCC xenografts.

Further studies revealed that ChA3 cytotoxicity is associated with the activity of stabilized HIF. This HIF dependency was confirmed using a variant of VHL that is unable to target HIFα for degradation. RCC4 cells expressing wild-type VHL exhibited resistance to ChA3 treatments, whereas cells with this mutated VHL, like VHL-deficient cells, remained sensitive, indicating HIF dependency. Glucose transporter 1 (GLUT1) is a well-known HIF transcriptional target and was used to analyze HIF activity. Further experiments associated ChA3 sensitivity with high Glut1 activity levels, indicating that the mechanism of action of ChA3 was dependent on HIF activity. Analysis of echinomycin and actinomycin D, which have comparable activity to ChA3, revealed similar preferential toxicity to VHL-deficient cells. Echinomycin was shown to bind to specific DNA sequences, including hypoxia-responsive elements, to exert its inhibitory effects (Kong et al. 2005). Therefore, the ChA3 family of antibiotics harbors the potential to act as novel small molecules for targeted therapy of VHL-deficient RCC. The use of ChA3 was tested in a small clinical trial, which was halted prematurely due to safety concerns. However, the findings of this in vitro study suggest that ChA3-related antibiotics and other derivatives may ultimately be adapted for clinical use.

11.4 STF-31 Selectively Kills VHL-Deficient Cells by Inhibiting Glucose Uptake

One major consequence of hypoxia and subsequent activation of HIF is the Warburg effect. During hypoxia, one means of adaptation is to switch from oxidative phosphorylation to aerobic glycolysis as the primary means of generating adenosine triphosphate (ATP). As a result of this adaptation, cancer cells, unlike their normal counterparts, become reliant on glucose, even under normal oxygenated conditions. This dependence on aerobic glycolysis represents another distinct feature of the HIF/VHL pathway that may be a potential target for therapy.

A chemical synthetic lethality screen of 64,000 small molecules identified a class of drugs, called the 3-series, exemplified by STF-31, which specifically kill VHL-deficient cells (Chan et al. 2011). The selectivity of STF-31 was verified by evaluating the drug sensitivity of multiple RCC cell lines lacking function VHL compared with their corresponding isogenic counterparts stably expressing a recombinant wild-type VHL. Necrotic cell death is induced in STF-31-treated VHL-deficient cells but not in treated cells expressing functional VHL. Overexpression of HIF in VHL-expressing cells is sufficient to impart sensitivity to STF-31, indicating that this drug is killing via the HIF pathway. Furthermore, knockdown of HIF with short hairpin RNA (shRNA) revealed that the mode of action by STF-31 was HIF-dependent. These findings suggested that characteristics imparted to cells by activation of the hypoxic response, either through low oxygen conditions or mutation of the upstream negative regulator VHL, can make cells selectively vulnerable to some cytotoxic drugs.

Further characterization revealed that altered metabolic pathways are important to conferring sensitivity to STF-31. HIF-mediated adaptation to low oxygen influences many different pathways, including glucose metabolism. VHL-deficient cells have higher glycolytic activity compared to wild-type VHL cells, a consequence of constitutive HIF stabilization. HIF activates the glycolytic pathway and results in the overexpression of glucose transporters. Renal cancer cells with mutated VHL, like a range of other cancers, are highly dependent on aerobic glycolysis for energy production. Unlike wild-type cells, cells lacking functional VHL cannot grow under conditions of glucose deprivation. Even in the presence of oxygen, HIF-dependent adaptations render VHL-deficient cells incapable of utilizing oxidative phosphorylation to overcome their dependence on glycolysis for energy production. The glycolytic activity was analyzed in cells treated with STF-31 by measuring intracellular lactate and extracellular acidification. Treatments with STF-31 significantly inhibited both of these in VHL-deficient cells, but did not affect the glycolytic activity in wild-type VHL cells. STF-31 inhibits glycolysis by impairing glucose uptake specifically in VHL-deficient cells. The decrease in glucose uptake by STF-31 resulted in a significant reduction in ATP production, indicating that STF-31 is killing VHL-deficient cells by inhibiting their primary mechanism of energy production.

Mechanistic studies revealed that STF-31 exploits the glucose addiction of tumor cells by impairing glucose uptake specifically in VHL-deficient cells. STF-31 inhibits glucose uptake by specifically blocking a glucose transporter that is selectively induced in HIF-activated cells. There are two main glucose transporters, GLUT1, the inducible, high-affinity transporter, and GLUT2, which is the responsible for basal glucose uptake (Pajor et al. 2008). Normal kidney cells with functional VHL typically express GLUT2 but very little GLUT1. In contrast, hypoxia or VHL deficiency induces GLUT1 and the suppression of GLUT2 expression. STF-31 selectively inhibits only the Glut1 glucose transporter by docking into the central core of the channel. The genetic inhibition of GLUT1 induces necrotic cell death only in VHL-deficient cells, mimicking the effects of the STF-31 treatment.

In vivo animal models confirmed the efficacy of the 3-series compounds for inhibiting tumor growth, with few side effects. STF-31 treatments can be monitored through fluorodeoxyglucose position emission tomography (FDG-PET). FDG-PET is based on the high glucose uptake presented in cancer cells and is used in cancer diagnosis to detect tumor cells. In mice treated with the 3-series, the 3-series was effective in reducing the tumor glucose uptake without altering brain uptake or hematocrit levels. STF-31 should be considered as a potential candidate not only for the treatment of renal carcinomas with inactive VHL but also for other cancers cells with high levels of GLUT1 and glycolytic dependency.

11.5 ELR510444 Inhibits Microtubule Stability and HIF

Recent work has demonstrated that the VHL gene encodes two isoforms of the protein, a full-length 30 kDa form that binds and promotes microtubule stability, and a smaller nuclear-localized variant responsible for degrading HIF (Hergovich et al. 2003). Since microtubules are essential for proliferation, modulating this activity of VHL would have the ability to inhibit proliferative pathways. An analysis of microtubule stability in cell lines expressing a variety of clinical mutants of VHL showed that the majority of the cells still contained intact microtubules, highlighting an attractive target for therapy.

ELR51044 (ELARA Pharmaceuticals), a novel compound originally described to disrupt microtubules through multiple mechanisms (Risinger et al. 2011), has also been shown to decrease HIF protein expression (Carew et al. 2012). The treatment of ELR51044 inhibited cell proliferation in several cancer cell lines, including breast (MDA-MB-231), melanoma (MDA-MB-435), and colon (HCT116)(Risinger et al. 2011). Treatment with this compound caused loss of microtubules, mitotic arrest, and aberrant spindles. ELR51044 was also able to displace the microtubule-binding compound, colchicine, indicating a direct effect on tubulin. Additionally, ELR51044 was unable to pass through a membrane in a permeability assay, indicating that the multidrug transporter cannot excrete this compound, thereby trapping the compound within the cell. When tested in xenograft models, ELR51044 was able to reduce the growth of MDA-MB-231 in a dose-dependent manner. The effects of ELR51044 were tested on RCC cell lines containing a variety of VHL mutations (Carew et al. 2012). ELR51044 was able to inhibit proliferation of all RCC cell lines, with increased toxicity to the VHL-deficient cell lines, indicating that the compound may be more efficacious in the context of increased HIF. Interestingly, ELR51044 was shown to not only depolymerize microtubles in the RCC cells and cause mitotic arrest, but also reduced HIF1 and HIF2 expression resulting in lower protein levels and reduced expression of the proangiogenic factor VEGF. When the wild-type VHL was ectopically expressed in the VHL-deficient cell lines, the toxicity of ELR51044 was reduced, indicating a dependence on VHL and HIF. This effect was reversed by stabilizing HIF1 levels with cobalt, thereby re-sensitizing the cells to ELR51044, which indicates that ELR51044 toxicity increases with higher levels of HIF1, a common characteristic of all solid tumors. When tested in mouse xenografts, ELR51044 was able to inhibit growth of the VHL-deficient cell lines (786-O and A498). An analysis of the ELR51044-treated tumors by immunohistochemistry revealed that the decrease in tumor growth was attributed to increased apoptosis, increased necrosis, decreased levels of VEGF, and fewer blood vessels. These results indicate that cells lacking VHL, such as RCCs, and cells with elevated HIF levels, such as solid hypoxic tumors, may be specifically susceptible to ELR51044. Taken together, these studies show the therapeutic potential of ELR51044 by illustrating its two-pronged attack on microtubules as well as HIF.

11.6 Stabilizing HIF for the Treatment of Anemia and Ischemia

In addition to regulating proangiogenic factors (VEGF) and glucose metabolism (GLUT1), HIF is also responsible for upregulating the red blood cell cytokine erythropoietin in response to hypoxia. Therefore, stabilizing HIF has the potential to treat chronic anemia and ischemia. Current clinical trials have focused on inhibiting the PHDs as a method to stabilize HIF and thereby increase endogenous erythropoietin for the treatment of anemia (reviewed in Yan et al. 2010). More recently, efforts have been made to destabilize the VHL/HIF interaction. This approach demonstrates the potential for the rational development of drugs to target and modulate the hypoxic response.

The first step in rational drug design was to identify the specific residues in the VHL and HIF that would need to be disrupted. Armed with a co-crystal structure of a HIF peptide-bound VHL, and the knowledge that VHL specifically recognizes the posttranslationally modified form of HIF, Buckley et al. used hydroxyl-proline as the foundation for targeted de novo drug design (Buckley et al. 2012a, b). During the drug optimization process, co-crystal structures and biophysical characterization revealed many novel properties of the VHL/HIF interaction as well as VHL/drug interaction. Several, mostly aromatic, residues in VHL (W88, Y98, S111, H115, and W117) interact with the HIF peptide as well as the prolyl-hydroxylated small molecules (Buckley et al. 2012a, b; Van Molle et al. 2012). Not surprisingly, this is a hotspot for clinical mutations that result in the dysregulation of HIF, indicating that competitive inhibition of this pocket can specifically interrupt the VHL/HIF interface while preserving other VHL interactions (Kaelin 2008). Interestingly, a comparative analysis of residues engaged by the HIF peptide and derivatives of the small molecule revealed that HIF1 does not interact with all possible residues in the binding pocket of VHL, specifically I566 in VHL is not engaged by HIF (Van Molle et al. 2012). Additionally, the characterization of the vast panel of drug derivatives revealed that future drug optimization should focus on modifications to the core prolyl-hydroxy pocket and right-hand side 2 (RHS2) as hot spot locations. Finally, the de novo design of these compounds achieved several important milestones important in protein–protein interaction inhibitions; these small molecules displaced water, engaged with cryptic interfaces, and explored novel interactions in VHL that were unused by HIF.

Although the herculean task of designing small molecules to replace an important protein–protein interaction was achieved, the functional and biological consequences of these compounds still need to be characterized. In order to translate from a proof of principal in de novo drug design to clinical application, several important experiments remain. First, these compounds need to be confirmed as HIF stabilizers, and the upregulation of erythropoietin needs to be verified. Additionally, intracellular consequences should also be characterized; these compounds should leave non-HIF pathways of VHL intact. Most importantly, and a common feature to all drugs, the oncogenic potential of HIF needs to be taken into consideration. Targeted application of these compounds should be excluded from any potential cancerous cells, and only locate to the bone marrow, the site of red blood cell production.

11.7 Targeting VHL-Deficient Cells Through HIF-Independent Mechanisms

In addition to the regulation of HIF, mutations in VHL can affect other oncogenic pathways. Specifically, VHL also regulates or interacts with pathways involved in the regulation of apoptosis, senescence, microtubules, extracellular matrix, adhesion, and autophagy. A number of these unexpected pathways interactions were revealed by drug and RNA interference screens that have provided alternative avenues for targeting VHL-deficient tumors.

One promising approach to targeting VHL-deficient tumors arises from a previously unknown pathway interaction—synthetic lethality between VHL and autophagy—that unexpectedly emerged from a chemical genetics screen. Interactions between a known target, such as VHL mutation, and a specific pathway are not always predictable when using an unbiased screen, such as a chemical synthetic lethal screen. Autophagy is an essential cellular process that regulates the turnover of organelles and long-live proteins to ensure homeostasis. Autophagy occurs in all eukaryotes in response to diverse biological stresses, such as starvation, hypoxia, and heat shock to promote survival. Under persistent stress conditions, autophagy can result in cell death (Mathew et al. 2007). During autophagy, portions of the cytoplasm and organelles are engulfed into double-membrane vesicles known as autophagosomes. Luminal components are degraded when membrane fusion occurs between the autophagosome with the lysosome, which contains hydrolases (Klionsky and Emr 2000).

Dysregulation of autophagy has been reported in diverse diseases, such as neuronal degeneration, infectious disease, and cancer (Levine 2008). The role of autophagy in these diseases is unclear. There are some reports suggesting a role of autophagy in cancer, but it is still uncertain whether autophagy drives survival or death pathways. Several genes involved in the autophagy process, such as beclin1 (*BECN*), are deleted in multiple types of cancer, suggesting roles as tumor suppressor genes (Aita et al. 1999). Recent reports have also described drug treatments able to induce autophagic cell death(Mathew et al. 2007, Jin and White 2007). Therefore, the regulation of autophagy is a possible therapeutic approach for cancer treatments.

11.8 STF-62247 Induces Autophagic Cell Death in VHL-Deficient Cells

The relevance of autophagy for the targeting of cells of the hypoxic phenotype became apparent when drug screening found a class of compounds (6-series) pyridyl aniline thiazoles derivatives (STF-62247) that were selectively toxic only to VHL-deficient cells (Turcotte et al. 2008). STF-62247 showed higher toxicity in RCC4 (inactive VHL) and SN12C-shVHL cells compared to RCC4/VHL and SN12C (wild-type VHL), respectively. The treatment of STF-622447 resulted in the accumulation of large intracytoplasmic vacuoles specifically in VHL-deficient cells, indicating that STF-62247 could induce cells to undergo autophagy. The addition of an autophagic inhibitor, 3-methyladenine, reverted the effects of STF-62247, indicating that autophagy is the mechanism through STF-62247 kills VHL-deficient cells. The knockdown of HIF did not alter the sensitivity of VHL-deficient cells to STF-64722, indicating that the mechanism of action was independent of HIF.

The fusion of autophagosomes with lysosomes, resulting in the acidification of the autophagosome, is a key step in autophagy-induced cell death. VHL-deficient cells treated with STF-62247 had higher acidification levels of autophagosomes compared to cells with wild-type VHL, indicating that the acidification of the autophagosomes is a precursor of autophagic cell death. When VHL-deficient cells were treated with nonfunctional analogs of STF-62247, there was no acidification detected and, correspondingly, no cell death, indicating a close relationship between acidification and autophagic cell death.

11.9 VHL-Deficient Cells Are Selectively Dependent on Several Kinases

Kinases play important roles in biology, are essential in cell-cycle regulation, and are frequently dysregulated in cancer. In renal cell carcinomas, the inactivation of VHL has been associated with the activation of several kinases that are important in the tumorigenesis process. Therefore, targeting these specific kinases could be a good strategy for developing new cancer therapies. Lentiviral short hairpin libraries (shRNA) have been used to test the loss of function screenings in mammalian cells. A lentiviral library targeting 88 different kinases was performed to analyze which kinases are essential for tumorigenesis when VHL is inactive (Bommi-Reddy et al. 2008). The screening was performed comparing the survival in two VHL-deficient renal carcinoma cells (RCC4 and 786-O) against their isogenic wild-type counterparts. Initial screening revealed five kinases (*CDK6, MAP2K1, MET, IRR, and HER4*) to be essential specifically in the VHL-deficient cell lines. Further, validations using small interfering RNAs (siRNAs) and immunoblotting confirmed the importance of three of these five kinases (CDK6, MET, and MAP2K1). To check the VHL dependency of inhibition of these kinases, a mutated VHL unable to degrade HIF was expressed ectopically in 786-O cells. This mutant VHL imparts resistance to the loss of some kinases, confirming that the sensitivity to the loss of CDK6, MET, and MAP2K1 kinases in VHL-deficient cells is HIF independent. The use of a drug that inhibits CDK4/6 kinase activity was used to confirm the HIF independency, and the selective dependency on of VHL-deficient cells on this kinase activity.

11.10 SP600125 Inhibits the JNK Pathway in VHL-Deficient RCC

In vitro and in vivo work using isogenic RCC cell lines, with mutant and wild-type forms of VHL, demonstrated that proliferation in VHL-deficient cell lines is dependent on the constitutively active c-Jun N-terminal kinase (JNK) pathway (An et al. 2013). In the absence of VHL, Card9 becomes dis-inhibited and the JNK pathway becomes activated via TRAF, TAKI1, and MKK4 which results in the activation of c-Jun and consequentially the upregulation of Twist. In vivo growth of RCCs was extremely abrogated when JNK was inhibited by shRNA, pharmacological inhibition (SP600125), and when wild-type VHL was ectopically expressed. Furthermore, immunohistochemical analysis of clear cell, chromophobe, papillary, and oncocytoma types of renal cancer confirmed that that hyper-activation of JNK was specifically upregulated in clear cell carcinomas, which is the most common type of renal cancer driven by loss of VHL. Therefore, SP600125 may act as a powerful inhibitor for the majority of ccRCC.

11.11 Conclusions

Targeting VHL and the hypoxic phenotype has been the focus of intense investigation for the treatment of cancer. A handful of small molecules and targets have been identified to specifically target cells that lack functional VHL. Interestingly, these exhibit a range of activities and mechanisms, some of which are dependent on HIF, and others that function independently of HIF. The diversity of the mechanisms of action demonstrates the broad range of the hypoxic phenotype. While these agents are still in preclinical development, they provide proof of concept that directly targeting the tumor microenvironment and exploiting multiple aspects of the hypoxic phenotype can be beneficial in the treatment of cancer.

References

Aita VM, Liang XH, Murty VV, Pincus DL, Yu W, Cayanis E, Kalachikov S, Gilliam TC, Levine B (1999) Cloning and genomic organization of beclin 1, a candidate tumor suppressor gene on chromosome 17q21. Genomics 59:59–65

An J, Liu H, Magyar CE, Guo Y, Veena MS, Srivatsan ES, Huang J, Rettig MB (2013) Hyperactivated JNK is a therapeutic target in pVHL-deficient renal cell carcinoma. Cancer Res 73:1374–1385

Bellmunt J, Szczylik C, Feingold J, Strahs A, Berkenblit A (2008) Temsirolimus safety profile and management of toxic effects in patients with advanced renal cell carcinoma and poor prognostic features. Ann Oncol 19:1387–1392

Bommi-Reddy A, Almeciga I, Sawyer J, Geisen C, Li W, Harlow E, Kaelin WG Jr, Grueneberg DA (2008) Kinase requirements in human cells: III. Altered kinase requirements in VHL-/- cancer cells detected in a pilot synthetic lethal screen. Proc Natl Acad Sci U S A 105:16484–16489

Buckley DL, Gustafson JL, Van Molle I, Roth AG, Tae HS, Gareiss PC, Jorgensen WL, Ciulli A, Crews CM (2012a) Small-molecule inhibitors of the interaction between the E3 ligase VHL and HIF1alpha. Angew Chem Int Ed Engl 51:11463–11467

Buckley DL, Van Molle I, Gareiss PC, Tae HS, Michel J, Noblin DJ, Jorgensen WL, Ciulli A, Crews CM (2012b) Targeting the von Hippel-Lindau E3 ubiquitin ligase using small molecules to disrupt the VHL/HIF-1alpha interaction. J Am Chem Soc 134:4465–4468

Carew JS, Esquivel JA, 2nd, Espitia CM, Schultes CM, Mulbaier M, Lewis JD, Janssen B, Giles FJ, Nawrocki ST (2012) ELR510444 inhibits tumor growth and angiogenesis by abrogating HIF activity and disrupting microtubules in renal cell carcinoma. PLoS One 7:e31120

Chan DA, Giaccia AJ (2007) Hypoxia, gene expression, and metastasis. Cancer Metastasis Rev 26:333–339

Chan DA, Sutphin PD, Nguyen P, Turcotte S, Lai EW, Banh A, Reynolds GE, Chi JT, Wu J, Solow-Cordero DE, Bonnet M, Flanagan JU, Bouley DM, Graves EE, Denny WA, Hay MP, Giaccia AJ (2011) Targeting GLUT1 and the Warburg effect in renal cell carcinoma by chemical synthetic lethality. Sci Transl Med 3:94ra70

Escudier B, Eisen T, Stadler WM, Szczylik C, Oudard S, Siebels M, Negrier S, Chevreau C, Solska E, Desai AA, Rolland F, Demkow T, Hutson TE, Gore M, Freeman S, Schwartz B, Shan M, Simantov R, Bukowski RM (2007) Sorafenib in advanced clear-cell renal-cell carcinoma. N Engl J Med 356:125–134

Hergovich A, Lisztwan J, Barry R, Ballschmieter P, Krek W (2003) Regulation of microtubule stability by the von Hippel-Lindau tumour suppressor protein pVHL. Nat Cell Biol 5:64–70

Jin S, White E (2007) Role of autophagy in cancer: management of metabolic stress. Autophagy 3:28–31

Kaelin WG (2007) Von hippel-lindau disease. Annu Rev Pathol 2:145–173

Kaelin WG Jr (2008) The von Hippel-Lindau tumour suppressor protein: O_2 sensing and cancer. Nat Rev Cancer 8:865–873

Kattan MW, Reuter V, Motzer RJ, Katz J, Russo P (2001) A postoperative prognostic nomogram for renal cell carcinoma. J Urol 166:63–67

Klionsky DJ, Emr SD (2000) Autophagy as a regulated pathway of cellular degradation. Science 290:1717–1721

Kong D, Park EJ, Stephen AG, Calvani M, Cardellina JH, Monks A, Fisher RJ, Shoemaker RH, Melillo G (2005) Echinomycin, a small-molecule inhibitor of hypoxia-inducible factor-1 DNA-binding activity. Cancer Res 65:9047–9055

Levine B, Kroemer G (2008) Autophagy in the pathogenesis of disease. Cell 132:27–42

Mathew R, Karantza-Wadsworth V, White E (2007) Role of autophagy in cancer. Nat Rev Cancer 7:961–967

Motzer RJ, Michaelson MD, Redman BG, Hudes GR, Wilding G, Figlin RA, Ginsberg MS, Kim ST, Baum CM, DePrimo SE, Li JZ, Bello CL, Theuer CP, George DJ, Rini BI (2006) Activity of SU11248, a multitargeted inhibitor of vascular endothelial growth factor receptor and platelet-derived growth factor receptor, in patients with metastatic renal cell carcinoma. J Clin Oncol 24:16–24

Motzer RJ, Escudier B, Oudard S, Hutson TE, Porta C, Bracarda S, Grunwald V, Thompson JA, Figlin RA, Hollaender N, Urbanowitz G, Berg WJ, Kay A, Lebwohl D, Ravaud A (2008) Efficacy of everolimus in advanced renal cell carcinoma: a double-blind, randomised, placebo-controlled phase III trial. Lancet 372:449–456

Ohh M, Park CW, Ivan M, Hoffman MA, Kim TY, Huang LE, Pavletich N, Chau V, Kaelin WG (2000) Ubiquitination of hypoxia-inducible factor requires direct binding to the beta-domain of the von Hippel-Lindau protein. Nat Cell Biol 2:423–427

Pajor AM, Randolph KM, Kerner SA, Smith CD (2008) Inhibitor binding in the human renal low- and high-affinity Na + /glucose cotransporters. J Pharmacol Exp Ther 324:985–991

Rini BI (2007) Vascular endothelial growth factor-targeted therapy in renal cell carcinoma: current status and future directions. Clin Cancer Res 13:1098–1106

Risinger AL, Westbrook CD, Encinas A, Mulbaier M, Schultes CM, Wawro S, Lewis JD, Janssen B, Giles FJ, Mooberry SL (2011) ELR510444, a novel microtubule disruptor with multiple mechanisms of action. J Pharmacol Exp Ther 336:652–660
Semenza GL (2002) HIF-1 and tumor progression: pathophysiology and therapeutics. Trends Mol Med 8:S62–S67
Sutphin PD, Chan DA, Li JM, Turcotte S, Krieg AJ, Giaccia AJ (2007) Targeting the loss of the von Hippel-Lindau tumor suppressor gene in renal cell carcinoma cells. Cancer Res 67:5896–5905
Turcotte S, Chan DA, Sutphin PD, Hay MP, Denny WA, Giaccia AJ (2008) A molecule targeting VHL-deficient renal cell carcinoma that induces autophagy. Cancer Cell 14:90–102
Van Molle I, Thomann A, Buckley DL, So EC, Lang S, Crews CM, Ciulli A (2012) Dissecting fragment-based lead discovery at the von Hippel-Lindau protein:hypoxia inducible factor 1alpha protein-protein interface. Chem Biol 19:1300–1312
Yan L, Colandrea VJ, Hale JJ (2010) Prolyl hydroxylase domain-containing protein inhibitors as stabilizers of hypoxia-inducible factor: small molecule-based therapeutics for anemia. Expert Opin Ther Pat 20:1219–1245

Chapter 12
Hypoxia and Radiation Therapy

Marianne Nordsmark, Jan Alsner, Morten Busk, Jens Overgaard and Michael R. Horsman

Abstract Hypoxia is a characteristic of most solid tumors. This makes hypoxia an obvious target for cancer therapy. Both in vitro and in vivo studies have identified a range of different approaches that can reduce or eliminate tumor hypoxia and improve radiation response. Several of these have been tested clinically and in particular in combination with radiotherapy targeting hypoxia has proven beneficial. In order to target hypoxia specifically, a range of clinically applicable techniques have been tested. These include measures of hypoxia either directly as physiological oxygen, using exogenous markers such as nitroimidazoles or endogenous molecular markers regulated by hypoxia. Substantial improvement in outcome following radiation therapy is obtained in particular in head and neck cancer; and in Denmark, the Danish head and neck cancer cooperative group (DAHANCA) has used the hypoxic cell radiosensitizer nimorazole in the treatment of head and neck cancer for several years. Currently, this concept has been adopted in confirmatory clinical trials worldwide as the strategy is refined using hypoxia gene classifiers to target the right patients. Thus, extensive preclinical and clinical studies are ongoing in this area, and hopefully it is a matter of time before the hypoxia problem is eliminated. This chapter reviews experimental and clinical evidence for the role of tumor hypoxia in radiotherapy.

Keywords Tumor hypoxia · Radiation therapy · Prognostic and predictive assays · Hypoxia gene classifier · Hypoxic modification · Radiation sensitizer · Bioreductive drugs

M. Nordsmark (✉) · J. Alsner · M. Busk · J. Overgaard · M. R. Horsman
Department of Experimental Clinical Oncology, Aarhus University Hospital,
Nørrebrogade 44, Bldg. 5, 8000 Aarhus C, Denmark
e-mail: marianne@oncology.dk

M. Nordsmark
Department of Oncology, Aarhus University Hospital, Aarhus, Denmark

G. Melillo (ed.), *Hypoxia and Cancer,* Cancer Drug Discovery and Development, DOI 10.1007/978-1-4614-9167-5_12,

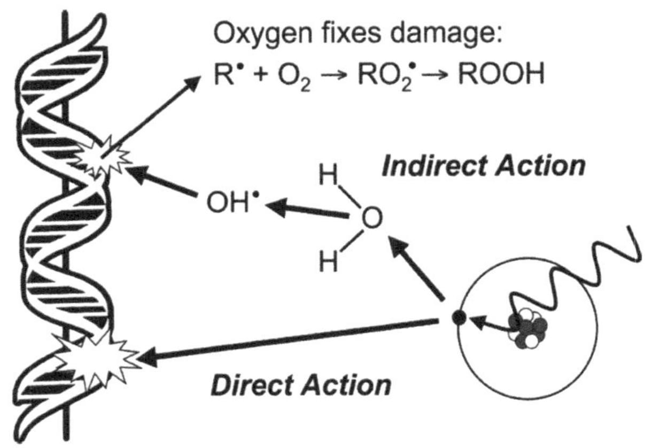

Fig. 12.1 Illustration of how the presence of oxygen is involved in repair of free radicals that are produced in the DNA by direct or indirect action of radiation. (Modified from Hall 1994)

12.1 Introduction

12.1.1 Tumor Hypoxia and Radiation Resistance

The significance of hypoxia in influencing outcome to radiotherapy clinically was first indicated from a study in which the radiation response of skin was markedly decreased if the blood flow to the irradiated area was reduced by compression (Schwarz 1909). Another early report showed that tissues in which blood flow was stimulated by diathermia had a more prominent response to radiation (Müller 1910). At that time, neither study actually attributed the effects to an oxygen dependency. Later, cell culture experiments showed that cells lacking oxygen (and glucose) were relatively insensitive to irradiation (Mottram 1936) and the role of oxygen deficiency as a major source of radiation resistance was hypothesized by Gray based on experimental observations in Chinese hamster cells irradiated in culture, showing that the radiation sensitivity declined at oxygen tensions below about 30 mmHg and declined steeply at very low oxygen tensions from 3 mmHg down to zero (Gray 1953).

As shown in Fig. 12.1, X-rays interact with biological material and create fast charged particles and free radicals that break chemical bonds and cause damage directly to the target molecule, mostly the DNA, or indirectly to other cellular molecules and from there to the critical target, named R*. The extent of this damage depends on the presence of oxygen because molecular oxygen reacts with R* to eventually form ROOH, a chemically stable composition that consolidates damage to the DNA and limits DNA repair mechanisms.

Radiobiologically relevant hypoxia can be determined in different ways in vitro and in vivo. In a classical assay, tumors in vivo are irradiated under normal air-breathing conditions or when the blood supply to the tumor is shut down by clamping. The tumors are then excised and the viability of the neoplastic cells is assessed by the clonogenic cell survival assay. The clamped dose–response curve represents total hypoxia. Air breathing produces a biphasic curve where the initial part of the survival curve is determined by the aerobic cells in the tumor. At a higher dose, the hypoxic tumor cells dominate the response and thus, the curve begins to parallel the clamped result. As the clamped curve represents 100 % hypoxia, the amount by which the curve in case of the air-breathing animals is displaced allows one to determine the percentage of hypoxia.

Two other ways include the clamped tumor growth delay assay, where measurements are taken of the time taken for tumors to reach a specific size after treatment, and the clamped tumor control assay, in which the percentage of animals showing local tumor control at a certain time after treatment is recorded (Moulder and Rockwell 1984; Horsman et al. 1993). This latter assay produces full radiation dose–response curves under air-breathing and fully anoxic (clamped tumor) conditions, and the hypoxic fractions can then be calculated from the displacement of the dose–response curves.

Evidence from exactly such classical radiobiology assays, used to determine hypoxic clonogenic radiation-resistant cells, suggests a possible link between hypoxia and cancer stem cells (Baumann et al. 2008) as stem cells possess clonogenic potential and hypoxia may affect cancer stem cell generation and maintenance through the upregulation of hypoxia-induced factors (Baumann et al. 2008; Hill et al. 2009). Preclinical studies have shown an inverse correlation between hypoxia and local tumor control after irradiation (Horsman et al. 1993; Yaromina et al. 2006), which could suggest that hypoxia may actually protect the cancer stem cells from the lethal effects of radiation. Using treatments that eliminate hypoxia should, therefore, be an effective method for enhancing the radiosensitivity of cancer stem cells.

12.2 Targeting Tumor Hypoxia in Patients During Radiotherapy

12.2.1 High-Oxygen-Content Gas Breathing

A simple method that was tested to eliminate the hypoxic cell population, and thereby improve the therapeutic potential of radiation therapy, involved allowing the tumor-bearing host to breathe high-oxygen-content gas mixtures before and during irradiation. Both oxygen and carbogen (95 % oxygen + 5 % carbon dioxide) breathing could substantially enhance the response of murine tumors to radiation (Du Sault 1963; Suit et al. 1972) and a superior effect was seen when the gasses were inspired under hyperbaric rather than normobaric conditions (Du Sault 1963; Suit et al. 1972). However, normobaric oxygen or carbogen improved the radiosensitizations quite substantially (Siemann et al. 1977; Rojas 1991; Grau et al. 1992; Mortensen et al. 2012).

The first clinical use of high-oxygen-content gas breathing was tested in the late 1960s (Churchill-Davidson 1968). These trials were fairly small, and suffered from the use of unconventional radiation fractionation schemes, but the effect of hyperbaric oxygen was superior to radiotherapy given in air, in particular when few and large fractions were applied (Churchill-Davidson 1968; Dische 1979; Dische et al. 1983).

Large multicenter clinical trials conducted by the UK Medical Research Council gave significant improvements in local tumor control and survival for both advanced head and neck and uterine cervix cancer (Dische 1979; Dische et al. 1993; Henk et al. 1977; Henk and Smith 1977; Watson et al. 1978; Overgaard 1989) but not in the case of bladder cancer (Overgaard 1989). The use of hyperbaric oxygen was stopped not only because of patient compliance and safety concerns but also because of access to chemical agents that mimic oxygen (Hypoxic Radiation Sensitizers, see below).

Experimental studies showed that the gas-breathing time factor during normobaric high-content-oxygen exposure was critical for the enhancement of radiation damage (Suit et al. 1972; Siemann et al. 1977; Rojas 1991; Chaplin et al. 1993). In clinical studies, breathing normobaric high-oxygen-content gas failed to show an improvement in radiation response (Bergsjø and Kolstad 1968; Rubin et al. 1979). This could possibly be due to the failure to achieve the optimum preirradiation gas-breathing time. Studies of carbogen breathing in head and neck cancer showed conflicting results; a benefit was reported when carbogen breathing was combined with accelerated radiation and nicotinamide (ARCON) (Kaanders et al. 2002b) but there was no improvement in radiation response (Mendenhall et al. 2005). The concept of ARCON included administration of accelerated radiotherapy, to overcome accelerated repopulation of tumor cells during fractionated radiotherapy, and nicotinamide to deal with fluctuating hypoxia (Petersen et al. 2001). Experimental studies had demonstrated that the vitamin B3 analog, nicotinamide, enhanced radiation damage in murine tumor models using both single and fractionated treatments (Horsman 1995). The ARCON therapy was tested in a phase II trial with promising local tumor control data (Kaanders et al. 2002c) followed by a phase III head and neck trial showing locoregional benefit from ARCON with acceptable toxicity, but no gain in overall survival (Janssens 2012a, b; Peters and Rischin 2012). A modified ARCON version with radiotherapy and concurrent carbogen and nicotinamide (CON) was tested in a phase III bladder cancer trial. In this study, toxicity was acceptable, and there was a modest improvement in cystoscopic tumor control and significant improvement in local relapse, risk of death, and overall survival in favor of radiotherapy and CON (Hoskin et al. 2010).

12.2.2 Hypoxia Radiation Sensitizers

Hypoxic cell sensitizers replaced oxygen-content gas breathing and became probably the most extensively investigated method for overcoming hypoxic cell radioresistance. It was found that the efficiency of radiosensitization was directly related to electron affinity (Adams and Cooke 1969). This led to in vitro studies demonstrating preferential radiosensitization of hypoxic cells by highly electron-affinic nitroaromatic compounds (Asquith et al. 1974; Adams et al. 1976). These chemical agents mimic oxygen but, unlike oxygen, they are slowly metabolized by the tumor cells and, therefore, by diffusion they reach all cells in the tumor.

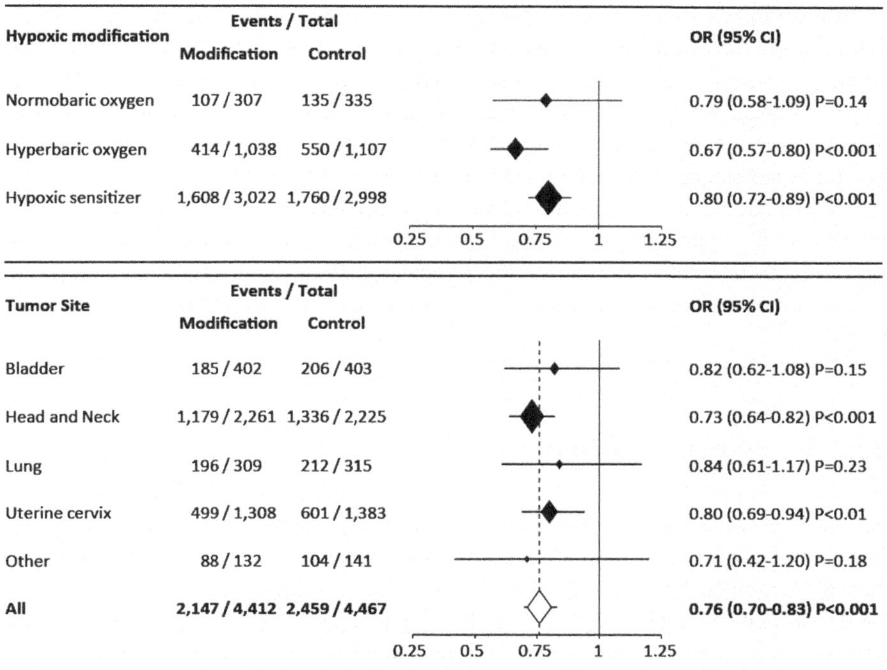

Fig. 12.2 Results from a meta-analysis of hypoxic modification trials by Overgaard 2007 and 2011. The effect of three different ways of modifying tumor hypoxia during radiotherapy, with significant effect of hyperbaric oxygen and hypoxic sensitizers. The gain of hypoxic modification during radiotherapy was most pronounced in head and neck cancer

The first clinical studies of radiosensitizers were conducted using metronidazole in brain tumors (Urtasun et al. 1976) followed by trials exploring the effect of misonidazole (Dische 1985; Overgaard et al. 1989; Overgaard 1994). Most of these trials failed to improve radiation response, although a benefit was seen in the DAHANCA 2 study (Overgaard et al. 1989). Unfortunately, dose-limiting toxicities of misonidazole were significant and misonidazole was replaced by more efficient or less toxic hypoxic sensitizers such as pimonidazole, etanidazole, and nimorazole. Among these studies, a European pimonidazole trial in uterine cervix cancer was disappointingly closed prematurely due to poor performance in the experimental arm (Dische et al. 1993). Two multicenter trials in head and neck cancer using etanidazole showed no benefit (Lee et al. 1995; Eschwége et al. 1997) but a highly significant, improved locoregional tumor control and disease-free survival was shown in the DAHANCA 5 study in patients with supraglottic and pharynx carcinomas treated with the less toxic drug nimorazole (Overgaard et al. 1998). A recent trial by the International Atomic Energy Agency (IAEA) with the 3-nitrotriazole compound, sanazole (AK-2123), in uterine cervical cancer also demonstrated a significant improvement in both local tumor control and overall survival (Dobrowsky et al. 2007).

As presented in Fig. 12.2, a meta-analysis of all randomized clinical studies in a range of tumor sites with hypoxic modification during radiotherapy showed significant improvement in locoregional tumor control (Overgaard 2007; Overgaard 2011). In particular, the administration of hypoxic radiosensitizers with primary radiotherapy showed that radiosensitizer modification of tumor hypoxia significantly improved the locoregional tumor control (the overall odds ratio was 0.76 (0.70–0.83). This treatment's benefit was mainly driven by the result of an improved response in head and neck cancer studies with an odds ratio of 0.73 (0.64–0.82) but to a lesser extent in bladder cancer, while no significant effect was observed in other tumor sites (cervix, lung, central nervous system, and esophagus). The overall gain was in the order of a 5–10 % improvement in local control, and although small, such gains are relevant to pursue because they seem to be associated with a similar improvement in survival.

12.2.3 Bioreductive Drugs

Another type of hypoxia-targeting drugs applied during radiotherapy are bioreductive drugs. These are agents which were relatively nontoxic to cells under normal oxygenated conditions, but when reduced showed to be more cytotoxic under hypoxia (Hall and Roizin-Towle 1975). This led to the development of various types of such bioreductive drugs that preferentially killed the radiation-resistant hypoxic tumor cell population. These drugs can be divided into three major groups: the quinones, nitroaromatics, and N-oxides (McKeown et al. 2007).

Among the quinones, the prototype bioreductive drug is mitomycin C (MMC), which has been used for many years in patients as a chemoradiosensitizer. It is activated by bioreduction to form products that cross-link DNA (Kennedy et al. 1980). In two randomized clinical trials in patients with squamous cell carcinoma of the head and neck, MMC improved radiation-induced local tumor control without any enhancement of radiation reactions in normal tissues (Weissberg et al. 1989; Haffty et al. 1993). However, in two other trials no major influence on response or survival was seen (Dobrowsky et al. 1995; Grau et al. 2003). This lack of response may be because the drug was only given once during the radiation schedule. Further, preclinical data show little differential between aerobic and hypoxic cell killing (Stratford and Stephens 1989; Hall 1994).

More efficient quinones such as porfiromycin and apaziquone (EO9) have been developed (McKeown et al. 2007). EO9 has only gone through phase I/II testing and back to the laboratory (Phillips et al. 2012), while porfiromycin was included in a prospective randomized trial in combination with radiation therapy in head and neck cancer but was found to be no better than MMC (Haffty et al. 2005).

Numerous efforts were made to find other nitroimidazole radiosensitizers that were effective as hypoxic cell cytotoxins. Along that line RSU 1069 was developed. This compound has the classic 2-nitroimidazole radiosensitizing properties, but an aziridine ring at the terminal end of the chain giving the molecule substantial potency as a hypoxic cell cytotoxin. Although the drug was found to have substantial activity in hypoxic cells in vitro and in vivo (Stratford et al. 1986), large animal studies indicated dose-limiting gastrointestinal toxicity. A less toxic prodrug, RB 6145, which is reduced to RSU 1069 in vivo, also showed potent antitumor activity in experimental systems but was dropped when toxicity studies revealed that it induced blindness in dogs. Other nitroaromatic compounds are in the pipeline, including NLCQ-1, CB1954, SN-23862, and PR-104 (McKeown et al. 2007).

A more promising group of bioreductive drugs is the organic nitroxides, of which the benzotriazene di-N-oxide, tirapazamine, is the lead compound. The parent moiety shows limited toxicity towards aerobic cells but after reduction under hypoxic conditions, a product is formed which has been shown to be highly toxic to cells in vitro and can substantially enhance radiation damage in tumors in vivo (Zeman et al. 1988). Most clinical studies have involved combining tirapazamine with chemotherapy, although there have been a few trials with radiation $\pm$ chemotherapy (McKeown et al. 2007). The results from the phase II trials generally showed promise, but in the few randomized trials that have been completed the results have been somewhat disappointing. Other N-oxides currently under development include chlorambucil N-oxide and N (banoxantrone), the latter being combined with radiation in a number of phase II trials (McKeown et al. 2007).

12.2.4 Hemoglobin and Erythropoietin

The potential benefit of increasing hemoglobin by blood transfusion prior to radiotherapy has been investigated in a number of studies (Thomas 2002). The first clinical investigation of this approach was in advanced squamous cell carcinoma of the uterine cervix (Evans and Bergsjø 1965). Transfusion to patients with low hemoglobin levels resulted in an increased tumor oxygen tension, as measured superficially in the tumor using first-generation oxygen-consuming electrodes. The same study was also the first to show that transfusion to a hemoglobin level of 11 g/dL or higher was significantly related to improved survival. A Canadian retrospective study of 605 cervix cancer patients showed that the negative influence of low hemoglobin on prognosis could be overcome by transfusion (Grogan et al. 1999). However, these observations were not supported by data from a prospective phase III trial, from the Danish Head and Neck Cancer (DAHANCA) study group, showing no benefit of transfusion in patients with low hemoglobin levels (Hoff et al. 2011a, b; Hoff 2012).

The concentration of hemoglobin can also be increased by stimulation with the hormone erythropoietin (EPO). Preclinical studies have shown that anemia in animals could be corrected by serial injection with EPO and that this EPO treatment also overcame the anemia-induced radiation resistance (Thews et al. 1998; Stuben et al. 2003). The concept of using EPO to correct anemia was tested in a number of clinical trials, and although low hemoglobin levels can be effectively and safely improved by EPO (Lavey and Dempsey 1993; Hoskin et al. 2009), a number of studies in patients undergoing treatment for head and neck cancer failed to show any benefit (Henke et al. 2003; Machtay et al. 2007; Overgaard et al. 2007; Hoskin et al. 2009). In fact, those patients who actually received EPO during radiation therapy did significantly worse than those patients who did not receive EPO and, as a result, all EPO and radiation trials have been stopped.

12.3 Measuring Hypoxia and Predicting Radiation Response

Methods to detect hypoxic treatment-resistant cells have continuously evolved. Tumor hypoxia can be assessed directly by measurements of oxygen partial pressure (pO_2) distributions with polarographic electrodes, or by the use of exogenous markers that are injected into the host and bind specifically to viable hypoxic tumor cells, or by assessments of endogenous markers that are genes/proteins regulated under hypoxia. Other more indirect approaches determine vascularization, perfusion, hemoglobin, or energy metabolism.

12.3.1 Direct Measures of Oxygen Partial Pressure

Direct measures of physiological oxygen in tumors became feasible with the introduction of the Eppendorf pO_2 Histograph, a polarographic oxygen-sensitive probe. This assay allows to sample hundreds of pO_2 values rapidly at multiple sites within the tumor (Nordsmark et al. 1994; Nordsmark 1997) in comparison with the old manually moved glass-sealed oxygen-consuming probes allowing only a few pO_2 values obtained 3–4 mm under the surface of a tumor (Kolstad 1963; Bergsjø and Kolstad1968; Gatenby et al. 1988). The Eppendorf pO_2 Histograph device is used preclinically and clinically for evaluation of tissue oxygen tension. A set of measurements can be obtained from a tumor and related normal tissue within 10–20 min and is generally well tolerated by patients even though it is invasive. It is relevant for the interpretation of such measurements to understand that the catchment range of one pO_2 value is between about 60 and 100 μm and is averaged over multiple cells that are malignant, normal; viable, or nonviable (Nordsmark et al. 1994; Nordsmark 1997).

Typically, human tumors are more hypoxic than normal tissues, and there is substantial inter- and intraheterogeneity. The most frequently reported parameters are the median pO_2, the proportion of pO_2 values ≤ 2.5 mmHg ($HP_{2.5}$), ≤ 5 mmHg (HP_5), ≤ 10 mmHg (HP_{10}), or the hypoxic subvolume, defined as HP_5 multiplied by the total tumor volume. The result of these parameters ($HP_{2.5}$, HP_5, and HP_{10}) are given as instantly printed values delivered by the Eppendorf pO_2 Histograph and they comply conveniently with the radiobiological rationale (Nordsmark et al. 1994).

The measurement of pO_2 distributions with polarographic electrodes is probably the most direct method for estimating tumor hypoxia and one that has showed to correlate with radiobiological hypoxic fractions based on direct analysis of the radiation response under normal and modified conditions in vivo in mouse C3H mammary carcinomas. Tumor hypoxia measured by the Eppendorf electrode was also correlated with tissue-based quantification of pimonidazole staining and the

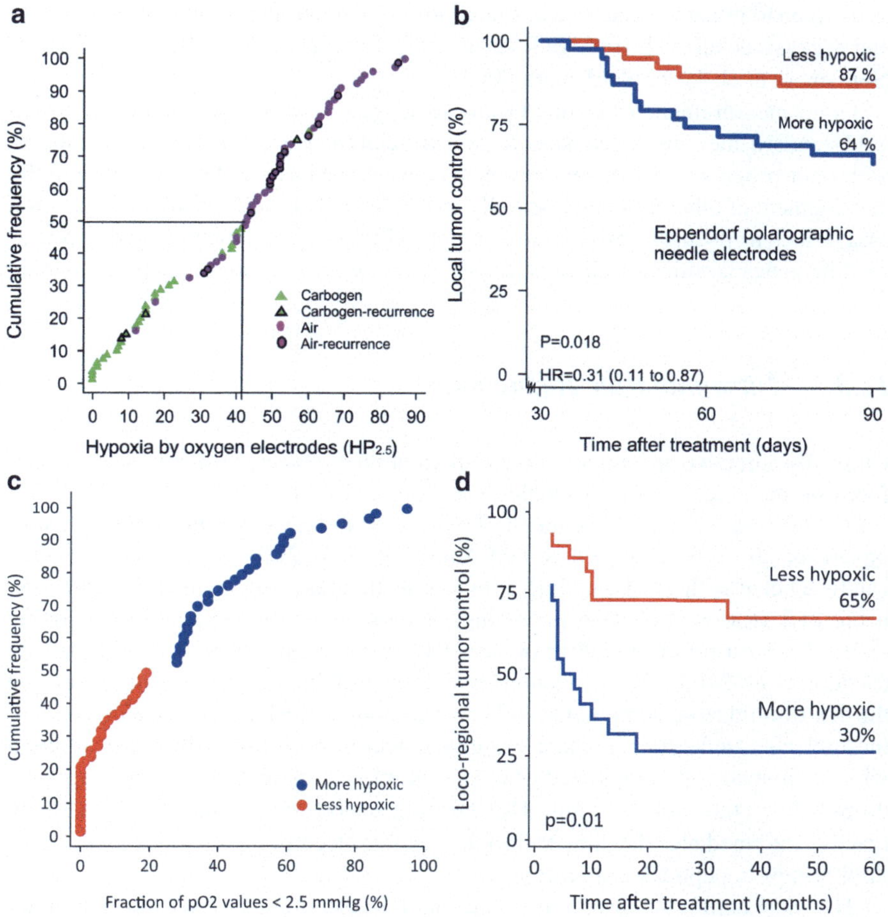

Fig. 12.3 **a** Cumulative frequency of mouse tumors as function of pO_2 values ≤ 2.5 mmHg measured by the Eppendorf electrodes under different gas-breathing conditions. **b** Local tumor control of tumors irradiated by 55 Gy single dose under different gas-breathing conditions and divided into "less hypoxic" and "more hypoxic" based on pO_2 values ≤ 2.5 mmHg measured by the Eppendorf electrodes. **c** Cumulative frequency of head and neck tumors as a function of pO_2 values ≤ 2.5 mmHg measured by the Eppendorf electrodes. **d** Locoregional tumor control in 63 head and neck cancer patients divided by median tumor pO_2 into "less hypoxic" and "more hypoxic." (**a**–**b** modified from Mortensen et al. 2011; **d** Nordsmark, with permission)

radiobiological hypoxic fraction in an experimental model. In a preclinical study using the mouse C3H model, it was shown that pretreatment pO_2 was predictable of tumor control probability following radiotherapy when tumors were grouped as more or less hypoxic (Fig. 12.3a–b). A similar observation was done in advanced head and neck cancer, where pretreatment oxygen measurements were used to classify patients into more hypoxic or less hypoxic and thereby predict local tumor control (Fig. 12.3c–d). Numerous such clinical studies have now clearly shown the ability of hypoxia to influence outcome in squamous cell carcinomas of the cervix and head and neck (Hoeckel et al. 1993; Fyles et al. 1998; Fyles et al. 2002; Brizel et al. 1997; Nordsmark et al. 1996; Nordsmark 2005).

Direct measurement in the tumor using an oxygen electrode is often referred to as a gold standard due to its direct measurement principle and its ability to identify patients with poor prognosis. These measurements have served as standard reference in the development of other new and hopefully more clinically applicable hypoxia-specific assays such as positron emission tomography (PET) hypoxia imaging and hypoxia-specific gene classifiers. Ironically, the device is no longer commercially available.

12.3.2 Nitroimidazole Reduction

Clinically attractive techniques to measure tumor hypoxia are available, with special focus on the detection of injectable exogenous tracers that accumulate in hypoxic cells. These tracers can subsequently be detected in biopsies using immunohistology (Raleigh et al. 1996; Olive and Aquino-Parsons 2004) or alternatively they can be labeled with a radionuclide prior to injection allowing noninvasive detection using PET (Rasey et al. 1996) and single-photon emission computed tomography (SPECT) (Urtasun et al. 1996) or detected by magnetic resonance spectroscopy (Seddon et al. 2003). The exogenous immunodetectable markers mainly belong to the 2-nitroimidazole compounds and enter cells by diffusion. Once inside the cell, the molecule undergoes a single electron reduction to form a nitro radical anion which is immediately reoxidized when pO_2 is sufficiently high. However, when pO_2 drops below a threshold ($\sim$ 10 mmHg) the molecule is further reduced to form highly reactive intermediates that bind to various macromolecules in the cells, in particular, thiol groups in peptides and protein.

The most commonly used nitroimidazole for detection of hypoxia is pimonidazole, which was originally developed as a hypoxia radiosensitizer under the name Ro 03-8799 (Williams et al. 1982). Although pimonidazole failed as a useful sensitizer, it was further developed as a marker of tissue hypoxia and is now commercially available as Hypoxyprobe™. Pimonidazole and related exogenous hypoxia probes (e.g., EF5) can be visualized on tissue sections prepared from biopsies or surgically resected tumor tissue from animals or patients previously administered with the probes. Since hypoxic cells are difficult to reach, tracer diffusion and accumulation is a slow process, and tracers are normally allowed to circulate for hours or even days before tissue sampling. During the staining procedure, the remaining unbound tracer (not hypoxia related) is washed away resulting in highly specific staining of viable cells with a pO_2 below $\sim$ 10 mmHg. Pimonidazole staining was correlated with the radiobiological hypoxic fraction in experimental models (Raleigh et al. 1999) and in accordance, locoregional control rates are lower in head and neck cancer patients with high uptake of pimonidazole than in patients with low uptake (Kaanders et al. 2002a). Recently, it was further demonstrated that regional control following hypoxia-targeting ARCON treatment (see Sect. 9.3.4) in head and neck cancer patients was improved specifically in patients with tumors with high retention of pimonidazole (Janessens et al. 2012a, b). In support of the usefulness of nitroimidazole-based compounds, (Evans et al. 2007) showed that elevated EF5 binding was significantly associated with poor outcome in patients with head and neck cancer receiving various treatments. In contrast, there was no correlation between pimonidazole and locoregional tumor control or overall survival in cervix carcinoma patients treated with radiotherapy (Nordsmark et al. 2006). Likewise in patients with bladder cancer treated with radiotherapy concurrent with carbogen and nicotinamide (CON), no clear correlations between pimonidazole staining and local control or metastases-free survival were observed (Hoskin et al. 2003), but conversely this could simply suggest that CON eliminated the negative influence of hypoxia effectively. Immune-detectable markers require a biopsy and provide an assessment of hypoxia in a small tumor area, which may result in sampling errors due to intratumoral heterogeneity in hypoxia. Weak or absent correlations between clinical endpoints and tracer retention quantification in small tumor subvolumes may indeed relate to this heterogeneity. Finally, invasive tissue-based analysis provides no three-dimensional (3-D) information on the distribution of hypoxia within the tumor, which is required for some treatments like hypoxic dose painting (see Chap. 9.3.5).

In order to overcome some of the limitations of invasive quantification of hypoxia marker retention, labeling 2-nitroimidazoles with positron emitters (in particular fluorine-18) has received a lot of attention. Such labeling allows noninvasive assessment of global as well as regional tumor tracer retention using PET scans. Tracer retention can be quantified in numerous ways but typically, tracer activity in the tumor is compared to tracer activity in a nonhypoxic reference tissue such as muscle or blood. However, slow distribution, uptake, and clearance of nitroimidazoles result in a significant presence of image-contaminating unbound tracer (unlike the invasive assays where it is washed away during staining), which leads to a relatively limited tissue contrast even 2–4 h after tracer administration when a typical scan is performed. The formation of unwanted metabolites, which varies substantially among different tracers, may further deteriorate image quality and hypoxia specificity. The low-contrast problem is further exacerbated by the relatively low resolution of PET (several millimeters in clinical scanners), which is much lower than the dimensions of the hypoxic structures that are studied. Taken together, these inherent limitations increase the risk of missing small hypoxic foci, especially in areas where hypoxia and necrosis are intermixed (Busk et al. 2008). Still, hypoxia imaging has shown some promising results (Horsman et al. 2012).

^{18}F-misonidazole (FMISO) was the first tracer developed for hypoxia PET imaging (Rasey et al. 1989). Despite its nonideal pharmacokinetics which results in low image contrast, FMISO has shown prognostic value in several studies in different tumor types including head and neck cancer (Kikuchi et al. 2011; Rajendran et al. 2006), renal cell carcinoma (Hugonnet et al. 2011), and non-small-cell lung cancer (NSCLC) (Eschmann et al. 2005). In addition, FMISO scans were able to identify those head and neck cancer patients who benefitted from treatment with the hypoxia-activated prodrug tirapazamine (Rischin et al. 2006). Intriguingly, a recent study showed that FMISO hypoxia PET conducted 1–2 weeks after initiation of treatment with radiotherapy was more strongly linked to therapeutic response than pretreatment scans (Zips et al. 2012).

Several second-generation hypoxia tracers with supposedly better pharmacokinetic characteristics than FMISO have been developed and are currently being tested. Fluoroazomycin arabinoside (FAZA) is less lipophilic than FMSIO and may therefore clear faster from nonhypoxic tissue. FAZA proved highly superior to FMISO in tumor-bearing mice in terms of tumor-to-reference tissue ratios (Piert et al. 2005), but unexpectedly FAZA was slightly inferior to FMISO in tumor-bearing rats (Sorger et al. 2003), suggesting that conclusions obtained in one species/model cannot uncritically be extrapolated to other species/models. A preclinical study showed that FAZA retention was able to predict tumor control probability following radiotherapy in tumor-bearing mice (Mortensen et al. 2011; Fig. 12.4a–b). Similarly, it was shown that FAZA could identify mice that benefit from hypoxia-directed radiochemotherapy using tirapzamine (Beck et al. 2007). FAZA is currently undergoing clinical testing, and it was recently shownthat FAZA retention was correlated to disease-free survival in patients treated with radiotherapy (Fig. 12.4c–d; Mortensen et al. 2012).

12.3.3 Molecular Markers, Gene Expression, and the Therapeutic Opportunity of Targeting Hypoxia During Radiotherapy

The endogenous markers of hypoxia include a number of molecular molecules regulated by low oxygen content. These can be measured from biopsy material using protein immunohistochemistry or gene expression, or proteins identified from blood samples. Although endogenous markers have been correlated with outcome to radiation therapy in some studies, it is not a universal finding. This probably reflects the fact that many of these endogenous markers are not hypoxic specific rather than being any indication that hypoxia does not play a role in influencing radiation response (Bussink et al. 2003). In an attempt to confirm how well these endogenous markers express radiobiologically relevant hypoxia, several studies that have compared tumor oxygenation with the expression level of the hypoxia marker hypoxia inducible factor-1α (HIF-1α) (Mayer et al. 2004; Haugland et al. 2002; Jankovic et al. 2006), carbonic anhydrase IX (CAIX) (Loncaster et al. 2001; Mayer et al. 2004; Hedley et al. 2003; Jankovic 2006), and Glut-1 (Mayer et al. 2005). These studies showed mixed observations in uterine cervix cancer. In head and neck cancer, oxygen tension was compared with HIF-1α, CAIX, osteopontin (OPN) plasma, and OPN tumor and showed no mutual correlation and had different prognostic values (Nordsmark et al. 2006; Nordsmark et al. 2007). In primary NSCLC, there was a correlation between the tumor/normal lung tissue pO_2 ratio and the expression of OPN measured in plasma by enzyme-linked immunosorbent assay (ELISA) ($r = 0.53$, $p = 0.02$) and CAIX measured by immohistochemistry, ($p = 0.006$) (Le et al. 2006).

Expression of CAIX has been related with poor outcome in some head and neck cancer studies (Buffa et al. 2004) but not in others (Nordsmark et al. 2007; Eriksen and Overgaard 2007).

OPN is a secreted arginine–glycine–aspartic acid-containing phosphoprotein that was correlated with tumor hypoxia (Le et al. 2003). OPN is inversely correlated

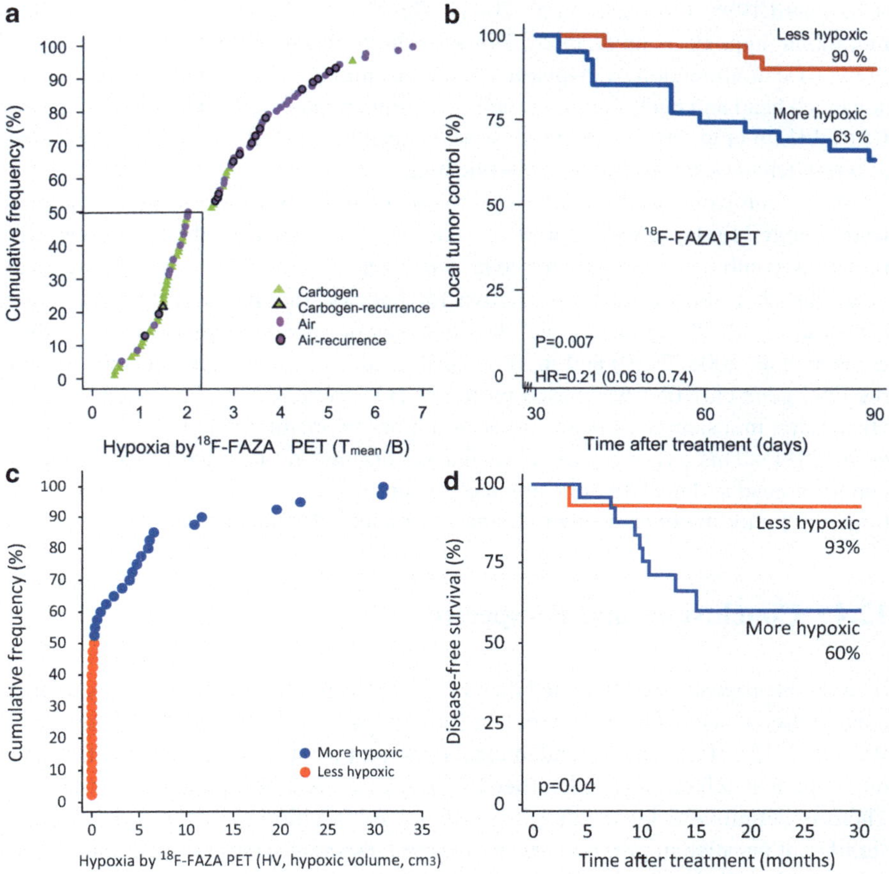

Fig. 12.4 **a** Cumulative frequency distribution of tumor hypoxia measured by ^{18}F FAZA PET in experimental mouse tumors under different gas-breathing conditions. **b** Local tumor control of tumors irradiated by 55 Gy single dose under different gas-breathing conditions and divided into "less hypoxic" and "more hypoxic" based on ^{18}F FAZA PET. **c** Cumulative frequency of 40 head and neck tumors as a function of hypoxia measured by ^{18}F FAZA PET prior to radiotherapy. **d** Disease-free survival of patients with hypoxic versus nonhypoxic tumors measured by ^{18}F FAZA PET (**a** and **b** modified from Mortensen et al. 2011; **c** and **d** adapted from Mortensen et al. 2012)

with Von Hippel–Lindau (VHL) gene expression, and the VHL protein drives the proteasome-mediated proteolysis of HIF-1α under normoxic conditions. In concordance with these results, there was a weak but significant correlation between median tumor pO_2 and plasma OPN in patients with advanced head and neck cancer (Nordsmark et al. 2007).

High levels of the plasma marker OPN were associated with poor local tumor control and survival after radiotherapy in 320 head and neck cancer patients who were included in a randomized double-blind placebo-controlled trial between radiation therapy alone and radiation combined with the hypoxia sensitizer nimorazole

(Overgaard 1998; Overgaard et al. 2005). This was the first proof of principle that modifiable hypoxia is predictable. However, high plasma OPN was not predictive of hypoxic modification by cisplatin and tirapazamin in patients with locoregionally advanced head and neck squamous cell carcinoma treated in the TROG 02.02 phase III trial (Lim et al. 2012), and more work is need before OPN is safely established as a predictive assay for hypoxic modification.

More promising results relate to a cohort of patients from the same randomized double-blind placebo-controlled trial between radiation therapy alone and radiation combined with the hypoxia sensitizer, nimorazole, where a hypoxia gene classifier showed to be predictive for hypoxic modification by nimoraxole (Toustrup et al. 2012). The gene classifier was developed stepwise in vitro (Soerensen et al. 2005, 2010) and in vivo (Busk et al. 2011a) and patient selection for the final gene classifier was based on tumor pO_2 measured by the Eppendorf pO_2 Histograph in a sample of head and neck tumors (Nordsmark et al. 2007; Toustrup et al. 2011). This hypoxia gene classifier is currently refined and tested in a new cohort of head and neck tumors and in other tumor sites as well, in order to advance targeting tumor hypoxia during radiotherapy on an individual patient basis.

12.4 Conclusions and Perspectives

It has been proposed that "targeted therapies" are the future of cancer therapy. In that context, hypoxia must be considered the ultimate target (Overgaard 2011; Peters and Rischin 2012). Hypoxia is a fundamental feature of most solid tumors, whether animal or human, which can be identified by a range of clinically applicable techniques. There is definitive evidence that its existence in specific tumor types will have a significant negative impact on cancer treatment, especially radiotherapy. Numerous preclinical studies have identified a range of different approaches that can reduce or eliminate tumor hypoxia and preferentially improve radiation response, and many of these approaches have undergone clinical testing and substantial improvements in outcome following radiation therapy. However, except for Denmark where the hypoxic cell radiosensitizer nimorazole is routinely used in the treatment of head and neck cancer, none of these hypoxic modifiers has become established as a standard therapy with radiation. Nevertheless, extensive preclinical and clinical studies are ongoing in this area, so hopefully, it is simply a matter of time before the hypoxia problem is eliminated.

Acknowledgments The authors would like to thank the following organizations for financial support: the Danish Cancer Society; the EC FP7 project METOXIA (project no. 222741); and CIRRO—the Lundbeck Foundation Center for Interventional Research in Radiation Oncology & the Danish Council for Strategic Research.

References

Adams GE, Cooke MS (1969) Int J Radiat Biol 15:457–471
Adams GE, Flockhart IR, Smithen CE et al (1976) Radiat Res 67, 9–20
Asquith JC, Watts ME, Patel K et al (1974) Radiat Res 60:108–118
Baumann M, Krause M, Hill R (2008) Nature Reviews Cancer 8:545–554
Beck R, Roper B, Carlsen JM et al (2007) Journal of Nuclear Medicine 48:973–80
Bergsjø P, Kolstad P (1968) Scand J Clin Lab Inv 106:167–171
Brizel DM, Sibley GS, Prosnitz LR et al (1997) Int J Radiat Oncol Biol Phys 38:285–289
Buffa FM, Bentzen SM, Daley FM, et al (2004) Clinical Cancer Research 10 (11):3745–3754
Busk M, Horsman MR, Overgaard J (2008) Acta Oncologica 47:1201–10
Busk M, Toustrup K, Soerensen BS, et al (2011) BMC Cancer 9(11):63
Bussink J, Kaanders JHAM, Kogel AJ van der (2003) Radiother Oncol 67:3–15
Chaplin DJ, Horsman MR, Siemann DW (1993) Br J Cancer 68:269–273
Churchill-Davidson I (1968) Front Radiat Ther Oncol 1:1–15
Dische S (1979) Br J Radiol 51:888–894
Dische S (1985) Radiother Oncol 3:97–115
Dische S, Anderson PJ, Sealy R et al (1983) Br J Radiol 56:251–255
Dische S, Machin D, Chassagne D (1993) Radiother Oncol 26:93–103
Dobrowsky W, Naude J, Dobrowsky E et al (1995) Acta Oncologica 34:270–272
Du Sault LA (1963) Br J Radiol 36:749–754
Dobrowsky W, Huigol NG, Jayatilake RS et al (2007) Radiother Oncol 82:24–29
Eriksen JG, Overgaard J (2007) Radiother Oncol 83(3):383–388
Eschmann SM, Paulsen F, Reimold M et al (2005) J Nucl Med 46:253–60
Eschwége F, Sancho-Garnier H, Chassagne D et al (1997) Int J Radiat Oncol Biol Phys 39:275–281
Evans JC, Bergsjø P (1965) Radiology 84:709–717
Evans SM, Du KL, Chalian AA et al (2007) Int J Radiat Oncol Biol Phys 69:1024–31
Fyles AW, Milosevic M, Wong R, et al (1998) Radiother Oncol 48(2):149–56
Fyles A, Milosevic M, Hedley D, et al (2002) J Clin Oncol 20(3):680–687
Gatenby RA, Kessler HB, Rosenblum JS et al (1988) Int J Radiat Oncol Biol Phys 14:831–838
Grau C, Horsman MR, Overgaard J (1992) Int J Radiat Oncol Biol Phys 22:415–419
Grau C, Agarwal JP, Jabeen K et al (2003) Radiother Oncol 67:17–27
Gray LH, Conger AD, Ebert M et al (1953) Br J Radiol 26:638–648
Grogan M, Thomas GM, Melamed I et al (1999) Cancer 86:1528–1536
Haffty BG, Son YH, Sasaki CT et al (1993) Int J Radiat Oncol Biol Phys 27:241–250
Haffty BG, Wilson LD, Son YH et al (2005) Int J Radiat Oncol Biol Phys 61:5–6
Hall EJ (1994) Radiobiology for the Radiobiologist, 4th edn. JB Lippincott, Philadelphia
Hall EJ, Roizin-Towle L (1975) Radiology 117:453–457
Haugland HK, Vukovic V, Pintilie M et al (2002) Int J Radiat Oncol Biol Phys 53:854–861
Hedley D, Pintilie M, Woo J et al (2003) Clin Cancer Res 9:5666–5674
Henk JM, Smith CW (1977 July) Lancet 104–105
Henk JM, Kunkler PB, Smith CW (1977 July) Lancet 101–103
Henke M, Laszig R, Rube C et al (2003) Lancet 362:1255–1260
Hill RP, Marie-Egyptienne DT, Hedley DW (2009) Semin Radiat Oncol 19:106–111
Hoeckel M, Knoop C, Schlenger K et al (1993) Radiother Oncol 26:45–50
Hoff CM, Lassen P, Eriksen JG, et al (2011a) Acta Oncologica 50:1006–1014
Hoff CM, Hansen HS, Overgaard M, et al (2011b) Radiother Oncol 98:28–33
Hoff CM (2012) Acta Oncologica 1–14
Horsman MR, Khalil AA, Nordsmark M, et al (1993) Radiother Oncol 28:69–71
Horsman MR (1995) Acta Oncologica 34:571–587
Horsman MR, Mortensen LS, Petersen JB, et al (2012) Nat Rev Clin Oncol 9(12):674–87
Hoskin PJ, Sibtain A, Daley FM, et al (2003) Br J Cancer 89:1290–7
Hoskin PJ, Robinson M, Slevin N et al (2009) J Clin Oncol 27:5751–5756

Hoskin PJ, Rojas A, Bentzen SM (2010) J Clin Oncol 33:4912–4918
Hugonnet F, Fournier L, Medioni J et al (2011) J Nucl Med 52:1048–55
Jankovic B, Aquino-Parsons quino-Parsons C, Raleigh JA et al (2006) Cytometry B Clin Cytom 70:45–55
Janssen HL, Haustermans KM, Sprong D et al (2002) Int J Radiat Oncol Biol Phys 54:1537–1549
Janssens GO, Rademakers SE, Terhaard CH, et al (2012a) J Clin Oncol 20, 30(15):1777–83
Janssens GO, Terhaard CH, Doornaert PA, Bijl HP, Ende P van den, Chin A, Pop LA, Kaanders JH (2012b) Int J Radiat Oncol Biol Phys 82(2):532–8
Kaanders JH, Wijffels KI, Marres HA et al (2002a) Cancer Res 62:7066–74
Kaanders JH, Bussink J, Kogel AJ van der (2002b) Lancet Oncology 3:728–737
Kaanders JHAM, Pop LAM, Marres HAM (2002c) Int J Radiat Oncol Biol Phys 52:769–778
Kennedy KA, Rockwell S, Sartorelli AC (1980) Cancer Res 40:2356–2360
Kikuchi M, Yamane T, Shinohara S et al (2011) Ann Nucl Med 25:625–33
Kolstad P (1963) Copenhagen Stockholm Göteborg, Scandinavian University Books, 1963
Lavey and Dempsey (1993)
Le QT, Sutpin PD, Raychaudhuri S et al (2003) Clin Cancer Res 9:59–67
Le QT, Chen E, Salim A, et al (2006) Clin Cancer Res 12(5):1507–1514
Lee DJ, Cosmatos D, Marcial VA et al (1995) Int J Radiat Oncol Biol Phys 32:567–576
Lim AM, Rischin D, Fisher R, et al (2012) Clin Cancer Res 18(1):301–7
Loncaster JA, Harris AL, Davidson SE et al (2001) Cancer Res 61:6394–6399
Mayer A, Wree A, Hockel M et al (2004) Cancer Res 64:5876–5881
Mayer A, Hockel M, Wree A et al (2005) Clin Cancer Res 11:2768–2773
Machtay M, Pajak T, Suntharalingam M et al (2007) Int J Radiat Oncol Biol Phys 69:1008–1017
Mendenhall WM, Morris CG, Amdur RJ, et al (2005) Cancer 104:333–337
McKeown SR, Cowen RL, Williams KJ (2007) Clin Oncol 19:427–442
Mortensen LS, Busk M, Nordsmark M, et al (2011) Radiother Oncol 99(3):418–423
Mortensen LS, Johansen J, Kallehauge J, et al (2012) Radiother Oncol 105(1):14–20
Mottram JC (1936) Br J Radiol 10:606–614
Moulder JE, Rockwell S (1984) Int J Radiat Oncol Biol Phys 10:695–712
Müller C (1910) Munchener Medizinische Wochenschrift 28:1490–1493
Nordsmark M, Bentzen SM, Overgaard J (1994) Acta Oncologica 33(4):383–389
Nordsmark M, Overgaard M, Overgaard J (1996) Radiother Oncol 41:31–39
Nordsmark M (1997) PHD thesis Faculty of Health Sciences. University of Aarhus 1:1–121
Nordsmark M, Bentzen SM, Rudat V et al (2005) Radiother Oncol 77:18–24
Nordsmark M, Loncaster J, Aquino-Parsons quino-Parsons C, et al (2006) Radiother Oncol 80(2):123–131
Nordsmark M, Eriksen JG, Gebski V et al (2007) Radiother Oncol 83:389–397
Olive PL, Aquino-Parsons C (2004) Semin Radiat Oncol 14:241–248
Overgaard J (1989) Int J Radiat Biol 56:801–811
Overgaard J (1994) Oncol Res 6:509–518
Overgaard J (2007) J Clin Oncol 25(26):4066–74
Overgaard J (2011) Radiother Oncol 100(1):22–32
Overgaard J, Hansen HS, Andersen AP et al (1989) Int J Radiat Oncol Biol Phys 16:1065–1068
Overgaard JS, Hansen H, Overgaard M et al (1998) Radiother Oncol 46:135–146
Overgaard J, Eriksen JG, Nordsmark M et al (2005) Lancet Oncol 6(10):757–64
Overgaard J, Hoff C, Sand HH (2007) Eur J Cancer 5 (Suppl 7)
Peters L, Rischin D (2012) J Clin Oncol 30(15):1741–3
Petersen C, Zips D, Krause M et al (2001) Int J Radiat Oncol Biol Phys 51:483–493
Piert M, Machulla HJ, Picchio M et al (2005) J Nucl Med 46:106–113
Phillips RM, Hendriks HR, Peters GJ (2012) Br J Pharmacol Apr 18
Rajendran JG, Schwartz DL, O'Sullivan J et al (2006) Clin Cancer Res 12:5435–41
Raleigh JA, Dewhirst MW, Thrall DE (1996) Semin Radiat Oncol 6:37–45
Raleigh JA, Chou SC, Arteel GE, et al (1999) Radiat Res 151(5):580–9
Rasey JS, Koh WJ, Grierson JR, et al (1989) Int J Radiat Oncol Biol Phys 17:985–91

Rasey JS, Koh WJ, Evans ML et al (1996) Int J Radiat Oncol Biol Phys 36:417–428
Rischin D, Hicks RJ, Fisher R et al (2006) J Clin Oncol 24:2098–104
Rouschop KM, Beucken T van den, Dubois L, et al (2010) J Clin Invest 120(1):127–141
Rojas A (1991) Radiother Oncol 20(Suppl 1):65–70
Roti Roti JL (2004) Int J Hyperther 201:109–114
Rubin P, Hanley J, Keys HM et al (1979) Int J Radiat Oncol Biol Phys 5:1963–1970
Schwarz G (1909) Munchener Medizinische Wochenschrift 24:1–2
Seddon BM, Payne GS, Simmons L et al (2003) Clin Cancer Res 9:5101–5112
Siemann DW, Hill RP, Bush RS (1977) Int J Radiat Oncol Biol Phys 2:903–911
Sorger D, Patt M, Kumar P et al (2003) Nucl Med Biol 30:317–26
Stratford IJ, Stephens MA (1989) Int J Radiat Oncol Biol Phys 16:973–976
Stratford IJ, O'Neill P, Sheldon PW et al (1986) Biochem Pharmacol 35:105–109
Stuben G, Pottgen C, Knuhmann K et al (2003) Int J Radiat Oncol Biol Phys 55:1358–1362
Suit HD, Marshall N, Woerner D (1972) Cancer 30:1154–1158
Sørensen BS, Hao J, Overgaard J et al (2005) Radiother Oncol 76:187–193
Sørensen BS, Toustrup K, Horsman MR et al (2010) Acta Oncologica 49:895–905
Thews O, Koenig R, Kelleher DK, Kutzner J, Vaupel P (1998) Br J Cancer 78:752–756
Thomas GM (2002) Oncology 63:19–28
Toustrup K, Sorensen BS, Nordsmark M, et al (2011) Cancer Res 71(17):5923–5931
Toustrup K, Sorensen BS, Lassen P et al (2012) Radiother Oncol 102:122–129
Urtasun R, Band P, Chapman, JD et al (1976) New Engl J Med 294:1364–1367
Urtasun RC, Parliament MB, McEwan AJ et al (1996) Br J Cancer 74(Suppl):S209–S212
Watson ER, Halnan KE, Dische S et al (1978) Br J Radiol 51:879–887
Weissberg JB, Son YH, Papac RJ et al (1989) Int J Radiat Oncol Biol Phys 17:3–9
Williams MV, Denekamp J, Minchinton AI, et al (1982) Int J Radiat Oncol Biol Phys 8:477–81
Yaromina A, Zips D, Thames HD et al (2006) Radiother Oncol 81:122–129
Zeman EM, Hirst VK, Lemmon MJ, Brown JM (1988) Radiother Oncol 12:209–218
Zips D, Zophel K, Abolmaali N et al (2012) Radiother Oncol 105:21–28

Chapter 13
Prodrug Strategies for Targeting Tumour Hypoxia

William R. Wilson, Kevin O. Hicks, Jingli Wang and Frederik B. Pruijn

Abstract Tumour hypoxia is a critically important, but elusive, target in cancer therapy; the importance of eliminating hypoxic cells is underscored by their central roles in tumour progression and resistance to cytotoxic chemotherapy and radiotherapy. While many molecular targets may offer synthetic lethal interactions with hypoxia, the development of prodrugs that are activated to generate cytotoxins by one-electron (1e) reduction in hypoxic cells represents a more direct approach. Although conceptually simple, significant challenges need to be overcome to achieve useful therapeutic activity. These include designing prodrugs able to diffuse efficiently into hypoxic tissue, avoidance of off-target (oxygen-insensitive) two-electron (2e) reduction, maximizing bystander effects from diffusion of the active metabolites to exploit severely hypoxic regions in tumours and (critically) the development of predictive biomarkers for what is in fact a multifaceted target (comprising hypoxia, activating reductases and the molecular determinants of sensitivity to the active metabolites). Here we provide an overview of recent progress towards these goals in the context of four classes of hypoxia-activated prodrugs (HAPs) in clinical trial or in advanced preclinical development, namely benzotriazine di-oxides (tirapazamine and SN30000), the dinitrobenzamide mustard (DNBM) PR-104, the phosphoramidate mustard TH-302 and nitrochloromethylbenzindoline (nitroCBI) prodrugs of potent DNA minor groove alkylators.

Keywords Hypoxia-activated prodrugs · Bioreductive prodrugs · Tirapazamine · SN30000 · PR-104 · TH-302 · Nitrochloromethylbenzindolines

13.1 Introduction: Hypoxia as a Therapeutic Target

Many tumours develop regions of severe hypoxia as a consequence of a structurally and functionally defective microvascular system (Dewhirst et al. 2008; Secomb et al. 2012; Vaupel and Mayer 2007); this notable feature of tumours has long been

W. R. Wilson (✉) · K. O. Hicks · J. Wang · F. B. Pruijn
Auckland Cancer Society Research Centre, The University of Auckland,
Private Bag 92019, Auckland 1142, New Zealand
e-mail: wr.wilson@auckland.ac.nz

G. Melillo (ed.), *Hypoxia and Cancer,* Cancer Drug Discovery and Development, DOI 10.1007/978-1-4614-9167-5_13,

considered a potential therapeutic target. The importance of hypoxia as a target is underscored by extensive evidence of its involvement in multiple interrelated aspects of tumour progression (many of them reviewed in this volume); these include suppression of cell death pathways including immune-mediated responses, the glycolytic switch in intermediary metabolism, increased resistance to extracellular acidosis, angiogenesis, recruitment of bone marrow-derived progenitor cells, genomic instability, reactive oxygen species signalling, activation of the epithelial–mesenchymal transition, adoption of stem-cell like phenotypes and enhanced invasion and metastasis (Chan and Giaccia 2007; Finger and Giaccia 2010; Hill et al. 2009). Despite the antiproliferative effects of severe hypoxia *in vitro*, in part mediated by decreases in deoxyribonucleotide triphosphate (dNTP) pools because of the oxygen dependence of ribonucleotidereductase (Norlund and Reichard 2006; Pires et al. 2010), hypoxia also seems to be associated with a proliferative phenotype in tumours (Chang et al. 2011; Mehta et al. 2011), possibly in part through its effect on receptor-mediated mitogenic signalling (Wang and Ohh 2010). Thus targeting of hypoxic cells may have potential to suppress the growth of primary tumours in some cases, as well as suppressing metastasis.

The case for considering hypoxia an important therapeutic target is further strengthened by the multiple contributions of hypoxia to resistance to radiotherapy and chemotherapy (Wilson and Hay 2011; Rohwer and Cramer 2011). The above considerations have generated much interest in molecular targets in hypoxic cells, a field that is the subject of many recent reviews (Ebbesen et al. 2009; Melillo 2007; Poon et al. 2009) although clinical exploration of such approaches is at an early stage. Some of the challenges in exploiting such targets include the difficulty of developing specific inhibitors of hypoxia-regulated transcription factors such as the hypoxia-inducible factor (HIF) family (Melillo 2007; Nordgren and Tavassoli 2011; Xia et al. 2012), concerns that HIF signalling is tumour-suppressive in some settings (Blouw et al. 2003; Keith et al. 2012; Mazumdar et al. 2010), the complexity of interaction between hypoxia-sensitive and hypoxia-independent regulation of this and other signalling pathways and the incomplete understanding of which targets would need to be inhibited in individual tumours to achieve synthetic lethality with hypoxia. However, important progress is being made in targeting carbonic anhydrase-IX which is encoded by an HIF-1-regulated gene and supports the survival of hypoxic cells in tumours (Dubois et al. 2011; McDonald et al. 2012; Poon et al. 2009; Chap.6 by the authors Pastorekova and Supran of this volume). In addition, other molecular targets required for survival of radioresistant hypoxic cells, such as IRE1 (Romero-Ramirez et al. 2005; Papandreou et al. 2011) or PERK (EIF2AK3) and its downstream signalling partners (Rouschop et al. 2010; Rouschop et al. 2013), provide attractive drug development opportunities.

The development of prodrugs that are activated selectively under hypoxia (hypoxia-activated prodrugs, HAPs) represents a superficially simpler challenge, and is the focus of this review. Importantly, HAPs offer strategies for exploiting hypoxia to kill more than just the hypoxic subpopulation. This can be achieved, for example, by multiple dosing of HAP to exploit cycling (fluctuating) hypoxia (Dewhirst 2009), thus eliminating cells as they transit through a hypoxic state (Denny and Wilson 1993; Kim and Brown 1994). An allied concept is to use an HAP with a long tumour residence time to exploit cycling hypoxia. Alternatively, an HAP can be used for hypoxia-dependent formation of an active drug that can kill well-oxygenated tumour cells. Such "bystander" effects can potentially be mediated by local diffusion of an active metabolite to adjacent cells at higher partial pressure of oxygen (pO_2, Denny and Wilson 1993), or by stable metabolites that are recycled when hypoxic cells die as is likely for AQ4, the active metabolite from banoxantrone (AQ4N, Patterson 2002).

For both molecular targeting and prodrug approaches, the identification of clinical settings in which hypoxia negatively impacts treatment outcomes is urgently needed. That evidence for hypoxia is a predictive factor is strongest in the case of definitive radiotherapy and chemoradiotherapy of head and neck squamous cell cancer (HNSCC; Nordsmark et al. 2005; Toustrup et al. 2011), and to a lesser extent in other radiation therapy contexts (Toustrup et al. 2012; Chap. 12 by the author Nordsmark of this volume). The current trend to hypofractionation in many radiotherapy settings lessens the opportunity for tumour reoxygenation during treatment and therefore further strengthens the case for hypoxia-targeted therapies (Brown et al. 2010; Carlson et al. 2011). Unfortunately, new drug development in a radiotherapy setting is a challenging undertaking (Harrington 2011), and has not yet been explored clinically with the most recent generation of HAPs. It is also disappointing that there is still little information on the predictive significance of hypoxia in clinical cancer chemotherapy. However, less invasive tools for monitoring tumour hypoxia that are now becoming available (Sect. 13.7.2) will hopefully stimulate investigation of this question.

In addition to viewing basal levels of hypoxia as a therapeutic target, there is evidence that some clinically important anticancer agents may induce additional hypoxia as a consequence of inhibiting tumour blood flow. Examples include the induction of hypoxia by bevacizumab in some breast cancers (Mehta et al. 2011), and by anti-angiogenic kinase inhibitors (Chang et al. 2007; Kruser et al. 2010; Paez-Ribes 2009), and vascular-disrupting agents (Siim et al. 2000; Taylor et al. 2012) in preclinical models. In radiation therapy, increases in hypoxia due to regrowth of tumours within the radiation field (Rofstad et al. 2005) have been implicated in recruitment of bone marrow-derived myeloid cells required for the irradiated tumour microvasculature to support tumour regrowth (Kioi et al. 2010). In addition, rapid inhibition of tumour blood flow and induction of hypoxia has recently been suggested to occur via ceramide signalling following the high radiation doses increasingly being used in hypofractionated radiotherapy protocols (Kolesnick et al. 2012). Such *therapy-induced hypoxia* is suspected to contribute to aggressiveness of tumours that escape during anti-angiogenic therapy (Paez-Ribes 2009; Sennino and McDonald 2012); coupled with the prospect that therapy-induced hypoxia may be more frequent, more severe and more predictable than basal hypoxia in some settings, it represents an attractive opportunity for hypoxia targeting (Rapsidara and Melillo 2012).

13.2 Hypoxia-activated Prodrugs

The concept of activating prodrugs selectively within tumours can be traced to the discovery that experimental tumours with high β-glucuronidase activity are sensitive to aniline mustard because they hydrolyse its glucuronide metabolite to regenerate

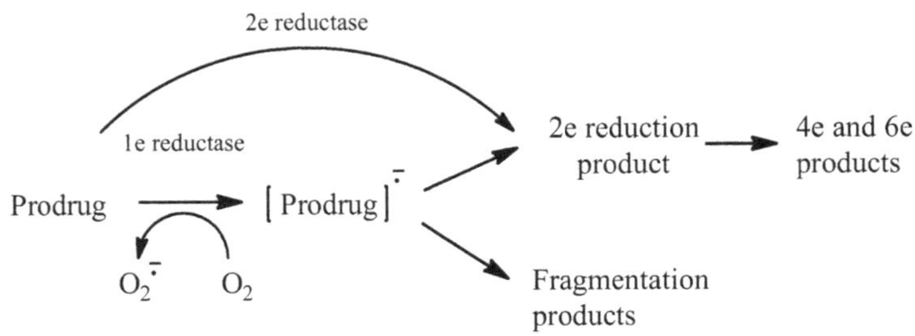

Fig. 13.1 Mechanism of hypoxia-selective activation of prodrugs. Selectivity for hypoxia typically depends on redox cycling in oxic cells, which suppresses formation of the initial prodrug radical. Under hypoxia, the latter can give rise to cytotoxins either by fragmentation (e.g. tirapazamine, SN30000, TH-302) or via further reduction to 4- and 6-electron reduction products (e.g. PR-104 and nitroCBIs)

the active nitrogen mustard within the tumour (Connors and Whisson 1966). This early observation sparked ongoing efforts to exploit high endogenous expression of prodrug-activating enzymes in tumours, including β-glucuronidase (Denny and Wilson 1998; Graaf et al. 2002). Recent efforts to exploit endogenous enzymes include activation of canfosfamide (Telcyta; TLK286, TER286) by glutathione S-transferase pi1 (GSTP1), potentially targeting drug-resistant cells that overexpress this glutathione-S-transferase isoform (Morgan et al. 1998; Rosario et al. 2000; Vergote et al. 2009). Further examples are attempts to exploit the high expression in some tumours of NADH:quinone oxidoreductase-1 (NQO-1/DT-diaphorase) with the quinone bioreductive prodrug apaziquone (EO9, Phillips et al. 2013) and NQO-2 with CB1954 (Knox and Chen 2004). The challenge of achieving sufficient activation (and sufficiently selective activation) using endogenous drug-metabolising enzymes has stimulated strategies for targeted delivery of exogenous enzymes with a variety of "directed enzyme prodrug therapy" (DEPT) systems, using antibodies (ADEPT, Bagshawe 1995), viruses (VDEPT, Grove et al. 1999) or tumour-tropic bacteria (BDEPT; Fox et al. 1996; Liu et al. 2008; Nuyts et al. 2002) as vectors.

Hypoxia is particularly attractive as a basis for selective activation of prodrugs using endogenous activating enzymes, given the prevalence of severe hypoxia in many tumour types (Vaupel et al. 2007) and the importance of hypoxic cells in resistance to conventional therapeutic agents. HAPs (also known as bioreductive prodrugs or hypoxic cytotoxins) have an extensive history, originating independently from observations that nitroimidazole radiosensitizers are selectively toxic to hypoxic tumour cells in culture (Hall and Roizin-Towle 1975; Mohindra and Rauth 1976) and from the development of quinone bioreductive alkylating agents by Sartorelli's group (Lin et al. 1972). These pioneering efforts represented something of an act of faith in that the existence of hypoxia in human tumours and its therapeutic significance was controversial at that time. While these issues have largely been resolved, the extent to which normal tissue hypoxia might compromise the therapeutic utility of HAPs is still unclear.

The mechanism by which most HAPs achieve selectivity for hypoxia is illustrated in Fig. 13.1. These compounds are substrates for oxidoreductases capable of catalysing one-electron (1e) transfer reactions, effecting reduction of the prodrug to a free radical. Oxygen at concentrations in the physiological (normoxic) range rapidly reoxidises the prodrug radical to the parent prodrug, generating superoxide. In the absence of oxygen, the steady-state concentration of the initial prodrug free radical increases markedly, facilitating its conversion to cytotoxic species. If these downstream cytotoxins are more potent than the unreduced prodrug, and the superoxide from redox cycling in oxic cells, hypoxia-selective cytotoxicity is achieved. The best-characterised 1e HAP oxidoreductase is NADPH:cytochrome P450 oxidoreductase (POR; Belcourt et al. 1996; Guise et al. 2007; Meng et al. 2012; Patterson et al. 1997; Patterson et al. 1998; Wang et al. 2012); other known 1e HAP reductases include nicotinamide adenine dinucleotide phosphate (NADPH)-dependent diflavinoxidoreductase 1 (NDOR1, Guise et al. 2012), methionine synthase reductase (MTRR, Guise et al. 2012), the nitric oxide synthase isoforms NOS1, NOS2A and NOS2B (Ask et al. 2003; Chandor et al. 2008), NADH-dependent cytochrome b5 reductase (CYB5R, Papadopoulou et al. 2003), thioredoxin reductase (Cenas et al. 2006), aldehyde oxidase (Tatsumi et al. 1986) and xanthine oxidase (Adams and Rickert 1995; Ueda et al. 2003). Most 1e reductases appear to have low substrate specificity, with the kinetics of reduction (and re-oxidation of the prodrug radical) determined largely by the one-electron reduction potential (E(1)) of the prodrug (Wardman 2001).Unfortunately there is little information on the contribution of any of the known 1e reductases at basal expression levels to HAP activity in tumour cells (even in model systems). There is also a dearth of information on whether differences in 1e reductase activity is an important determinant of HAP sensitivity of hypoxic cells in human tumours, although the limited data available point to substantial variation between cell lines (Guise et al. 2012; Wang et al. 2012) and between individual human tumours (Evans et al. 2000; Patterson et al. 1997).

A complicating factor in targeting hypoxia with HAP is that many of these prodrugs can also be activated by oxygen-insensitive two-electron (2e) oxidoreductases; by catalysing concerted addition of 2e (usually via hydride transfer from NAD(P)H), these enzymes bypass the initial oxygen-sensitive prodrug radical. This results in cytotoxicity independent of hypoxia when the active cytotoxin is the 2e reduced metabolite or is downstream of this. The best-studied 2e reductase is NQO-1 which, in common with other 2e reductases, has a narrower substrate specificity than the 1e reductases. This competing 2e reduction by NQO-1 is most pronounced for quinone bioreductive prodrugs (including mitomycin C), many of which have poor hypoxic selectivity as a consequence. However, NQO-1 is highly expressed in some tumours as a result of redox signalling through Keap1/Nrf2 (Cullinan and Diehl 2006), and thus represents a classical endogenous enzyme/prodrug strategy. The overexpression of NQO-1 in bladder carcinomas has led to the evaluation of the NQO-1 prodrug apaziquone (EO9) in two phase III trials in this context (Phillips et al. 2013); although both trials failed their primary endpoint of reduced tumour recurrence at 2 years, when the trials were pooled there was a significant delay in recurrence ($p = 0.017$) (http://www.businesswire.com/news/home/20120405005360/en/Spectrum-Pharmaceuticals-Announces-Results-Apaziquone-Phase-3).

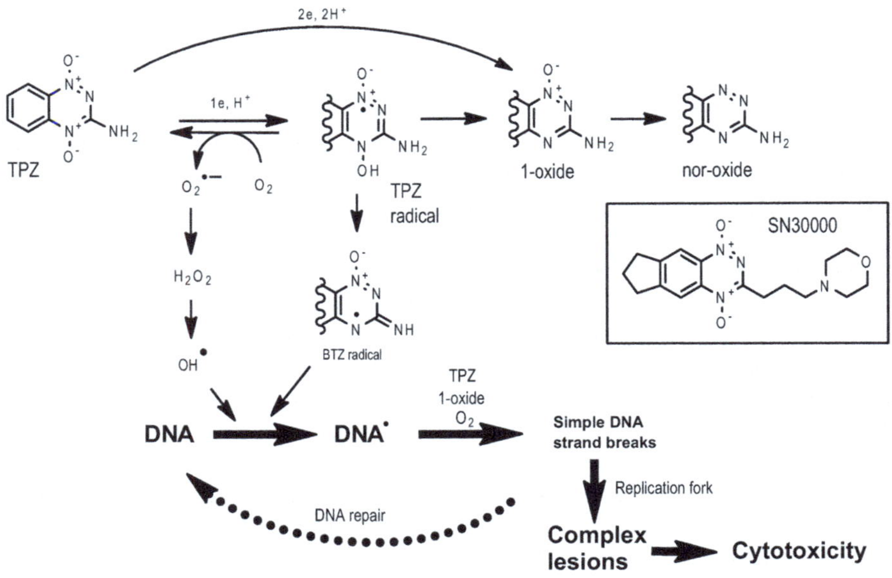

Fig. 13.2 Mechanism of action of benzotriazine HAPs inferred from studies of TPZ and its analogue SN30000. Net enzymatic 1e reduction is inhibited in the presence of oxygen by reoxidation of the initial radical by O_2. In the absence of O_2, the radical lifetime is extended and it undergoes spontaneous dismutation to mono-oxide metabolites, in particular the 1-oxide which may undergo further 2e reduction to the nor-oxide. The 1-oxide may also be formed by enzymatic 2e reduction which is oxygen insensitive. Downstream radicals (OH·and BTZ radicals) cause oxidative damage of DNA which may be further oxidised by oxygen or parent prodrug or 1-oxide (under hypoxia) with subsequent production of complex DNA lesions. Simple DNA breaks may be repaired by base excision repair and non-homologous end-joining pathways, with complex lesions undergoing homologous repair. (Modified from Hunter et al. 2012 to which the reader is referred for further details)

The following sections outline four classes of HAPs currently in clinical trial or which are well advanced in preclinical development. A number of reviews describe earlier HAPs (Ahn and Brown 2007; Chen and Hu 2009; Denny et al. 1996; McKeown et al. 2007; Patterson 2002; Rauth et al. 1998; Rockwell et al. 2009; Sartorelli 1988; Stratford and Workman 1998; Wardman 2001; Wilson and Hay 2011; Workman and Stratford 1993); even though not clinically successful, many of these compounds have illustrated important concepts and have influenced the ongoing development of the field.

13.3 Benzotriazine di-N-oxides (Tirapazamine and SN30000)

13.3.1 Overview

The most intensively studied HAP, tirapazamine (TPZ; Fig. 13.2), shows high selectivity for hypoxia cells in culture through bioreductive metabolism to an oxidising free radical that elicits complex DNA damage. It neatly sidesteps the problem of 2e reduction; although it is a substrate for NQO-1, the 2e (and 4e) reduction products have low cytotoxic potency. An additional feature differentiating TPZ from other HAPs is that it is activated at relatively high oxygen concentrations, potentially allowing it to eliminate HIF-stabilised and radioresistant cells that are only moderately hypoxic. Despite signs of clinical activity in early phase trials, TPZ has not achieved clinical registration, perhaps in part because the pivotal trials have not been conducted with predictive biomarker support to confirm the presence of the target (hypoxia and appropriate reductase activity). However, even in preclinical models with extensive hypoxia, TPZ is less selective for hypoxic cells than in cell culture as a consequence of limited penetration into severely hypoxic tissue. The second-generation benzotriazine di-oxide (TPZ analogue) SN30000 provides a greater hypoxic cell killing than TPZ in tumour xenografts, in part because of improved tissue penetration. Planned clinical development of SN30000, through Cancer Research UK, seeks to leverage experience with TPZ and make appropriate use of predictive biomarkers to identify potentially responsive patients.

13.3.2 Discovery and Mechanism of Action of TPZ

The aromatic N-oxide TPZ (3-amino-1,2,4-benzotriazine 1,4-dioxide, Fig. 13.2) represents a class of aromatic N-oxide HAPs based on the benzotriazine moiety (Brown 1993; Brown 2010), which were first developed as pesticides and antifungals. TPZ in particular was shown to have marked selectivity for cancer cells exposed under hypoxic conditions (Zeman et al. 1986), and many studies have demonstrated that TPZ and analogues have a hypoxic cytotoxicity ratio (HCR) of 50–300 *in vitro* using both clonogenic and growth inhibition endpoints (Hay et al. 2003; Koch 1993; Zeman et al. 1989). Early studies showed that under anoxia TPZ was rapidly metabolised, mainly to its 1-oxide which was found to be non-cytotoxic, suggesting that cytotoxicity was due to an intermediate radical (Baker et al. 1988; Laderoute et al. 1988). Animal studies combining TPZ with radiotherapy confirmed selective killing of hypoxic tumour cells and indicated the potential to overcome radioresistance (Brown and Lemmon 1990; Zeman et al. 1988). An analogue development programme demonstrated that cytotoxicity, bioreduction (measured by oxygen consumption due to redox cycling in oxic cells) and LD_{50} in mice of neutral compounds were dependent on E(1) but that the HCR reaches a maximum at E(1) low enough for acceptable host toxicity (Zeman et al. 1989). These trends were also seen in a further analogue series (Hay et al. 2003), and several studies have shown that cell killing under anoxia is directly related to the amount of TPZ metabolised (Hicks et al. 2003; Siim and van Zijl 1996).

TPZ undergoes enzyme-catalysed 1e reduction (illustrated in Fig. 13.2) to a TPZ radical which is efficiently reoxidised to TPZ in the presence of O_2, producing a superoxide radical and downstream reactive oxygen species. At low oxygen concentrations the radical lifetime is substantially increased allowing its spontaneous conversion to oxidising radicals capable of damaging DNA (Anderson et al. 2003; Daniels and Gates 1996). The oxidising radical(s) producing DNA damage are still not completely characterised; evidence supports the cytotoxic OH radical (Chowdhury et al. 2007; Daniels and Gates 1996), the benzotriazinyl radical (BTZ, Fig. 13.2, Anderson et al. 2003; Shinde et al. 2009) and aryl radicals (Shinde et al. 2010; Yin et al. 2012). Metabolism by 2e reductases does not contribute to cytotoxicity as it bypasses the critical prodrug radical, producing the much less toxic 1-oxide (SR4317) and nor-oxide (SR4330) metabolites (Fig. 13.2). POR has been identified as an important 1e reductase for TPZ (Fitzsimmons et al. 1994; Patterson et al. 1997; Patterson et al. 1998) and analogues (Wang et al. 2012) with iNOS also contributing (Chinje et al. 2003). Early studies also suggested that unidentified nuclear reductase(s) may be particularly important in the hypoxic cytotoxicity of TPZ (Evans et al. 1998), but the 1e reductases responsible for activation in human tumours remain poorly characterised.

It has been proposed that DNA radicals produced by the oxidising cytotoxic radicals can be further oxidised by TPZ itself (Daniels et al. 1998; Jones and Weinfeld 1996) or, more efficiently, by its 1-oxide metabolite SR4317 (Hwang et al. 1999) to DNA breaks and other DNA lesions. Marked potentiation of the anoxic cytotoxicity of TPZ by high concentrations of SR4317 has been observed in a study designed to exploit this phenomenon (Siim et al., 2004), although this might also be accounted for by displacement of the redox equilibrium between the TPZ radical and benzotriazinyl radical by SR4317 (Anderson et al. 2006). Complex damage (single-strand breaks, base damage and DNA–protein crosslinks) produced by TPZ under anoxia has been detected by comet assay (Olive et al. 1996; Siim and van Zijl 1996). This damage, which includes the formation of covalently bound topoisomerase complexes (Peters and Brown 2002), interferes with the DNA replication fork leading to sensitivity of cells deficient in homologous recombination repair (Evans et al. 2008; Hunter et al. 2012). Recent evidence shows that multiple DNA repair pathways are responsible for resolution of this complex DNA damage, with different repair profiles under aerobic and hypoxic conditions (Hunter et al. 2012). DNA damage by TPZ and analogues elicits γH2AX induction (Olive et al. 2004; Wang et al. 2012), which correlates with cytotoxicity, hypoxia and reductase expression (Wang et al. 2012).

A unique feature of TPZ is that solution oxygen concentrations required to suppress its cytotoxicity are generally 10–30-fold higher than for other HAPs (Koch 1993) with oxygen concentrations required to reduce cytotoxic potency by half (K_{O2}) of 1–3 μM O_2 (Hicks et al. 2004; Koch 1993). This results in significant killing of cells at intermediate oxygen concentration, considered to be the most important in fractionated radiation protocols (Wouters and Brown 1997) and provides good complementarity with radiation (Hicks et al. 2004), for which K_{O2} values for radioresistance are of the order of 4 μM O_2 (Koch et al. 1984; Whillans and Hunt 1982).

13.3.3 Preclinical Efficacy

TPZ enhances cell kill in tumour xenografts in mice in combination with both single high-dose radiation (Zeman et al. 1988) and fractionated radiation protocols (Brown and Lemmon 1991); in the latter study no additional sensitivity of normal tissues in the radiation field was observed. TPZ was also tested in combination with other chemotherapeutic agents and, in particular, shows a hypoxia-dependent synergistic interaction with cisplatin in cell culture and in preclinical tumour models (Dorie and Brown 1993, 1994) with maximum enhancement when TPZ is given 2 hours before cisplatin. This interaction appears to result from inhibition of cisplatin–DNA crosslink repair (Kovacs et al. 1999). TPZ was found to undergo extensive metabolism to SR4317 in mice after intraperitoneal (IP) or intravenous (IV) injection (Graham et al. 1997; Walton and Workman 1993). Major toxicities after high single doses of TPZ were weight loss, pilo-erection, hyperactivity and tremors (Graham et al. 1997), while bone marrow suppression was the major finding with daily dosing (Spiegel et al. 1993).

13.3.4 Clinical Trials

TPZ has been extensively studied in clinical trials for 15 years and has provided evidence for the utility of targeting hypoxia (see below) but has generally provided disappointing results in late-stage trials (for reviews, see Ghatage and Sabagh 2012; McKeown et al. 2007; Reddy and Williamson 2009). Early phase I trials showed TPZ to be well tolerated up to doses of 390–450 mg/m^2 in single agent studies (Senan et al. 1997) and 260 mg/m^2 in combination with cisplatin (Johnson et al. 1997). Major toxicities included nausea, vomiting, diarrhoea, weight loss and muscle cramps, with reversible hearing loss and tinnitus being often dose limiting. Visual disturbances have also been observed, but without the permanent retinal damage reported in mice (Lee and Wilson 2000). TPZ dose reduction has been necessary in combination with chemotherapy, especially to minimise acute hearing loss (Reddy and Williamson 2009). Major toxicities reported in chemotherapy combinations include nausea and vomiting, severe diarrhoea, fatigue, anaemia and febrile neutropaenia (Ghatage and Sabagh 2012; Reddy and Williamson 2009). These trials have focused mainly on head and neck (Rischin et al. 2005; Rischin et al. 2010b), non-small-cell lung (Sandler et al. 2000; Shepherd et al. 2000; von Pawel et al. 2000; Williamson et al. 2005) and cervical carcinomas (Aghajanian et al. 1997; Covens et al. 2006; Craighead et al. 2000; DiSilvestro et al. 2012; Maluf et al. 2006; Rischin et al. 2010a) and have concentrated on the synergy with radiation and cisplatin (Ghatage and Sabagh 2012; McKeown et al. 2007; Reddy and Williamson 2009). Generally promising indications with these combinations in early phase trials have not been confirmed in randomised phase III trials, although the initial phase III trial (CATAPULT I) of TPZ with cisplatin in advanced non-small cell lung cancer (NSCLC) demonstrated a significant increase in response rate and overall survival (OS) relative to cisplatin alone (von Pawel et al. 2000). However, etoposide was superior to TPZ in combination with cisplatin in the NSCLC (CATAPULT II) trial, and the recent HeadSTART trial of TPZ with chemoradiation in advanced head and neck cancer failed to meet its primary endpoint of OS (Rischin et al. 2010b). The latter study was significantly compromised by substandard radiotherapy (Peters et al. 2010); exclusion of patients

Table 13.1 Primary site failure in 92 randomised patients categorised for tumour hypoxia by ^{18}F-misonidazole or ^{18}F-FAZA PET imaging at the Peter MacCallum Centre during clinical trials of TPZ in combination with fractionated radiotherapy and cisplatin. (After Brown 2010)

PET hypoxia status	Treatment		*p*-value
	RT + cis + 5FU	RT + cis + TPZ	
Non-hypoxic	2/27	3/21	NS
Hypoxic	8/18	0/26	0.0002
p-value	0.008	NS	

with non-compliant treatment plans identified a trend in favour of the TPZ arm (HR 0.74, 95 % CI 0.53–1.04, $p = 0.067$) for time to locoregional failure. In addition, a retrospective analysis of potential serological markers for hypoxia suggested an advantage for TPZ in patients with high hepatocyte growth factor and interleukin-8 (Le et al. 2012).

The lack of patient selection for hypoxia, the therapeutic target, may be an important reason for the failure to demonstrate efficacy of TPZ in phase III studies. When tumours were categorised as hypoxic using [^{18}F]-fluoromisonidazole positron emission tomography (PET) there was significant benefit in locoregional control by radiotherapy in the patients treated with TPZ/cisplatin, compared to 5FU/cisplatin, while no advantage was seen for TPZ in non-hypoxic tumours (Rischin et al. 2006); the updated experience including patients imaged for hypoxia with [^{18}F]-FAZA is shown in Table 13.1. This suggests both that hypoxia targeting has potential to benefit patients and that PET imaging with hypoxia markers can be used to identify TPZ-responsive patients. In summary, despite promising indictors of efficacy, the lack of consistent results at phase III trials with unselected patients, accompanied by significant toxicities, has caused clinical development of TPZ to cease.

13.3.5 Extravascular Penetration of TPZ

In addition to the possible reasons for failure in clinical trials mentioned above, it became apparent that the selectivity of TPZ for hypoxic cells was higher in two-dimensional (2D monolayer) cell cultures than in 3D spheroid cultures or in tumour xenografts. Durand and Olive (1992) used fluorescence-activated cell sorting (FACS) to demonstrate that the difference in TPZ sensitivity between the outer cells (at 5 % O_2) and the inner (severely hypoxic) cells in spheroids was more than an order of magnitude lower than in monolayer culture; when spheroids were exposed under anoxia, little killing was observed in cells at a depth of $> 100\ \mu m$. This was attributed to poor extravascular transport due to metabolic consumption of TPZ during its diffusion. This low differential was also observed *in vivo* using FACS to sort cells based on distance from blood vessels (Durand and Olive 1997). More directly, diffusion studies using multicellular layer (MCL) cultures showed a significant limitation of TPZ transport under anoxic conditions (Hicks et al. 1998; Hicks et al. 2003; Kyle and

Minchinton 1999). MCLs (also known as multi-layered cell cultures, MCCs) are a model for the tumour extravascular compartment, in which cells are grown on porous support membranes (Cowan et al. 1996; Minchinton et al. 1997) to form diffusion-limited structures to a thickness up to several hundred micrometres. MCLs have many features in common with multicellular spheroids and avascular tumour nodules, including central hypoxia (Hicks et al. 1998). Their planar structure is particularly useful as drug and metabolite diffusion can be measured using compound-specific analytical techniques such as Liquid Chromatography–Tandem Mass Spectrometry (LC–MS/MS); use of this model in the context of drug transport in tumours has been reviewed elsewhere (Hicks 2008; Minchinton and Tannock 2006).

Using measured transport parameters of TPZ in HT29 MCL (diffusion coefficient under aerobic conditions and rate constant for anoxic metabolism), and cytotoxic potency in anoxic cell cultures, we developed a spatially resolved pharmacokinetic/pharmacodynamic (SR-PKPD) model which demonstrated a large decrease in TPZ concentration with distance into the MCL. This model was able to predict the observed resistance when cells were exposed as intact MCL compared to stirred suspensions (Hicks et al. 2003). The SR-PKPD model was then extended to tumours using the measured oxygen dependence of TPZ metabolism and cytotoxicity (Hicks et al. 2004) and a mapped microvascular network (Hicks et al. 2006) with measured TPZ plasma pharmacokinetic (PK) as input. The model predicted that log cell kill of TPZ in the hypoxic region is markedly reduced relative to that achievable if there were no extravascular transport limitation. The SR-PKPD model was then used to successfully predict the activity of 15 TPZ analogues in tumour xenografts based on measured diffusion coefficients, rates of metabolism, anoxic cytotoxicity and plasma PK (Hicks et al. 2006).

13.3.6 Development of Second-Generation TPZ Analogues

There has been substantial work on the development of TPZ analogues (Delahoussaye et al. 2003; Kelson et al. 1998; Minchinton et al. 1992), with our own programme based on improving extravascular transport using the above SR-PKPD model to guide optimisation (Hay et al. 2007a, b, 2008). A key hypothesis was that improving extravascular transport would lead to a higher therapeutic ratio, based on the assumption that penetration into hypoxic regions of tumours limits anti-tumour activity, but penetration into normal tissues does not limit toxicity. We also sought to improve anoxic potency relative to solubility (a limitation for TPZ; the initial phase I required long infusion times for dose escalation (Senan et al. 1997)), and hypoxic selectivity *in vitro*, but retain the favourable features of TPZ such as lack of toxicity of the 2e reduction product and activity at intermediate oxygen concentrations. To achieve this an initial "toolbox" of compounds was developed that were used to define relationships between physicochemical properties (log*P*, pKa, MW, hydrogen bond donors and acceptors) and tissue diffusion coefficient (Pruijn et al. 2005, 2008) and between E(1) and anoxic potency (Hay et al. 2003).

The SR-PKPD model, reviewed recently (Hicks 2012), was then used to predict compounds with maximal *in vivo* selectivity and prioritise them for testing in tumour xenografts (Hay et al. 2007a, b, 2008). Measured tissue diffusion coefficients increased with increasing log$P$ (pH 7.4), decreased with increasing MW and number of hydrogen bond donors and acceptors, and covered a ca. 100-fold range depending on cell line (Pruijn et al. 2005, 2008). Rates of reductive metabolism in the analogue series also varied by $>$ 100-fold (Hicks et al. 2010), allowing a wide range of properties to be investigated. Sixteen of the 18 analogues predicted to be active by the SR-PKPD model were found to be active in initial HT29 xenograft screens and two (SN29751 and SN30000) had improved therapeutic activity relative to TPZ in a range of tumour xenografts when combined with single dose or fractionated radiotherapy (Hicks et al. 2010).

The lead compound from this optimisation, SN30000, is currently in preclinical development by Cancer Research UK under a licence from the University of Auckland. SN30000 has improved extravascular transport with a threefold higher diffusion coefficient than TPZ in HT29 and SiHa MCL (Hicks et al. 2010) due to a higher log$P$ (0.5 vs -0.34 at pH 7.4) and fewer hydrogen bond donors (0 vs 2). Higher anoxic potency for SN30000 than TPZ is seen in cell lines (Hicks et al. 2010), which partly reflects its higher E(1) (-399 mV vs -456 mV) leading to more rapid bioreductive metabolism. However, the increased rate of metabolism is more than offset by its higher diffusion coefficient; good penetration of SN30000 has been recently confirmed in a further MCL model (HCT116 cells) with only a small transport impediment under hypoxia (Yin and Hicks, unpublished data). In addition, SN30000 is at least five times more soluble than TPZ due to the morpholino sidechain, which is a sufficiently weak base that it does not confer any greater resistance at low extracellular pH than seen for TPZ.

13.3.7 Mechanism of Action of SN30000 and Response Biomarkers

The SN30000 mechanism of action is analogous to that of TPZ in all respects investigated to date (Fig. 13.2). Under anoxia it is metabolised to the corresponding 1-oxide and nor-oxide which, as for TPZ, are much less cytotoxic than the parent prodrug (Hicks et al. 2010). There is a strong correlation between SN30000 and TPZ, across a panel of cell lines, in relation to anoxic cytotoxicity (Hicks et al. 2010) and rates of anoxic reduction (Wang et al. 2012) suggesting their mechanistic similarity. In addition, the lesions responsible for SN30000 cytotoxicity are repaired by the same pathways as for TPZ with homologous recombination repair prominent (Evans et al. 2008; Hunter et al. 2012). Further, when reductase activity and pO_2 is modulated in a common genetic background, the relationship between γH2AX and clonogenic killing is the same for both TPZ and SN30000 (Wang et al. 2012). This DNA damage response is partially localised to hypoxic regions in HCT116 xenografts as defined by EF5 immunostaining, although a significant response is

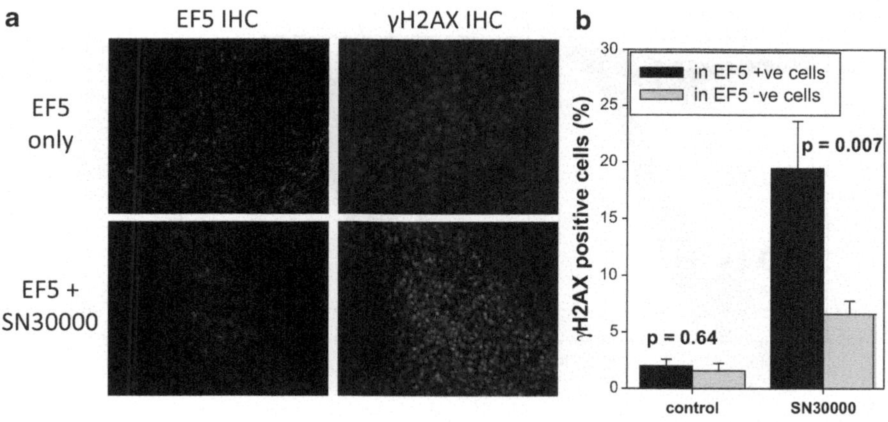

Fig. 13.3 DNA damage response (γH2AX induction) in hypoxic regions of an orthotopic triple negative breast cancer xenograft. Mice bearing MDA-MB-231-D3H2LN mammary fat pad tumours were dosed with the 2-nitromidazole hypoxia marker EF5 at 60 mg/kg with or without SN30000 at 200 mg/kg, ip, 2 h before tumour removal. **a** Fluorescent immunohistochemistry (*IHC*) on frozen sections. Images were obtained with a Leica fluorescent microscope using a 20 × objective. **b** Flow cytometry evaluation of dual staining for EF5 and γH2AX. Values are mean and errors are standard error of the mean (SEM) for four EF5-only (control) mice and 13 mice treated with EF5 plus SN30000

also seen in EF5-negative cells (Wang et al. 2012) which may reflect the higher K_{O2} value of SN30000 than EF5. Similar data are illustrated for a triple negative breast cancer xenograft in Fig. 13.3. The DNA damage response as defined by γH2AX or 53BP1 focalisation may provide a useful response biomarker for proof of concept in early phase clinical trials.

Under aerobic conditions substantial oxygen consumption, H_2O_2 production and aerobic cytotoxicity are observed in cells with high POR activity (Hunter et al. 2012). The significance of redox cycling in aerobic cells, and potential effects on mitochondrial function as demonstrated for TPZ (Wouters et al. 2001), warrant further investigation in relation to normal tissue toxicity. This redox biology, and associated biomarkers, might also well be of significance in tumour cells at intermediate oxygen concentrations.

Although not a response biomarker in the conventional sense, the above SR-PKPD model also has a role in making an early assessment of the utility of SN30000 during its clinical development by enabling comparison with TPZ once the human plasma PK of SN30000 is known. To illustrate this, in Fig. 13.4a tumour cell killing by TPZ at the human maximum tolerated dose (MTD) of 390 mg/m^2 for 3-weekly dosing (Graham et al. 1997) is predicted for a "virtual" patient with a tumour having sensitivity equivalent to HT29 xenografts. This model uses a previously published tumour capillary network and HT29 MCL transport and cytotoxicity parameters (Hicks et al. 2006). Cell kill is predicted to be heterogeneous due to the poor penetration of TPZ and variable intercapillary distances in the microvascular network as previously described (Hicks et al. 2006). The model predicts that for a single dose of TPZ there

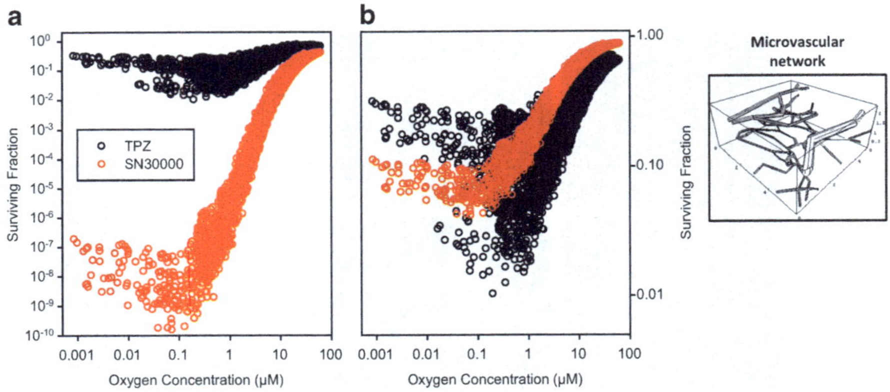

Fig. 13.4 Spatially resolved pharmacokinetic/pharmacodynamic (SR-PKPD) modelling of tumour cell killing by TPZ and SN30000 in the illustrated microvascular network (Hicks et al. 2006) using *in vitro* transport and cytotoxicity parameters for HT29 tumours. **a** Predicted cell killing assuming plasma PK of both compounds is the same as TPZ at its human MTD of 390 mg/m^2 (area under the concentration-time curve (AUC) 132 μM.h, Graham et al. 1997). **b** Prediction for SN30000 at a dose that gives the same average cell killing (0.86 log cell kill) in the radiobiologically hypoxic region (< 4 μM O_2) as TPZ at its MTD (redrawn from panel **a**). The SN30000 plasma AUC required for this dose (isoeffective for hypoxic cells, but less cytotoxic to oxic cells) is predicted to be 41 % of that for TPZ

would be 0.86 logs of cell kill averaged across the radiobiologically hypoxic region (< 4 μM O_2). By comparison, Fig. 13.4a also shows the prediction for SN30000 if the same plasma PK as TPZ were attained, resulting in the predicted log cell kill of 5.8 in the same hypoxic regions. Assuming PK scales linearly with dose, in Fig. 13.4b an SN30000 plasma AUC 41 % that of TPZ is predicted to give the same average cell killing in the hypoxic region as TPZ (0.86 logs). The much less heterogeneous relationship between O_2 and cell kill due to improved transport is evident and also results in an increase in cell kill in the most hypoxic (< 1 μM O_2) zone compared to TPZ, with greater selectivity for hypoxic versus oxic cells in the tumour. Since the ratio of rates of bioreductive metabolism between TPZ and SN30000 is similar across all cell lines tested (Wang et al. 2012), as it is for cytotoxic potency (Hicks et al. 2010), it is expected that this favourable AUC prediction would apply similarly to other tumours. The improvement in extravascular transport of SN30000 relative to TPZ can be appreciated by the calculation of penetration half distances into a planar anoxic tissue element; these half distances are 68 vs 42 μm for HT29 MCLs, and 106 vs 77 μm for SiHa MCLs (both calculated from data in Hicks et al. 2010) and 91 vs 50 μm for HCT116 calculated from Yin and Hicks (unpublished data). If the plasma AUC of SN30000 at MTD in humans exceeds that for TPZ, as it does in mice and rats (Hicks et al. 2010), the above SR-PKPD modelling gives some confidence that SN30000 will target hypoxic cells in human tumours more successfully than TPZ.

13.4 Dinitrobenzamide Mustards (PR-104)

13.4.1 Overview

The nitro compound PR-104 is a clinical stage HAP which embodies a distinctly different prodrug strategy from the benzotriazine di-oxides. PR-104 is a phosphate ester (conferring high aqueous solubility) which is rapidly hydrolysed *in vivo* to the corresponding alcohol, PR-104A. The active metabolites generated by bioreduction of PR-104A are reactive nitrogen mustard crosslinking agents (hydroxylamine PR-104H and amine PR-104M) which are stable enough to diffuse from hypoxic zones. PR-104A requires more extreme hypoxia than TPZ/SN30000 for activation in cell culture. This profile (activation under severe hypoxia with bystander killing of cells at higher pO_2) potentially provides a greater tumour selectivity than for HAPs activated under moderate (physiological) hypoxia. However, PR-104A is also a substrate for 2e reductive activation by aldo–ketoreductase 1C3 (AKR1C3), which may contribute to its dose-limiting myelotoxicity (including prolonged thrombocytopaenia) in humans. Its clinical development is currently focused on settings in which both hypoxia and AKR1C3 might be exploited, particularly in advanced leukaemias.

13.4.2 Discovery of Dinitrobenzamide Mustard HAPs

PR-104 originated from the proposal that the hypoxia-selective metabolic reduction of a strongly electron-withdrawing nitro group ($\sigma_p = 0.78$) to an electron-donating hydroxylamine ($\sigma_p = -0.45$) or amine ($\sigma_p = -0.66$) could be used as an electronic switch to activate a latent nitrogen mustard moiety on the same aromatic ring (Denny and Wilson 1986; Siim et al. 1997). It had long been known that the reactivity of nitrogen mustards is controlled by the electron density on the nitrogen (Panthananickal et al. 1979), and the *in vitro* cytotoxicity of *p*-nitro- and *p*-aminoaniline chloromustards was shown to differ by over 4 orders of magnitude (Palmer et al. 1990b). The relatively low reduction potentials of early mononitro examples (Palmer et al. 1990a, 1990b) led to the evaluation of more readily reduced dinitrobenzamide mustards (DNBMs), in which both the second nitro group and the carboxamide provide additional electron withdrawal. The 2,4-dinitrobenzamide-5-chloromustard SN23862 (Fig. 13.5, cmpd**1**) provided modest hypoxia-selective cytotoxicity in culture (Palmer et al. 1992), which was shown to be due to hypoxia-dependent DNA crosslinking (Palmer et al. 1994).

Further exploration of structure–activity relationships showed that two other regioisomer patterns, with a *meta* relationship between the nitro groups as in the 2,4-dinitrobenzamide-5-mustards, also provided hypoxic selectivity (Palmer et al. 1996). The more soluble basic analogues of these compounds (e.g. Fig. 13.5, cmpd**2** and **3**) showed appreciable killing of hypoxic cells in KHT tumours (Palmer et al. 1994, 1996), but had almost as much activity against oxic cells. This raised the

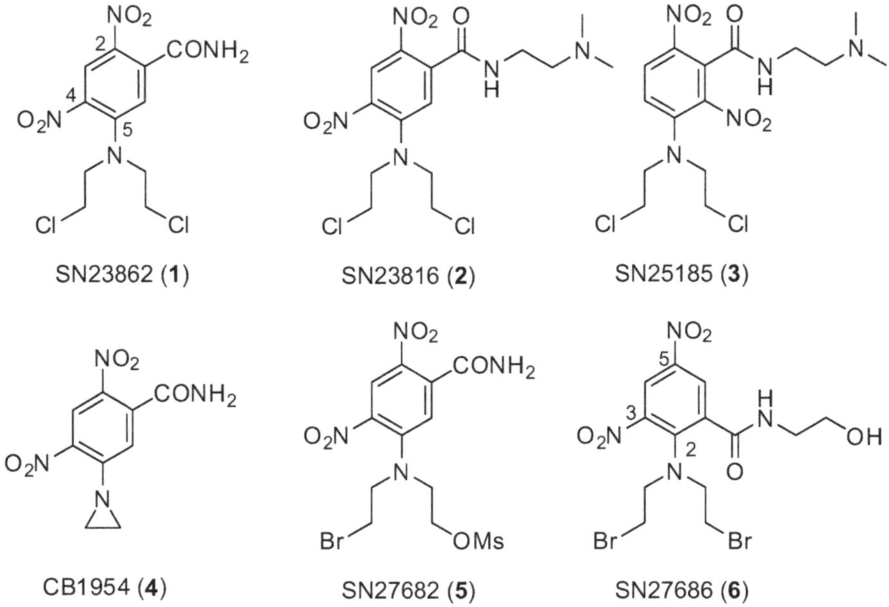

Fig. 13.5 Structures of dinitrobenzamide prodrugs investigated as prodrugs for activation by endogenous 1e reductases in hypoxic cells, and by bacterial 2e reductases for DEPT applications

important question whether activity against oxic cells reflects a bystander effect. Subsequent studies with MCL co-cultures of A459 cells with high and low 1e reductase activity confirmed the existence of a bystander effect from SN23862 (**1**) and SN23816 (**2**) within this model tissue, under uniform anoxia, while no such bystander killing could be detected for TPZ (Wilson et al. 2007). However, quantifying the contribution of metabolite diffusion to the monotherapy activity of this and other HAPs continues to be a difficult challenge as discussed below.

13.4.3 DNBMs as Substrates for Bacterial Nitroreductases

These early DNBM prodrugs proved to be much better substrates for the *Escherichia coli nfsB* nitroreductase (Anlezark et al. 1995) than for endogenous reductases in hypoxic tumour cell lines, so the focus shifted to optimization for *nfsB* in DEPT contexts. The mustard SN23862 (**1**) is superior to its well-studied aziridine analogue CB1954 (**4**) as a substrate for *nfsB* (Anlezark et al. 1995). Although the two prodrugs are structurally very similar, the mechanism of activation by nitroreduction is fundamentally different; in the case of CB1954 the cytotoxicity of the reduced metabolites is primarily due to the formation of a second reactive centre which generates a bifunctional DNA crosslinking agent (Knox et al. 1991), whereas for SN23862 (**1**) the electronic activation of the latent mustard moiety is responsible (Helsby et al. 2003).

Advantages of SN23862 (**1**) for *nfs B*DEPT include its lack of activation by NQO-1 (Palmer et al. 1992; Palmer et al. 1994) and the regioselectivity of its reduction; although reduction at the 4-nitro group is thermodynamically favoured (Palmer et al. 1995), *nfsB* selectively forms the potently cytotoxic 2-hydroxylamino metabolite (Anlezark et al. 1995) which is not prone to the spontaneous intramolecular alkylation which compromises potency of the 4-hydroxylamine (Palmer et al. 1995). The bystander effect due to activation of SN23862 (**1**) by *nfsB* is also greater than that from CB1954 (Wilson et al. 2002) because of superior tissue diffusion of the 2-amine metabolite of SN23862 (**1**) relative to the CB1954 metabolites (Helsby et al. 2004).

A systematic exploration of structure–activity relationships for SN23862 analogues as *nfsB* substrates showed greater cytotoxic potency and *nfsB* selectivity for asymmetric mustards with both halogen and mesylate leaving groups, and demonstrated the major role of lipophilicity as a determinant of bystander effect efficiency in MCL studies (Atwell et al. 2007). Several analogues from this study showed curative activity against *nfsB*-expressing xenografts, including marked activity of the bromo/mesylate mustard SN27682 (**5**) in tumours in which only 10 % of the cells expressed *nfsB*, indicating a strong bystander effect. A related 3,5-dinitrobenzamide-2-bromomustard SN27686 (**6**) was shown to be a highly effective *nfsB* prodrug and to provide an efficient bystander effect in MCLs (Singleton et al. 2007). The 3,5-dinitrobenzamide-2-mustard PR-104 (see Sect. 13.4.4) markedly enhanced antitumour activity of *nfsB* targeted to hypoxic/necrotic regions of SiHa tumours using *nfsB*-armed spores of *Clostridium sporogenes* as biological vectors (Liu et al. 2008). The NADPH-selective *E. coli* nitroreductase *nfsA*, which is distantly related to *nfsB*, was also shown to activate very efficiently all three DNBM bromomustard regioisomers with *meta*-disposed nitro groups (Vass et al. 2009). Evaluation of optimized *nfsA* variants with DNBM prodrugs is ongoing (Patterson et al. 2010).

13.4.4 Mechanism of Action and Preclinical Pharmacology of PR-104

Stimulated in part by the marked activity of DNBMs with mixed leaving groups as *nfsB* prodrugs (e.g. **5**), an extensive survey for hypoxia-selective cytotoxicity identified the 3,5-dinitrobenzamide-2-mustard PR-104 (Fig. 13.6) as a promising HAP candidate. PR-104 is a readily water-soluble phosphate ester that is rapidly hydrolysed to the corresponding alcohol (PR-104A) *in vivo* (Patel et al. 2007; Patterson et al. 2007); this "pre-prodrug" strategy overcomes formulation problems with the relatively insoluble DNBMs as either HAP or DEPT prodrugs. (The same pre-prodrug approach is also used by the phosphate ester of SN27686 (**6**), which is active against H1299 xenografts in conjunction with the systemically administered *nfsB*-armed conditionally replicating adenovirus ONYX^{NTR} (Singleton et al. 2007)). Unlike TPZ, PR-104A is as active against cells in anoxic spheroids as it is against single cells in anoxic stirred suspensions suggesting efficient tissue penetration (and/or bystander killing at high cell density (Patterson et al. 2007); more efficient MCL penetration by PR-104A than TPZ has been demonstrated (Hicks et al. 2007). PR-104A

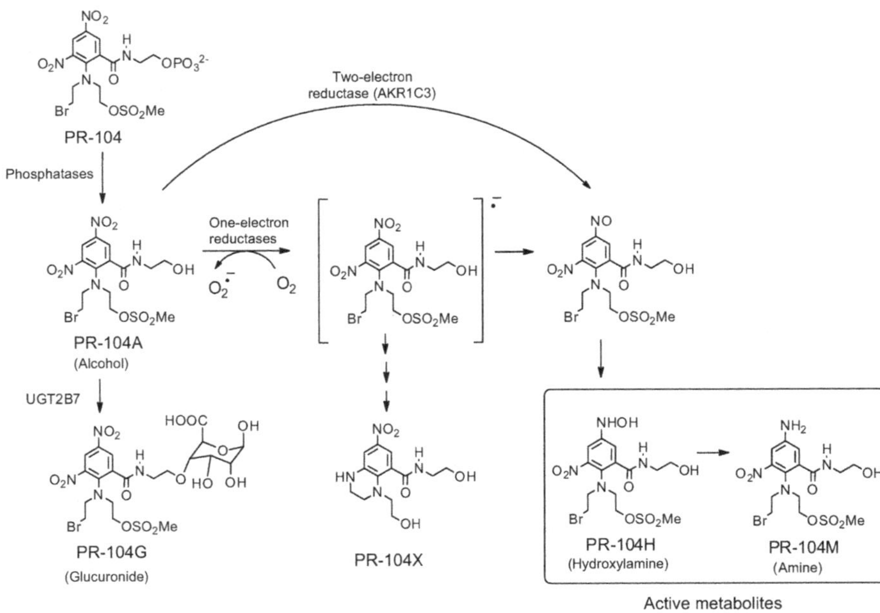

Fig. 13.6 Structure and mechanism of activation of the 3,5-dinitrobenzamide-2-mustard PR-104. The phosphate ester of PR-104A is rapidly hydrolysed to alcohol PR-104A *in vivo*. Reduction of the electron-withdrawing nitro group of PR-104A to the electron-donating hydroxylamine (PR-104H) or amine (PR-104M) activates the nitrogen mustard moiety. One-electron reductases form a nitro radical that is reoxidised by oxygen to confer hypoxic selectivity as for TPZ/SN30000. Two-electron reduction by aldo–ketoreductase 1C3 can bypass the nitro radical to effect oxygen-insensitive activation. Inactive metabolites include the *O*-glucuronide PR-104G and products (e.g. PR-104X) derived from reduction of the 3-nitro group leading to intramolecular alkylation

shows HCR values (20 % O_2/anoxia IC_{50} ratio) in the range of 6–160 in human tumour cell lines (Patterson et al. 2007). This wide variation largely reflects differences in aerobic sensitivity to PR-104A, subsequently shown to be due to reduction by a variably expressed oxygen-insensitive 2e reductase, AKR1C3 (Guise et al. 2010).

Under anoxia, 1e reduction of PR-104A is catalysed by POR, although siRNA and antisense studies have shown that this is not the sole hypoxia-dependent PR-104A reductase in SiHa cells (Guise et al. 2007), and multi-allelic knockout of the *POR* gene in SiHa or HCT116 cells with zinc finger nucleases has little effect on PR-104A metabolism or cytotoxicity (Su et al. 2012). Other diflavin reductases (methionine synthase reductase (MTRR), novel diflavinoxidoreductase 1 (NDOR1) and inducible nitric-oxide synthase (NOS2A)) are also able to activate PR-104A under hypoxia (Guise et al. 2012), although the contribution of these enzymes to PR-104A activation in human tumours is unknown.

Both AKR1C3 and the hypoxia-dependent 1e reductases reduce the 5-nitro group of PR-104A (*para* to the mustard), generating the same hydroxylamine (PR-104H) and amine (PR-104M) metabolites (Fig. 13.6); the 3-nitro group is also reduced in hypoxic SiHa cells as demonstrated by the formation of the ring-closed product PR-104X (Patterson et al. 2007), but the monofunctional alkylating agent intermediates in this pathway are expected to be non-toxic (Palmer et al. 1995); the contribution of 3-nitro reduction has not been studied further, although this stable metabolite could be of interest as a potential biomarker. The bifunctional activated mustards PR-104H and PR-104M are predominantly responsible for DNA interstrand crosslinking and cytotoxicity in hypoxic cells, as demonstrated in studies using single-cell gel electrophoresis (comet assay, Singleton et al. 2009) and DNA-repair-deficient cell lines (Gu et al. 2009).

Importantly, the comparison of the cytotoxicity of PR-104A and TPZ in stirred SiHa tumour cell suspensions equilibrated at different gas-phase oxygen concentrations, with steady-state O_2 concentrations monitored in solution using fibre optic probes, showed a K_{O2} value for PR-104A tenfold lower than TPZ (0.13 μM O_2 vs 1.3 μM, respectively; Hicks et al. 2007). This is consistent with other studies showing that cytotoxicity and/or metabolism of nitroaromatic compounds of similar reduction potential is very sensitive to oxygen (Koch 1993; Siim et al. 1994; Taylor and Rauth 1982). PR-104A differs from these earlier compounds in generating relatively stable cytotoxic metabolites that appear to provide a substantial bystander effect (see below), and therefore embodies the concept of exploiting severe (pathological) hypoxia to kill cells in adjacent regions at higher pO_2 (Wilson et al. 2007).

PR-104 has marked activity against radiobiologically hypoxic cells in human tumour xenografts, as demonstrated by its activity when administered immediately after irradiation (Hicks et al. 2007; Patterson et al. 2007), but in these and other studies (Benito et al. 2011; Guise et al. 2010; Houghton et al. 2011) it also shows substantial monotherapy activity in some xenograft models. The latter studies both noted a possible relationship between AKR1C3 expression and monotherapy activity of PR-104, and forced expression of AKR1C3 in two cell lines with low basal activity (H1299 and HCT116) sensitised the resulting xenografts to PR-104 (Guise et al. 2010). In addition, dosing of tumour-bearing mice with the AKR1C3 inhibitor naproxen suppressed concentrations of PR-104H/M and tumour cell killing in H460 xenografts (Gu et al. 2011a), which have high endogenous AKR1C3 expression (Guise et al. 2010). These observations suggested the potential for PR-104 to exploit expression of this 2e reductase, which is upregulated by the Nrf2 transcription factor and is highly expressed in some human tumours (Guise et al. 2010).

Although AKR1C3 is likely to contribute to the monotherapy anti-tumour activity of PR-104 in some contexts, it is obviously important for its clinical development (especially for the development of predictive biomarkers) to understand to what extent other mechanisms might play a role. Systemic metabolism to PR-104H and PR-104M is predominantly hepatic in mice, although the enzymes responsible appear to be neither oxygen sensitive nor paralogous to AKR1C3 (Gu et al. 2011a). Plasma concentrations of PR-104H and PR-104M are independent of the presence of tumours (Gu et al. 2011a) with AUC values (both active metabolites combined) 15–20 % of that for PR-104A in CD-1 nude mice (Gu et al. 2010), 2 % in SCID mice (Houghton et al. 2011), 3 % in NIH-III nude mice (Foehrenbacher et al. 2012) and ~8 % in cancer patients (Gu et al. 2010; McKeage et al. 2011). This systemic activation appeared not to correlate with myelotoxicity in patients (McKeage et al. 2011), although this preliminary evaluation does not preclude a role in either toxicity or anti-tumour activity. Perhaps of greater interest is the potential contribution of bystander effects arising from hypoxic metabolism in tumours, in relation to both monotherapy activity and killing of moderately hypoxic cells (at O_2 concentrations above the PR-104A K_{O2}) which may be critically important for radioresistance and chemoresistance. Several lines of evidence demonstrate that such a bystander effect must exist, including the measurable efflux of PR-104H and PR-104M from anoxic cells in culture (Patterson et al. 2007) and underprediction of PR-104 monotherapy activity (and its activity with radiation) in SiHa tumours using SR-PKPD models analogous to those used successfully for TPZ/SN30000 (i.e. without incorporation of a bystander effect, Hicks et al. 2007). We have recently extended these models to incorporate diffusion of bystander metabolites explicitly (Foehrenbacher et al. 2013) in order to develop a better understanding of the contribution of AKR1C3-dependent, hypoxic and systemic metabolism of PR-104A to anti-tumour activity in different settings. The latter study has shown that ~30-50 % of the activity of PR-104A in SiHa and HCT116 xenografts, both as monotherapy and with radiation, can be ascribed to bystander effects.

13.4.5 Clinical Development of PR-104

Phase I clinical trials of PR-104 monotherapy in patients with advanced solid tumours identified the MTD as 1,100 mg/m^2 on a 3-weekly schedule (Jameson et al. 2010), and 675 mg/m^2 when given on days 1, 8 and 15 every 28 days (McKeage et al. 2011). The plasma PK of PR-104A in the first cycle, in both studies, provided PR-104A AUC values in a range consistent with activity against hypoxic cells in xenograft models although would not be expected to provide major monotherapy activity. Minor reductions in tumour size were seen in some patients in both phase I studies, but no objective responses were observed. Myelotoxicity was the dose-limiting toxicity (DLT) with both schedules. Particularly for weekly dosing, thrombocytopenia was problematic with a delayed onset (nadir 6–10 weeks from start of treatment) and slow and/or incomplete recovery leading to the suggestion that its clinical use might be restricted to short courses of treatment such as combination with radiotherapy or in myeloablative regimes with haematological stem cell support (McKeage et al. 2011).

Pharmacokinetic parameters in humans were well predicted by allometric scaling from mice and rats. (Allometrically scaled clearance in dogs was faster (Patel et al. 2011), which has been ascribed to the high rate of *O*-glucuronidation of PR-104A which decreases in the order dogs > humans > rodents (Gu et al. 2011b)). However, at the 3-weekly MTD of 1,100 mg/m^2 PR-104, the plasma AUC of PR-104A was only ~25 % of that achieved in mice at MTD (Patel et al. 2011). The greater sensitivity to myelosuppression in humans was suggested (Patel et al. 2011) to reflect activation by AKR1C3, which is expressed in human myeloid progenitors (Birtwistle et al. 2009; Down et al. 2010) but appears to have no orthologue in mice (Gu et al. 2011a; Velica et al. 2009). Consistent with this, a preliminary study has shown PR-104A to be more cytotoxic to human than mouse myeloid progenitors under aerobic conditions *in vitro*, and that its activity against human myeloid progenitors is sensitive to the AKR1C3 inhibitor naproxen and insensitive to hypoxia (Down et al. 2010).

Despite the possible involvement of AKR1C3 in the normal tissue toxicity of PR-104, its very high expression in some tumours (Guise et al. 2010) has led to investigation in tumour types that highly express this enzyme. A small study of 14 patients with hepatocellular carcinoma, in combination with sorafenib, demonstrated one partial response and stable disease in six patients but was discontinued when greater toxicity was observed than in the above monotherapy studies; subsequent evaluation of plasma PK showed a statistically significant twofold increase in PR-104A exposure than in previous (non-HCC) patients at the same dose levels, suggesting that the elevated toxicity in these patients might reflect compromised PR-104A clearance because of impaired liver function (Abou-Alfa et al. 2011). Based on evidence for high AKR1C3 expression in some NSCLCs (Guise et al. 2010), and a phase Ib study defining the MTD of PR-104 as 770 mg/m^2 with docetaxel (60 or 75 mg/m^2) with prophylactic G-CSF on a 21-day cycle (McKeage et al. 2012), a randomized phase II trial of docetaxel (75 mg/m^2) vs docetaxel (60 mg/m^2) with PR-104 (770 mg/m^2) was undertaken in relapsed stage IIIb/IV NSCLC (clinicaltrials.gov ID: NCT00862134). The study was closed early based on an interim assessment of 38 patients that showed cumulative thrombocytopenia in the PR-104/docetaxel arm and no evidence of a higher objective response rate.

Evidence for high expression of AKR1C3 at the messenger RNA (mRNA) level in acute myeloid leukaemia (AML, Birtwistle et al. 2009), coupled with emerging evidence for hypoxia in advanced haematological malignancies based on studies with 2-nitroimidazole markers (Azab et al. 2012; Benito et al. 2011; Frolova et al. 2012; Jensen et al. 2000; Martin et al. 2011; Matsunaga et al. 2012), has also led to preclinical and clinical evaluation of PR-104 monotherapy in human leukaemias. Notably, PR-104 decreased leukaemic cell burden and prolonged survival in multiple acute lymphoblastic leukaemia (ALL) xenograft models, including curative activity (maintained complete responses) in an AKR1C3-positive ALL at PR-104 doses providing plasma PK achievable in solid tumour oncology patients (Benito et al. 2011). The rationale for PR-104 in this setting includes the very low expression in human leukaemias of UDP-glucuronosyltransferase 2B7, the isoform responsible for detoxification of PR-104A by glucuronidation (Gu et al. 2011b). A recently completed phase I trial of PR-104 in AML/ALL (clinicaltrials.gov ID:NCT01037556) has achieved dose escalation to 3 and 4 g/m^2; although myelosuppression, sometimes prolonged, was again an issue, clear evidence of clinical responses (Arana Yi et al. 2013) suggests that PR-104 could have utility in a subgroup that is prospectively defined using predictive biomarkers for AKR1C3, hypoxia and other determinants of alkylating agent sensitivity.

13.5 Phosphoramidate Mustards (TH-302)

13.5.1 Overview

TH-302, from Threshold Pharmaceuticals, is the HAP that is now most advanced in clinical development. As a hypoxia-activated nitrogen mustard, TH-302 is superficially similar to PR-104A but is not a substrate for activation by AKR1C3 or other

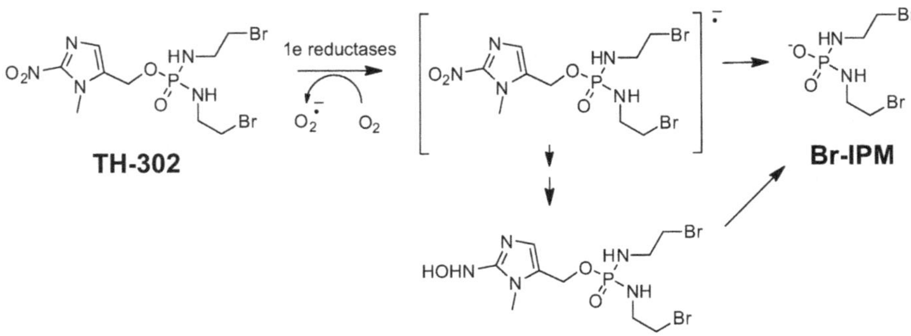

Fig. 13.7 Mechanism of activation of TH-302, which fragments when reduced to release the DNA crosslinking agent Br-IPM under hypoxia, as assessed from radiolytic reduction studies. (Based on Meng et al. 2012)

known 2e reductases, and exploits a different strategy for activating the nitrogen mustard moiety involving fragmentation of reduced metabolites to release the reactive cytotoxin bromo-isophosforamide mustard (Br-IPM; Fig. 13.7). Similarly to PR-104A, the activation of TH-302 requires severe hypoxia, appears to elicit bystander killing and generates DNA crosslinks that are responsible for hypoxic cytotoxicity. Extensive preclinical studies have shown anticancer efficacy as a monotherapy agent and in combination with chemotherapy/radiation therapy. Promising results from phase II trials in advanced soft-tissue sarcoma (in combination with doxorubicin) and in advanced pancreatic cancer (in combination with gemcitabine) have led to ongoing phase III trials in these disease settings.

13.5.2 Mechanism of Action and Preclinical Pharmacology of TH-302

Earlier investigations of nitroheterocyclic prodrugs of phosphoramidate mustards (Borch et al. 2000; Borch et al. 2001) paved the way for development of achiral analogues, with the 2-nitroimidazole TH-302 demonstrating high hypoxic selectivity in H460 cultures (Duan et al. 2008). A subsequent study confirmed the marked hypoxic selectivity of TH-302 across a panel of 32 human cancer cell lines, with HCR (IC_{50} air/IC_{50} anoxia) values of 11–600 (Meng et al. 2012). This study also used steady-state and pulse radiolysis methods to demonstrate fragmentation of the initial TH-302 free radical, showing this to be a major mechanism of release of the reactive toxin Br-IPM as illustrated in Fig. 13.7. Hypoxia-dependent DNA crosslinking in H460 cells was demonstrated by the comet assay, and the hypersensitivity to TH-302 of oxic or hypoxic cells with defects in homologous recombination repair (*xrcc3, BRCA1, BRCA2*) and fanconi anaemia (*FANCA*) genes or the incision step of nucleotide excision repair (*ercc4/xpf*$^{-/-}$) confirmed that crosslinks are responsible for hypoxia cytotoxicity (as also demonstrated by TH-302 hypersensitivity of *rad51d* knockouts, Hunter et al. 2012). In addition to DNA crosslinking, inhibition of thioredoxin reductase and glutathione reductase after high-dose TH-302 (300 mg/kg) has been proposed to contribute to anti-tumour activity in a murine hepatoma model (Li et al. 2010). Importantly, TH-302 cytotoxicity in H460 cells was more sensitively inhibited by oxygen than for TPZ; the gas-phase O_2 concentrations required for a tenfold increase in IC_{50} were between 0.1 and 0.6 % O_2 for TH-302, but 5–10 % O_2 for TPZ (Meng et al. 2012). The same study provided evidence for a bystander effect when TH-302 is activated by the bacterial 2e reductase *nfsA* under oxic conditions, in MCL co-cultures, although it has not been shown that the bacterial 2e reductase *nfsA* generates the same metabolite(s) as human 1e reductases.

POR has been identified as one of the human 1e reductases able to activate TH-302 under anoxia, with increased cytotoxicity in POR-overexpressing SiHa cells (Meng et al. 2012). POR overexpression in SiHa increased TH-302 cytotoxicity even more under aerobic than anoxic conditions (Meng et al. 2012), as also observed in Chinese hamster ovary (CHO) cell lines (Hunter et al. 2012); interestingly, diphenyliodonium did not inhibit the aerobic cytotoxicity of TH-302 in parental SiHa cells (Meng et al. 2012), suggesting that this may not be mediated by flavoproteins. It is not yet clear to what extent the very large variation in sensitivity across cell lines (900-fold under anoxia and > 200-fold under aerobic conditions, Meng et al. 2012) reflects differences in metabolic activation. Given that the above radiolytic reduction studies (Meng et al. 2012) indicate fragmentation to release Br-IPM from the 1e radical, presumably any 2e reduction would be of little significance, which would mean that TH-302 combines attractive features of PR-104A (low K_{O2} and bystander effect) and TPZ/SN30000 (lack of cytotoxicity from 2e reduction). However, the radiolytic reduction data do not preclude a contribution from fragmentation of a downstream hydroxylamine or amine as originally proposed for this class (Borch et al. 2000), which would be more consistent with the marked activation of TH-302 by the 2e reductase *nfsA* in hyperoxic HCT116 MCLs (Meng et al. 2012).

The pharmacokinetics, metabolism and excretion of TH-302 have been characterised in non-human species (Jung et al. 2012a, b, c), demonstrating rapid systemic clearance, urinary and biliary excretion, extensive metabolism by conjugation with glutathione and measurable concentrations of Br-IPM in plasma. Assuming that Br-IPM is generated intracellularly by bioreduction, the latter observation suggests that Br-IPM is able to efflux from cells and is thus consistent with a bystander effect from hypoxic activation of TH-302 in tumours. In all these respects TH-302 is broadly similar to PR-104 (Gu et al. 2010) except for the extensive *O*-glucuronidation of the primary alcohol of PR-104A in dogs and humans (Gu et al. 2010, 2011b).

Extensive preclinical efficacy studies have been reported in a broad range of xenografts where TH-302 was administered as either monotherapy or in combination with chemotherapy or radiation therapy. TH-302 demonstrated activity as a single agent and in combination with bortezomib in an extensively hypoxic multiple myeloma model (Hu et al. 2010, 2013). In a recent study of 11 xenograft models (Sun et al. 2012), monotherapy activity of TH-302 was shown to be hypoxia-dependent, based on multiple lines of evidence: (1) tumour growth delay correlated with hypoxia as determined by pimonidazole staining in different xenograft models; (2) tumour responses to TH-302 in H460 tumours were increased by breathing low oxygen gas and suppressed by 95 % O_2 (Fig. 13.8); (3) hypoxic fractions were significantly reduced 48 h after a single dose of TH-302 at 150 mg/kg in three different xenograft models;

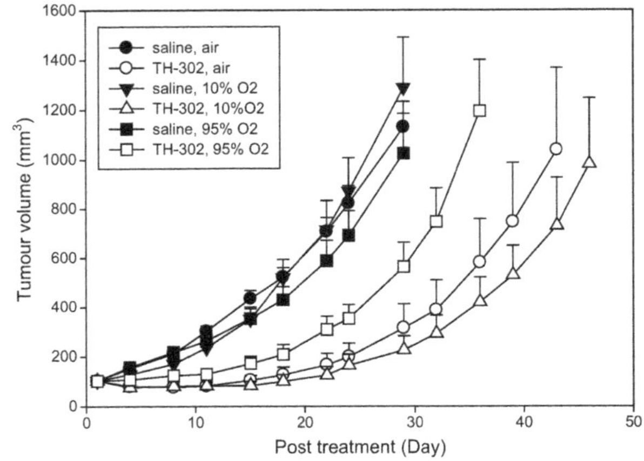

Fig. 13.8 Monotherapy activity of TH-302 in H460 xenografts is sensitive to modification of tumour hypoxia. Mice were treated with TH-302 (50 mg/kg, daily × 10) with the indicated respiratory gases for 30 min before and 2 h after treatment. Values are mean and SEM for 10 mice. (Redrawn from Sun et al. 2012)

(4) TH-302-induced DNA damage measured by γH2AX 6 hr after treatment was localised in or near hypoxic regions defined by pimonidazole staining, but extended to pimonidazole-negative regions 24 hr post treatment. The authors interpreted this time dependence as evidence for a hypoxia-dependent bystander effect, although this is difficult to reconcile with the very short plasma half-life of TH-302 (∼ 10 min) in nude mice (Jung et al. 2012c) and high reactivity of Br-IPM. An alternative explanation (not inconsistent with a bystander effect) might be that less efficient TH-302 activation at higher O_2 concentrations results in slower induction of a DNA damage response.

As might be expected for a hypoxia-targeting prodrug, TH-302 also provided increased efficacy when combined with a broad range of chemotherapeutic agents in xenograft models. This was originally noted for combination with gemcitabine in MiaPaca-2 xenografts (Duan et al. 2008). In a comprehensive study involving 11 xenograft models and seven different chemotherapy agents (Liu et al. 2012), combination with TH-302 was well tolerated and in each case showed a trend to greater anti-tumour activity compared to chemotherapy alone; this increase was statistically significant for 4 of 15 combinations. Given the well-established importance of hypoxia in radioresistance, a combination with radiation would also be a rational strategy. A study currently in progress (Lohse et al. 2012) suggests activity of TH-302 (50 mg/kg; d1,5,9) combined with radiation (2 Gy; d2,3,4) in two of three highly hypoxic, patient-derived pancreatic cancer xenograft lines.

13.5.3 Clinical Trials of TH-302

Subsequent to Investigational New Drug (IND) filing in 2007, Threshold Pharmaceuticals has undertaken a broad TH-302 clinical development programme (http://www.thresholdpharm.com/sec/clinical_pipeline). As of September 2013, four trials had completed, nine were recruiting and two were pending; together these will evaluate over 2,000 patients with TH-302 as monotherapy or in combination with other agents. Initial phase I trials of TH-302 monotherapy against advanced solid

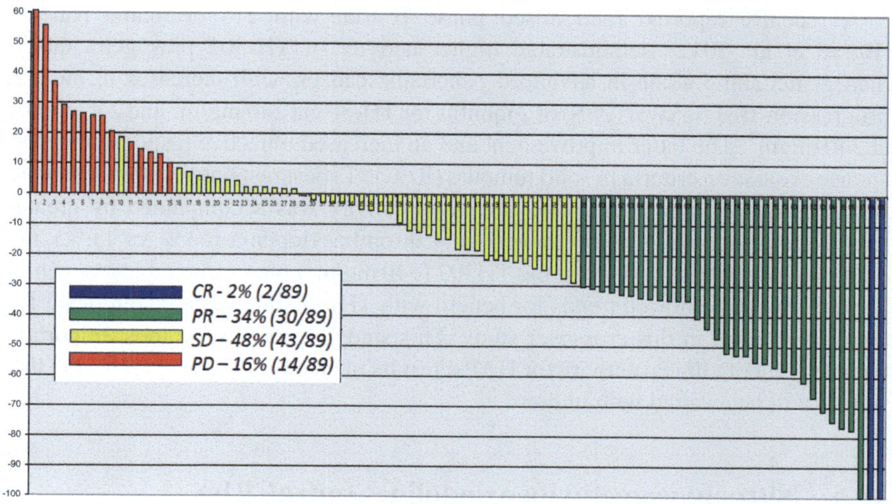

Fig. 13.9 A waterfall plot of the percent change in the sum of the longest diameters of the target lesions from baseline to nadir (*Y*-axis) in a phase II study of TH-302 (300 mg/m^2, day 1 and 8 of a 21-day cycle) in combination with doxorubicin (75 mg/m^2 on day 1, 2 hr after TH-302, for a maximum of six cycles) in advanced soft-tissue sarcoma. (From http://www.thresholdpharm.com/pdf/chawa_ctos_annual_10-11.pdf)

tumours, using a 30–60 min infusion, established an MTD of 575 mg/m^2 when dosed on days 1, 8 and 15 with a 28-day cycle and 670 mg/m^2 once every 3 weeks (Weiss et al. 2011). Unlike dogs and rats where the DLT was mainly hematologic, skin and mucosal toxicities were dose limiting in humans on the weekly schedule and fatigue and vaginitis/proctitis on the 3-weekly schedule. Two out of 37 partial responses were achieved with the weekly schedule, and stable disease in 27/57 patients across the two arms of the study. Median plasma terminal half-lives of TH-302 and its active toxin Br-IPM were 0.81 and 0.70 h, respectively, independent of dose, and plasma concentrations of Br-IPM were approximately 1–2 % of those for TH-302.

In contrast to TH-302 monotherapy, haematological toxicities represent the DLT for TH-302 in combination with gemcitabine, docetaxel or pemetrexed in advanced solid tumours (Borad et al. 2010) and with doxorubicin in soft-tissue sarcoma (Ganjoo et al. 2011). Common adverse events also included fatigue, nausea, skin rash and mucosal toxicity. In these phase 1b studies, MTDs of TH-302 were moderately lower than for the monotherapy regimens. The TH-302 MTD was 340 mg/m^2 when given 2 hr before gemcitabine (1 g/m^2 on days 1, 8 and 15 of a 28-day cycle), 340 mg/m^2 with docetaxel (75 mg/m^2 given on day 1), 480 mg/m^2 with pemetrexed (500 mg/m^2 on day 1) and 300 mg/m^2 with doxorubicin (75 mg/m^2 on day 1 with TH-302 on days 1 and 8 of a 21-day cycle, with prophylactic growth factor support). A non-randomised phase II study of TH-302 with doxorubicin in advanced soft-tissue sarcoma demonstrated a high objective response rate (Fig. 13.9) and a median OS of 17.5 months (Chawla et al. 2011). This promising activity led to a randomised phase III trial in the same indication, which opened in September 2011 and is expected to enrol 620 patients.

A recently reported randomised phase II trial with 214 evaluable patients (Borad et al. 2012) demonstrated higher activity of TH-302 plus gemcitabine than gemcitabine alone in advanced pancreatic cancer, with increases in median progression-free survival (PFS) of 2 months for TH-302 at 240 mg/m^2 and 2.4 months at 340 mg/m^2. The latter improvement and an increased objective response rate (response evaluation criteria in solid tumours (RECIST) best response) in this arm were both statistically significant. This increased activity was accompanied by greater haematological toxicity with grade 3 or 4 thrombocytopenia (63 % vs 11 %) and neutropaenia (60 % vs 31 %) in the TH302 (340 mg/m^2) plus gemcitabine vs gemcitabine arms. OS showed trends for benefit with TH-302 but was not significant for either dose level in this crossover study. This study represents the most important recent signal of clinical activity for HAPs; results of an ongoing phase III trial in this setting will be awaited with interest.

13.6 Nitrochloromethylbenzindolines (nitroCBIs)

13.6.1 Overview

NitroCBIs are HAPs derived from natural products that bind in the minor groove of DNA and selectively alkylate N3 of adenine. Although the natural products are extremely potent cytotoxins with broad-spectrum anti-tumour activity, and a unique mechanism of action, they failed in clinical trials because of lack of tumour selectivity. The nitroCBIs have retained the desired properties of the natural products while adding a layer of tumour selectivity. The formulation properties have also been dramatically improved by virtue of incorporating a phosphate ester moiety. The active and stable aminoCBI metabolites of nitroCBIs appear to confer efficient bystander effects, and nitroCBIs exhibit anti-tumour activity in xenograft models at well-tolerated doses. At present, preclinical studies are aimed at selecting a lead compound for clinical development.

13.6.2 NitroCBIs are Derived from Natural Products

The cyclopropylindolines form a family of anti-tumour antibiotics (including CC-1065, duocarmycin SA and yatakemycin; Fig. 13.10) that alkylate in a sequence-selective manner at the N3 position of adenine in the minor groove of DNA (illustrated in Fig. 13.11) and are exceptionally potent cytotoxins (Boger and Johnson 1996; Tichenor and Boger 2008). Four synthetic analogues of the natural products (adozelesin, carzelesin, bizelesin and KW-2189) have been evaluated clinically because of their excellent broad-spectrum anti-tumour activity in preclinical models (Kobayashi et al. 1994; Li et al. 1992), but all proved to be severely myelotoxic in patients and lacked anti-tumour activity at the low doses that could be administered systemically (Cristofanilli et al. 1998; Markovic et al. 2002; Pavlidis et al. 2000; Pitot et al. 2002;

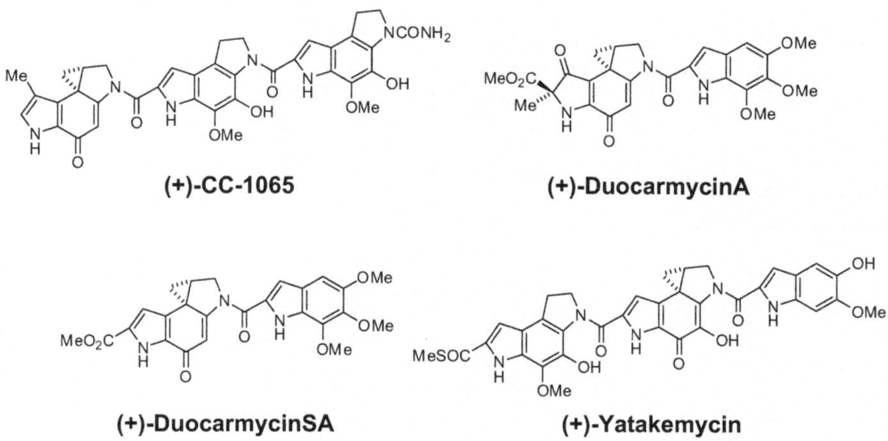

Fig. 13.10 Potently cytotoxic cyclopropylindoline natural products that provide the starting point for development of the nitrochloromethylbenzindoline (nitroCBI) class of HAPs

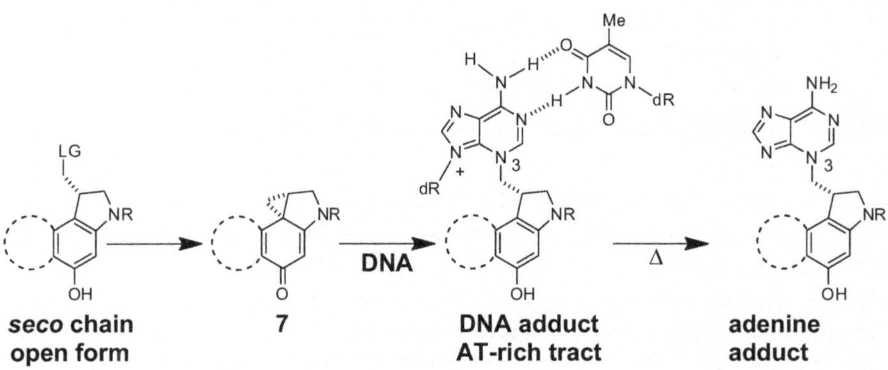

Fig. 13.11 Mechanism of DNA alkylation by cycloprolylindolines (**7**) and their *seco* open chain precursors. LG denotes leaving group, e.g. Cl, Br

Schwartz et al. 2003). These observations provided a strong rationale to explore an HAP approach to increase tumour selectivity and thereby improve the therapeutic index.

Early studies identified the minimum potent pharmacophore as the cyclopropylindoline unit **7** and showed that the free OH group in the *seco* chain open form (Fig. 13.11) was not obligatory for DNA alkylation, suggesting that it could be replaced with other electron-donating substituents (Boger et al. 1990). Indeed, replacing the phenol with an amino substituent resulted in potent cytotoxic analogues (Tercel et al. 1996b) shown capable of similar DNA sequence-selective alkylation (Gieseg et al. 1999; Tercel et al. 1999a; Tercel et al. 1999b). This suggested the potential for prodrugs in which the amino group is replaced with a strongly electron-withdrawing nitro group; a nitro analogue was shown to be masked as a cytotoxin and to be activated by an aerobic *E. coli* nitroreductase (*nfsB*) resulting in an increase in cytotoxicity of several hundredfold (Tercel et al. 1996a).

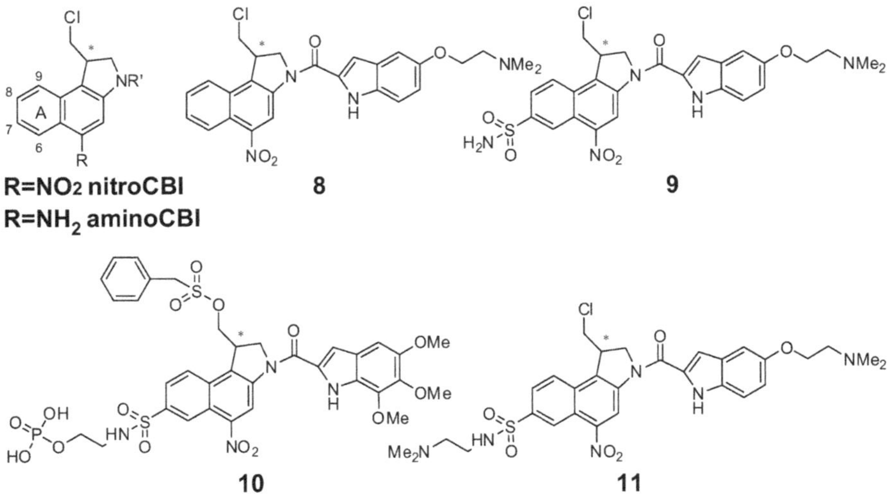

Fig. 13.12 Generic structures of nitro- and aminoCBIs, with A-ring numbering, and structures of nitroCBI analogues **8** and **9** and nitroCBI phosphate pre-prodrug **10** and **11** evaluated as hypoxia-activated prodrugs. Asterisks mark the chiral centre

13.6.3 Initial Mechanistic Studies with NitroCBIs

A number of nitro-substituted alkylating subunits based on **1** were then investigated to find potential substrates for the 1e reductases in hypoxic tumour cells; this identified the nitroCBIs (Fig. 13.12), exemplified by cmpd**8**, as having hypoxic/oxic cytotoxicity differentials of up to 300-fold following short-term (4 h) exposure in a Chinese hamster cell line, although hypoxic selectivity of **8** was more modest in human HT29 cells (Wilson et al. 2009). The mechanism was confirmed by demonstration of selective formation of the aminoCBI metabolite in hypoxic cells, and the corresponding N3 adenine DNA adduct which was detected by LC-MS following thermal depurination of DNA from nitroCBI-treated hypoxic (but not oxic) cells (Wilson et al. 2009). In a later study, it was found that the oxic and hypoxic cell survival curves for cmpd**12a** (see Fig. 13.13) were superimposable when plotted against the *measured* extracellular concentration of the corresponding aminoCBI metabolite suggesting that even the low-potency aerobic cytotoxicity of nitroCBIs is due to nitroreduction (Tercel et al. 2011b). Given that aminoCBIs are relatively stable in aqueous medium and quite hydrophobic it was not surprising to find evidence for a bystander effect in high-density 3D MCL cultures with human cancer cells overexpressing the *E. colinfsB* aerobic nitroreductase (Wilson et al. 2009). Further indication of a bystander effect (i.e. an active metabolite diffusing from the cells in which it is generated to kill surrounding cell populations) was obtained in studies with cmpd**9** with 3D spheroid cultures (Tercel et al. 2009).

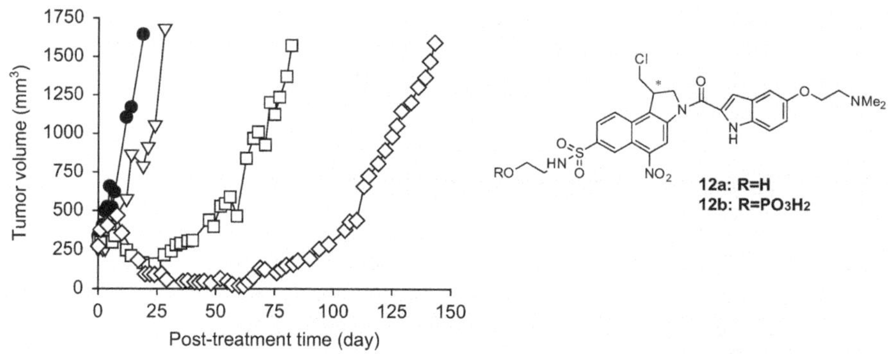

Fig. 13.13 Anti-tumour activity of the phosphate pre-prodrug **12b** of nitroCBI **12a** by growth delay assay. Mice bearing SiHa xenografts were treated on days 1, 2 and 3 with cmpd**12b** (42 μmol/kg, i.v.) and/or local tumour irradiation (5 Gy, cobalt-60). The curves represent the tumour volume of the median mouse in each group; circles: control; triangles: **12b** alone; squares: radiation alone; diamonds: **12b** 5 min after radiation. Median times to quadrupling of the initial treatment volume were 16, 24, 77 and 128 days, respectively. (Redrawn from Tercel et al. 2011a)

13.6.4 Improving Hypoxia Selectivity of NitroCBIs

Initial studies with cmpd**8** demonstrated relatively modest hypoxic selectivity against human tumour cell lines. To improve hypoxic selectivity, structure–activity relationships of the nitroCBIs were explored systematically through substitution on the A-ring (Stevenson et al. 2012; Tercel et al. 2009; Tercel et al. 2010), the nature of the leaving group and the side chain that binds non-covalently to the minor groove of DNA (Ashoorzadeh et al. 2011; Stevenson et al. 2011; Stevenson et al. 2012). One limiting factor for hypoxic selectivity was the low E(1) of the A-ring unsubstituted nitroCBIs, which results in low rates of enzymatic reduction and thus low hypoxic cytotoxic potency. By placing an electron-withdrawing substituent in each of the available positions on the A-ring it was found that substituent effect on E(1) was strongest in the 7-position (Tercel et al. 2009). NitroCBI analogues with a basic dimethylamino minor groove binding side chain and with an electron-withdrawing substituent in the 7-position of the A-ring with H-bond donor capacity (e.g. sulfonamide or carboxamide) were the most selective of all compounds surveyed (Tercel et al. 2009). One analogue (cmpd**9**) was 300- and 500-fold more potent than PR-104A and tirapazamine, respectively, in a panel of 11 human tumour cell lines, and displayed equal or greater hypoxic selectivity in almost every case (Tercel et al. 2009). Subsequently, it was found that analogues with sulphonate leaving groups did not require a basic minor groove binding side chain for good hypoxic selectivity, e.g. cmpd**10** showed exquisite activity against hypoxic tumour cells *in vivo* (Ashoorzadeh et al. 2011; Stevenson et al. 2012). The enzymology of nitroCBI reductive metabolism has not yet been fully characterised. However, forced overexpression of human POR resulted in increased cytotoxicity under hypoxic but not aerobic conditions of nitroCBI analogues, making this the first human enzyme identified as being capable of hypoxia-dependent nitroCBI activation (Ashoorzadeh et al. 2011; Tercel et al. 2009). In mouse liver S9 fractions nitroCBI analogues were cleanly converted into the corresponding aminoCBI under hypoxic conditions with no evidence for reductive metabolism under aerobic conditions, which indicates these HAPs are not substrates for hepatic 2e reductases (Tercel et al. 2009; Tercel et al. 2010).

13.6.5 Improving Aqueous Solubility of NitroCBIs

Many nitroCBI analogues have low aqueous solubility which compromised their testing in tumour models; in some cases low solubility even precluded reaching on-scale cytotoxicity during *in vitro* cell-based assays (Tercel et al. 2009; Wilson et al. 2009). Replacing the neutral side chain with a basic dimethylamino moeity modestly improved solubility, which was further improved by introducing a second basic substituent via a side chain on the A-ring although only by threefold (Tercel et al. 2010). This analogue, **11,** did not show hypoxia-selective cell kill in human tumour xenografts in mice, but it did pave the way for further analogue development via other solubilising substituents on the A-ring, which also built on the finding that sulphonamide and carboxamide substituents in the 7-position gave potent hypoxia-selective nitroCBIs. Using the same phosphate ester pre-prodrug approach as for PR-104 increased the aqueous solubility by > 1,000-fold (Stevenson et al. 2011; Tercel et al. 2011a). The phosphate nitroCBI pre-prodrug **12b** demonstrated hypoxia-selective anti-tumour activity in a tumour growth delay assay with fractionated radiotherapy (Fig. 13.13, Tercel et al. 2011a). Cmpd**12b** was well tolerated and caused only minimal and transient body weight loss, with favourable plasma pharmacokinetics demonstrating rapid hydrolysis of the phosphate to the corresponding alcohol **12a** (Tercel et al. 2011a).

A number of phosphate nitroCBIs were synthesised and tested for hypoxia-selective anti-tumour activity by combining with a high-dose radiotherapy, using an *ex vivo* clonogenic assay. Broad-spectrum activity was found in xenograft models representing cancers from different tissue origins (Stevenson et al. 2011; Stevenson et al. 2012; Tercel et al. 2011a). In contrast, administration of the alkylating aminoCBI metabolite itself resulted in reversible body weight loss at its MTD, which was at least 24-fold lower than the corresponding nitroCBI phosphate **12b**, and was completely ineffective against hypoxic cells in the tumours (Tercel et al. 2011a). Thus, the prodrug approach conferred tumour selectivity, most likely through hypoxia-selective formation of the aminoCBI alkylating metabolite. In several studies modest but statistically significant single-agent activity in tumours was observed in an *ex vivo* clonogenic assay, indicative of a bystander effect (Stevenson et al. 2011; Stevenson et al. 2012; Tercel et al. 2011a, b).

13.6.6 Chirality of NitroCBIs

The nitroCBIs have a chiral centre (Fig. 13.12). All the naturally occurring duocarmycins are *S*-enantiomers, which are much more potent cytotoxins than the

R-enantiomers, reflecting differences in DNA alkylation (Boger et al. 1994a, b, 1996; Tichenor et al. 2008). Like the original natural products, the cytotoxic potency of aminoCBIs is greater for the *S*- than the *R*-enantiomer (Atwell et al. 1998, 1999; Gieseg et al. 1999). Enantiopure nitroCBI enantiomers were prepared by chiral chromatographic separation of a synthetic intermediate (Tercel et al. 2011b). In SiHa cell cultures the *S*-enantiomer of nitroCBI analogue **12a** was more cytotoxic than the *R*-enantiomer, while hypoxia selectivity was similar for both enantiomers and for the racemate (Tercel et al. 2011b). Consistent with this, no enantioselectivity was observed for enzymatic reduction of nitro- to aminoCBI. At equimolar doses of the corresponding phosphate pre-prodrugs, all anti-tumour activity of the racemate **12b** could be attributed to the *S*-enantiomer component in an *ex vivo* clonogenic assay using SiHa tumour xenografts (Tercel et al. 2011b). This correspondence between *in vitro* and *in vivo* activity of the *S*- and *R*-enantiomers further confirms the mechanism of action of the anti-tumor activity of the nitroCBIs.

13.6.7 Current Status of Development of NitroCBIs

It is clear that nitroCBIs are potent HAPs with a unique mechanism of action and broad spectrum of biological activity. Via judicious design of analogues it has been possible to harness the desired anti-tumour properties of the natural products and develop water-soluble prodrugs that confer hypoxic selectivity, and therefore tumour selectivity. Current studies are exploring the toxicology of this novel class, in small and large animal models, to select a preferred candidate for clinical development.

13.7 Future Directions for HAP Development

13.7.1 HAPs for Selective Inhibition of Molecular Targets

The HAPs described above all generate DNA-reactive cytotoxins as their primary mechanism of cytotoxicity. While activation of these prodrugs within tumours via hypoxia-dependent metabolism offers potential for improved tumour selectivity relative to conventional cytotoxics, some activation within normal tissues is inevitable (whether due to normal tissue hypoxia or the activity of oxygen-insensitive reductases such as NQO-1 or AKR1C3 as discussed above). Hepatic activation, in particular, is an important challenge given peri-portal hypoxia (Arteel et al. 1995; Gu et al. 2011a) and high activity of many oxidoreductases in the liver. Extra-tumoural bioreductive generation of DNA-reactive cytotoxins is almost certainly the major contributor to DLTs (especially myelotoxicity) that overlap the toxicity profile of widely used cytotoxic chemotherapy drugs. Thus addition of HAPs to current standard of care presents a significant challenge for clinical development of these agents. For this reason, future HAP will ideally offer a very different normal tissue toxicity profile.

In this respect, an attractive strategy is to exploit HAP chemistry to generate inhibitors of molecular targets, utilising hypoxia to impose additional tumour selectivity on compounds with relatively benign toxicity profiles. One example is the development of HAPs that are metabolised under hypoxia to generate inhibitors of O^6-alkylguanine DNA alkyltransferase (AGT; Baumann et al. 2011; Penketh et al. 2012; Zhu et al. 2011), potentially overcoming the lack of tumour selectivity of such agents as sensitisers of temozolomide and related monoalkylating agents. A second example is provided by HAPs of inhibitors of CHK1 (Cazares-Korner et al. 2013), which has an important role in DNA damage responses under severe hypoxia and after reoxygenation (Olcina et al. 2010). In both these cases, there is also a risk of potentiating normal tissue toxicity of chemotherapy if HAP activation is not strictly confined to tumours. In contrast, the HAP SN29966 utilises an earlier-reported (Tercel et al. 2001; Wilson et al. 1998) 4-nitromidazole 'trigger unit' with a quaternary ammonium linker to generate an irreversible pan-ErbB kinase inhibitor under hypoxia (Patterson et al. 2009; Smaill et al. 2012), which may avoid such problems. The related HAP PR-610 is metabolised to a pan-ErbB inhibitor that is effective against the T790M epidermal growth factor receptor (EGFR) mutation frequently responsible for erlotinib resistance in NSCLC, and is currently in phase I clinical trial through Proacta Inc (www.proacta.com). Clearly, these kinds of HAP strategies are also potentially applicable to many other molecular targets in tumours.

13.7.2 Predictive Biomarkers

The most pressing need in the clinical development of HAPs is to discover and validate biomarkers that can be used to identify hypoxic tumours prospectively, using accessible (and minimally invasive) technology. As discussed in relation to TPZ (Sect. 3.4), the failure to identify the target has been a glaring omission in late-stage clinical trials of HAP to date; no well-powered randomised trial has yet utilised predictive biomarkers to identify potentially responsive tumours. Direct measurement of oxygen in tumours using polarographic electrodes has been historically important in documenting the role of hypoxia in disease progression and therapeutic resistance (e.g. Nordsmark et al. 2005), but less invasive approaches able to better sample the whole tumour are urgently needed (Le and Courter 2008; Horsman et al. 2012). There is much interest in the development of hypoxia biomarkers based on exogenously administered 2-nitroimidazole bioreductive probes for imaging hypoxia (Mees et al. 2009), and endogenous biomarkers including mRNA expression signatures (Jubb et al. 2010; Buffa et al. 2010; Toustrup et al. 2011, 2012) and HIF-1-regulated proteins such as carbonic anhydrase IX that can be imaged non-invasively (Ahlskog et al. 2009; Hoeben et al. 2010).

However, the use of hypoxia biomarkers to predict HAP sensitivity addresses only one aspect of what is in fact a complex therapeutic target. Activation of these prodrugs requires not only hypoxia but also expression of the required 1e reductases in hypoxic cells. In addition, these hypoxic cells (and surrounding cells at higher pO_2, if a bystander effect is to be exploited) need to express the molecular profile required for sensitivity to the active metabolites. While all three elements (hypoxia, reductase and determinants of intrinsic sensitivity) might be amenable to molecular profiling, this is not a realistic prospect at this time, especially when the identities of the 1e reductases in human tumours are poorly understood. In this respect, it would obviously be highly desirable to utilise biomarkers that themselves reflect multiple elements of this multifaceted target. An example is provided by our recent demonstration that the hypoxic bioreductive activation of the 2-nitroimidazole-imaging agent EF5 is highly correlated with that of SN30000 (and TPZ) across a panel of human tumour cell lines (Wang et al. 2012). Thus, as well as reporting on hypoxia, PET imaging with [^{18}F]-EF5 (and presumably other 2-nitroimidazole probes) has the potential to report activity of the 1e reductases for SN30000 in hypoxic regions of tumours, even though the identities of these are poorly understood.

The other side of the same coin is that 2-nitroimidazole probes may be less than ideal for imaging hypoxia itself; their dependence on 1e reductase activity means that hypoxic tumours with low reductase activity could be misclassified as well oxygenated. This points to the need for matching predictive biomarkers, mechanistically, with therapeutics so that the biomarker/therapeutic pair can be jointly validated during clinical development of HAPs. However, there is also an overarching need for validated tests for hypoxia in order to identify clinical settings in which patients are most at risk of failure as a result of tumour hypoxia (Peters et al. 2012), and for focusing clinical development of HAPs in these settings.

References

Abou-Alfa GK, Chan SL, Lin CC et al (2011) PR-104 plus sorafenib in patients with advanced hepatocellular carcinoma. Cancer Chemother Pharmacol 68:539–545

Adams PC, Rickert DE (1995) Metabolism of [C-14]1,3-dinitrobenzene by rat small intestinal mucosa *in vitro*. Drug Metab Dispos 23:982–987

Aghajanian C, Brown C, O'flaherty C et al (1997) Phase I study of tirapazamine and cisplatin in patients with recurrent cervical cancer. Gynecol Oncol 67:127–130

Ahlskog JK et al (2009) Human monoclonal antibodies targeting carbonic anhydrase IX for the molecular imaging of hypoxic regions in solid tumours. Br J Cancer 101:645–657 (2009)

Ahn GO, Brown M (2007) Targeting tumors with hypoxia-activated cytotoxins. Front Biosci 12:3483–3501

Anderson RF, Shinde SS, Hay MP et al (2003) Activation of 3-amino-1,2,4-benzotriazine 1,4-dioxide antitumor agents to oxidizing species following their one-electron reduction. J Am Chem Soc 125:748–756

Anderson RF, Shinde SS, Hay MP et al (2006) Potentiation of the cytotoxicity of the anticancer agent tirapazamine by benzotriazine N-oxides: the role of redox equilibria. J Am Chem Soc 128:245–249

Anlezark GM, Melton RG, Sherwood RF et al (1995) Bioactivation of dinitrobenzamide mustards by an E. coli B nitroreductase. Biochem Pharmacol 50:609–618

Arana Yi GYA, Borthakur G, Thall PF et al (2013) Final report of phase I/II study of PR104, a hypoxia-activated pro-drug, in relapsed/refractory acute leukemia. Abstract #115667: 2013 ASCO Annual Meeting

Arteel GE, Thurman RG, Yates JM et al (1995) Evidence that hypoxia markers detect oxygen gradients in liver: pimonidazole and retrograde perfusion of rat liver. Br J Cancer 72:889–895
Ashoorzadeh A, Atwell GJ, Pruijn FB et al (2011) The effect of sulfonate leaving groups on the hypoxia-selective toxicity of nitro analogs of the duocarmycins. Bioorg Med Chem 19: 4851–4860
Ask K, Dijols S, Giroud C et al (2003) Reduction of nilutamide by NO synthases: implications for the adverse effects of this nitroaromatic antiandrogen drug. Chem Res Toxicol 16:1547–1554
Atwell GJ, Tercel M, Boyd M et al (1998) Synthesis and cytotoxicity of 5-amino-1-(chloromethyl)-3-[(5,6,7-trimethoxyindol-2-yl)carbonyl]-1,2-dihydro-3H-benz[e]indole (amino-seco-CBI-TMI) and related 5-alkylamino analogues: New DNA minor groove alkylating agents. J Org Chem 63 9414–9420
Atwell GJ, Milbank JJ, Wilson WR et al (1999) 5-Amino-1-(chloromethyl)-1,2-dihydro-3H-benz[e]indoles: relationships between structure and cytotoxicity for analogues bearing different DNA minor groove binding subunits. J Med Chem 42:3400–3411
Atwell GJ, Yang S, Pruijn FB et al (2007) Synthesis and Structure-Activity Relationships for 2,4-Dinitrobenzamide-5 mustards as Prodrugs for the *Escherichia coli nfsB* Nitroreductase in Gene Therapy. J Med Chem 50:1197–1212
Azab AK, Hu J, Quang P et al (2012) Hypoxia promotes dissemination of multiple myeloma through acquisition of epithelial to mesenchymal transition-like features. Blood 119:5782–5794
Bagshawe KD (1995) Antibody-directed enzyme prodrug therapy: a review. Drug Dev Res 34: 220–230
Baker MA, Zeman EM, Hirst VK et al (1988) Metabolism of SR 4233 by Chinese hamster ovary cells: basis of selective hypoxic cytotoxicity. Cancer Res 48:5947–5952
Baumann RP, Ishiguro K, Penketh PG et al (2011) KS900: a hypoxia-directed, reductively activated methylating antitumor prodrug that selectively ablates O(6)-alkylguanine-DNA alkyltransferase in neoplastic cells. Biochem Pharmacol 81:1201–1210
Belcourt MF, Hodnick WF, Rockwell S et al (1996) Differential toxicity of mitomycin C and porfiromycin to aerobic and hypoxic Chinese hamster ovary cells overexpressing human NADPH:cytochrome c (P-450) reductase. Proc Natl Acad Sci U S A 93:456–460
Benito J, Shi Y, Szymanska B et al (2011) Pronounced hypoxia in models of murine and human leukemia: high efficacy of hypoxia-activated prodrug PR-104. PLoS ONE 6:e23108
Birtwistle J, Hayden RE, Khanim FL et al (2009) The aldo-keto reductase AKR1C3 contributes to 7,12-dimethylbenz(a)anthracene-3,4-dihydrodiol mediated oxidative DNA damage in myeloid cells: Implications for leukemogenesis. Mutat Res 662:67–74
Blouw B, Song H, Tihan T et al (2003) The hypoxic response of tumors is dependent on their microenvironment. Cancer Cell 4:133–146
Boger DL, Johnson DS (1996) CC-1065 and the duocarmycins: Understanding their biological function through mechanistic studies. Angew Chem Int Ed Engl 35:1438–1474
Boger DL, Han NH, Tarby CM et al (1996) Synthesis, chemical properties, and preliminary evaluation of substituted CBI analogs of CC-1065 and the duocarmycins incorporating the 7-cyano-1,2,9,9a-tetrahydrocyclopropa[*c*]benz[*e*]indol-4-one alkylation subunit: Hammett quantitation of the magnitude of electronic effects on functional reactivity. J Org Chem 61:4894–4912
Boger DL, Yun W (1994) CBI-TMI: synthesis and evaluation of a key analog of the duocarmycins. Validation of a direct relationship between chemical solvolytic stability and cytotoxic potency and confirmation of the structural features responsible for the distinguishing behavior of enantiomeric pairs of agents. J Am Chem Soc 116:7996–8006
Boger DL, Ishizaki T, Zarrinmayeh H et al (1990) Duocarmycin-pyrindamycin DNA alkylation properties and identification, synthesis, and evaluation of agents incorporating the pharmacophore of the duocarmycin-pyrindamycin alkylation subunit. Identification of the CC-1065-duocarmycin common pharmacophore. J Am Chem Soc 112:8961–8971
Boger DL, Johnson DS, Yun W (1994) (+)- and ent-(−)-Duocarmycin SA and (+)- and ent-(−)-N-BOC-DSA DNA alkylation properties. Alkylation site models that accommodate the offset

AT-rich adenine N3 alkylation selectivity of the enantiomeric agents. J Am Chem Soc 116:1635–1656
Borad M, Infante JR, Mita AC et al (2010) Complete phase 1B study of TH-302 in combination with gemcitabine, docetaxel or pemetrexed. The 35th ESMO Congress in Milan, Italy, 8–12 Oct
Borad MJ, Reddy R, bahary N et al (2012) TH-302 plus gemcitabine vs. gemcitabine in patients with untreated advanced pancreatic adenocarcinoma. European Society for Medical Oncology (ESMO) 2012 Congress, Vienna, Austria, 28th Sep–2nd Oct
Borch RF, Liu J, Schmidt JP et al (2000) Synthesis and evaluation of nitroheterocyclic phosphoramidates as hypoxia-selective alkylating agents. J Med Chem 43:2258–2265
Borch RF, Liu J, Joswig C et al (2001) Antitumor activity and toxicity of novel nitroheterocyclic phosphoramidates. J Med Chem 44:74–77
Brown JM (1993) SR 4233 (tirapazamine): a new anticancer drug exploiting hypoxia in solid tumours. Br J Cancer 67:1163–1170
Brown M (2010) Henry S. Kaplan distinguished scientist award lecture 2007. The remarkable yin and yang of tumour hypoxia. Int J Radiat Biol 86:907–917
Brown JM, Lemmon MJ (1990) Potentiation by the hypoxic cytotoxin SR 4233 of cell killing produced by fractionated irradiation of mouse tumors. Cancer Res 50:7745–7749
Brown JM, Lemmon MJ (1991) Tumor hypoxia can be exploited to preferentially sensitize tumors to fractionated irradiation. Int J Radiat Oncol Biol Phys 20:457–461
Brown JM, Diehn M, Loo BW (2010) Stereotactic ablative radiotherapy should be combined with a hypoxic cell radiosensitizer. Int J Radiat Oncol Biol Phys 78:323–327
Buffa FM, Harris AL, West CM et al (2010) Large meta-analysis of multiple cancers reveals a common, compact and highly prognostic hypoxia metagene. Br J Cancer 102:428–435
Carlson DJ, Keall PJ, Loo BW Jr et al (2011) Hypofractionation results in reduced tumor cell kill compared to conventional fractionation for tumors with regions of hypoxia. Int J Radiat Oncol Biol Phys 79:1188–1195
Cazares-Korner C, Pires IM, Swallow ID et al (2013) CH-01 is a hypoxia-activated prodrug that sensitizes cells to hypoxia/reoxygenation through inhibition of Chk1 and Aurora A. ACS Chem Biol 8:1451–1459
Cenas N, Prast S, Nivinskas H et al (2006) Interactions of nitroaromatic compounds with the mammalian selenoprotein thioredoxin reductase and the relation to induction of apoptosis in human cancer cells. J Biol Chem 281:5593–5603
Chan DA, Giaccia AJ (2007) Hypoxia, gene expression, and metastasis. Cancer Metastasis Rev 26:333–339
Chandor A, Dijols S, Ramassamy B et al (2008) Metabolic activation of the antitumor drug 5-(Aziridin-1-yl)-2,4-dinitrobenzamide (CB1954) by NO synthases. Chem Res Toxicol 21: 836–843
Chang Q, Jurisica I, Do T et al (2011) Hypoxia predicts aggressive growth and spontaneous metastasis formation from orthotopically grown primary xenografts of human pancreatic cancer. Cancer Res 71:3110–3120
Chang YS, Adnane J, Trail PA et al (2007) Sorafenib (BAY 43-9006) inhibits tumor growth and vascularization and induces tumor apoptosis and hypoxia in RCC xenograft models. Cancer Chemother Pharmacol 59:561–574
Chawla SP, Ganjoo KN, Adkins D et al (2011) A phase 2 study of TH-302 in combination with doxorubicin in advanced soft tissue sarcoma. Connective Tissue Oncology Society (CTOS) Annual Meeting, Chicago, USA 26–29 Oct
Chen Y, Hu L (2009) Design of anticancer prodrugs for reductive activation. Med Res Rev 29:29–64
Chinje EC, Cowen RL, Feng J et al (2003) Non-nuclear localized human NOSII enhances the bioactivation and toxicity of tirapazamine (SR4233) *in vitro*. Mol Pharmacol 63:1248–1255
Chowdhury G, Junnotula V, Daniels JS et al (2007) DNA strand damage product analysis provides evidence that the tumor cell-specific cytotoxin tirapazamine produces hydroxyl radical and acts as a surrogate for O(2). J Am Chem Soc 129:12870–12877

Connors TA, Whisson ME (1966) Cure of mice bearing advanced plasma cell tumours with aniline mustard: the relationship between glucuronidase activity and tumour sensitivity. Nature 210:866–867
Covens A, Blessing J, Bender D et al (2006) A phase II evaluation of tirapazamine plus cisplatin in the treatment of recurrent platinum-sensitive ovarian or primary peritoneal cancer: a Gynecologic Oncology Group study. Gynecol Oncol 100:586–590
Cowan DS, Hicks KO, Wilson WR (1996) Multicellular membranes as an *in vitro* model for extravascular diffusion in tumours. Br J Cancer Suppl 27:S28–S31
Craighead PS, Pearcey R, Stuart G (2000) A phase I/II evaluation of tirapazamine administered intravenously concurrent with cisplatin and radiotherapy in women with locally advanced cervical cancer. Int J Radiat Oncol Biol Phys 48:791–795
Cristofanilli M, Bryan WJ, Miller LL et al (1998) Phase II study of adozelesin in untreated metastatic breast cancer. Anticancer Drugs 9:779–782
Cullinan SB, Diehl JA (2006) Coordination of ER and oxidative stress signaling: the PERK/Nrf2 signaling pathway. Int J Biochem Cell Biol 38:317–332
Daniels JS, Gates KS (1996) DNA cleavage by the antitumor agent 3-amino-1,2,4-benzotriazine 1,4-dioxide (SR4233): Evidence for involvement of hydroxyl radical. J Am Chem Soc 118: 3380–3385
Daniels JS, Gates KS, Tronche C et al (1998) Direct evidence for bimodal DNA damage induced by tirapazamine. Chem Res Toxicol 11:1254–1257
Delahoussaye YM, Hay MP, Pruijn FB et al (2003) Improved potency of the hypoxic cytotoxin tirapazamine by DNA-targeting. Biochem Pharmacol 65:1807–1815
Denny WA, Wilson WR (1986) Considerations for the design of nitrophenyl mustards as agents with selective toxicity for hypoxic tumor cells. J Med Chem 29:879–887
Denny WA, Wilson WR (1993) Bioreducible mustards: a paradigm for hypoxia-selective prodrugs of diffusible cytotoxins (HPDCs). Cancer Metastasis Rev 12:135–151
Denny WA, Wilson WR (1998) The design of selectively-activated anti-cancer prodrugs for use in antibody-directed and gene-directed enzyme-prodrug therapies. J Pharm Pharmacol 50:387–394
Denny WA, Wilson WR, Hay MP (1996) Recent developments in the design of bioreductive drugs. Br J Cancer 74(Suppl XXVII):S32–S38
Dewhirst MW (2009) Relationships between cycling hypoxia, HIF-1, angiogenesis and oxidative stress. Radiat Res 172:653–665
Dewhirst MW, Cao Y, Moeller B (2008) Cycling hypoxia and free radicals regulate angiogenesis and radiotherapy response. Nat Rev Cancer 8:425–437
DiSilvestro P, Ali S, Peter C et al (2012) A Gynecologic Oncology Group phase III randomized trial of weekly cisplatin and radiation versus cisplatin and tirapazamine and radiation in stage IB2, IIA, IIIB and IVA cervical carcinoma limited to the pelvis. Gynecol Oncol 125(Suppl 1):S3–S4
Dorie MJ, Brown JM (1993) Tumor-specific, schedule-dependent interaction between tirapazamine (SR 4233) and cisplatin. Cancer Res 53:4633–4636
Dorie MJ, Brown JM (1994) Potentiation of the anticancer effect of cisplatin by the hypoxic cytotoxin tirapazamine. In: Vaupel PW, Kelleher DK, Gunderoth M (eds) Tumor oxygenation, Fischer-Verlag, Stuttgart, pp 125–135
Down J, Parmar K, D'Andrea AD et al (2010) Aldo-Keto Reductase 1C3 as a novel target for the aerobic activation of the prodrug PR-104A in human hematopoietic cells. Proceedings of 56th Annual Meeting Radiation Research Society, September 25–29, 2010, Maui, Hawaii, abstract number PS7 95
Duan JX, Jiao H, Kaizerman J et al (2008) Potent and highly selective hypoxia-activated achiral phosphoramidate mustards as anticancer drugs. J Med Chem 51:2412–2420
Dubois L, Peeters S, Lieuwes NG et al (2011) Specific inhibition of carbonic anhydrase IX activity enhances the *in vivo* therapeutic effect of tumor irradiation 9. Radiother Oncol 99:424–431
Durand RE, Olive PL (1992) Evaluation of bioreductive drugs in multicell spheroids. Int J Radiat Oncol Biol Phys 22:689–692

Durand RE, Olive PL (1997) Physiologic and cytotoxic effects of tirapazamine in tumor-bearing mice. Radiat Oncol Investig 5:213–219

Ebbesen P, Pettersen EO, Gorr TA et al (2009) Taking advantage of tumor cell adaptations to hypoxia for developing new tumor markers and treatment strategies. (Review) (375 refs). Journal of Enzyme Inhibition Medicinal Chemistry 24(Suppl 1):1–39

Evans JW, Yudoh K, Delahoussaye YM et al (1998) Tirapazamine is metabolized to its DNA-damaging radical by intranuclear enzymes. Cancer Res 58:2098–2101

Evans SM, Hahn S, Pook DR et al (2000) Detection of hypoxia in human squamous cell carcinoma by EF5 binding. Cancer Res 60:2018–2024

Evans JW, Chernikova SB, Kachnic LA et al (2008) Homologous recombination is the principal pathway for the repair of DNA damage induced by tirapazamine in mammalian cells. Cancer Res 68:257–265

Finger EC, Giaccia AJ (2010) Hypoxia, inflammation, and the tumour microenvironment in metastatic disease. Cancer Met Rev 29:285–293

Fitzsimmons SA, Lewis AD, Riley RJ et al (1994) Reduction of 3-amino-1,2,4-benzotriazine-1,4-di-N-oxide (tirapazamine, WIN 59075, SR 4233) to a DNA-damaging species: a direct role for NADPH:cytochrome P450 oxidoreductase. Carcinogenesis 15:1503–1510

Foehrenbacher A, Patel K, Abbattista M et al (2013) The role of bystander effects in the antitumor activity of the hypoxia-activated prodrug PR-104. Front Oncol 3:263

Fox ME, Lemmon MJ, Mauchline ML et al (1996) Anaerobic bacteria as a delivery system for cancer gene therapy: *in vitro* activation of 5-fluorocytosine by genetically engineered clostridia. Gene Ther 3:173–178

Frolova O, Samudio I, Benito JM et al (2012) Regulation of HIF-1alpha signaling and chemoresistance in acute lymphocytic leukemia under hypoxic conditions of the bone marrow microenvironment. Cancer Biol Ther 13:858–870

Ganjoo KN, Cranmer LD, Butrynski JE et al (2011) A phase I study of the safety and pharmacokinetics of the hypoxia-activated prodrug TH-302 in combination with doxorubicin in patients with advanced soft tissue sarcoma. Oncology 80:50–56

Ghatage P, Sabagh H (2012) Is there a role for tirapazamine in the treatment of cervical cancer? Expert Opin Drug Metab Toxicol 8:1589–1597

Gieseg MA, Matejovic J, Denny WA (1999) Comparison of the patterns of DNA alkylation by phenol and amino seco- CBI-TMI compounds: use of a PCR method for the facile preparation of single end-labelled double-stranded DNA. Anticancer Drug Des 14:77–84

Graaf MM, Boven E, Scheeren HW et al (2002) Beta-glucuronidase-mediated drug release. Curr Pharm Des 8:1391–1403

Graham MA, Senan S, Robin H Jr et al (1997) Pharmacokinetics of the hypoxic cell cytotoxic agent tirapazamine and its major bioreductive metabolites in mice and humans: retrospective analysis of a pharmacokinetically guided dose-escalation strategy in a phase I trial. Cancer Chemother Pharmacol 40:1–10

Grove JI, Searle PF, Weedon SJ et al (1999) Virus-directed enzyme prodrug therapy using CB1954. Anticancer Drug Des 14:461–472

Gu Y, Patterson AV, Atwell GJ et al (2009) Roles of DNA repair and reductase activity in the cytotoxicity of the hypoxia-activated dinitrobenzamide mustard PR-104A. Mol Cancer Ther 8:1714–1723

Gu Y, Atwell GJ, Wilson WR (2010) Metabolism and excretion of the novel bioreductive prodrug PR-104 in mice, rats, dogs and humans. Drug Metab Dispos 38:498–508

Gu Y, Guise CP, Patel K et al (2011a) Reductive metabolism of the dinitrobenzamide mustard anticancer prodrug PR-104 in mice. Cancer Chemother Pharmacol 67:543–555

Gu Y, Tingle MD, Wilson WR (2011b) Glucuronidation of anticancer prodrug PR-104A: Species differences, identification of human UDP-glucuronosyltransferases and implications for therapy. J Pharmacol Exp Ther 337:692–702

Guise CP, Wang A, Thiel A et al (2007) Identification of human reductases that activate the dinitrobenzamide mustard prodrug PR-104A: a role for NADPH:cytochrome P450 oxidoreductase under hypoxia. Biochem Pharmacol 74:810–820

Guise CP, Abbattista M, Singleton RS et al (2010) The bioreductive prodrug PR-104A is activated under aerobic conditions by human aldo-keto reductase 1C3. Cancer Res 70:1573–1584

Guise CP, Abbattista MR, Tipparaju SR et al (2012) Diflavin oxidoreductases activate the bioreductive prodrug PR-104A under hypoxia. Mol Pharmacol 81:31–40

Hall EJ, Roizin-Towle L (1975) Hypoxic sensitizers: radiobiological studies at the cellular level. Radiology 117:453–457

Harrington KJ (2011) Guidelines for preclinical and early phase clinical assessment of novel radiosensitisers. Br J Cancer 105:628–639

Hay MP, Gamage SA, Kovacs MS et al (2003) Structure-activity relationships of 1,2,4-benzotriazine 1,4-dioxides as hypoxia-selective analogues of tirapazamine. J Med Chem 46:169–182

Hay MP, Hicks KO, Pruijn FB et al (2007a) Pharmacokinetic/pharmacodynamic model-guided identification of hypoxia-selective 1,2,4-benzotriazine 1,4-dioxides with antitumor activity: the role of extravascular transport. J Med Chem 50:6392–6404

Hay MP, Pchalek K, Pruijn FB et al (2007b) Hypoxia-selective 3-alkyl 1,2,4-benzotriazine 1,4-dioxides: the influence of hydrogen bond donors on extravascular transport and antitumor activity. J Med Chem 50:6654–6664

Hay MP, Hicks KO, Pchalek K et al (2008) Tricyclic [1,2,4]triazine 1,4-dioxides as hypoxia selective cytotoxins. J Med Chem 51:6853–6865

Helsby NA, Wheeler SJ, Pruijn FB et al (2003) Effect of nitroreduction on the alkylating reactivity and cytotoxicity of the 2,4-dinitrobenzamide-5-aziridine CB 1954 and the corresponding nitrogen mustard SN23862: distinct mechanisms of bioreductive activation. Chem Res Toxicol 16:469–478

Helsby NA, Ferry DM, Patterson AV et al (2004) 2-amino metabolites are key mediators of CB 1954 and SN23862 bystander effects in nitroreductase GDEPT. Br J Cancer 90:1084–1093

Hicks KO (2008) Three-dimensional tissue cultures. In: Schwab M (ed) Encyclopedia of Cancer, Springer-Verlag, Berlin, pp 2956–2961

Hicks KO (2012) Introducing drug transport early in the design of hypoxia selective anticancer agents using a mathematical modelling approach. In: D'Onofrio A, Cerrai P, Gandolfi A (eds) New challenges for cancer systems biomedicine. Springer, Italy, pp 337–354

Hicks KO, Fleming Y, Siim BG et al (1998) Extravascular diffusion of tirapazamine: effect of metabolic consumption assessed using the multicellular layer model. Int J Radiat Oncol Biol Phys 42:641–649

Hicks KO, Pruijn FB, Sturman JR et al (2003) Multicellular resistance to tirapazamine is due to restricted extravascular transport: a pharmacokinetic/pharmacodynamic study in HT29 multicellular layer cultures. Cancer Res 63:5970–5977

Hicks KO, Siim BG, Pruijn FB et al (2004) Oxygen dependence of the metabolic activation and cytotoxicity of tirapazamine: implications for extravascular transport and activity in tumors. Radiat Res 161:656–666

Hicks KO, Pruijn FB, Secomb TW et al (2006) Use of three-dimensional tissue cultures to model extravascular transport and predict *in vivo* activity of hypoxia-targeted anticancer drugs. J Natl Cancer Inst 98:1118–1128

Hicks KO, Myint H, Patterson AV et al (2007) Oxygen dependence and extravascular transport of hypoxia-activated prodrugs: comparison of the dinitrobenzamide mustard PR-104A and tirapazamine. Int J Radiat Oncol Biol Phys 69:560–571

Hicks KO, Siim BG, Jaiswal JK et al (2010) Pharmacokinetic/pharmacodynamic modeling identifies SN30000 and SN29751 as tirapazamine analogues with improved tissue penetration and hypoxic cell killing in tumors. Clin Cancer Res 16:4946–4957

Hill RP, Marie-Egyptienne DT, Hedley DW (2009) Cancer stem cells, hypoxia and metastasis. Semin Radiat Oncol 19:106–111

Hoeben BA, Kaanders JH, Franssen GM et al (2010) PET of Hypoxia with ^{89}Zr-Labeled cG250-F(ab')$_2$ in Head and Neck Tumors...etc. J Nucl Med 51:1076–1083
Horsman MR, Mortensen LS, Petersen JB (2012) Imaging hypoxia to improve radiotherapy outcome. Nat Rev Clin Oncol 9:674–687
Houghton PJ, Lock R, Carol H et al (2011) Initial testing of the hypoxia activated prodrug PR-104 by the Pediatric Preclinical Testing Program. Pediatr Blood Cancer 57:443–453
Hu J, Handisides DR, Van Valckenborgh E (2010) Targeting the multiple myeloma hypoxic niche with TH-302, a hypoxia-activated prodrug. Blood 116:1524–1527
Hunter FW, Wang J, Patel R et al (2012) Homologous recombination repair-dependent cytotoxicity of the benzotriazine di-N-oxide CEN-209: Comparison with other hypoxia-activated prodrugs. Biochem Pharmacol 83:574–585
Hu J, Van Valckenborgh E, Xu D et al (2013) Synergistic induction of apoptosis in multiple myeloma cells by bortezomib and hypoxia-activated prodrug TH-302, *in vivo* and *in vitro*. Mol Cancer Ther 12:1763–1773
Hwang JT, Greenberg MM, Fuchs T et al (1999) Reaction of the hypoxia-selective antitumor agent tirapazamine with a C1'-radical in single-stranded and double-stranded DNA. Biochemistry (Mosc) 38:14248–14255
Jameson MB, Rischin D, Pegram M et al (2010) A phase I trial of PR-104, a nitrogen mustard prodrug activated by both hypoxia and aldo-keto reductase 1C3, in patients with solid tumors. Cancer Chemother Pharmacol 65:791–801
Jensen PO, Mortensen BT, Hodgkiss RJ et al (2000) Increased cellular hypoxia and reduced proliferation of both normal and leukaemic cells during progression of acute myeloid leukaemia in rats. Cell Prolif 33:381–395
Johnson CA, Kilpatrick D, von Roemeling R et al (1997) Phase I trial of tirapazamine in combination with cisplatin in a single dose every 3 weeks in patients with solid tumors. J Clin Oncol 15: 773–780
Jones GD, Weinfeld M (1996) Dual action of tirapazamine in the induction of DNA strand breaks. Cancer Res 56:1584–1590
Jubb AM, Buffa FM, Harris AL (2010) Assessment of tumour hypoxia for prediction of response to therapy and cancer prognosis. J Cell Mol Med 14:18–29
Jung D, Jiao H, Duan JX et al (2012a) Metabolism and excretion of TH-302 in dogs. Xenobiotica 42:687–700
Jung D, Jiao H, Duan JX et al (2012b) Metabolism, pharmacokinetics and excretion of a novel hypoxia activated cytotoxic prodrug, TH-302, in rats. Xenobiotica 42:372–388
Jung D, Lin L, Jiao H et al (2012c) Pharmacokinetics of TH-302: a hypoxically activated prodrug of bromo-isophosphoramide mustard in mice, rats, dogs and monkeys. Cancer Chemother Pharmacol 69:643–654
Keith B, Johnson RS, Simon MC (2012) HIF1apha and HIF2alpha: sibling rivalry in hypoxic tumor growth and progression. Nature Rev Cancer 12:9–22
Kelson AB, McNamara JP, Pandey A et al (1998) 1,2,4-Benzotriazine 1,4-dioxides. An important class of hypoxic cytotoxins with antitumor activity. Anticancer Drug Des 13:575–592
Kim IH, Brown JM (1994) Reoxygenation and rehypoxiation in the SCCVII mouse tumor. Int J Radiat Oncol Biol Phys 29:493–497
Kioi M, Vogel H, Schultz G et al (2010) Inhibition of vasculogenesis, but not angiogenesis, prevents the recurrence of glioblastoma after irradiation in mice. J Clin Invest 120:694–705
Knox RJ, Chen S (2004) Quinone reductase-mediated nitro-reduction: clinical applications. Methods Enzymol 382:194–221
Knox RJ, Friedlos F, Marchbank T et al (1991) Bioactivation of CB 1954: reaction of the active 4-hydroxylamino derivative with thioesters to form the ultimate DNA-DNA interstrand crosslinking species. Biochem Pharmacol 42:1691–1697
Kobayashi E, Okamoto A, Asada M et al (1994) Characteristics of antitumor activity of KW-2189, a novel water-soluble derivative of duocarmycin, against murine and human tumors. Cancer Res 54:2404–2410
Koch CJ (1993) Unusual oxygen concentration dependence of toxicity of SR-4233, a hypoxic cell toxin. Cancer Res 53:3992–3997

Koch CJ, Stobbe CC, Bump EA (1984) The effect on the Km for radiosensitization at 0 degree C of thiol depletion by diethylmaleate pretreatment: quantitative differences found using the radiation sensitizing agent misonidazole or oxygen. Radiat Res 98:141–153

Kolesnick R, Haimovitz-Friedman A, Fuks Z (2012) Coupling of endothelial cell dysfunction to tumor stem cell demise. In: Abstracts pf the 58th annual meeting of the Radiation Research Society, Puerto Rico, 30 Sept–3 Oct 2012 (abstract #S102).

Kovacs MS, Hocking DJ, Evans JW et al (1999) Cisplatin anti-tumour potentiation by tirapazamine results from a hypoxia-dependent cellular sensitization to cisplatin. Br J Cancer 80:1245–1251

Kruser TJ, Wheeler DL, Armstrong EA et al (2010) Augmentation of radiation response by motesanib, a multikinase inhibitor that targets vascular endothelial growth factor receptors. Clin Cancer Res 16:3639–3647

Kyle AH, Minchinton AI (1999) Measurement of delivery and metabolism of tirapazamine to tumour tissue using the multilayered cell culture model. Cancer Chemother Pharmacol 43:213–220

Laderoute K, Wardman P, Rauth AM (1988) Molecular mechanisms for the hypoxia-dependent activation of 3-amino- 1,2,4-benzotriazine-1,4-dioxide (SR 4233). Biochem Pharmacol 37: 1487–1495

Le QT, Courter D (2008) Clinical biomarkers for hypoxia targeting. Cancer Metastasis Rev 27: 351–362

Le QT, Fisher R, Oliner KS et al (2012) Prognostic and predictive significance of plasma HGF and IL-8 in a phase III trial of chemoradiation with or without tirapazamine in locoregionally advanced head and neck cancer. Clin Cancer Res 18:1798–1807

Lee AE, Wilson WR (2000) Hypoxia-dependent retinal toxicity of bioreductive anticancer prodrugs in mice. Toxicol Appl Pharmacol 163:50–59

Li LH, DeKoning TF, Kelly RC et al (1992) Cytotoxicity and antitumor activity of carzelesin, a prodrug cyclopropylpyrroloindole analogue. Cancer Res 52:4904–4913

Li S, Zhang J, Li J et al (2010) Inhibition of both thioredoxin reductase and glutathione reductase may contribute to the anticancer mechanism of TH-302. Biol Trace Elem Res 136:294–301

Lin AJ, Cosby LA, Shanky CW et al (1972) Potential bioreductive alkylating agents. I. Benzoquinone derivatives. J Med Chem 15:1247–1252

Liu SC, Ahn GO, Kioi M et al (2008) Optimised Clostridium-directed enzyme prodrug therapy improves the antitumor activity of the novel DNA crosslinking agent PR-104. Cancer Res 68:7995–8003

Liu Q, Sun JD, Wang J et al (2012) TH-302, a hypoxia-activated prodrug with broad *in vivo* preclinical combination therapy efficacy: optimization of dosing regimens and schedules. Cancer Chemother Pharmacol 69:1487–1498

Lohse I, Rasowski J, Cao PJ et al (2012) Targeting tumor hypoxia in patient-derived pancreatic xenografts using TH-302. AACR Pancreatic Cancer Meeting, Lake Taho, June 18–20

Maluf FC, Leiser AL, Aghajanian C et al (2006) Phase II study of tirapazamine plus cisplatin in patients with advanced or recurrent cervical cancer. Int J Gyne Cancer 16:1165–1171

Markovic SN, Suman VJ, Vukov AM et al (2002) Phase II trial of KW2189 in patients with advanced malignant melanoma. Am J Clin Oncol 25:308–312

Martin SK, Diamond P, Gronthos S et al (2011) The emerging role of hypoxia, HIF-1 and HIF-2 in multiple myeloma. Leukemia 25:1533–1542

Matsunaga T, Imataki O, Torii E et al (2012) Elevated HIF-1alpha expression of acute myelogenous leukemia stem cells in the endosteal hypoxic zone may be a cause of minimal residual disease in bone marrow after chemotherapy. Leuk Res 36:e122-e124

Mazumdar J, Hickey MM, Pant DK et al (2010) HIF-2alpha deletion promotes Kras-driven lung tumor development. Proc Natl Acad Sci U S A 107:14182–14187

McDonald PC, Winum JY, Supuran CT et al (2012) Recent developments in targeting carbonic anhydrase IX for cancer therapeutics. Oncotarget 3:84–97

McKeage MJ, Gu Y, Wilson WR et al (2011) A phase I trial of PR-104, a pre-prodrug of the bioreductive prodrug PR-104A, given weekly to solid tumour patients. BMC Cancer 11:432

McKeage MJ, Jameson MB, Ramanathan RK et al (2012) PR-104 a bioreductive pre-prodrug combined with gemcitabine or docetaxel in a phase Ib study of patients with advanced solid tumours. BMC Cancer 12:496

McKeown SR, Cowen RL, Williams KJ (2007) Bioreductive drugs: from concept to clinic. Clin Oncol 19:427–442

Mees G, Dierckx R, Vangestel C, Van de Wiele C et al (2009) Molecular imaging of hypoxia with radiolabelled agents. Eur J Nucl Med Mol Imag 36:1674–1686

Mehta S, Hughes NP, Buffa FM et al (2011) Assessing early therapeutic response to bevacizumab in primary breast cancer using magnetic resonance imaging and gene expression profiles. J Natl Cancer Inst Monogr 2011(43):71–74

Melillo G (2007) Targeting hypoxia cell signalling for cancer therapy. Cancer Metastasis Rev 26:341–352

Meng F, Evans JW, Bhupathi D et al (2012) Molecular and cellular pharmacology of the hypoxia-activated prodrug TH-302. Mol Cancer Ther 11:740–751

Minchinton AI, Tannock IF (2006) Drug penetration in solid tumours. Nat Rev Cancer 6:583–592

Minchinton AI, Lemmon MJ, Tracy M et al (1992) Second-generation 1,2,4-benzotriazine 1,4-di-N-oxide bioreductive anti- tumor agents: pharmacology and activity *in vitro* and *in vivo*. Int J Radiat Oncol Biol Phys 22:701–705

Minchinton AI, Wendt KR, Clow KA et al (1997) Multilayers of cells growing on a permeable support. An *in vitro* tumour model. Acta Oncol 36:13–16

Mohindra JK, Rauth AM (1976) Increased cell killing by metronidazole and nitrofurazone of hypoxic compared to aerobic mammalian cells. Cancer Res 36:930–936

Morgan AS, Sanderson PE, Borch RF et al (1998) Tumor efficacy and bone marrow-sparing properties of TER286, a cytotoxin activated by glutathione S-transferase. Cancer Res 58:2568–2575

Nordgren IK, Tavassoli A (2011) Targeting tumour angiogenesis with small molecule inhibitors of hypoxia inducible factor. Chem Soc Rev 40:4307–4317

Nordsmark M, Bentzen SM, Rudat V et al (2005) Prognostic value of tumor oxygenation in 397 head and neck tumors after primary radiation therapy. An international multi-center study. Radiother Oncol 77:18–24

Norlund P, Reichard P (2006) Ribonucleotide reductases. Ann Rev BioChemist 75:681–706

Nuyts S, Van Mellaert L, Theys J et al (2002) Clostridium spores for tumour-specific drug delivery. Anticancer Drugs 13:115–125

Olcina M, Lecane PS, Hammond EM (2010) Targeting hypoxic cells through the DNA damage response. Clin Cancer Res 16:5620–5629

Olive PL, Vikse CM, Banath JP (1996) Use of the comet assay to identify cells sensitive to tirapazamine in multicell spheroids and tumors in mice. Cancer Res 56:4460–4463

Olive PL, Banath JP, Sinnott LT (2004) Phosphorylated histone H2AX in spheroids, tumors, and tissues of mice exposed to etoposide and 3-amino-1,2,4-benzotriazine-1,3-dioxide. Cancer Res 64:5363–5369

Paez-Ribes M (2009) Antiangiogenic therapy elicits malignant progression of tumors to Increased local invasion and distant metastasis. Cancer Cell

Palmer BD, Wilson WR, Denny WA (1990a) Nitro analogues of chlorambucil as potential hypoxia-selective anti- tumour drugs. Anticancer Drug Des 5:337–349

Palmer BD, Wilson WR, Pullen SM et al (1990b) Hypoxia-selective antitumor agents. 3. Relationships between structure and cytotoxicity against cultured tumor cells for substituted N, N-bis(2- chloroethyl)anilines. J Med Chem 33:112–121

Palmer BD, Wilson WR, Cliffe S et al (1992) Hypoxia-selective antitumor agents. 5. Synthesis of water-soluble nitroaniline mustards with selective cytotoxicity for hypoxic mammalian cells. J Med Chem 35:3214–3222

Palmer BD, Wilson WR, Atwell GJ et al (1994) Hypoxia-selective antitumor agents. 9. Structure-activity relationships for hypoxia-selective cytotoxicity among analogues of 5-[N, N-bis(2-chloroethyl)amino]-2,4-dinitrobenzamide. J Med Chem 37:2175–2184

Palmer BD, Zijl P van, Denny WA et al (1995) Reductive chemistry of the novel hypoxia-selective cytotoxin 5-[*N*, *N*- bis(2-chloroethyl)amino]-2,4-dinitrobenzamide. J Med Chem 38:1229–1241
Palmer BD, Wilson WR, Anderson RF et al (1996) Hypoxia-selective antitumor agents. 14. Synthesis and hypoxic cell cytotoxicity of regioisomers of the hypoxia-selective cytotoxin 5-[*N*, *N*- bis(2-chloroethyl)amino]-2,4-dinitrobenzamide. J Med Chem 39:2518–2528
Panthananickal A, Hansch C, Leo A (1979) Structure-activity relationship of aniline mustards acting against B-16 melanoma in mice. J Med Chem 22:1267–1269
Papadopoulou MV, Ji M, Rao MK et al (2003) Reductive activation of the nitroimidazole-based hypoxia-selective cytotoxin NLCQ-1 (NSC 709257). Oncology Res 14:21–29
Papandreou I, Denko NC, Olson M et al (2011) Identification of an Ire1alpha endonuclease specific inhibitor with cytotoxic activity against human multiple myeloma. Blood 117:1311–1314
Patel K, Lewiston D, Gu Y et al (2007) Analysis of the hypoxia-activated dinitrobenzamide mustard phosphate prodrug PR-104 and its alcohol metabolite PR-104A in plasma and tissues by liquid chromatography-mass spectrometry. J Chromatogr B Analyt Technol Biomed Life Sci 856: 302–311
Patel K, Choy SF, Hicks KO et al (2011) A combined pharmacokinetic model for the hypoxia-targeted prodrug PR-104A in humans, dogs, rats and mice predicts species differences in clearance and toxicity. Cancer Chemother Pharmacol 67:1145–1155
Patterson LH (2002) Bioreductively activated antitumor N-oxides: the case of AQ4N, a unique approach to hypoxia-activated cancer chemotherapy. Drug Metab Rev 34:581–592
Patterson AV, Saunders MP, Chinje EC et al (1997) Overexpression of human NADPH:cytochrome c (P450) reductase confers enhanced sensitivity to both tirapazamine (SR 4233) and RSU 1069. Br J Cancer 76:1338–1347
Patterson AV, Saunders MP, Chinje EC et al (1998) Enzymology of tirapazamine metabolism: a review. Anticancer Drug Des 13:541–573
Patterson AV, Ferry DM, Edmunds SJ et al (2007) Mechanism of action and preclinical antitumor activity of the novel hypoxia-activated DNA crosslinking agent PR-104. Clin Cancer Res 13:3922–3932
Patterson AV, Jaiswal J, Syddall SP et al (2009) Cellular metabolism, murine pharmacokinetics and preclinical antitumor activity of SN29966, a novel hypoxia-activated irreversible pan-HER inhibitor. Mol Cancer Ther 8:Abstract B76-doi: 10.1158/1535–7163.
Patterson AV, Smaill JB, Ackerley DF et al (2010) Bacterial nitroreductase enzymes and methods relating thereto. Provisional patent NZ586849
Pavlidis N, Aamdal S, Awada A et al (2000) Carzelesin phase II study in advanced breast, ovarian, colorectal, gastric, head and neck cancer, non-Hodgkin's lymphoma and malignant melanoma: a study of the EORTC early clinical studies group (ECSG). Cancer Chemother Pharmacol 46:167–171
Penketh PG, Shyam K, Baumann RP et al (2012) A strategy for selective O6-alkylguanine-DNA alkyltransferase depletion under hypoxic conditions. Chem Biol Drug Des 80:279–290
Peters KB, Brown JM (2002) Tirapazamine: a hypoxia-activated topoisomerase ii poison. Cancer Res 62:5248–5253
Peters L, Rischin D (2012) Elusive goal of targeting tumor hypoxia for therapeutic gain. J Clin Oncol 30:1741–1743
Peters LJ, O'Sullivan B, Giralt J et al (2010) Critical impact of radiotherapy protocol compliance and quality in the treatment of advanced head and neck cancer: Results from TROG 02.02. J Clin Oncol 28:2996–3001
Phillips RM, Hendriks HR, Peters GJ (2013) EO9 (Apaziquone): from the clinic to the laboratory and back again. Br J Pharmacol 168:11–18
Pires IM, Bencokova Z, Milani M et al (2010) Effects of acute versus chronic hypoxia on DNA damage responses and genomic instability. Cancer Res 70:925–935
Pitot HC, Reid JM, Sloan JA et al (2002) A Phase I study of bizelesin (NSC 615291) in patients with advanced solid tumors. Clin Cancer Res 8:712–717
Poon E, Harris AL, Ashcroft M (2009) Targeting the hypoxia-inducible factor (HIF) pathway in cancer. Expert Rev Mol Med 11:e26

Pruijn FB, Sturman JR, Liyanage HDS et al (2005) Extravascular transport of drugs in tumor tissue: effect of lipophilicity on diffusion of tirapazamine analogs in multicellular layer cultures. J Med Chem 48:1079–1087

Pruijn FB, Patel K, Hay MP et al (2008) Prediction of tumour tissue diffusion coefficients of hypoxia-activated prodrugs from physicochemical parameters. Aust J Chem 61:687–693

Rapisarda A, Melillo G (2012) Overcoming disappointing results with antiangiogenic therapy by targeting hypoxia. Nat Rev Clin Oncol 9:378–390

Rauth AM, Melo T, Misra V (1998) Bioreductive therapies: an overview of drugs and their mechanisms of action. Int J Radiat Oncol Biol Phys 42:755–762

Reddy SB, Williamson SK (2009) Tirapazamine: a novel agent targeting hypoxic tumor cells. Expert Opinion InvestDrugs 18:77–87

Rischin D, Peters L, Fisher R et al (2005) Tirapazamine, cisplatin, and radiation versus fluorouracil, cisplatin, and radiation in patients with locally advanced head and neck cancer: a randomized phase II trial of the Trans-Tasman Radiation Oncology Group (TROG 98.02). J Clin Oncol 23:79–87

Rischin D, Hicks RJ, Fisher R et al (2006) Prognostic significance of [18F]-misonidazole positron emission tomography-detected tumor hypoxia in patients with advanced head and neck cancer randomly assigned to chemoradiation with or without tirapazamine: a substudy of Trans-Tasman Radiation Oncology Group Study 98.02. J Clin Oncol 24:2098–2104

Rischin D, Narayan K, Oza AM et al (2010a) Phase 1 study of tirapazamine in combination with radiation and weekly cisplatin in patients with locally advanced cervical cancer. Int J Gynecol Cancer 20:827–833

Rischin D, Peters LJ, O'Sullivan B et al (2010b) Tirapazamine, cisplatin, and radiation versus cisplatin and radiation for advanced squamous cell carcinoma of the head and neck (TROG 02.02, HeadSTART): a phase III trial of the Trans-Tasman Radiation Oncology Group. J Clin Oncol 28:2989–2995

Rockwell S, Dobrucki IT, Kim EY et al (2009) Hypoxia and radiation therapy: past history, ongoing research, and future promise. Curr Mol Med 9:442–458

Rofstad EK, Mathiesen B, Henriksen K et al (2005) The tumor bed effect: increased metastatic dissemination from hypoxia-induced up-regulation of metastasis-promoting gene products. Cancer Res 65:2387–2396

Rohwer N, Cramer T (2011) Hypoxia-mediated drug resistance: Novel insights on the functional interaction of HIFs and cell death pathways. Drug Resist Updat 14:191–201

Romero-Ramirez L, Cao H, Nelson D et al (2005) XBP1 is essential for survival under hypoxic conditions and is required for tumour growth. Cancer Res 64:5943–5947

Rosario LA, O'Brien ML, Henderson CJ et al (2000) Cellular response to a glutathione S-transferase P1-1 activated prodrug. Mol Pharmacol 58:167–174

Rouschop KM, van den Beuken T, Dubois L et al (2010) The unfolded protein response protects human tumor cells during hypoxia through regulation of the autophagy genes MAP1LC3B and ATG5. J Clin Invest 120:127–141

Rouschop KM, Dubois LJ, Keulers TG et al (2013) PERK/eIF2alpha signaling protects therapy resistant hypoxic cells through induction of glutathione synthesis and protection against ROS. Proc Natl Acad Sci U S A 110:4622–4627

Sandler AB, Nemunaitis J, Denham C et al (2000) Phase III trial of gemcitabine plus cisplatin versus cisplatin alone in patients with locally advanced or metastatic non-small-cell lung cancer. (see comment). J Clin Oncol 18:122–130

Sartorelli AC (1988) Therapeutic attack of hypoxic cells of solid tumors: Presidential address. Cancer Res 48:775–778

Schwartz GH, Patnaik A, Hammond LA et al (2003) A phase I study of bizelesin, a highly potent and selective DNA-interactive agent, in patients with advanced solid malignancies. Ann Oncol 14:775–782

Secomb TW, Dewhirst MW, Pries AR (2012) Structural adaptation of normal and tumour vascular networks. Basic Clin Pharmacol Toxicol 110:63–69

Senan S, Rampling R, Graham MA et al (1997) Phase I and pharmacokinetic study of tirapazamine (SR 4233) administered every three weeks. Clin Cancer Res 3:31–38
Sennino B, McDonald DM (2012) Controlling escape from angiogenesis inhibitors. Nat Rev Cancer 12:699–709
Shepherd F, Koschel G, von Pawel J et al (2000) Comparison of Tirazone (tirapazamine) and cisplatin vs. etoposide and cisplatin in advanced non-small cell lung cancer (NSCLC): final results of the international phase III CATAPULT II Trial. Lung Cancer 29(Suppl 1):28, abstract
Shinde SS, Hay MP, Patterson AV et al (2009) Spin trapping of radicals other than the *OH radical upon reduction of the anticancer agent tirapazamine by cytochrome P450 reductase. J Am Chem Soc 131:14220–14221
Shinde SS, Maroz A, Hay MP et al (2010) Characterization of radicals formed following enzymatic reduction of 3-substituted analogues of the hypoxia-selective cytotoxin 3-amino-1,2,4-benzotriazine 1,4-dioxide (tirapazamine). J Am Chem Soc 132:2591–2599
Siim BG, Atwell GJ, Wilson WR (1994) Oxygen dependence of the cytotoxicity and metabolic activation of 4- alkylamino-5-nitroquinoline bioreductive drugs. Br J Cancer 70:596–603
Siim BG, van Zijl PL, Brown JM (1996) Tirapazamine-induced DNA damage measured using the comet assay correlates with cytotoxicity towards hypoxic tumour cells *in vitro*. Br J Cancer 73:952–960
Siim BG, Denny WA, Wilson WR (1997) Nitro reduction as an electronic switch for bioreductive drug activation. Oncol Res 9:357–369
Siim BG, Laux WT, Rutland MD et al (2000) Scintigraphic imaging of the hypoxia marker (99m)technetium-labeled 2,2'-(1,4-diaminobutane)bis(2-methyl-3-butanone) dioxime (99mTc-labeled HL-91; prognox): noninvasive detection of tumor response to the antivascular agent 5,6-dimethylxanthenone-4-acetic acid. Cancer Res 60:4582–4588
Siim BG, Pruijn FB, Sturman JR et al (2004) Selective potentiation of the hypoxic cytotoxicity of tirapazamine by its 1-N-oxide metabolite SR 4317. Cancer Res 64:736–742
Singleton DC, Li D, Bai SY et al (2007) The nitroreductase prodrug SN28343 enhances the potency of systemically administered armed oncolytic adenovirus ONYX-411(NTR). Cancer Gene Ther 14:953–967
Singleton RS, Guise CP, Ferry DM et al (2009) DNA crosslinks in human tumor cells exposed to the prodrug PR-104A: relationships to hypoxia, bioreductive metabolism and cytotoxicity. Cancer Res 69:3884–3891
Smaill JB, Jaiswal JK, Abbattista MR et al (2012) Mechanism of action of the hypoxia-activated irreversible pan-HER inhibitor SN29966. In: Chicago, IL, USA
Spiegel JF, Spear MA, Brown JM (1993) Toxicology of daily administration to mice of the radiation potentiator SR 4233 (WIN 59075). Radiother Oncol 26:79–81
Stevenson RJ, Denny WA, Ashoorzadeh A et al (2011) The effect of a bromide leaving group on the properties of nitro analogs of the duocarmycins as hypoxia-activated prodrugs and phosphate pre-prodrugs for antitumor therapy. Bioorg Med Chem 19:5989–5998
Stevenson RJ, Denny WA, Tercel M et al (2012) Nitro seco analogues of the duocarmycins containing sulfonate leaving groups as hypoxia-activated prodrugs for cancer therapy. J Med Chem 55:2780–2802
Stratford IJ, Workman P (1998) Bioreductive drugs into the next millennium. Anticancer Drug Des 13:519–528
Su J, Guise CP, Smaill JB et al (2012) NADPH: cytochrome P450 oxidoreductase (POR)-specific zinc finger nucleases demonstrate a limited role for POR in hypoxic activation of bioreductive prodrugs. Tumor microenvironment and cellular stress: Signalling, metabolism, imaging and therapeutic targets, Crete, 3–6 Oct 2012
Sun JD, Liu Q, Wang J et al (2012) Selective tumor hypoxia targeting by hypoxia-activated prodrug TH-302 inhibits tumor growth in preclinical models of cancer. Clin Cancer Res 18:758–770
Tatsumi K, Kitamura S, Narai N (1986) Reductive metabolism of aromatic nitro compounds including carcinogens by rabbit liver preparations. Cancer Res 46:1089–1093

Taylor YC, Rauth AM (1982) Oxygen tension, cellular respiration, and redox state as variables influencing the cytotoxicity of the radiosensitizer misonidazole. Radiat Res 91:104–123
Taylor M, Billiot F, Marty V et al (2012) Reversing resistance to vascular-disrupting agents by blocking late mobilization of circulating endothelial progenitor cells. Cancer Discov 2:434–449
Tercel M, Denny WA, Wilson WR (1996a) A novel nitro-substituted seco-CI—application as a reductively activated adept prodrug. Bioorg Med Chem Lett 6:2741–2744
Tercel M, Denny WA, Wilson WR (1996b) Nitrogen and sulfur analogues of the seco-co alkylating agent—synthesis and cytotoxicity. Bioorg Med Chem Lett 6:2735–2740
Tercel M, Gieseg MA, Denny WA et al (1999a) Synthesis and cytotoxicity of amino-seco-DSA: an amino analogue of the DNA alkylating agent duocarmycin SA. J Org Chem 64:5946–5953
Tercel M, Gieseg MA, Milbank JB et al (1999b) Cytotoxicity and DNA interaction of the enantiomers of 6-amino-3- (chloromethyl)-1-[(5,6,7-trimethoxyindol-2-yl)carbonyl]indo- line (amino-seco-CI-TMI). Chem Res Toxicol 12:700–706
Tercel M, Lee AE, Hogg A et al (2001) Hypoxia-selective antitumor agents. 16. Nitroarylmethyl quaternary salts as bioreductive prodrugs of the alkylating agent mechlorethamine. J Med Chem 44:3511–3522
Tercel M, Atwell GJ, Yang S et al (2009) Hypoxia-activated prodrugs: substituent effects on the properties of nitro seco-1,2,9,9a-tetrahydrocyclopropa[c]benz[e]indol-4-one (nitroCBI)prodrugs of DNA minor groove alkylating agents. J Med Chem 52:7258–7272
Tercel M, Yang S, Atwell GJ et al (2010) Hypoxic selectivity and solubility-investigating the properties of A-ring substituted nitro seco-1,2,9,9a-tetrahydrocyclopropa[c]benz[e]indol-4-ones (nitroCBIs) as hypoxia-activated prodrugs for antitumor therapy. Bioorg Med Chem 18:4997–5006
Tercel M, Atwell GJ, Yang S et al (2011a) Selective treatment of hypoxic tumor cells *in vivo*: phosphate pre-prodrugs of nitro analogues of the duocarmycins. Angew Chem 50:2606–2609
Tercel M, Lee HH, Yang S et al (2011b) Preparation and antitumour properties of the enantiomers of a hypoxia-selective nitro analogue of the duocarmycins. ChemMedChem 6:1860–1871
Tichenor MS, Boger DL (2008) Yatakemycin: total synthesis, DNA alkylation, and biological properties. Nat Prod Rep 25:220–226
Toustrup K, Sorensen BS, Nordsmark M et al (2011) Development of a hypoxia gene expression classifier with predictive impact for hypoxic modification of radiotherapy in head and neck cancer. Cancer Res 71:5923–5931
Toustrup K, Sorensen BS, Alsner J et al (2012) Hypoxia gene expression signatures as prognostic and predictive markers in head and neck radiotherapy. Semin Radiat Oncol 22:119–127
Ueda O, Kitamura S, Ohashi K et al (2003) Xanthine oxidase-catalysed metabolism of 2-nitrofluorene, a carcinogenic air pollutant, in rat skin. Drug Metab Dispos 31:367–372
Vass SO, Jarrom D, Wilson WR et al (2009) E. coli *nfsA*: an alternative nitroreductase for prodrug activation gene therapy in combination with CB1954. Br J Cancer 100:1903–1911
Vaupel P, Mayer A (2007) Hypoxia in cancer: significance and impact on clinical outcome. Cancer Metastasis Rev 26:225–239
Vaupel P, Hockel M, Mayer A (2007) Detection and characterization of tumor hypoxia using pO2 histography. Antioxid Redox Signal 9:1221–1235
Velica P, Davies NJ, Rocha PP et al (2009) Lack of functional and expression homology between human and mouse aldo-keto reductase 1C enzymes: implications for modelling human cancers. Mol Cancer 8:121–132
Vergote I, Finkler N, del Campo J et al (2009) Phase 3 randomised study of canfosfamide (Telcyta, TLK286) versus pegylated liposomal doxorubicin or topotecan as third-line therapy in patients with platinum-refractory or -resistant ovarian cancer. Eur J Cancer 45:2324–2332
von Pawel J, von Roemeling R, Gatzemeier U et al (2000) Tirapazamine plus cisplatin versus cisplatin in advanced non-small-cell lung cancer: a report of the international CATAPULT I study group. Cisplatin and tirapazamine in subjects with advanced previously untreated non-small-cell lung tumors. J Clin Oncol 18:1351–1359

Walton MI, Workman P (1993) Pharmacokinetics and bioreductive metabolism of the novel benzotriazine di-N-oxide hypoxic cell cytotoxin tirapazamine (WIN 59075; SR 4233; NSC 130181) in mice. J Pharmacol Exp Ther 265:938–947

Wang Y, Ohh M (2010) Oxygen-mediated endocytosis in cancer. J Cell Mol Med 14:496–503

Wang J, Foehrenbacher A, Su J et al (2012) The 2-nitroimidazole EF5 is a biomarker for oxidoreductases that activate bioreductive prodrug CEN-209 under hypoxia. Clin Cancer Res 18:1684–1695

Wardman P (2001) Electron transfer and oxidative stress as key factors in the design of drugs selectively active in hypoxia. Curr Med Chem 8:739–761

Weiss GJ, Infante JR, Chiorean EG et al (2011) Phase 1 study of the safety, tolerability and pharmacokinetics of TH-302, a hypoxia-activated prodrug, in patients with advanced solid malignancies. Clin Cancer Res 17:2997–3004

Whillans DW, Hunt JW (1982) A rapid-mixing comparison of the mechanisms of radiosensitization by oxygen and misonidazole in CHO cells. Radiat Res 90:126–141

Williamson SK, Crowley JJ, Lara PN Jr et al (2005) Phase III trial of paclitaxel plus carboplatin with or without tirapazamine in advanced non-small-cell lung cancer: Southwest Oncology Group Trial S0003. J Clin Oncol 23:9097–9104

Wilson WR, Ferry DM, Tercel M et al (1998) Reduction of nitroarylmethyl quaternary ammonium prodrugs of mechlorethamine by radiation. Radiat Res 149:237–245

Wilson WR, Hay MP (2011) Targeting hypoxia in cancer therapy. Nat Rev Cancer 11:393–410

Wilson WR, Pullen SM, Hogg A et al (2002) Quantitation of bystander effects in nitroreductase suicide gene therapy using three-dimensional cell cultures. Cancer Res 62:1425–1432

Wilson WR, Hicks KO, Pullen SM et al (2007) Bystander effects of bioreductive drugs: potential for exploiting pathological tumor hypoxia with dinitrobenzamide mustards. Radiat Res 167: 625–636

Wilson WR, Stribbling SM, Pruijn FB et al (2009) Nitro-chloromethylindolines: hypoxia-activated prodrugs of potent adenine *N*3 DNA minor groove alkylators. Mol Cancer Ther 8:2903–2913

Workman P, Stratford IJ (1993) The experimental development of bioreductive drugs and their role in cancer therapy. Cancer Metastasis Rev 12:73–82

Wouters BG, Brown JM (1997) Cells at intermediate oxygen levels can be more important than the "hypoxic fraction" in determining tumor response to fractionated radiotherapy. Radiat Res 147:541–550

Wouters BG, Delahoussaye YM, Evans JW et al (2001) Mitochondrial dysfunction after aerobic exposure to the hypoxic cytotoxin tirapazamine. Cancer Res 61:145–152

Xia Y, Choi HK, Lee K (2012) Recent advances in hypoxia-inducible factor (HIF)-1 inhibitors. Eur J Med Chem 49:24–40

Yin J, Glaser R, Gates KS (2012) On the reaction mechanism of tirapazamine reduction chemistry: unimolecular N-OH homolysis, stepwise dehydration, or triazene ring-opening. Chem Res Toxicol 25:634–645

Zeman EM, Brown JM, Lemmon MJ et al (1986) SR-4233: a new bioreductive agent with high selective toxicity for hypoxic mammalian cells. Int J Radiat Oncol Biol Phys 12:1239–1242

Zeman EM, Hirst VK, Lemmon MJ et al (1988) Enhancement of radiation-induced tumor cell killing by the hypoxic cell toxin SR 4233. Radiother Oncol 12:209–218

Zeman EM, Baker MA, Lemmon MJ et al (1989) Structure-activity relationships for benzotriazine di-N-oxides. Int J Radiat Oncol Biol Phys 16:977–981

Zhu R, Liu MC, Luo MZ et al (2011) 4-nitrobenzyloxycarbonyl derivatives of O(6)-benzylguanine as hypoxia-activated prodrug inhibitors of O(6)-alkylguanine-DNA alkyltransferase (AGT), which produces resistance to agents targeting the O-6 position of DNA guanine. J Med Chem 54: 7720–7728

Chapter 14
Exploiting "Hif Addiction" For Cancer Therapy

Nicole D. Fer and Annamaria Rapisarda

Abstract Intratumor hypoxia is a common and unifying characteristic of solid tumors that are associated with altered cellular metabolism, an invasive and metastatic phenotype, as well as resistance to radiation and chemotherapy. The discovery of hypoxia-inducible factor-1 (HIF-1), a transcription factor critically involved in cellular responses to hypoxia and tumor progression, has provided a molecular target of the hypoxic tumor microenvironment that has been exploited for the development of novel cancer therapeutics. However, HIF-1 expression in human cancers is focal and heterogeneous, strongly arguing against the clinical efficacy of single agent HIF-1 inhibitors and opening the door to combinational therapies. A therapy-induced hypoxic tumor microenvironment, and the subsequent activation of HIF-1-dependent pathways, may create an "HIF addiction" that will represent the perfect setting to fully exploit HIF-1 inhibitors. Interestingly, therapies targeting tumor vasculature have been shown to increase intratumor hypoxia in some instances and hypoxia has been suggested to mediate several mechanisms of resistance to antiangiogenic therapies, providing a rationale for combination therapies with HIF-1 inhibitors. In this chapter, we discuss the potential of exploiting a therapy-induced "HIF addiction" by using known HIF inhibitors to improve clinical efficacy keeping in mind that patient selection is essential.

Keywords Hypoxia · HIF-1 · Tumor microenvironment · Antiangiogenic agents · Vascular disrupting agents · Combination therapies · Cancer therapeutics

14.1 Introduction

The disparity between increased oxygen consumption by proliferating tumor cells and decreased delivery from an aberrant vasculature generates areas of low oxygen concentration (hypoxia) within solid tumors. Hypoxia is implicated in several

A. Rapisarda (✉) · N. D. Fer
Tumor Microenvironment Laboratory, SAIC-Frederick Inc.,
Frederick National Laboratory for Cancer Research, Frederick,
MD 21702-1201, USA
e-mail: rapisardaa@mail.nih.gov

G. Melillo (ed.), *Hypoxia and Cancer,* Cancer Drug Discovery and Development, DOI 10.1007/978-1-4614-9167-5_14,

aspects of cancer cell biology, including maintenance of the cancer stem cell microenvironment and the "stemness" phenotype; selection of clonal populations resistant to apoptosis; downregulation of proapoptotic genes; induction of autophagy and the anaerobic metabolic switch (Majmundar et al. 2010; Mazumdar et al. 2009; Graeber et al. 1996; Erler et al. 2004; Rouschop et al. 2010; Semenza 2011). Moreover, intratumor hypoxia contributes to genome instability by suppressing DNA-repair pathways; producing reactive oxygen species; and contributing to resistance to radiotherapy and chemotherapy (Bristow and Hill 2008; Guzy et al. 2005; Semenza 2003).

The identification of hypoxia-inducible factor (HIF-1) as the main mediator of hypoxia-dependent response has provided a selective molecular target associated with hypoxia and has led to the discovery and development of targeted therapies exploiting the hypoxic tumor microenvironment (Semenza 2003; Melillo 2007). HIF-1 is a heterodimeric transcription factor consisting of a constitutively expressed β-subunit (HIF-1β) and of an oxygen-regulated α-subunit (HIF-1α or HIF-2α; Semenza 2008). HIF-α expression has been shown in the majority of human cancers and their metastases, where they are associated with patient mortality and poor response to treatment (Semenza 2010). However, even conceptually, HIF-targeting approaches used as single agents may not achieve clinically relevant results due to fundamental aspects of HIF biology: the focal and heterogeneous expression of HIF-1α in solid tumors and the reliance of cancer cells on HIF-1-independent pathways in oxygenated areas. Moreover, it has been shown that the effects of HIF-1α downregulation by an inducible short hairpin RNA (shRNA)system is transient and tumor stage-dependent in xenografts and that tumors quickly adapted to the loss of HIF-1 and well-established large tumors were resistant to HIF-1 inhibition (Li et al. 2006). Indeed, HIF inhibitors may be more effective in a context in which survival of tumor cells may be largely and more consistently relying on HIF-1-dependent mechanisms, such as in combination with therapies that induce a hypoxic stress of the tumor microenvironment, therefore creating a HIF addiction.

An example may be vascular regression caused by the administration of the antiangiogenic and vascular-targeting agents, which has been shown to be associated with a concomitant increase in intratumor hypoxia (Rapisarda et al. 2009a; Rapisarda and Melillo 2009b; Rapisarda and Melillo 2012; Tozer et al. 2005). To date five drugs targeting the vascular endothelial growth factor (VEGF) pathway are approved by the Food and Drug Adminstration (FDA), alone or in combination with chemotherapy; however, despite the promise that has developed from these studies with some patients leading to improved progression-free survival when compared to standard therapy, a significant number of patients do not respond to antiangiogenic therapy or quickly develop resistance to it (Ebos and Kerbel 2011). Several mechanisms of resistance to this therapy have been proposed; one being the induction of intratumor hypoxia which occurs as a result of vessel pruning and extensive inhibition of neoangiogenesis (Rapisarda 2012).

In this chapter, we discuss the potential of exploiting the intratumor hypoxic microenvironment or "HIF addiction" produced as an adaptive response to antiangiogenic agents and vascular-disrupting agents (VDAs) by using known HIF inhibitors, which could generate a powerful combination and thus increase the effectiveness of the overall treatment. In addition, we explore the importance of patient selection based on the therapy being administered and the relative HIF expression.

14.2 Hypoxic Tumor Microenvironment

The tumor microenvironment is a critical component of tumor progression and subsequent response to treatment. Each tumor type is a heterogeneous collection of genetic and epigenetic alterations; however, a unifying characteristic of human cancers is the presence of hypoxic regions, a critical feature of the tumor microenvironment (Brown and Giaccia 1998; Harris 2002; Denko 2008; Pouyssegur et al. 2006; Vaupel 2010). Hypoxia is involved in a number of characteristic events in cancer cell biology and several biological functions associated with tumor progression, including maintenance of the cancer stem cell microenvironment and the "stemness" phenotype; selection of clonal populations that are resistant to apoptosis; downregulation of proapoptotic genes; induction of autophagy; the anaerobic metabolic switch; angiogenesis and vasculogenesis; the epithelial–mesenchymal transition; invasiveness and metastasis; and suppression of immune surveillance (Majmundar et al. 2010; Mazumdar et al. 2009; Graeber et al. 1996; Erler et al. 2004; Rouschop et al. 2010; Semenza 2011; Hirota and Semenza 2006; Li et al. 2006; Jiang et al. 2011; DeBrock et al. 2011; Yotnda et al. 2010). In addition, intratumor hypoxia is a major contributing factor to poor patient outcomes in a variety of tumor types by contributing to genome instability through the suppression of DNA-repair pathways and the generation of reactive oxygen species as well as by contributing to resistance to radiotherapy and chemotherapy (Semenza 2007; Bristow and Hill 2008; Guzy et al. 2005). Indeed, hypoxia is a common prognosticator of poor patient response in a variety of tumor types.

Tumor cells adjust to a hypoxic microenvironment by modifying their transcriptional program. HIF is a family of transcription factors that mediates the majority of hypoxia-dependent responses responsible for the adaptation to hypoxia in tumor cells. HIF-1 consists of a constitutively expressed β-subunit (HIF-1β) and an oxygen-regulated α-subunit (HIF-1α and HIF-2α) that are responsible for the activation of a number of target genes related to the pathways mentioned previously (Semenza 2008; Qing and Simon 2009). Increased HIF-α expression is also connected with a variety of genetic alterations that inactivate tumor suppressor genes or activate oncogenic pathways, as well as activation of receptor tyrosine kinase-dependent signaling pathways (Semenza 2003). Overexpression of HIF-1α and or HIF-2α has been shown in a variety of human cancers, such as lung, breast, and prostate cancer, and is directly associated with poor patient prognosis and treatment failure (Semenza 2010).

The overwhelming evidence supporting the key role of hypoxia in the tumor microenvironment is consistent with the idea that hypoxia is an attractive target for cancer therapy and, therefore, several strategies to target hypoxic cells and hypoxia-dependent responses are being explored in clinical trials (Onnis et al. 2009; Wilson and Hay 2011).

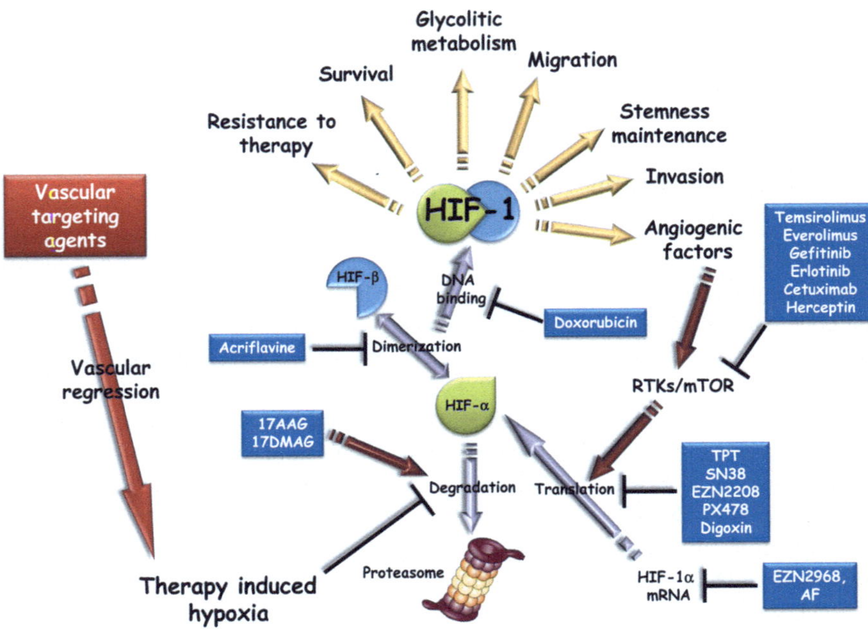

Fig. 14.1 Schematic representation of the interaction between the HIF-1 pathway and vascular targeting agents. HIF-1 inhibitors, with potential mechanisms of action, are indicated

14.2.1 HIF Inhibitors

Regardless of the challenges associated with targeting transcription factors, numerous attempts have been made to identify HIF-1 inhibitors based mostly on high throughput screening assays designed to identify small molecules affecting HIF-1 expression and/or transcriptional activity (Rapisarda et al. 2002; Melillo 2007). Small molecules have been identified that target various points of HIF regulation (Fig. 14.1 and Table 14.1, the table is intended to be illustrative rather than comprehensive), but whether and how these agents might be exploited in the clinical setting remains poorly defined. HIF-α expression is mainly controlled at the level of protein degradation and translation. HIF-1α is a client protein of heat shock protein 90 (Hsp90); the Hsp90 inhibitors 17-N-allylamino-17-demethoxygeldanamycin (17AAG) and17-dimethylaminoethylamino-17-demethoxygeldanamycin (17 DMAG) have been shown to increase HIF 1α protein degradation (Isaacs et al. 2002; Mabjeesh et al. 2002; Porter et al. 2009). Several agents reduce HIF-1α translation, including the topoisomerase I inhibitors topotecan and EZN-2208 (a pegylated form of SN38), the small molecule PX 478, cardiac glycosides as well as the mammalian target of rapamycin (mTOR) inhibitors temsirolimus and everolimus, and several receptor tyrosine kinase inhibitors (Rapisarda et al. 2004a; Rapisarda et al. 2004b; Sapra et al. 2011; Koh et al. 2008; Zhang et al. 2008; Wan et al. 2006; Del Bufalo et al. 2006).

Table 14.1 HIF inhibitors in clinical development with potential mechanism of action

Agent	Target	Mechanism of HIF inhibition	Reference
Temsirolimus, Everolimus	mTOR	HIF-1α translation	(Del Bufalo et al. 2006; Wan et al. 2006)
Silibilin	mTOR/p70S6K/4E-BP1	HIF-1α translation	(Garcia-Maceira et al. 2009)
Nelfinavir, Amprenavir	AKT	HIF-1α translation	(Pore et al. 2006a)
Topotecan, EZN-2208	Topoisomerase 1	HIF-1α translation	(Rapisarda et al. 2004a; Rapisarda et al. 2004b; Sapra et al. 2008)
2ME2, ENMD1198	Microtubule	HIF-1α translation	(Mabjeesh et al. 2003; LaVallee et al. 2008)
Gefitinib, Erlotinib, Cetuximab	EGFR	HIF-1α translation	(Pore et al. 2006b; Luwor et al. 2005)
Trastuzumab	Her2/Neu	HIF-1α translation	(Koukourakis et al. 2003)
Imatinib	Bcr-Abl, c-Kit, PDGFR-β	HIF-1α translation	(Litz and Krystal 2006)
PX-478	Unknown	HIF-1α translation	(Koh et al. 2008)
Digoxin (cardiac glycosides)	Unknown	HIF-1α translation	(Zhang et al. 2008)
17AAG 17DMAG	Hsp90	HIF-1α degradation	(Mabjeesh et al. 2002; Porter et al. 2009; Issacs et al. 2002)
SCH66336	Farnesyltransferase	HIF-1α degradation	(Han et al. 2005)
EZN-2968	HIF-1α mRNA	HIF-1α mRNA degradation	(Greenberger et al. 2008)
FK228, Vorinostat	HDAC	HIF-1α degradation, transcriptional activity	(Kong et al. 2006; Lee et al. 2003)
Amphotericin B	FIH-1	HIF-1α degradation, transcriptional activity	(Yeo et al. 2006)
Acriflavine	HIF-α PAS-B domain	HIF-α/HIF-β dimerization	(Lee et al. 2009b)
Echinomycin, Doxorubicin	DNA binding	HIF-1α DNA binding	(Kong et al. 2005; Lee et al. 2009a)
Bortezomib	Proteasome	HIF-1α transcriptional activity	(Kaluz et al. 2006)
Aminoflavone	AhR	HIF-1α mRNA processing	(Terzuli et al. 2010)

Additional strategies to inhibit HIF-1 include targeting its messenger RNA (mRNA) using either antisense oligonucleotides (EZN 2968) or the aryl hydrocarbon receptor (AhR) ligand aminoflavone by inhibiting HIF-α and HIF-β dimerization and by inhibiting HIF-1 DNA-binding activity using echinomycin or anthracyclines (such as doxorubicin and daunorubicin; Greenberger et al. 2008; Terzuoli et al. 2010; Kong et al. 2005; Lee et al. 2009a; Lee et al. 2009b).

It is interesting to point out that the majority of known HIF-1 inhibitors target signaling pathways that not only are involved in HIF-1 regulation but also affect distinct cellular functions hampering the pharmacological validation of HIF-1 as a therapeutic target. Indeed, the anticancer activity observed with these agents might be dependent on inhibition of multiple targets, including but not limited to HIF-1.

14.3 HIF-1 Inhibitors as Single Agents or Combination Strategies: Creating the Addiction

Despite the fact that HIF-1 inhibition has been convincingly reported in several experimental tumor models, expression of HIF-1α in the majority of solid tumors is focal and heterogeneous making the validation of target inhibition in the clinical setting challenging and strongly arguing for the development of combinatorial therapy with agents that increase the dependency of tumor cells on HIF 1 (HIF-1 addiction). Interestingly, vascular regression, caused by the administration of the antiangiogenic and vascular-targeting agents, has been shown to be associated with a concomitant increase in intratumor hypoxia, providing the rationale for combination therapies with HIF-1 inhibitors (Rapisarda and Melillo 2009b; Tozer et al. 2005; Rapisarda and Melillo 2012).

14.3.1 Antiangiogenic Therapies and Vascular Disruptive Agents

Regardless of the expectations, clinical trials using antiangiogenic therapies emphasized that a significant number of patients do not respond to these agents or quickly develop resistance to it, strongly suggesting that combination strategies aimed at hampering resistance mechanism will improve the therapeutic outcome (Rapisarda and Melillo 2012). However, the consequences of antiangiogenic therapies on the tumor microenvironment are poorly understood and a current matter of discussion with at least two contrasting hypotheses having been suggested that are not mutually exclusive and could be cancer and context dependent: vascular normalization and vascular regression (Jain 2005; Goel et al. 2011; Kerbel and Folkman 2002). Clinical data to support the vascular normalization notion suggest that after careful use of the antiangiogenic therapy, there is a potential therapeutic window of tumor normalization that can be exploited to improve chemotherapy delivery and increased effectiveness (Ribatti 2011). Conversely, it has been recently shown that treatment with sunitinib and bevacizumab increases cancer stem cells due to the generation of intratumor hypoxia, thus potentially limiting treatment efficacy (Conely et al. 2012). In addition, clinical studies have found that increased intratumor hypoxia occurs in patients with non-small-cell lung cancer and hepatocellular carcinoma after treatment with bevacizumab and a recent study has reported the acquisition of an invasive phenotype in patients with glioblastoma who have developed multifocal recurrence of tumors during the course of antiangiogenic therapy (Smit et al. 2011; Yopp et al. 2011; Narayana et al. 2009).

In contrast to antiangiogenic therapies, VDAs increase vascular permeability and decrease blood flow, sometimes inducing loss of vascular endothelial cells by apoptosis. Such processes can lead to a catastrophic vascular failure and the development of massive intratumoral hypoxia and necrosis. However, a viable rim of tumor tissue invariably remains from which tumor regrowth rapidly resumes (Tozer et al. 2005; Siemann and Horsman 2009). Interestingly, it has been shown that treatment with VDAs causes acute systemic mobilization and homing of bone marrow-derived circulating endothelial precursor (CEP) cells that could promote tumor regrowth in xenografts (Shaked et al. 2006). This effect is coupled with a rapid increase in circulating plasma VEGF, stromal-derived factor-1 (SDF-1), and granulocyte colony-stimulating factor (G-CSF) levels (Shaked et al. 2009). Evidence for rapid elevations in circulating plasma G-CSF, VEGF, and SDF-1 was also observed in patients treated with the VDA combretastatin-A4 phosphate (Farace et al. 2007). Notably, G-CSF has been shown to induce HIF-1α (Liu et al. 2010), and VEGF and SDF-1 are HIF-dependent genes (Hirota and Semenza 2006), highlighting the importance of HIF-dependent responses following the treatment with VDAs and providing the rationale for combination therapies with HIF inhibitors.

14.4 Therapy-Induced Intratumor Hypoxia: Exploiting Hif Addiction

Treatment with antiangiogenic agents and VDAs can result in the development of extensive "therapy-induced" hypoxic regions within tumors and activation of cancer cells survival and escape mechanism through the expression of HIF-1-dependent genes. HIF-addicted cancer cells are, conceptually, more sensitive to HIF inhibitors; therefore, the efficacy of these agents might be improved in the presence of therapy-induced intratumor hypoxia. Moreover, given that HIF-1-dependent genes have important roles in multiple mechanisms implicated in the resistance to anti-VEGF therapies, combining antiangiogenic and HIF-1-targeting agents might provide good clinical results. Indeed, it has been shown that bevacizumab increases intratumor hypoxia and HIF-dependent genes in glioma xenografts and that inhibition of HIF-1α by low-dose daily topotecan in a hypoxic-stressed tumor microenvironment resulted in a more pronounced antitumor effect, relative to either agent alone (Rapisarda et al. 2009c). The effects on tumor growth were associated with significantly decreased HIF-1 transcriptional activity and reduced tumor cell proliferation, consistent with the hypothesis that targeting HIF-1α activity may abrogate compensatory pathways required for cancer cell survival (Rapisarda et al. 2009c). Additional studies have also shown that genetic disruption of both HIF-1α and HIF-2α expression in colon cancer xenografts improves tumor response to sunitinib (a multitargeted receptor tyrosine kinase inhibitor; Burkitt et al. 2009). This effect was mediated by a marked decrease in tumor angiogenesis and perfusion, through inhibition of multiple pro-angiogenic factors, as well as a marked decrease in tumor cell proliferation. In addition, the combination of bevacizumab and irinotecan (another topoisomerase I inhibitor that also inhibits HIF-1) has shown clinical benefit in glioblastoma patients with a 6-month overall survival of 62–77 % (Vredenburgh et al. 2007; Chen et al. 2007). Moreover, in 2012, Moroney et al. reported in a phase I study where bevacizumab and temsirolimus were combined with liposomal doxorubicin in patients with advanced malignancies, a median duration of partial response/complete response for all patients who achieved tumor responses of 9 months, thus warranting further clinical evaluation of the potential for this regimen. More recently, a study examining hypoxia-mediated sorafenib resistance and the use of EF24 (compound structurally similar to curcumin) to overcome the resistance in hepatocellular carcinoma provided evidence of a synergistic effect against metastasis both *in vivo* and *in vitro* using this combination (Liang et al. 2013). It has also been suggested that the inhibition of HIF-2α and not HIF-1α might be more beneficial in several tumor types (Raval et al. 2005; Franovic et al. 2009); however, the possibility to exploit HIF-2α as a target is hampered by the current lack of HIF-2α-specific inhibitors.

Using antiangiogenic agents or VDAs as single-agent therapies has proven to be somewhat disappointing, while HIF inhibitors are, conceptually, not going to be good candidates for single-agent treatment; however, the use of antiangiogenic agents or VDAs to create a therapy-induced stressed hypoxic microenvironment could result in an HIF addiction that will sensitize cancer cells to HIF inhibitors.

14.4.1 Patient Selection: Key to Successful Combination Therapy

As with the development of any cancer treatment, patient selection is essential in designing successful combination therapies. With the concepts presented in this chapter, the need for biomarkers that can direct the selection of patients for whom combined targeting of intratumor hypoxia and vasculature will be most beneficial is crucial. Currently, the most common detection system used to identify hypoxic regions in patient samples is binding of chemical hypoxia markers, such as pimonidazole or EF5, partnered with the expression and colocalization of HIF-1α and hypoxia target genes, such as CAIX (Jankovic et al. 2006). Other methods of measuring hypoxia in patients include imaging techniques such as positron emission tomography (PET) scans, blood oxygenation level-dependent (BOLD) imaging, or dynamic contrast-enhanced (DCE) m*agnetic resonance imaging* (MRI). Using these methods in combination with endogenous markers of hypoxia-responsive pathways provide a better idea in selecting patients for specific therapies. In a target-driven pilot study recently conducted using oral topotecan in patients with advanced solid tumors, Kummar et al. (2011) selected patients based on the expression of HIF-1α in archival tissue ($\geq 10\,\%$ of cells showing positive nuclear staining for HIF-1α) and then used the expression of HIF-1α and HIF-1 target genes, VEGF and glucose transporter-1 (GLUT-1), along with DCE-MRI, to assess HIF-1α modulation in response to treatment and changes in tumor blood flow and permeability. Indeed, it was reported that topotecan decreased HIF-1α expression, HIF-dependent genes and tumor blood flow, and permeability in advanced tumors (Kummar et al. 2011).

Several biomarkers that may predict responses to antiangiogenic therapies have also been explored, but with mixed results; therefore providing that identifying valid biomarkers to successfully select patients will be very challenging to develop (Murukesh et al. 2010). Further studies are required to better define the patient population that may benefit from hypoxia-targeted therapy in combination with antiangiogenic agents or possibly VDAs by pairing the techniques described above, as well as future biomarker development.

14.5 Conclusion

In the complex biology of human cancers, the role of HIF-1 in resistance to therapy and tumor progression has been extensively validated; therefore, HIF-1 inhibition appears as a logical therapeutic strategy. Many small molecule inhibitors of HIF-1 have been described, several of which have entered early clinical development. However, we still have limited understanding of when, and to what extent, the inhibition of HIF-1 in cancer patients may be effective. Based on HIF-1 biology, it is conceivable that induction of an HIF-addictive phenotype in cancer cells will be the ideal setting to exploit HIF inhibitors. Therefore, we discussed the potential of combination therapies where the creation of an HIF-addicted tumor microenvironment by using antiangiogenic agents or VDAs can be exploited and thus provide an enhanced clinical response. The main challenge in using this type of combinational therapy will be the development of robust predictive biomarkers for patient selection.

Acknowledgements The authors would like to thank Beverly A. Teicher for helpful discussions. This project has been funded in whole or in part with Federal funds from the National Cancer Institute, National Institutes of Health, under Contract No. HHSN261200800001E. The content of this publication does not necessarily reflect the views or policies of the Department of Health and Human Services, nor does mention of trade names, commercial products, or organizations imply endorsement by the US government. This research was supported, in part, by the Developmental Therapeutics Program in the Division of Cancer Treatment and Diagnosis of the National Cancer Institute, NIH.

References

Bristow RG, Hill RP (2008) Hypoxia and metabolism. Hypoxia, DNA repair and genetic instability. Nat Rev Cancer 8:180

Brown JM, Giaccia AJ (1998) The unique physiology of solid tumors: opportunities (and problems) for cancer therapy. Cancer Res 58:1408

Burkitt K et al (2009) Targeting both HIF-1 and HIF-2 in human colon cancer cells improves tumor response to sunitinib treatment. Mol Cancer Ther 8:1148
Chen W et al (2007) Predicting treatment response of malignant gliomas to bevacizumab and irinotecan by imaging proliferation with [18F] fluorothymidine positron emission tomography: a pilot study. J Clin Oncol 25:4714
Conely S et al (2012) Antiangiogenic agents increase breast cancer stem cells via the generation of tumor hypoxia. Proc Natl Acad Sci USA 109:2784
DeBrock K et al (2011) Antiangiogenic therapy, hypoxia and metastasis: risky liaisons, or not? Nat Rev Clin Oncol 8:393
Del Bufalo D et al (2006) Antiangiogenic potential of the mammalian target of rapamycin inhibitor temsirolimus. Cancer Res 66:5549
Denko NC (2008) HIF1 and glucose metabolism in the solid tumour. Nat Rev Cancer 8:705
Ebos JM, Kerbel RS (2011) Antiangiogenic therapy: impact on invasion, disease progression, and metastasis. Nat Rev Clin Oncol 8:210
Erler JT et al (2004) Hypoxia-mediated down-regulation of Bid and Bax in tumors occurs via hypoxia-inducible factor 1-dependent and –independent mechanisms and contributes to drug resistance. Mol Cell Biol 24:2875
Farace F et al (2007) Vascular disrupting therapy-induced mobilization of circulating endothelial progenitor cells. Ann Oncol 18:1421
Franovic A et al (2009) Human cancers converge at the HIF-2alpha oncogenic axis. Proc Natl Acad Sci USA 106:21306
García-Maceira P, Mateo J (2009) Silibinin inhibits hypoxia-inducible factor- 1alpha and mTOR/p70S6K/4E-BP1 signalling pathway in human cervical and hepatoma cancer cells: implications for anticancer therapy. Oncogene 28:313
Goel S et al (2011) Normalization of the vasculature for treatment of cancer and other diseases. Physiol Rev 91:1071
Graeber TG et al (1996) Hypoxia-mediated selection of cells with diminished apoptotic potential in solid tumors. Nature 379:88
Greenberger LM et al (2008) A RNA antagonist of hypoxia-inducible factor-1alpha, EZN-2968, inhibits tumor cell growth. Mol Cancer Ther 7:3598
Guzy RD et al (2005) Mitochondrial complex III is required for hypoxia-induced ROS production and cellular oxygen sensing. Cell Metab 1:401
Han JY et al (2005) Hypoxia-inducible factor-1alpha and antiangiogenic activity of farnesyltransferase inhibitor SCH66336 in human areodigestive tract cancer. J Natl Cancer Inst 97:1272
Harris AL (2002) Hypoxia—a key regulatory factor in tumour growth. Nat Rev Cancer 2:38
Hirota K, Semenza GL (2006) Regulation of angiogenesis by hypoxia-inducible factor 1. Crit Rev Oncol Hematol 59:15
Isaacs JS et al (2002) Hsp90 regulates a von Hippel Lindau-independent hypoxia-inducible factor-1 alpha-degradative pathway. J Biol Chem 277:29936
Jain RK (2005) Antiangiogenic therapy for cancer: current and emerging concepts. Oncology 19:7
Jankovic B et al (2006) Comparison between pimonidazole binding, oxygen electrode measurements, and expression of endogenous hypoxia markers in cancer of the uterine cervix. Cytometry B Clin Cytom 70:45
Jiang J et al (2011) EMT: a new vision of hypoxia promoting cancer progression. Cancer Biol Ther 11:714
Kaluz S, Kaluzova M, Stanbridge EJ (2006) Proteasomal inhibition attenuates transcriptional activity of hypoxia-inducible factor 1 (HIF-1) via specific effect on the HIF-1α C-terminal activation domain. Mol Cell Biol 26:5895
Kerbel RS, Folkman J (2002) Clinical translation of angiogenesis inhibitors. Nat Rev Cancer 2:727
Koh et al (2008) Molecular mechanisms for the activity of PX-478, an antitumor inhibitor of the hypoxia-inducible factor-1alpha. Mol Cancer Ther 7:90

Kong D et al (2005) Echinomycin, a small-molecule inhibitor of hypoxia-inducible factor-1 DNA-binding activity. Cancer Res 65:9047

Kong X et al (2006) Histone deacetylase inhibitors induce VHL and ubiquitin-independent proteasomal degradation of hypoxia-inducible factor 1 alpha. Mol Cell Biol 26:2019

Koukourakis MI et al (2003) The effect of trastuzumab/docatexel combination on breast cancer angiogenesis: dichotomus effect predictable by the HIF1 alpha/VEGF pre-treatment status? Anticancer Res 23:1673

Kummar S et al (2011) Multihistology, target-driven pilot trial of oral topotecan as an inhibitor of hypoxia-inducible factor-1α in advanced solid tumors. Clin Cancer Res 17:5123

LaVallee TM et al (2008) Significant antitumor activity *in vivo* following treatment with the microtubule agent ENMD-1198. Mol Cancer Ther 7:1472

Lee K et al (2009a) Anthracycline chemotherapy inhibits HIF-1 transcriptional activity and tumor-induced mobilization of circulating angiogenic cells. Proc Natl Acad Sci USA 106:2353

Lee K et al (2009b) Acriflavine inhibits HIF-1 dimerization, tumor growth, and vascularization. Proc Natl Acad Sci U S A 106:17910

Lee MY et al (2003) Inhibition of hypoxia-induced angiogenesis by FK228, a specific histone deacetylase inhibitor, via suppression of HIF-1alpha activity. Biochem Biophys Res Commun 300:241

Li B et al (2006) VEGF and PIGF promote adult vaculogenesis by enhancing EPC recruitment and vessel formation at the site of tumor neovascularisation. FASEB J 20:1495

Liang Y et al (2013) Hypoxia-mediated sorafenib resistance can be overcome by EF24 through VHL-dependent HIF-1α inhibition in hepatocellular carcinoma. Hepatology. doi: 10.1002/hep.26224

Litz J, Krystal GW (2006) Imatinib inhibits c-Kit-induced hypoxia-inducible factor-1activity and vascular endothelial growth factor expression in small cell lung cancer cells. Mol Cancer Ther 5:1415

Liu SP et al (2010) Granulocyte colony-stimulating factor activating HIF-1alpha acts synergistically with erythropoietin to promote tissue plasticity. PLoS One 5:10093

Luwor RB et al (2005) The antiepidermal growth factor receptor monoclonal antibody cetuximab/C225 reduces hypoxia-inducible factor-1 alpha, leading to transcriptional inhibition of vascular endothelial growth factor expression. Oncogene 24:4433

Mabjeesh NJ et al (2002) Geldanamycin induces degradation of hypoxia-inducible factor 1alpha protein via proteasome pathway in prostate cancer cells. Cancer Res 62:2478

Mabjeesh NJ et al (2003) 2ME2 inhibits tumor growth and angiogenesis by disrupting mircotubules and dysregulating HIF. Cancer Cell 3:363

Majumundar AJ, Wong WJ, Simon MC (2010) Hypoxia-inducible factors and the response to hypoxic stress. Mol Cell 40:294

Mazumdar J, Dondeti V, Simon MC (2009) Hypoxia-inducible factors in stem cells and cancer. Eur J Nucl Med Mol Imaging 36:1674

Melillo G (2007) Hypoxia-inducible factor 1 inhibitors. Methods Enzymol 435:385

Moroney J et al (2012) Phase I study of the antiangiogenic antibody bevacizumab and the mTOR/hypoxia-inducible factor inhibitor temsirolimus combined with liposomal doxorubicin: tolerance and biological activity. Clin Cancer Res 18:5796

Murukesh N, Dive C, Jayson GC (2010) Biomarkers of angiogenesis and their role in the development of VEGF inhibitors. Br J Cancer 102:8

Narayana A et al (2009) Antiangiogenic therapy using bevacizumab in recurrent high-grade glioma: impact on local control and patient survival. J Neurosurg 110:173

Onnis B et al (2009) Development of HIF-1 inhibitors for cancer therapy. J Cell Mol Med 13:2780

Pore N et al (2006a) HIV protease inhibitors decrease VEGF/HIF-1expression and angiogenesis in glioblastoma cells. Neoplasia 8:889

Pore N et al (2006b) EGFR tyrosine kinase inhibitors decrease VEGF expression by both hypoxia-inducible factor (HIF)-1-independent and HIF-1-dependent mechanisms. Cancer Res 66:3197

Porter JR et al (2009) Ansamycin inhibitors of Hsp90: nature's prototype for anti-chaperone therapy. Curr Top Med Chem 9:1386
Pouysségur J, Dayan F, Mazure NM (2006) Hypoxia signalling in cancer and approaches to enforce tumour regression. Nature 441:2006
Qing G, Simon MC (2009) Hypoxia inducible factor-2alpha: a critical mediator of aggressive tumor phenotypes. Curr Opin Genet Dev 19:60
Rapisarda A, Melillo G (2009b) Role of the hypoxic tumor microenvironment in the resistance to anti-angiogenic therapies. Drug Resist Updat 12:74
Rapisarda A, Melillo G (2012) Overcoming disappointing results with antiangiogenic therapy by targeting hypoxia. Nat Rev Clin Oncol 9:378
Rapisarda A et al (2002) Identification of small molecule inhibitors of hypoxia-inducible factor 1: mechanism and therapeutic implications. Cancer res 62: 4316
Rapisarda A et al (2004a) Topoisomerase I-mediated inhibition of hypoxia-inducible factor 1: mechanism and therapeutic implications. Cancer Res 64:1475
Rapisarda A et al (2004b) Schedule-dependent inhibition of hypoxia-inducible factor 1alpha accumulation, angiogenesis, and tumor growth by topotecan in U251-HRE glioblastoma xenografts. Cancer Res 64:6845
Rapisarda A et al (2009a) Antiangiogenic agents and HIF-1 inhibitors meet at the crossroads. Cell Cycle 8:4040
Rapisarda A et al (2009c) Increased antitumor activity of bevacizumab in combination with hypoxia inducible factor-1 inhibition. Mol Cancer Ther 8:1867
Raval RR et al (2005) Contrasting properties of hypoxia-inducible factor 1 (HIF-1) and HIF-2 in von Hippel-Lindau-associated renal cell carcinoma. Mol Cell Biol 25:5675
Ribatti D (2011) Editorial: overview of antiangiogenic therapies in haematological malignancies. Curr Cancer Drug Targets 11:995
Rouschop KM et al (2010) The unfolded protein response protects human tumor cells during hypoxia through regulation of the autophagy genes MAP1LC3B and ATG5. J Clin Invest 120:127
Sapra P et al (2008) Novel delivery of SN38 markedly inhibits tumor growth in xenografts, including a camptothecin-11-refractory model. Clin Cancer Res 14:1888
Sapra P et al (2011) Potent and sustained inhibition of HIF-1α and downstream genes by a polyethyleneglycol-SN38 conjugate, EZN-2208, results in anti-angiogenic effects. Angiogenesis 14:245
Semenza GL (2003) Targeting HIF-1 for cancer therapy. Nat Rev Cancer 3:721
Semenza GL (2007) Hypoxia and cancer. Cancer Metastasis Rev 26:223
Semenza GL (2008) Hypoxia-inducible factor 1 and cancer pathogenesis. IUBMB Life 60(9):591–7
Semenza GL (2010) Defining the role of hypoxia-inducible factor 1 in cancer biology and therapeutics. Oncogene 29:625
Semenza GL (2011) Regulation of metabolism by hypoxia-inducible factor 1. Cold Sping Harb Symp Quant Biol
Shaked Y et al (2006) Therapy-induced acute recruitment of circulating endothelial progenitor cells to tumors. Science 313:1785
Shaked Y et al (2009) Contribution of granulocyte colony-stimulating factor to the acute mobilization of endothelial precursor cells by vascular disrupting agents. Cancer Res 69:7524
Siemann DW, Horsman MR (2009) Vascular targeted therapies in oncology. Cell Tissue Res 335:241
Smit EF et al (2011) Effects of the antiangiogenic drug bevacizumab on tumor perfusion and drug delivery of 11C-labeled docetaxel in patients with non-small cell lung cancer (NSCLC): implications for scheduling of antiangiogenic agents. In: Abstracts of the 2011 ASCO annual meeting. Illinois, 3–7 June, Chicago
Terzuoli E et al (2010) Aminoflavone, a ligand of the aryl hydrocarbon receptor, inhibits HIF-1alpha expression in an AhR-independent fashion. Cancer Res 70:6837
Tozer GM, Kanthou C, Baguley BC (2005) Disrupting tumour blood vessels. Nat Rev Cancer 5:423
Vaupel P (2010) Metabolic microenvironment of tumor cells: a key factor in malignant progression. Exp Oncol 32:125

Vredenburgh JJ et al (2007) Bevacizumab plus irinotecan in recurrent glioblastoma multiforme. J Clin Oncol 25:4722

Wan X et al (2006) CCI-779 inhibits rhabdomyosarcoma xenograft growth by an antiangiogenic mechanism linked to the targeting of mTOR/Hif-1alpha/VEGF signalling. Neoplasia 8:394

Wilson WR, Hay MP (2011) Targeting hypoxia in cancer therapy. Nat Rev Cancer 11:393

Yeo EJ et al (2006) Amphotericin B blunts erythropoietic response to hypoxia by reinforcing FIH-mediated repression of HIF-1. Blood 107:916

Yopp AC et al (2011) Antiangiogenic therapy for primary liver cancer: correlation of changes in dynamic contrast-enhanced magnetic resonance imaging with tissue hypoxia markers and clinical response. Ann Surg Oncol 18:2192

Yotnda P et al (2010) Hypoxic tumors and their effect on immune cells and cancer therapy. Methods Mol Biol 651:1

Zhang H et al (2008) Digoxin and other cardiac glycosides inhibit HIF-1alpha synthesis and block tumor growth. Proc Natl Acad Sci U S A 105:19579

Chapter 15
Clinical Exploitation of Hypoxia

Geoff S Higgins and Adrian L Harris

Abstract This chapter will review previous attempts at overcoming tumour hypoxia using treatments such as oxygen mimetics. The potential to exploit the hypoxic environment for synthetic lethality approaches will be explored along with ongoing attempts to alter tumour hypoxia by manipulating the tumour vasculature.

The possible causes for the limited efficacy of previous attempts at overcoming tumour hypoxia will be discussed along with suggestions as to how future studies may be able to better utilise existing approaches. These include improved biomarker identification of patients likely to benefit from hypoxic manipulation using techniques such as gene expression profiling and functional imaging in order to enrich trial populations with appropriate patients. The use of dynamic monitoring of changes in hypoxia occurring throughout treatment will be needed to exploit microenvironmental changes induced by therapy.

Keywords Radiotherapy trials · Hypoxia metagene · Hypoxia imaging · Contextual synthetic lethality · Vascular normalisation

15.1 Hypoxia Modification in Radiotherapy

Tumour hypoxia is well recognised as conferring an adverse prognosis in many different tumour types. Although this poor prognosis can be seen in patients treated with either surgery or chemotherapy, attempts to overcome tumour hypoxia have predominantly been performed in combination with radiotherapy treatment.

The understanding that cells are much more sensitive to ionising radiation in the presence of oxygen than in its absence has been recognised for over 50

G. S. Higgins (✉)
Gray Institute for Radiation Oncology and Biology, Department of Oncology, Oxford University, Oxford, UK
e-mail: Geoffrey.higgins@oncology.ox.ac.uk

A. L. Harris
Molecular Oncology Laboratories, Department of Oncology, Oxford University, Oxford, UK
e-mail: patoadrianharris@oncology.ox.ac.uk; aharris.lab@imm.ox.ac.uk

G. Melillo (ed.), *Hypoxia and Cancer*, Cancer Drug Discovery and Development, DOI 10.1007/978-1-4614-9167-5_15,

years (Gray et al. 1953; Wright and Howard-Flanders 1957; Palcic et al. 1982). Multiple approaches have been adopted to try to overcome this. Initial attempts at improving tumour hypoxia involved patients breathing high-oxygen content gas under hyperbaric oxygen (HBO) conditions. A Cochrane review of all clinical trials performed with HBO concluded that there was some evidence that HBO improved local control rates in cervix and head and neck cancer, but that these benefits may only arise with unconventional fractionation schemes. Importantly, they found that HBO was associated with significant adverse effects, including oxygen toxic seizures and severe tissue radiation injury (Bennett et al. 2012).

Other strategies have previously been employed to try to increase tumour oxygen levels. One particular approach has been to try to increase the oxygen that can be delivered to the tumour by increasing patient's haemoglobin levels. Anaemia is recognised as conferring a poor prognosis in many different tumour types (Gordon 2002). Although the association between anaemia and poor prognosis may simply reflect more advanced disease, previous studies have suggested that anaemia is an independent prognostic factor in patients receiving radiotherapy treatment (Warde et al. 1998). As a result of this, multiple small studies have investigated whether reversing anaemia during radiotherapy treatment results in an improvement in clinical outcomes. These studies have been unable to consistently show that treating anaemia with blood transfusions results in an improvement in patient outcomes. Although some studies show an improvement in outcomes, other studies have shown that blood transfusions may have a positively detrimental effect on outcome, possibly by inducing mild immunosuppression (Varlotto and Stevenson 2005). Alternative means of correcting anaemia without using blood transfusions have also been pursued. Most notably, several trials have investigated the possibility of improving anaemia and therefore response to radiotherapy, by treating head and neck cancer patients with erythropoietin β. A meta-analysis of these trials suggested that the addition of erythropoietin to radiotherapy treatment caused a statistically significant detrimental effect on overall survival (Lambin et al. 2009). The mechanism for this is not clear, but since erythropoietin receptors are known to be expressed on tumour cells, it is possible that erythropoietin treatment causes adverse clinical outcomes by inducing tumour cell progression and proliferation (Henke et al. 2006).

A more promising approach to overcoming radioresistance associated with tumour hypoxia has been to use 'oxygen mimetics'. The origins to this area of research began with the identification that electron-affinic compounds sensitised hypoxic cells to radiation (Adams and Cooke 1969).

Nimorazole has been the most widely investigated of these drugs and was the subject of a large phase III study (DAHANCA 5) in Danish patients with cancer of the pharynx or supraglottic larynx. This study showed that the combined treatment of nimorazole and radiotherapy was associated with a statistically significant improvement in locoregional control compared with patients treated with radiotherapy alone (Overgaard et al. 1998). Despite these trial results, nimorazole is rarely used in routine practice outwith Denmark (Wardman 2007). The reasons for this are unclear but may be due to the trial failing to show that the addition of nimorazole resulted in an improvement in patient survival (Overgaard et al. 1998). No other hypoxic radiosensitiser is currently in routine, widespread use.

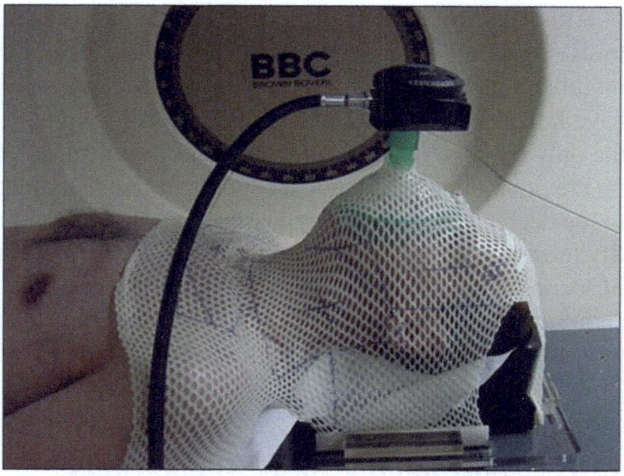

Fig. 15.1 Carbogen treatment. A patient receiving radiotherapy treatment for a head and neck cancer is immobilised with perspex mesh in order for the radiotherapy to be delivered accurately. Carbogen is breathed via an anaesthetic mask placed directly over the patient's mouth. The carbogen is breathed for 5 minutes prior to radiotherapy and throughout the duration of the radiotherapy treatment. (Picture courtesy of Professor Hans Kaanders)

It has been suggested that the disappointing results associated with attempts to correct tumour hypoxia are due to the fact that these treatments act exclusively against chronic 'diffusion-limited' hypoxia (Kaanders et al. 2002a). More recent studies have used the vitamin B3 derivative, nicotinamide, in an attempt to reverse 'perfusion-limited' acute hypoxia, by preventing transient fluctuations in tumour blood flow (Horsman 1995). In particular, nicotinamide has been used as part of the accelerated radiation, carbogen and nicotinamide (ARCON) studies. Carbogen (95 % oxygen, 5 % carbon dioxide) breathing commences 5 minutes prior to radiotherapy and is continued throughout (Fig. 15.1). Nicotinamide is administered orally, 90 minutes prior to radiotherapy. The rationale for this approach is that nicotinamide can reduce the intermittent closure of blood vessels, thereby decreasing acute hypoxia, whilst carbogen breathing can increase the oxygen partial pressure in tissues and reduce chronic hypoxia.

Phase II studies investigating the effects of ARCON have previously been undertaken in head and neck (Kaanders et al. 2002a) and bladder cancer patients (Hoskin et al. 2009). A recent randomised phase III trial found that ARCON improved 2 year and 5 year regional control rates in patients with laryngeal carcinoma compared with accelerated radiotherapy (AR) alone (88 and 86 % for AR vs. 95 and 93 % for ARCON, respectively; HR 0.46; 95 %CI, 0.22–0.97; $p = 0.04$) (Janssens et al. 2012). The study, however, did not show significant improvements in laryngeal preservation, local control, disease-free survival, overall survival, treatment-related toxicity or functional outcomes (Janssens et al. 2012). A translational sub-study of 79 patients defined the primary tumours as either well oxygenated or hypoxic based on the fraction of cells that had pimonidazole staining using a cut-off value of 2.6 %. Regional control in the group with a high hypoxic fraction was significantly improved with ARCON compared with AR treatment (100 % vs. 55 %, respectively; $p = 0.01$). There was no difference observed in the group with a low hypoxic fraction (96 % for ARCON vs. 92 % for AR; $p = 0.7$). This finding may be central to the limited benefits observed in most hypoxic modification studies.

Although a recent meta-analysis of squamous cell carcinoma of the head and neck trials suggested a small but statistically significant survival benefit from hypoxic modification (Overgaard 2011), the improvements seen from hypoxic sensitisation have failed to live up to expectations. To date, and only in Denmark, nimorazole is the only hypoxic radiosensitiser in routine clinical use (Overgaard et al. 2003).

Previous studies have not accounted for the heterogeneity that exists with regard to the degree and extent of hypoxia within tumours and have not attempted to identify those individuals with significantly hypoxic tumours who have the greatest risk of locoregional failure. Future trials may be able to demonstrate larger, clinically significant benefits from hypoxic modification if they are sufficiently enriched with those individuals who may potentially benefit the most from hypoxic modification.

15.2 Non-Imaging-Based Biomarkers

Recent research has focussed on the use of several different biomarkers as a means of identifying tumour hypoxia without having to obtain direct oxygen concentration measurements. A common hypoxia metagene is one such approach that has recently been described (Buffa et al. 2010). Established hypoxia-regulated genes (seeds) were used to generate hypoxia co-expression cancer networks from which core high-connectivity genes were used to derive a 51-gene signature. This common hypoxia metagene was shown to have strong prognostic significance in several independent data sets and showed greater prognostic potential compared with separate hypoxia signatures.

The common hypoxia metagene appears to have significant overlap with a gene expression classifier, subsequently published by a separate group, to identify hypoxia in head and neck tumours (Toustrup et al. 2011). This 15-gene signature was shown to predict response to nimorazole treatment (Toustrup et al. 2012). Three hundred and twenty-three biopsy samples from patients recruited to the DAHANCA 5 study were classified as either ‘less’ or ‘more’ hypoxic. As previously discussed, DAHANCA 5 demonstrated superior locoregional control in head and neck cancer patients treated with radiotherapy who were randomised to nimorazole treatment rather than placebo alone (Overgaard et al. 1998). One hundred and fourteen of the samples (35 %) were classified as being ‘more’ hypoxic. These patients had a significant, pronounced benefit from hypoxic modification with nimorazole compared with placebo in terms of locoregional control (5-year control rates of 49 % vs. 18 %, respectively; $p = 0.001$) and disease-specific survival (48 % vs 30 %; $p = 0.04$). Patients with ‘less’ hypoxic tumours had no significant benefit from nimorazole treatment and their clinical outcomes were similar to those patients with ‘more’ hypoxic tumours treated with nimorazole.

Osteopontin (OPN), a secretory phosphoglycoprotein, has previously been shown to be inversely correlated with the expression of the Von Hippel Lindau protein which modulates the expression of hypoxia-inducible factor-1 alpha (HIF-1 alpha)

(Le et al. 2003). OPN is also inversely associated with tumour oxygenation in head and neck cancers as measured by Eppendorf probes. A retrospective analysis of the DAHANCA 5 data showed that OPN could be used to identify patients who would benefit from hypoxic modification with nimorazole (Overgaard et al. 2005). Plasma OPN, however, was not shown to confer a poor prognosis or predict response to the bioreductive agent tirapazamine in the large TROG 02.02 phase III trial and the clinical significance of this biomarker is therefore currently unclear (Lim et al. 2012).

Carbonic anhydrase 9 (CA-9) is an extracellular, membrane-tethered, enzyme that regulates intracellular and extracellular pH in hypoxia (De Simone and Supuran 2010). It is one of the most inducible genes by the HIF-1 alpha pathway. Inhibition of this enzyme by drugs, antibodies or small interfering RNA (siRNA) reduces survival in hypoxia (Supuran 2012; Dubois et al. 2011; Perez-Sayans et al. 2012). As it can be readily assessed by immunohistochemistry, and is stable compared to HIFs, it has been widely investigated as a biomarker for hypoxia. Its messenger RNA (mRNA) expression is included in the signatures described above. Although clearly linked to poor prognosis as an independent factor in many tumour types, the data relating expression assessed by immunohistochemistry to response in randomised trials of radiotherapy modulation is limited (Beasley et al. 2001; Chia et al. 2001; Hoskin et al. 2003; De Schutter et al. 2005; Brennan et al. 2006; Sherwood et al. 2007; Kyndi et al. 2008; Schrijvers et al. 2008). Although some studies report that high CA-9 is associated with poor outcome after radiotherapy, others show no relationship (Eriksen and Overgaard 2007). Yet others have found that CA-9 is associated with a good outcome following ARCON treatment (Jonathan et al. 2006).

CA-9 and glut-1 expression have been correlated with an injectable hypoxia marker, pimonidazole, and all have been correlated with Eppendorf electrode measurements of hypoxia in cancer of the cervix (Kaanders et al. 2002b; Loncaster et al. 2001; Vordermark and Brown 2003). In the 13 reports reviewed (Vordermark and Brown 2003), HIF-1 alpha overexpression was associated with poor outcome in nine studies. CA-9 was an adverse factor in cervix, head and neck and lung cancer in some reports but not in other head and neck cancer reports. Not one predicted for poor survival in lung, cervix and colorectal cancer emphasising the heterogeneity and difficulty in scoring histology reports, the marked variability and lack of concordance between studies.

In head and neck cancers, at least three studies have been carried out to assess the benefit of continuous hyperfractionated accelerated radiotherapy (CHART) versus standard radiotherapy. They have shown that several biomarkers and pathways could predict for a better effect of CHART, although overall the study yielded little significant difference. These included a low proliferation index (Buffa et al. 2004), high epidermal growth factor (EGF) receptor expression (Bentzen et al. 2005) and low expression of CA-9 or HIF-2 (Buffa et al. 2004; Koukourakis et al. 2006). Thus, defining subgroups using three different markers suggested improvement in benefit, which was missed in the main study. However, these have not been further replicated in studies so far but are rational candidates for further analysis.

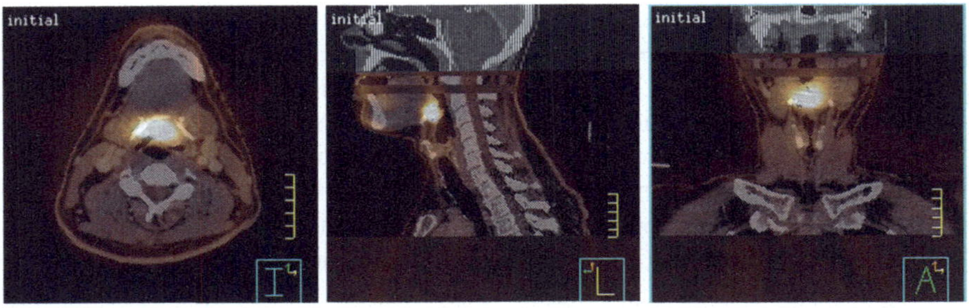

Fig. 15.2 Co-registered CT and ^{18}F-Misonidazole PET images in transaxial, sagittal and coronal projections from a patient with a base of tongue tumour allow the identification of hypoxic subregions suitable for radiotherapy 'boost' planning. (Reproduced from (Hendrickson et al. 2011) with permission from Elsevier)

15.3 Imaging Biomarkers of Hypoxia

As discussed in previous chapters, significant advances in imaging tumour hypoxia have been made in recent years. Several positron emission tomography (PET) tracers such as ^{18}F-Misonidazole, ^{18}F-fluoroazomycin arabinoside (^{18}F-FAZA) and Cu(II)-diacetyl-bis(N^4-methylthiosemicarbazone) (Cu-ATSM) appear to have significant potential clinical use and their use may become increasingly widespread (Chitneni et al. 2011). In the absence of trial data, directly comparing the efficacy of these markers, it is currently difficult to ascertain which marker has the greatest clinical value. At present, ^{18}F-Misonidazole remains the most extensively studied hypoxia-imaging agent.

Technical developments such as the implementation of intensity-modulated radiotherapy treatment (IMRT) have enabled the delivery of a non-uniform dose of radiation to tumour volumes. This, combined with the progress made in hypoxia imaging, means that co-registered PET/computed tomography (CT) images can be used to define hypoxic subregions suitable for radiotherapy 'boost' planning (Fig. 15.2). Delivering a higher dose of radiation to radioresistant, hypoxic areas may be able to improve local tumour control. Although acute hypoxia may result in some variability in repeat ^{18}F-Misonidazole PET scans, a significant proportion of patients can be shown to have reproducible regions of hypoxia on PET imaging suggestive of chronic hypoxia (Nehmeh et al. 2008). It has been shown that it is feasible to deliver radiotherapy boosts to such regions (Fig. 15.3).

The prognostic use of ^{18}F-Misonidazole PET has been explored previously. A small 40-patient study, in patients with either non-small cell lung cancer or head and neck cancer, has previously suggested that pre-radiotherapy ^{18}F-Misonidazole PET imaging could predict response to radiotherapy treatment based on the ratio of retained ^{18}F-Misonidazole within the tumour compared to that seen in surrounding normal tissues (Eschmann et al. 2005).

The Trans-Tasman Radiation Oncology Group found that patients with hypoxic tumours treated with a chemoradiation regimen containing the hypoxic cytotoxin tirapazamine had lower rates of locoregional failure than similar patients treated with conventional chemoradiotherapy treatment (Rischin et al. 2006). This raises

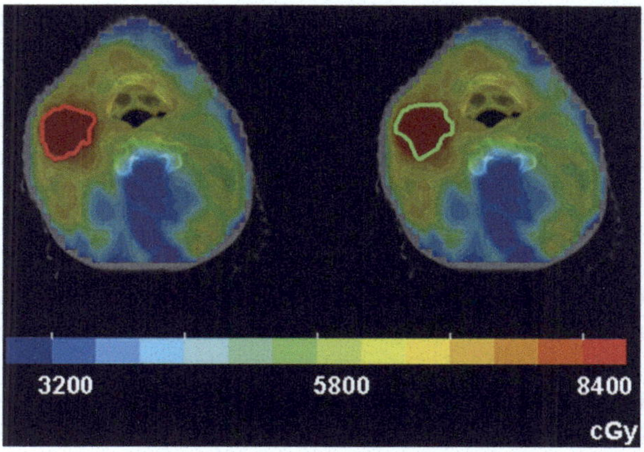

Fig. 15.3 IMRT dose distributions shown in colour wash superimposed on the hypoxic volumes from two sequential ^{18}F-Misonidazole PET scans showing reproducible regions of chronic hypoxia. The hypoxic target volume was defined on the first scan (outlined in *red*). A boost dose of 14 Gy was delivered to this region in addition to a prescription dose of 70 Gy. The hypoxic volume identified on the second ^{18}F-Misonidazole PET scan (outlined in *green*) was well covered by the 84-Gy region. (Reproduced from (Lin et al. 2008) with permission from Elsevier)

the possibility that molecular imaging of baseline hypoxia may be an effective way of identifying those patients likely to benefit from treatments aimed at modifying tumour hypoxia. Ongoing research will need to confirm the benefits of such an approach and identify whether the initially hypoxic regions should be boosted or whether the radiotherapy plan should be adapted to changes in hypoxic volumes.

Future clinical trials attempting to overcome tumour hypoxia need to better utilise appropriate biomarkers such as the hypoxia metagene or hypoxic-imaging studies. This will enable better selection of patients likely to benefit from hypoxic manipulation and exclude those patients who are unlikely to derive any benefit.

15.4 Evolving Therapeutic Approaches to Overcome Hypoxia

To date, attempts at overcoming tumour hypoxia have had only modest effects on clinical outcomes. In the near future it is likely that clinical improvements are most likely to come from exploiting either 'contextual' synthetic lethality or 'normalisation' of the tumour vasculature.

15.5 'Contextual' Synthetic Lethality

It has become increasingly apparent that hypoxia mediates several changes in the DNA repair capacity of tumour cells. Mismatch repair (MMR) is an essential pathway that repairs errors which arise in newly synthesised DNA and is required in order to maintain genomic integrity (Jiricny 2006). Hypoxia downregulates the expression of several key components of MMR such as MLH1 (Mihaylova et al. 2003), MSH2 (Shahrzad et al. 2005) and MSH6 (Rodriguez-Jimenez 2008) which could potentially render hypoxic cells more sensitive to certain cytotoxic agents which cause DNA damage that is repaired by MMR. These include topoisomerase inhibitors such as camptothecin and etoposide (Jacob et al. 2001) and certain alkylating agents such as CCNU and mitomycin C (Fiumicino et al. 2000).

Probably the greatest therapeutically exploitable defect induced by hypoxia relates to a reduction in homologous recombination (HR) efficiency. DNA double-strand breaks (DSBs) are critical DNA lesions generated when the two complementary strands of the DNA double helix are broken simultaneously, at sites that are sufficiently close to one another that base pairing and chromatin structure are insufficient to keep the two DNA ends juxtaposed. The critical importance of being able to repair DSBs has resulted in the evolution of two distinct DNA repair processes: non-homologous end-joining (NHEJ) and HR. NHEJ is a potentially error-prone mechanism for the repair of DSBs throughout the cell cycle without the requirement for a homologous template (Helleday et al. 2007). HR uses homologous DNA from a sister chromatid as a template to perform error-free repair of DNA DSBs. The need for a sister chromatid means that HR occurs only in G2 and S phases of the cell cycle (Helleday et al. 2007).

The term synthetic lethality has been used to describe the observation that mutations in two genes leads to cell death, whereas a mutation in only one of these genes does not. Homozygous loss of breast cancer related gene (BRCA) renders cells exquisitely sensitive to the single-strand break (SSB) repair protein PARP1 (Farmer et al. 2005; Bryant et al. 2005), whilst heterozygous loss of BRCA has no effect on sensitivity to PARP. This phenomenon occurs because inhibition of PARP1 results in increased SSBs which are converted to DSBs when they encounter a replication fork. Since these collapsed forks require HR for repair (Arnaudeau et al. 2001), deficiency in HR due to BRCA loss results in synthetic lethality (Bryant et al. 2005). This increase in sensitivity to PARP inhibition has subsequently been shown to occur following the loss of many other proteins involved in HR such as RAD51, RAD54, FANCD2, FANCA, CHK1 and CHK2 (McCabe et al. 2006).

Under hypoxic conditions, HR efficiency is significantly reduced due to decreased expression of key genes in the HR pathway such as *Rad51* (Bindra et al. 2004), *BRCA1* and *BRCA2* (Meng et al. 2005; Chan et al. 2008). In keeping with this reduction in HR efficiency, hypoxic tumour cells have been shown to have increased sensitivity to PARP inhibitor treatment both in vitro (Hegan et al. 2010) and in vivo (Chan et al. 2010). This contextual synthetic lethality may enable PARP inhibitors, in combination with either chemotherapy or radiotherapy to target hypoxic tumour cells which would otherwise be resistant to treatment.

It is also possible that CA-9 inhibition may increase the sensitivity of hypoxic tumour cells to radiotherapy. CA-9 inhibition should increase extracellular pH and decrease intracellular pH, thus producing additional toxicity to cells surviving the hypoxic environment to make them more susceptible to radiation. It has recently been shown that low molecular weight inhibitors of CA-9 can potentiate radiotherapy in tumour cells expressing CA-9. CA-9 knockdown was shown to cause radiosensitisation, but this could not be further modified by the use of a CA-9 inhibitor (Dubois et al. 2011). Since CA-9 can be detected by antibody scanning and as there are several blocking antibodies and inhibitors in preclinical development, this is a target for the near future. It may be that using biomarkers on tumour sections or imaging would then be relevant to the modulation (Askoxylakis et al. 2010).

Tumour radiosensitisation may also be possible if induction of the hypoxia response could be modified. For example, HIF-2 alpha regulates many enzymes involved in protection against free radical stress (Hervouet et al. 2008). Since much of the DNA damage caused by radiotherapy treatment is mediated by free radical formation, HIF-2 alpha may result in tumour radioresistance. Although this has not been studied directly, it is well recognised that renal cancers, which have high HIF-2 alpha expression, are resistant to radiotherapy compared to many other tumour types.

15.6 Vascular 'Normalisation' Using Direct Inhibitors of Angiogenesis

Tumour blood vessels are typically grossly abnormal anatomically and physiologically. The vasculature has a chaotic, tortuous and hyperpermeable structure (Nagy et al. 2009) which results in increased interstitial fluid pressure as well as tumour hypoxia.

Anti-angiogenesis treatments, such as the vascular endothelial growth factor (VEGF) monoclonal antibody bevacizumab, were developed on the rationale that disrupting new blood vessel formation would restrict tumours' access to nutrients thereby rendering them dormant. Although, clinical trial data has shown that bevacizumab monotherapy in the metastastic setting is associated with objective response rates as low as 3.3 % (Giantonio et al. 2007), bevacizumab has Food and Drug Administration (FDA) approval for combined use with chemotherapy treatment in metastatic colorectal cancer and metastatic non-small cell lung cancer (Shojaei 2012). These decisions were based on two key studies which found that the addition of bevacizumab was associated with a median overall survival benefit of 2.1 months in metastatic colorectal and non-small cell lung cancer (Giantonio et al. 2007; Sandler et al. 2006). It has been argued that the synergistic effect seen with a combination of bevacizumab and chemotherapy therapy may result from 'vascular normalisation' (Goel et al. 2011). It is proposed that, rather than destroying tumour vessels, bevacizumab may work by restoring the normal structure and function of the tumour vasculature, thereby normalising the tumour microenvironment and reducing hypoxia. Such changes may improve the tumour delivery of systemically administered anti-cancer agents as well as increasing tumour radiosensitivity. This 'normalisation' may occur due to anti-VEGF treatment restoring the balance between pro- and anti-angiogenic signalling. Prolonged or excessive doses of bevacizumab are held to cause vascular regression thereby reversing the process of vascular normalisation and resulting in only a transient period of vascular normalisation.

The vascular normalisation hypothesis is supported by data from mouse xenograft models in which bevacizumab was shown to increase pericyte coverage of vessels and decrease tumour microvessel density during the first 4 days of treatment (Dings et al. 2007). In keeping with these findings, tumour oxygen levels increased significantly during this period. However, these physiological changes were transient and from treatment day 5 tumour oxygenation in treated mice decreased significantly below that of control mice.

Tumour growth delay was increased when the tumours were irradiated during the period of normalisation compared with other periods. It is currently unclear whether the dosing and scheduling of bevacizumab can be adjusted in patients in such a way as to prolong the duration of vascular normalisation in order to increase the efficacy of treatments such as chemotherapy and radiotherapy.

15.7 Vascular 'Normalisation' Using Indirect Inhibitors of Angiogenesis

In contrast to the transient normalisation seen with bevacizumab, disruption of oncogenic signalling has recently been shown to cause a more durable, pronounced normalisation. One such example of this is the EGFR-Ras-PI3K-AKT pathway which is activated in many cancers and has been targeted by several different clinical compounds. Pre-clinical data using mice bearing human tumour xenografts has shown that inhibitors against EGFR (gefitinib), Ras (L778,123), PI3K (PI-103) and AKT (nelfinavir) induce a significant and prolonged increase in tumour blood flow and tumour perfusion and a marked decrease in tumour hypoxia (Qayum et al. 2009). These changes were observed at drug concentrations below that required to induce tumour regression and therefore appear to be independent of any anti-proliferative effect.

More recent work with the selective PI3K inhibitor BKM120 and the dual PI3K/mTOR inhibitor BEZ235 has shown similar effects on the tumour microenvironment. Xenograft models were used to show that these drugs induced structural changes in the vasculature and reductions in tumour hypoxia which persisted for over 7 weeks from the start of drug treatment. These alterations in the tumour microenvironment were associated with significant delays in tumour growth delay experiments following exposure to single treatment irradiation (Fokas et al. 2012). These agents are currently being tested in clinical trials (Bendell et al. 2012). It is likely that, if these findings can be reproduced in the clinical setting, these drugs will be able to significantly increase the effectiveness of both chemotherapy and radiotherapy treatment.

15.8 Angiogenesis Inhibition as Opposed to Normalisation

VEGF and other angiogenic factors are well recognised as being induced by hypoxia. High angiogenesis could relate to well-oxygenated tumours, except that high tumour vascularity is often torturous, leaky and ineffective. Low angiogenic vasculature is also associated with hypoxia and radioresistance and it is likely that moderate- to well-differentiated vasculature will be most important, with vasculature of either extreme associated with poor outcome, as has been reported in some more correlative studies (Koukourakis et al. 2013).

Currently there are over 40 ongoing studies combining radiotherapy with anti-angiogenic drugs and in many cases also chemotherapy (Kleibeuker et al. 2012). The vast majority of phase I and phase II studies are therefore unlikely to answer the critical questions about combination therapy efficacy and are really developing safety aspects. Although a priori reducing angiogenesis might be thought to increase hypoxia, it will depend on the state of the vasculature as described above. Furthermore, our group has shown that radiotherapy appears to sensitise the vasculature and it has been well described that there can be an inhibitory effect of radiation through Smad signalling (Imaizumi et al. 2010).

We treated xenografts with an antibody that blocks human and mouse delta-like 4 signalling, so this would affect both tumour vasculature and potentially stem cells in the tumour. This showed dramatic enhancement of radiation therapy. Histological examination revealed major areas of necrosis developing with the combined therapy (Liu et al. 2011). We therefore hypothesised that the radiosensitisation was due to the combination of radiotherapy with an inhibitory drug which blocked the critical pathway required for vascular stabilisation.

A wide variety of interactions with angiogenic inhibitors have been reported. Some inhibitors show effects when given with radiotherapy. Others, such as the VEGF receptor inhibitor PTK787, show effects only when given after radiotherapy (Kleibeuker et al. 2012). This is a complex area and there are many problems to be solved, particularly with the drugs, the schedule and the trial design. It is clear that the positive effects of the combinations may warrant translation into the clinic, particularly the delta-like 4 inhibition.

15.9 Dynamic Monitoring of Therapy

It is well recognised that radiation can produce rapid changes in gene expression and, by killing tumour cells, may affect packing density and re-vascularisation. The tumour is therefore changing its biology continuously during radiation therapy. However, those changes may induce resistance, changes in the microenvironment and include an induction in angiogenic signalling pathways and activation of AKT.

In order to know which patients and which tumours should be modulated, it will be important to know hypoxia at baseline and during treatment. To fully evaluate tumour hypoxia, a combination marker including measurement of the hypoxic volume by imaging and assessment of hypoxia gene response by gene profiling will probably be required.

The individualisation and neutralisation of hypoxia for synergistic toxicity or to overcome its biological effects are likely to depend not just on the baseline assessment, but on early response analysis. Evidence to support this early response comes from several sources, but particularly in breast cancer where the effects of aromatase inhibitor treatment on tumour gene expression have been correlated with long-term outcome in a series of elderly patients given aromatase inhibitor treatment before surgery (Mackay et al. 2007; Ellis et al. 2008). The baseline gene expression profiles, or Ki67, were not predictive of outcome at 3 years, but the gene profiles and changing Ki67 at just 2 weeks after the start of treatment correlated with long-term outcome at 3 years. Similar demonstration of the early vascular heterogeneity of tumour response to anti-VEGF therapy can be detected at 2 weeks, associated with major changes in gene expression (Mehta et al. 2011). This provides a model for the type of studies that we now need to do on hypoxia and its modulation.

A key principle of radiotherapy is that hypoxic tumour regions show improved oxygenation over time. As better-oxygenated cells respond to treatment and die, the less well-oxygenated cells then receive a better oxygen supply because less oxygen is consumed by those cells initially closer to the blood vessels. This phenomenon of 'tumour rexoygenation' can be assessed by undertaking serial ^{18}F-Misonidazole PET scans throughout the treatment. Emerging evidence suggests that ^{18}F-Misonidazole PET undertaken 1 or 2 weeks into the treatment may be a promising way of identifying patients whose tumours contain residual areas of hypoxia, and who may derive most benefit from hypoxia modification or dose-escalated treatment (Zips et al. 2012).

Tumours that show evidence of reoxygenation are likely to be more sensitive to radiotherapy, but those that do not are likely to benefit from hypoxia modulation or radiation dose escalation. As improved hypoxia imaging becomes available, it will be critical to integrate these not only into conventional radiation schedules but also into modified fractionation treatments and combination therapies.

15.10 Summary

Previous attempts at modifying tumour hypoxia have met with limited success. This is at least partly due to an inability to focus such interventions upon the population of patients most likely to benefit. The use of validated biomarkers to enrich hypoxia modification trials with those patients most likely to benefit is likely to demonstrate more clinically meaningful benefits for both established and novel interventions.

References

Adams GE, Cooke MS (1969) Electron-affinic sensitization. I. A structural basis for chemical radiosensitizers in bacteria. Int J Radiat Biol Relat Stud Phys Chem Med 15(5):457–471

Arnaudeau C, Lundin C, Helleday T (2001) DNA double-strand breaks associated with replication forks are predominantly repaired by homologous recombination involving an exchange mechanism in mammalian cells. J Mol Biol 307(5):1235–1245

Askoxylakis V, Garcia-Boy R, Rana S, Kramer S, Hebling U, Mier W, Altmann A, Markert A, Debus J, Haberkorn U (2010) A new peptide ligand for targeting human carbonic anhydrase IX, identified through the phage display technology. PloS One 5(12):e15962

Beasley NJ, Wykoff CC, Watson PH, Leek R, Turley H, Gatter K, Pastorek J, Cox GJ, Ratcliffe P, Harris AL (2001) Carbonic anhydrase IX, an endogenous hypoxia marker, expression in head and neck squamous cell carcinoma and its relationship to hypoxia, necrosis, and microvessel density. Cancer Res 61(13):5262–5267

Bendell JC, Rodon J, Burris HA, Jonge M de, Verweij J, Birle D, Demanse D, De Buck SS, Ru QC, Peters M, Goldbrunner M, Baselga J (2012) Phase I, dose-escalation study of BKM120, an oral pan-Class I PI3K inhibitor, in patients with advanced solid tumors. Journal of clinical oncology: official journal of the American Society of. Clin Oncol 30(3):282–290

Bennett MH, Feldmeier J, Smee R, Milross C (2012) Hyperbaric oxygenation for tumour sensitisation to radiotherapy. Cochrane Database Syst Rev 4:CD005007

Bentzen SM, Atasoy BM, Daley FM, Dische S, Richman PI, Saunders MI, Trott KR, Wilson GD (2005) Epidermal growth factor receptor expression in pretreatment biopsies from head and neck squamous cell carcinoma as a predictive factor for a benefit from accelerated radiation therapy in a randomized controlled trial. Journal of clinical oncology: official journal of the American Society of. Clin Oncol 23(24):5560–5567

Bindra RS, Schaffer PJ, Meng A, Woo J, Maseide K, Roth ME, Lizardi P, Hedley DW, Bristow RG, Glazer PM (2004) Down-regulation of Rad51 and decreased homologous recombination in hypoxic cancer cells. Mol Cell Biol 24(19):8504–8518

Brennan DJ, Jirstrom K, Kronblad A, Millikan RC, Landberg G, Duffy MJ, Ryden L, Gallagher WM, O'Brien SL (2006) CA IX is an independent prognostic marker in premenopausal breast cancer patients with one to three positive lymph nodes and a putative marker of radiation resistance. Clinical cancer research: an official journal of the American Association for. Cancer Res 12(21):6421–6431

Bryant HE, Schultz N, Thomas HD, Parker KM, Flower D, Lopez E, Kyle S, Meuth M, Curtin NJ, Helleday T (2005) Specific killing of BRCA2-deficient tumours with inhibitors of poly(ADP-ribose) polymerase. Nature 434(7035):913–917

Buffa FM, Bentzen SM, Daley FM, Dische S, Saunders MI, Richman PI, Wilson GD (2004) Molecular marker profiles predict locoregional control of head and neck squamous cell carcinoma in a randomized trial of continuous hyperfractionated accelerated radiotherapy. Clinical cancer research: an official journal of the American Association for. Cancer Res 10(11):3745–3754

Buffa FM, Harris AL, West CM, Miller CJ (2010) Large meta-analysis of multiple cancers reveals a common, compact and highly prognostic hypoxia metagene. Br J Cancer 102(2):428–435

Chan N, Koritzinsky M, Zhao H, Bindra R, Glazer PM, Powell S, Belmaaza A, Wouters B, Bristow RG (2008) Chronic hypoxia decreases synthesis of homologous recombination proteins to offset chemoresistance and radioresistance. Cancer Res 68(2):605–614

Chan N, Pires IM, Bencokova Z, Coackley C, Luoto KR, Bhogal N, Lakshman M, Gottipati P, Oliver FJ, Helleday T, Hammond EM, Bristow RG (2010) Contextual synthetic lethality of cancer cell kill based on the tumor microenvironment. Cancer Res 70(20):8045–8054

Chia SK, Wykoff CC, Watson PH, Han C, Leek RD, Pastorek J, Gatter KC, Ratcliffe P, Harris AL (2001) Prognostic significance of a novel hypoxia-regulated marker, carbonic anhydrase IX, in invasive breast carcinoma. Journal of clinical oncology: official journal of the American Society of. Clin Oncol 19(16):3660–3668

Chitneni SK, Palmer GM, Zalutsky MR, Dewhirst MW (2011) Molecular imaging of hypoxia. Journal of nuclear medicine: official publication. Society of Nuclear Medicine 52(2):165–168

De Schutter H, Landuyt W, Verbeken E, Goethals L, Hermans R, Nuyts S (2005) The prognostic value of the hypoxia markers CA IX and GLUT 1 and the cytokines VEGF and IL 6 in head and neck squamous cell carcinoma treated by radiotherapy + /- chemotherapy. BMC Cancer 5:42

De Simone G, Supuran CT (2010) Carbonic anhydrase IX: Biochemical and crystallographic characterization of a novel antitumor target. Biochimica et biophysica acta 1804(2):404–409

Dings RP, Loren M, Heun H, McNiel E, Griffioen AW, Mayo KH, Griffin RJ (2007) Scheduling of radiation with angiogenesis inhibitors anginex and Avastin improves therapeutic outcome via vessel normalization. Clinical cancer research: an official journal of the American Association for. Cancer Res 13(11):3395–3402

Dubois L, Peeters S, Lieuwes NG, Geusens N, Thiry A, Wigfield S, Carta F, McIntyre A, Scozzafava A, Dogne JM, Supuran CT, Harris AL, Masereel B, Lambin P (2011) Specific inhibition of carbonic anhydrase IX activity enhances the in vivo therapeutic effect of tumor irradiation. Radiotherapy and oncology: journal of the European Society for Therapeutic Radiology and. Oncology 99(3):424–431

Ellis MJ, Tao Y, Luo J, A'Hern R, Evans DB, Bhatnagar AS, Chaudri Ross HA, Kameke A von, Miller WR, Smith I, Eiermann W, Dowsett M (2008) Outcome prediction for estrogen receptor-positive breast cancer based on postneoadjuvant endocrine therapy tumor characteristics. J Natl Cancer Inst 100(19):1380–1388

Eriksen JG, Overgaard J (2007) Lack of prognostic and predictive value of CA IX in radiotherapy of squamous cell carcinoma of the head and neck with known modifiable hypoxia: an evaluation of the DAHANCA 5 study. Radiotherapy and oncology: journal of the European Society for Therapeutic Radiology and. Oncology 83(3):383–388

Eschmann SM, Paulsen F, Reimold M, Dittmann H, Welz S, Reischl G, Machulla HJ, Bares R (2005) Prognostic impact of hypoxia imaging with 18F-misonidazole PET in non-small cell lung cancer and head and neck cancer before radiotherapy. Journal of nuclear medicine: official publication. Society of Nuclear Medicine 46(2):253–260

Farmer H, McCabe N, Lord CJ, Tutt AN, Johnson DA, Richardson TB, Santarosa M, Dillon KJ, Hickson I, Knights C, Martin NM, Jackson SP, Smith GC, Ashworth A (2005) Targeting the DNA repair defect in BRCA mutant cells as a therapeutic strategy. Nature 434(7035):917–921

Fiumicino S, Martinelli S, Colussi C, Aquilina G, Leonetti C, Crescenzi M, Bignami M (2000) Sensitivity to DNA cross-linking chemotherapeutic agents in mismatch repair-defective cells in vitro and in xenografts. Int JCancer 85(4):590–596

Fokas E, Im JH, Hill S, Yameen S, Stratford M, Beech J, Hackl W, Maira SM, Bernhard EJ, McKenna WG, Muschel RJ (2012) Dual inhibition of the PI3K/mTOR pathway increases tumor radiosensitivity by normalizing tumor vasculature. Cancer Res 72(1):239–248

Giantonio BJ, Catalano PJ, Meropol NJ, O'Dwyer PJ, Mitchell EP, Alberts SR, Schwartz MA, Benson AB 3rd (2007) Bevacizumab in combination with oxaliplatin, fluorouracil, and leucovorin (FOLFOX4) for previously treated metastatic colorectal cancer: results from the Eastern Cooperative Oncology Group Study E3200. J Clin Oncol 25(12):1539–1544

Goel S, Duda DG, Xu L, Munn LL, Boucher Y, Fukumura D, Jain RK (2011) Normalization of the vasculature for treatment of cancer and other diseases. Physiol Rev 91(3):1071–1121

Gordon MS (2002) Managing anemia in the cancer patient: old problems, future solutions. Oncologist 7(4):331–341

Gray LH, Conger AD, Ebert M, Hornsey S, Scott OC (1953) The concentration of oxygen dissolved in tissues at the time of irradiation as a factor in radiotherapy. Br J Radiol 26(312):638–648

Hegan DC, Lu Y, Stachelek GC, Crosby ME, Bindra RS, Glazer PM (2010) Inhibition of poly(ADP-ribose) polymerase down-regulates BRCA1 and RAD51 in a pathway mediated by E2F4 and p130. Proc Natl Acad Sci USA. 107(5):2201–2206

Helleday T, Lo J, Gent DC van, Engelward BP (2007) DNA double-strand break repair: from mechanistic understanding to cancer treatment. DNA Repair (Amst) 6(7):923–935

Hendrickson K, Phillips M, Smith W, Peterson L, Krohn K, Rajendran J (2011) Hypoxia imaging with [F-18] FMISO-PET in head and neck cancer: potential for guiding intensity modulated radiation therapy in overcoming hypoxia-induced treatment resistance. Radiother Oncol 101(3):369–375

Henke M, Mattern D, Pepe M, Bezay C, Weissenberger C, Werner M, Pajonk F (2006) Do erythropoietin receptors on cancer cells explain unexpected clinical findings? J Clin Oncol 24(29):4708–4713

Hervouet E, Cizkova A, Demont J, Vojtiskova A, Pecina P, Franssen-van Hal NL, Keijer J, Simonnet H, Ivanek R, Kmoch S, Godinot C, Houstek J (2008) HIF and reactive oxygen species regulate oxidative phosphorylation in cancer. Carcinogenesis 29(8):1528–1537

Horsman MR (1995) Nicotinamide and other benzamide analogs as agents for overcoming hypoxic cell radiation resistance in tumours. A review. Acta Oncol 34(5):571–587

Hoskin P, Rojas A, Saunders M (2009) Accelerated radiotherapy, carbogen, and nicotinamide (ARCON) in the treatment of advanced bladder cancer: mature results of a Phase II nonrandomized study. Int J Radiat Oncol Biol Phys 73(5):1425–1431

Hoskin PJ, Sibtain A, Daley FM, Wilson GD (2003) GLUT1 and CAIX as intrinsic markers of hypoxia in bladder cancer: relationship with vascularity and proliferation as predictors of outcome of ARCON. Bri J Cancer 89(7):1290–1297

Imaizumi N, Monnier Y, Hegi M, Mirimanoff RO, Ruegg C (2010) Radiotherapy suppresses angiogenesis in mice through TGF-betaRI/ALK5-dependent inhibition of endothelial cell sprouting. PloS One 5(6):e11084

Jacob S, Aguado M, Fallik D, Praz F (2001) The role of the DNA mismatch repair system in the cytotoxicity of the topoisomerase inhibitors camptothecin and etoposide to human colorectal cancer cells. Cancer Res 61(17):6555–6562

Janssens GO, Rademakers SE, Terhaard CH, Doornaert PA, Bijl HP, Ende P van den, Chin A, Marres HA, Bree R de, Kogel AJ van der, Hoogsteen IJ, Bussink J, Span PN, Kaanders JH (2012) Accelerated Radiotherapy With Carbogen and Nicotinamide for Laryngeal Cancer: Results of a Phase III Randomized Trial. Journal of clinical oncology: official journal of the American Society of. Clin Oncol 30(15):1777–1783

Jiricny J (2006) The multifaceted mismatch-repair system. Nat Rev Mol Cell Biol 7(5):335–346

Jonathan RA, Wijffels KI, Peeters W, Wilde PC de, Marres HA, Merkx MA, Oosterwijk E, Kogel AJ van der, Kaanders JH (2006) The prognostic value of endogenous hypoxia-related markers for head and neck squamous cell carcinomas treated with ARCON. Radiother Oncol 79(3):288–297

Kaanders JH, Pop LA, Marres HA, Bruaset I, Hoogen FJ van den, Merkx MA, Kogel AJ van der (2002a) ARCON: experience in 215 patients with advanced head-and-neck cancer. Int J Radiat Oncol Biol Phys 52(3):769–778

Kaanders JH, Wijffels KI, Marres HA, Ljungkvist AS, Pop LA, Hoogen FJ van den, Wilde PC de, Bussink J, Raleigh JA, Kogel AJ van der (2002b) Pimonidazole binding and tumor vascularity predict for treatment outcome in head and neck cancer. Cancer Res 62(23):7066–7074

Kleibeuker EA, Griffioen AW, Verheul HM, Slotman BJ, Thijssen VL (2012) Combining angiogenesis inhibition and radiotherapy: a double-edged sword. Drug Resist Updat 15(3):173–182

Koukourakis MI, Bentzen SM, Giatromanolaki A, Wilson GD, Daley FM, Saunders MI, Dische S, Sivridis E, Harris AL (2006) Endogenous markers of two separate hypoxia response pathways (hypoxia inducible factor 2 alpha and carbonic anhydrase 9) are associated with radiotherapy failure in head and neck cancer patients recruited in the CHART randomized trial. Journal of clinical oncology: official journal of the American Society of. Clin Oncol 24(5):727–735

Koukourakis MI, Giatromanolaki A, Sivridis E, Gatter KC, Harris AL (2013) High DLL4 expression in tumour-associated vessels predicts for favorable radiotherapy outcome in locally advanced squamous cell head-neck cancer (HNSCC). Angiogenesis 16(2):343–351

Kyndi M, Sorensen FB, Knudsen H, Alsner J, Overgaard M, Nielsen HM, Overgaard J (2008) Carbonic anhydrase IX and response to postmastectomy radiotherapy in high-risk breast cancer: a subgroup analysis of the DBCG82 b and c trials. Breast Cancer Res 10(2):R24

Lambin P, Ramaekers BL, Mastrigt GA van, Van den Ende P, Jong J de, De Ruysscher DK, Pijls-Johannesma M (2009) Erythropoietin as an adjuvant treatment with (chemo) radiation therapy for head and neck cancer. Cochrane Database Syst Rev 8(3):CD006158

Le QT, Sutphin PD, Raychaudhuri S, Yu SC, Terris DJ, Lin HS, Lum B, Pinto HA, Koong AC, Giaccia AJ (2003) Identification of osteopontin as a prognostic plasma marker for head and neck squamous cell carcinomas. Clinical cancer research: an official journal of the American Association for. Cancer Res 9(1):59–67

Lim AM, Rischin D, Fisher R, Cao H, Kwok K, Truong D, McArthur GA, Young RJ, Giaccia A, Peters L, Le QT (2012) Prognostic significance of plasma osteopontin in patients with locoregionally advanced head and neck squamous cell carcinoma treated on TROG 02.02 phase III trial. Clinical cancer research: an official journal of the American Association for. Cancer Res 18(1):301–307

Lin Z, Mechalakos J, Nehmeh S, Schoder H, Lee N, Humm J, Ling CC (2008) The influence of changes in tumor hypoxia on dose-painting treatment plans based on 18F-FMISO positron emission tomography. Int J Radiat Oncol Biol Phys 70(4):1219–1228
Liu SK, Bham SA, Fokas E, Beech J, Im J, Cho S, Harris AL, Muschel RJ (2011) Delta-like ligand 4-notch blockade and tumor radiation response. J Natl Cancer Inst 103(23):1778–1798
Loncaster JA, Harris AL, Davidson SE, Logue JP, Hunter RD, Wycoff CC, Pastorek J, Ratcliffe PJ, Stratford IJ, West CM (2001) Carbonic anhydrase (CA IX) expression, a potential new intrinsic marker of hypoxia: correlations with tumor oxygen measurements and prognosis in locally advanced carcinoma of the cervix. Cancer Res 61(17):6394–6399
Mackay A, Urruticoechea A, Dixon JM, Dexter T, Fenwick K, Ashworth A, Drury S, Larionov A, Young O, White S, Miller WR, Evans DB, Dowsett M (2007) Molecular response to aromatase inhibitor treatment in primary breast cancer. Breast cancer Res 9(3):R37
McCabe N, Turner NC, Lord CJ, Kluzek K, Bialkowska A, Swift S, Giavara S, O'Connor MJ, Tutt AN, Zdzienicka MZ, Smith GC, Ashworth A (2006) Deficiency in the repair of DNA damage by homologous recombination and sensitivity to poly(ADP-ribose) polymerase inhibition. Cancer Res 66(16):8109–8115
Mehta S, Hughes NP, Buffa FM, Li SP, Adams RF, Adwani A, Taylor NJ, Levitt NC, Padhani AR, Makris A, Harris AL (2011) Assessing early therapeutic response to bevacizumab in primary breast cancer using magnetic resonance imaging and gene expression profiles. J Natl Cancer Inst Monogr 2011(43):71–74
Meng AX, Jalali F, Cuddihy A, Chan N, Bindra RS, Glazer PM, Bristow RG (2005) Hypoxia down-regulates DNA double strand break repair gene expression in prostate cancer cells. Radiotherapy and oncology: journal of the European Society for Therapeutic Radiology and. Oncology 76(2):168–176
Mihaylova VT, Bindra RS, Yuan J, Campisi D, Narayanan L, Jensen R, Giordano F, Johnson RS, Rockwell S, Glazer PM (2003) Decreased expression of the DNA mismatch repair gene Mlh1 under hypoxic stress in mammalian cells. Mol Cell Biol 23(9):3265–3273
Nagy JA, Chang SH, Dvorak AM, Dvorak HF (2009) Why are tumour blood vessels abnormal and why is it important to know? Bri J Cancer 100(6):865–869
Nehmeh SA, Lee NY, Schroder H, Squire O, Zanzonico PB, Erdi YE, Greco C, Mageras G, Pham HS, Larson SM, Ling CC, Humm JL (2008) Reproducibility of intratumor distribution of (18)F-fluoromisonidazole in head and neck cancer. Int J Radiat Oncol Biol Phys 70(1):235–242
Overgaard J (2011) Hypoxic modification of radiotherapy in squamous cell carcinoma of the head and neck–a systematic review and meta-analysis. Radiotherapy and oncology: journal of the European Society for Therapeutic Radiology and. Oncology 100(1):22–32
Overgaard J, Hansen HS, Overgaard M, Bastholt L, Berthelsen A, Specht L, Lindelov B, Jorgensen K (1998) A randomized double-blind phase III study of nimorazole as a hypoxic radiosensitizer of primary radiotherapy in supraglottic larynx and pharynx carcinoma. Results of the Danish Head and Neck Cancer Study (DAHANCA) Protocol 5–85. Radiotherapy and oncology: journal of the European Society for Therapeutic Radiology and. Oncology 46(2):135–146
Overgaard J, Hansen HS, Specht L, Overgaard M, Grau C, Andersen E, Bentzen J, Bastholt L, Hansen O, Johansen J, Andersen L, Evensen JF (2003) Five compared with six fractions per week of conventional radiotherapy of squamous-cell carcinoma of head and neck: DAHANCA 6 and 7 randomised controlled trial. Lancet 362(9388):933–940
Overgaard J, Eriksen JG, Nordsmark M, Alsner J, Horsman MR (2005) Plasma osteopontin, hypoxia, and response to the hypoxia sensitiser nimorazole in radiotherapy of head and neck cancer: results from the DAHANCA 5 randomised double-blind placebo-controlled trial. Lancet Oncol 6(10):757–764
Palcic B, Brosing JW, Skarsgard LD (1982) Survival measurements at low doses: oxygen enhancement ratio. Br J Cancer 46(6):980–984
Perez-Sayans M, Garcia-Garcia A, Scozzafava A, Supuran CT (2012) Inhibition of V-ATPase and carbonic anhydrases as interference strategy with tumor acidification processes. Curr Pharm Des 18(10):1407–1413

Qayum N, Muschel RJ, Im JH, Balathasan L, Koch CJ, Patel S, McKenna WG, Bernhard EJ (2009) Tumor vascular changes mediated by inhibition of oncogenic signaling. Cancer Res 69(15):6347–6354

Rischin D, Hicks RJ, Fisher R, Binns D, Corry J, Porceddu S, Peters LJ (2006) Prognostic significance of [18F]-misonidazole positron emission tomography-detected tumor hypoxia in patients with advanced head and neck cancer randomly assigned to chemoradiation with or without tirapazamine: a substudy of Trans-Tasman Radiation Oncology Group Study 98.02. J Clin Oncol 24(13):2098–2104

Rodriguez-Jimenez FJ, Moreno-Manzano V, Lucas-Dominguez R, Sanchez-Puelles JM (2008) Hypoxia causes downregulation of mismatch repair system and genomic instability in stem cells. Stem Cells 26(8):2052–2062

Sandler A, Gray R, Perry MC, Brahmer J, Schiller JH, Dowlati A, Lilenbaum R, Johnson DH (2006) Paclitaxel-carboplatin alone or with bevacizumab for non-small-cell lung cancer. N Engl J Med 355(24):2542–2550

Schrijvers ML, Laan BF van der, Bock GH de, Pattje WJ, Mastik MF, Menkema L, Langendijk JA, Kluin PM, Schuuring E, Wal JE van der (2008) Overexpression of intrinsic hypoxia markers HIF1alpha and CA-IX predict for local recurrence in stage T1-T2 glottic laryngeal carcinoma treated with radiotherapy. Int J Radiat Oncol Biol Phys 72(1):161–169

Shahrzad S, Quayle L, Stone C, Plumb C, Shirasawa S, Rak JW, Coomber BL (2005) Ischemia-induced K-ras mutations in human colorectal cancer cells: role of microenvironmental regulation of MSH2 expression. Cancer Res 65(18):8134–8141

Sherwood BT, Colquhoun AJ, Richardson D, Bowman KJ, O'Byrne KJ, Kockelbergh RC, Symonds RP, Mellon JK, Jones GD (2007) Carbonic anhydrase IX expression and outcome after radiotherapy for muscle-invasive bladder cancer. Clin Oncol (Royal College of Radiologists (Great Britain)) 19(10):777–783

Shojaei F (2012) Anti-angiogenesis therapy in cancer: current challenges and future perspectives. Cancer lett 320(2):130–137

Supuran CT (2012) Inhibition of carbonic anhydrase IX as a novel anticancer mechanism. World J Clin Oncol 3(7):98–103

Toustrup K, Sorensen BS, Nordsmark M, Busk M, Wiuf C, Alsner J, Overgaard J (2011) Development of a hypoxia gene expression classifier with predictive impact for hypoxic modification of radiotherapy in head and neck cancer. Cancer Res 71(17):5923–5931

Toustrup K, Sorensen BS, Lassen P, Wiuf C, Alsner J, Overgaard J (2012) Gene expression classifier predicts for hypoxic modification of radiotherapy with nimorazole in squamous cell carcinomas of the head and neck. Radiotherapy and oncology: journal of the European Society for Therapeutic Radiology and. Oncology 102(1):122–129

Varlotto J, Stevenson MA (2005) Anemia, tumor hypoxemia, and the cancer patient. Int J Radiat Oncol Biol Phys 63(1):25–36

Vordermark D, Brown JM (2003) Endogenous markers of tumor hypoxia predictors of clinical radiation resistance? Strahlenther Onkol 179(12):801–811

Warde P, O'Sullivan B, Bristow RG, Panzarella T, Keane TJ, Gullane PJ, Witterick IP, Payne D, Liu FF, McLean M, Waldron J, Cummings BJ (1998) T1/T2 glottic cancer managed by external beam radiotherapy: the influence of pretreatment hemoglobin on local control. Int J Radiat Oncol Biol Phys 41(2):347–353

Wardman P (2007) Chemical radiosensitizers for use in radiotherapy. Clin Oncol (R Coll Radiol) 19(6):397–417

Wright EA, Howard-Flanders P (1957) The influence of oxygen on the radiosensitivity of mammalian tissues. Acta radiol 48(1):26–32

Zips D, Zophel K, Abolmaali N, Perrin R, Abramyuk A, Haase R, Appold S, Steinbach J, Kotzerke J, Baumann M (2012) Exploratory prospective trial of hypoxia-specific PET imaging during radiochemotherapy in patients with locally advanced head-and-neck cancer. Radiother Oncol J Eur Soc Therap Radiol Oncol 105(1):21–28

Index

G. Melillo (ed.), *Hypoxia and Cancer,* Cancer Drug Discovery and Development, DOI 10.1007/978-1-4614-9167-5,